Springer Series in Solid-State Sciences

Volume 201

The Springer Series in Solid-State Sciences features fundamental scientific books prepared by leading and up-and-coming researchers in the field. They strive to communicate, in a systematic and comprehensive way, the basic principles as well as new developments in theoretical and experimental solid-state science.

We welcome submissions for monographs or edited volumes from scholars across this broad domain. Topics of current interest include, but are not limited to:

- Semiconductors and superconductors
- Quantum phenomena
- Spin physics
- Topological insulators
- Multiferroics
- Nano-optics and nanophotonics
- Correlated electron systems and strongly correlated materials
- Vibrational and electronic properties of solids
- Spectroscopy and magnetic resonance

Maxim Yu. Kagan · Kliment I. Kugel ·
Alexander L. Rakhmanov · Artem O. Sboychakov

Electronic Phase Separation in Magnetic and Superconducting Materials

Recent Advances

Springer

Maxim Yu. Kagan
Donostia International Physics Center
San Sebastian, Spain

Kliment I. Kugel
Institute for Theoretical and Applied Electrodynamics
Russian Academy of Sciences
Moscow, Russia

Alexander L. Rakhmanov
Institute for Theoretical and Applied Electrodynamics
Russian Academy of Sciences
Moscow, Russia

Artem O. Sboychakov
Institute for Theoretical and Applied Electrodynamics
Russian Academy of Sciences
Moscow, Russia

ISSN 0171-1873 ISSN 2197-4179 (electronic)
Springer Series in Solid-State Sciences
ISBN 978-3-031-55466-7 ISBN 978-3-031-55467-4 (eBook)
https://doi.org/10.1007/978-3-031-55467-4

This Springer imprint is published by the registered company Springer Nature Switzerland AG
The registered company address is: Gewerbestrasse 11, 6330 Cham, Switzerland

Paper in this product is recyclable.

Preface

We consider the nanoscale electronic phase separation in a wide class of different materials, mostly in strongly correlated electron systems. The phase separation turns out to be quite ubiquitous manifesting itself in different situations, where the itineracy of charge carriers competes with their tendency toward localization. The latter is often related to some specific type of magnetic ordering, e.g., antiferromagnetic in manganites and low-spin states in cobaltites. The interplay between the localization-induced lowering of potential energy and metallicity (which provides the gain in the kinetic energy) favors an inhomogeneous ground state such as nanoscale ferromagnetic droplets in an antiferromagnetic insulating background.

The present book deals with the advances in the subject of electronic phase separation and formation of different types of nanoscale ferromagnetic (FM) metallic droplets (FM polarons or ferrons) in antiferromagnetically ordered (AFM), charge-ordered (CO), or orbitally ordered (OO) insulating matrices, as well as the colossal magnetoresistance (CMR) effect and tunneling electron transport in the nonmetallic phase-separated state of complex magnetic oxides. It also touches upon the compounds with spin-state transitions, inhomogeneous phase-separated state in strongly correlated multiband systems, and electron polaron effect. A special attention is paid to the systems with the imperfect Fermi surface nesting such as chromium alloys, iron-based pnictides, and AA stacked graphene bilayers.

The second part of the book deals with the novel superconducting materials and electronic phase separation in them. In this context, first of all, we should mention the results related to the formation of inhomogeneous space-separated Fermi–Bose mixture of paired (bosonic) and normal (fermionic) BiO_6 octahedral clusters in bismuth oxides BaKBiO. Another important result here concerns the possibility of the formation of the paired (order parameter) droplets in the matrix of the unpaired (normal) states in the low-density attracting electron system in the presence of strong random potential.

We analyze also the droplet formation, which arises in superconducting and superfluid (Bose-condensed) systems ranging from ultracold Fermi gases and mixtures, cuprates, and metallic hydrogen, as well as in the excitonic and plasmonic systems.

We briefly comment on the phase separation in two-dimensional electron gas and on disorder effects in the Dirac and Weyl systems.

In Conclusion, we discuss some additional problems and provide an extensive reference list based on the results of different groups in this popular subject worldwide.

We hope our book will be useful for a broad scientific community ranging from quantum condensed matter physicists (both theorists and experimentalists) to solid-state and inorganic chemists and technologists.

Moscow, Russia
San Sebastian, Spain
September 2023

Maxim Yu. Kagan
Kliment I. Kugel
Alexander L. Rakhmanov
Artem O. Sboychakov

Acknowledgements

Maxim Yu. Kagan acknowledges the support and hospitality of the Donostia International Physics Center (DIPC), San Sebastian, Spain during his regular visits in the years 2018–2023.

We are grateful to S. Aksenov, B. Aronzon, N. Bebenin, A. Bianconi, E. Burovski, R. Ikhsanov, D. Khomskii, A. Kuznetsov, E. Kuznetsov, E. Mazur, A. Menushenkov, Ya. Rodionov, A. Rozhkov, V. Silkin, S. Streltsov, V. Tripathi, and A. Turlapov for helpful discussions.

Contents

Chapter 1
Introduction. Spontaneously Formed Nanoscale Inhomogenieties in Different Materials

Abstract We provide an introduction to the subject of phase separation and briefly recollect very different systems (ranging from quantum gases and magnetic oxides to high T_c superconductors and rotating neutron stars), where the phase separation takes place.

The phase separation is a universal and ubiquitous phenomenon, which is present not only in electronic liquid in metals, semiconductors [1], heavy-fermion [2, 3], and high-T_c superconductors, but also in different types of mixtures (Fermi–Bose, Fermi–Fermi, and Bose–Bose mixtures), in bismuthates (BaKBiO), liquid and solid solutions of ^{3}He in ^{4}He, mixtures of ^{6}Li and ^{7}Li, ^{40}K and ^{87}Rb isotopes, in spin-polarized (imbalanced) Fermi gases, and in two kinds of Bose gases in magnetic traps and on optical lattices, as well as in low-dimensional organic compounds and in mixed states (different pasta-like structure) in nuclear physics [4–14]. In condensed matter physics, it usually manifests itself either in the form of electronic phase separation (with the formation of paramagnetic spin bags or strings, ferrons, stripes [15–17], etc.) or the phase separation in ionic subsystems (e.g., charge density wave formation in superconducting cuprates such as LaBaCuO and LaSrCuO for some specific values of hole doping [4–6, 18, 19]). While the latter phenomena usually require a reconstruction of the crystal lattice or formation of the superstructures (e.g., doubling of the lattice period), the former one does not necessarily correspond to the instability of the crystal lattice. In our book, we are dealing with the electronic phase separation. Moreover, we will mostly consider the nanoscale phase separation in the electronic subsystem and only briefly discuss other types of electronic phase separation (such as a total phase separation giving rise to two large clusters, formation of stripes, etc.). We will also focus on strongly correlated electron systems with a strong interplay between kinetic and potential energy of electrons. The gain in kinetic energy favors delocalization and metallicity in the electron subsystem. At the same time, the potential energy promotes localization and insulating behavior (as in the Hubbard model at strong onsite Coulomb repulsion U or in the $t - J$ model in the limit of large J/t ratios [20–22]).

We stress again that the instability of the ground state with respect to the nanoscale phase separation is an important feature characteristic of numerous

M. Yu. Kagan et al., *Electronic Phase Separation in Magnetic and Superconducting Materials*, Springer Series in Solid-State Sciences 201,
https://doi.org/10.1007/978-3-031-55467-4_1

strongly correlated electron systems. It corresponds, in particular, to the creation of small ferromagnetic (FM) droplets inside different insulating host materials. The host materials can be antiferromagnetic [23–30], charge-ordered (CO) [31–34], or orbitally ordered (OO) ones [35, 36]. The FM droplets are randomly distributed within these hosts and are usually metallic ones. The shape of the droplets is usually a spherical or an ellipsoidal one [23, 24, 27]. We can say that the droplets are indeed real nanoscale objects looking like raisins in a cake. Their radius is of the order of $R = 10$–15 Å, which corresponds to 2–4 interatomic distances). Inside droplets, there are typically 30–100 local spins, which are FM oriented owing in fact to one additional conduction electron or hole. Hence, an FM droplet exhibits a large magnetic moment of the order of 15–50 Bohr magnetons. Usually, the phase-separated state in complex magnetic oxides (such as manganites, cobaltites, and nickelates) is a superparamagnetic one. This fact is due to the residual magneto-dipole interaction between the FM droplets. Meanwhile, the electron transport here is governed by the tunneling of conduction electrons between the neighboring metallic droplets through the tunneling barriers, which are formed in the insulating host. The tunneling usually corresponds to the regime of Coulomb blockade [37–45], and it is spin dependent, which could be important for the applications in spintronics.

In this book, we will consider different types of free and bound magnetic polarons arising in AFM, CO, and OO matrices both in the isotropic and anisotropic cases on regular (cubic, quadratic) and frustrated (triangular) lattices. We analyze the thermodynamic and transport (kinetic) characteristics of the phase-separated state including magnetization and magnetic susceptibility, resistivity, magnetoresistance, and the spectrum of $1/f$-noise in it [46–49].

As we have already mentioned, in cobaltites, for example, the phase separation is usually accompanied by the transitions between high- and low-spin states [50]. For cobaltites, the electronic phase separation was considered not only in 3D and 2D systems but in quasi-one-dimensional magnets such as $BaCoO_3$ as well (see [51–54]). These papers deal with the creation of magnetic polarons in spin chains with Heisenberg AFM interaction upon doping. For such 1D systems, it was possible to demonstrate the formation of rather long-range spin distortions around the polaronic core. The long-range tail of slowly decaying spin distortions (usually they decay in a power-law fashion) is also formed around the ferromagnetic core of the magnetic polaron bound by donor impurity. This type of spin distortions arise in manganites at very low doping levels in 2D and 3D cases in the model, which combines Hund's rule and Heisenberg-type magnetic interactions (as well as the one-site magnetic anisotropy) with electron–ion Coulomb attraction potential between Sr (or Ba) impurity ion and conduction electron [55, 56].

Another type of interesting mesoscopic phenomena in strongly correlated electron systems corresponds to the electron polaron effect (EPE). EPE usually takes place in two-band systems with strongly differing band masses. This situation is typical of many mixed valence electron systems [57–62], including uranium-based heavy-fermion compounds. The most general model, which should be used here, is a two-band Hubbard model with one narrow band [63–67]. The EPE usually describes the strong enhancement of a heavy mass due to many-body effects.

They are predominantly related to the dressing of a heavy particle in a coat of low-energy (infrared) electron–hole pairs of light particles and lead to a strong additional narrowing of the heavy particle bandwidth. The EPE has a lot of similarities with the physics of infrared divergences [68–70] for the Brownian motion of a heavy particle in the Fermi liquid of light particles [71, 72]). Another close analogy here is that with the Anderson physics of the orthogonality catastrophe [73, 74], which describes a 1D chain of N electrons in the presence of one impurity in the system.

The electron polaron effect not necessarily leads to the nanoscale phase separation (this scenario is also possible, at least, at the level of the strong density redistribution for the large initial mismatch between the densities of heavy and light components in agreement with the predictions of [75]).

At the same time, in all the cases (with and without phase separation), the electron polaron effect leads to the anomalous temperature dependence of the resistivity in 3D and especially in quasi-2D (layered) case providing possible alternative explanation to the mechanism of spin-dependent tunneling conductivity when we try to explain resistivity characteristics in manganese silicides [76] or layered manganites. In manganese silicides MnSi and in the stoichiometric alloys $Mn_{1-x}Fe_xSi$, we have a very peculiar power-law (almost of the square root type) dependence of the resistivity $R(T)$ (for temperatures in the range from 30 to 250 K) and magnetization $M(H)$ (in moderate and high magnetic fields from 2–4 T up to 50 T) in the paramagnetic phase.

We generalize the manifestation of the electron polaron effect (EPE) in the micro-contact with deep and shallow traps specified by the model with the quadrupole structure of four quantum dots (QD) [77–79]. In this model two central QDs play the role of a deep two-level trap, while two other QDs at the periphery of the quadrupole QD structure play the role of the shallow traps with one level in them. In the framework of the nonequilibrium Keldysh Green's function technique [80], we calculate the electric current in this system both in the isotropic case and in the anisotropic case, where the physical picture is close to the EPE.

In our book, we also consider the role of a more standard electron–phonon polaron effect as well as a long-range part of Coulomb interaction in the formation of different inhomogeneities and nanoscale phase-separated structures. Usually the electron polaron effect and polaron effect of the phonon origin play in concert leading to the multiplicative effect for the bandwidth narrowing. In this connection, it is important to mention the giant oxygen isotope effects in manganites [81–83] (see also [84] and references therein) and in cobaltites [85], which are closely related both to the polaron effect and phase separation. Moreover, the oxygen isotope exchange can even induce the metal–insulator transition in manganites [86–88]. Electron–phonon (electron–ion) interactions are responsible for the creation of different types of superstructures in manganites and other Jahn–Teller systems [89–91]. In manganites, two conduction bands (delocalized orbitals) of e_g electrons are additionally split by sufficiently large Jahn–Teller gap, while three localized t_{2g} orbitals form the local spin $S = 3/2$ at them in accordance with the generalized Hund's rule.

For the physics of phase separation and the formation of different anisotropic structures (stripes, zig-zags, spin ladders, etc.), a very important role is played by the orbital degrees of freedom (often related to the Jahn–Teller-type effects), as well as by the relativistic (magneto-dipole or spin-orbit) interactions. The anisotropic magneto-dipole interaction could be also a source for the formation of anisotropic elongated structures (such as stripes) in many strongly correlated systems [89]. Another possibility for a strongly anisotropic situation and formation of stripes arises when we take into account the interplay between nanoscale phase separation and charge ordering. At the same time, the formation of zig-zags and spin ladders [92, 93] takes place even when we consider stable crystallographic distortions.

An important role of orbital degrees of freedom also manifests itself in multiorbital systems such as transition metals compounds with several e_g and t_{2g} orbitals for d electrons [35]. In such a case, it is possible to formulate a spin–orbital model, which adequately describes entangled spin–orbital excitations in the system [94]. The phase separation in Jahn–Teller systems with localized and itinerant electrons was analyzed in detail in [95].

In the first part of the book, we consider also the systems with the imperfect nesting such as chromium-based alloys, iron-based superconductors, stacked AA graphene bilayers, hydrogenized graphene, and graphene-based heterostructures [96]. These systems are inherently unstable toward the electronic phase separation [97–100] already in the framework of the simplest Rice model [101], which was originally proposed for the description of spin density waves in chromium-based alloys. This model captures the band structure effects in itinerant antiferromagnetism and contains repulsive Coulomb interaction between almost spherical electron and hole pockets of different radii (imperfect nesting).

An idealized graphene monolayer and an idealized AB graphene bilayer are also the subjects of anomalous Kohn–Luttinger type of superconductivity [102–104]. The same type of anomalous superconductivity and nanoscale phase separation with the formation of small metallic droplets inside the CO matrix (similar to phase separation in the Verwey model [105, 106] for manganites) occurs in the strong-coupling limit of the Shubin–Vonsovsky model [107] for strong Coulomb interaction at nearest-neighbor sites, $V \gg t$, and $n = 1/2$ (near the quarter filling) [108]. The Shubin–Vonsovsky model can be important for cuprates, iron-based superconductors, and idealized monolayer and bilayer graphene. The electron polaron effects are clearly pronounced in monolayer and bilayer graphene as well as in the other quasi-2D and quasi-3D (Weyl) semimetals with the Dirac spectrum such as bismuth, for example, [109–111]. It is interesting here to study more thoroughly a combined effect of strong Coulomb correlations and disorder for the conductivity in this class of materials.

One of the main goals of our book is to describe the recent progress in the subject of nanoscale phase separation. It is quite surprising that a lot of intriguing results on this subject obtained during the last 15 years have not been reflected yet in comprehensive review articles. Indeed, on the wave of the so-called "manganite boom" at the end of 90s, there appeared several reviews and books [23–25, 112–114] related mostly to the phase separation in manganites, and nearly nothing after that. In the first part of the book, we are trying to fill somehow such evident "blank spot".

The second part of the book concerns the novel superconducting materials and electronic phase separation in them. In this context, first of all, we should mention the results related to the formation of the inhomogeneous space-separated

Fermi–Bose mixture of paired (bosonic) and normal (fermionic) BiO_6 octahedral clusters in BaKBiO bismuth oxides.

The idea of a spatially separated Fermi–Bose mixture of finite boson clusters or complexes (containing compact electron or hole pairs) in a host medium involving unpaired (Fermi liquid) states was first formulated in [9] with the aim to explain the mechanism of superconductivity and the nature of the electron transport in the normal state of $BaKBiO_3$ compounds. The formation of the long-range order and of the macroscopic wave function, which characterize the superconducting state in $Ba_{1-x}K_xBiO_3$ are due to the tunneling of local electron pairs from one BiO_6 boson cluster to the neighboring BiO_6 cluster through the tunneling barrier formed by normal fermionic clusters. It is important to stress that the components of the Fermi–Bose mixture at the metallic concentrations in the range of $0.37 < x < 0.5$ are separated in the real space, but not in the energy space [9, 10].

Developing this set of ideas we study the inhomogeneous Fermi–Bose mixture in the interference device described by the Aharonov–Bohm ring [115] with the superconducting wire in a topologically nontrivial state serving as a bridge in this structure [116–118]. We analyze Majorana bound modes [119–121] on the ends of the wire and consider the manifestation of different resonances in the conductance of the Aharonov–Bohm ring. We show that the collapse of the Fano resonance [122] in this system is related to an increase in the length of the superconducting bridge when the binding energy for the Majorana end modes tends to zero.

Another important result here deals with the possibility of the formation of the droplets of the order parameter in the matrix of the unpaired normal states in the low-density attracting two-dimensional electron system [123] in the presence of strong random potential specified by the diagonal disorder [124–126]. In this system in the framework of the seminal Bogoliubov–de Gennes approach, we can illustrate the formation of inhomogeneous states of spatially separated Fermi–Bose mixture of Cooper pairs and unpaired electrons. These states involve superconducting bosonic droplets of different sizes in a host matrix of the unpaired normal electronic states. At very small densities, the droplets of the order parameter contain only one or two bielectron pairs. When the density of electrons is increased the formation of a large percolation cluster takes place and insulator–superconductor transition occurs in our 2D system [4, 5]. These results are important both for constructing of the global phase diagram of the system and for the understanding of the nature of the phase transition between the superconducting and localized (insulating) states in a quasi-two-dimensional thin film of a dirty metal. In a more practical sense, they are also interesting for applications in superconducting nanoelectronics and for the implementation of superconducting qubits in granular media.

In the second part of the book, we analyze also the droplets formation, which arise in superconducting and superfluid (Bose-condensed) systems ranging from imbalanced ultracold Fermi gases [127, 128] and mixtures [129–132] to high-T_c cuprates and metallic hydrogen [133, 134], as well as to the excitonic [135–142] and plasmonic systems [143–147].

Note that the formation of droplets is typical for collapsing excitonic [148–150]and plasmonic systems with the attractive exciton–exciton and plasmon–plasmon interactions, as well as for metallic hydrogen in shock-wave experiments near the onset of the first-order phase transition between the liquid and crystalline phases.

In the last chapter before the conclusions, we briefly comment on the phase separation in two-dimensional electron gas [151] and disorder effects in the Dirac and Weyl systems. We discuss also the beautiful and intensively debated problem of the chiral anomaly [152–154] specified by the EB term in the longitudinal negative magnetoresistance [155–157] of the topological insulators, such as $Bi_{1-x}Sb_x$ and its possible suppression by the large number of impurities in the regime of the high quantizing magnetic fields and low temperatures.

Finally, we discuss some additional actual problems in the physics of phase separation which are not totally resolved until nowadays and provide an extensive reference list based on the results of different groups in this popular subject worldwide.

References

1. L.V. Keldysh, Phys. Status Solidi (a) **164**(1), 3 (1997)
2. A. H. Castro Neto, B.A. Jones, Phys. Rev. B **62**, 14975 (2000)
3. P. Chandra, P. Coleman, J.A. Mydosh, V. Tripathi, Nature **417**(6891), 831 (2002)
4. C. Castellani, C.Di Castro., M. Grilli, Phys. Rev. Lett. **75**(7), 4650 (1995)
5. C. Castellani, C. Di Castro., M. Grilli, J. Phys. Chem. Solids **59**(10-12), 1694 (1998)
6. A. Perali, C. Castellani, C. Di Castro, M. Grilli, Phys. Rev. B **54**, 16216 (1996)
7. J. Zaanen, O. Gunnarsson, Phys. Rev. B **40**, 7391 (1989)
8. M.W. Zwierlein, A. Schirotzek, C.H. Schunck, W. Ketterle, Science **311**(5760), 492 (2006)
9. A.P. Menushenkov, K.V. Klementev, A.V. Kuznetsov, M.Yu. Kagan, J. Exp. Theor. Phys. **93**(3), 615 (2001)
10. A.P. Menushenkov, A.V. Kuznetsov, K.V. Klementiev, M.Yu. Kagan, J. Supercond. Novel Magn. **29**(3), 701 (2016)
11. E.P. Bashkin, A.E. Meyerovich, Adv. Phys. **30**(1), 1 (1981)
12. M.Yu. Kagan, Phys. Usp. **37**(1), 69 (1994)
13. T. Maruyama, T. Tatsumi, D.N. Voskresensky, T. Tanigawa, S. Chiba, Phys. Rev. C **72**, 015802 (2005)
14. T. Maruyama, T. Tatsumi, D.N. Voskresensky, T. Tanigawa, T. Endo, S. Chiba, Phys. Rev. C **73**, 035802 (2006)
15. A. Kampf, J.R. Schrieffer, Phys. Rev. B **41**, 6399 (1990)
16. L.N. Bulaevskii, E.L. Nagaev, D.I. Khomskii, Sov. Phys. JETP **27**, 836 (1968)
17. W.F. Brinkman, T.M. Rice, Phys. Rev. B **2**, 1324 (1970)
18. J.M. Tranquada, B.J. Sternlieb, J.D. Axe, Y. Nakamura, S. Uchida, Nature **375**(6532), 561 (1995)
19. A. Bianconi, N.L. Saini, A. Lanzara, M. Missori, T. Rossetti, H. Oyanagi, H. Yamaguchi, K. Oka, T. Ito, Phys. Rev. Lett. **76**, 3412 (1996)
20. J. Hubbard, Proc. Roy. Soc. A **205**, 238 (1963)
21. V.J. Emery, S.A. Kivelson, H.Q. Lin, Phys. Rev. Lett. **64**, 475 (1990)
22. M.Yu. Kagan, T.M. Rice, J. Phys.: Condens. Matter **6**(20), 3771 (1994)
23. E.L. Nagaev, Phys. Rep. **346**(6), 387 (2001)
24. E. Dagotto, T. Hotta, A. Moreo, Phys. Rep. **344**(1), 1 (2001)
25. M.Yu. Kagan, K.I. Kugel, Phys. Usp. **44**(6), 553 (2001)
26. M.Yu. Kagan, M.V. Mostovoy, D.I. Khomskii, Physica B **284-288**(Part 2), 1209 (2000)
27. M.Yu. Kagan, A.V. Klaptsov, I.V. Brodskii, K.I. Kugel, A.O. Sboichakov, A.L. Rakhmanov, Phys. Usp. **46**(8), 851 (2003)

28. M.Yu. Kagan, A.V. Klaptsov, I.V. Brodsky, K.I. Kugel, A.O. Sboychakov, A.L. Rakhmanov, J. Phys. A: Math. Gen. **36**(35), 9155 (2003)
29. M.Yu. Kagan, K.I. Kugel, A.L. Rakhmanov, K.S. Pazhitnykh, J. Phys.: Condens. Matter **18**(48), 10905 (2006)
30. M.Yu. Kagan, *Modern Trends in Superconductivity and Superfluidity*, Lecture Notes in Physics, vol. 874. (Springer, Netherlands, 2014)
31. M.Yu. Kagan, K.I. Kugel, D.I. Khomskii, J. Exp. Theor. Phys. **93**(2), 415 (2001)
32. M.Yu. Kagan, K.I. Kugel, A.L. Rakhmanov, D.I. Khomskii, Low Temp. Phys. **27**(8), 601 (2001)
33. M.Yu. Kagan, K.I. Kugel, D.I. Khomskii, A.L. Rakhmanov, Physica C **364-365**(Supplement C), 643 (2001)
34. J. Lorenzana, C. Castellani, C. Di Castro, Phys. Rev. B **64**, 235128 (2001)
35. K.I. Kugel, D.I. Khomskii, Sov. Phys. Usp. **25**, 231 (1982)
36. K.I. Kugel, A.L. Rakhmanov, A.O. Sboychakov, D.I. Khomskii, Phys. Rev. B **78**(15), 155113 (2008)
37. A.L. Rakhmanov, K.I. Kugel, Ya.M. Blanter, M.Yu. Kagan, Phys. Rev. B **63**(17), 174424 (2001)
38. A.L. Rakhmanov, K.I. Kugel, Ya.M. Blanter, M.Yu. Kagan, Phys. Metals Metallography **918**(Suppl. 1), 207 (2001)
39. A.O. Sboychakov, A.L. Rakhmanov, K.I. Kugel, M.Yu. Kagan, I.V. Brodsky, J. Exp. Theor. Phys. **95**(4), 753 (2002)
40. A.O. Sboychakov, A.L. Rakhmanov, K.I. Kugel, M.Yu. Kagan, I.V. Brodsky, J. Magn. Magn. Mater. **258**(Supplement C), 296 (2003)
41. K.I. Kugel, A.L. Rakhmanov, A.O. Sboychakov, M.Yu. Kagan, I.V. Brodsky, A.V. Klaptsov, J. Exp. Theor. Phys. **98**(3), 572 (2004)
42. A.O. Sboychakov, A.L. Rakhmanov, K.I. Kugel, M.Yu. Kagan, I.V. Brodsky, J. Phys.: Condens. Matter **15**(10), 1705 (2003)
43. Y. Imry, *Introduction to Mesoscopic Physics*, 2nd edn. (Oxford University Press, 2008)
44. D.V. Averin, K.K. Likharev, in *Mesoscopic Phenomena in Solids*, ed. by B.L. Altshuler, P.A. Lee, R.A. Webb (Elsevier, Amsterdam, 1992)
45. C.I.M. Beenakker, S.K. Bootsma, in *Frontiers in Nanoscale Science of Micron/Submicron Devices*, ed. by A.P. Jauho, E.V. Buzaneva (Kluwer Academic Publishers, Dordrecht, 1996)
46. S. Kogan, *Electronic Noise and Fluctuations in Solids* (Cambridge University Press, Cambridge, 1996)
47. P. Dutta, P.M. Horn, Rev. Mod. Phys. **53**, 497 (1981)
48. B.I. Shklovskii, A.L. Efros, *Electronic Properties of Doped Semiconductors, Springer Series in Solid-State Sciences* (Springer, Berlin, 1984)
49. V. Podzorov, M. Uehara, M.E. Gershenson, T.Y. Koo, S.W. Cheong, Phys. Rev. B **61**, R3784 (2000)
50. A.O. Sboychakov, K.I. Kugel, A.L. Rakhmanov, D.I. Khomskii, Phys. Rev. B **80**, 024423 (2009)
51. I. González, J. Castro, D. Baldomir, A.O. Sboychakov, A.L. Rakhmanov, K.I. Kugel, Phys. Rev. B **69**, 224409 (2004)
52. A.O. Sboychakov, A.L. Rakhmanov, K.I. Kugel, I. González, J. Castro, D. Baldomir, Phys. Rev. B **72**, 014438 (2005)
53. V. Pardo, J. Rivas, D. Baldomir, Appl. Phys. Lett. **86**(20), 202507 (2005)
54. J. Castro, I. González, D. Baldomir, Eur. Phys. J. B **39**(4), 447 (2004)
55. S.L. Ogarkov, M.Yu. Kagan, A.O. Sboychakov, A.L. Rakhmanov, K.I. Kugel, Phys. Rev. B **74**, 014436 (2006)
56. K.I. Kugel, A.L. Rakhmanov, A.O. Sboychakov, M.Yu. Kagan, S.L. Ogarkov, Physica B **403**(5), 1353 (2008)
57. D.M. Newns, N. Read, Adv. Phys. **36**(6), 799 (1987)
58. P. Coleman, Phys. Rev. B **35**, 5072 (1987)

59. A.M. Tsvelik, *Quantum Field Theory in Condensed Matter Physics*, 2nd edn. (Cambridge University, Cambridge, 2007)
60. C.M. Varma, Physica B **378-380**(Supplement C), 17 (2006)
61. P. Fulde, *Electron Correlations in Molecules and Solids*, Springer Series in Solid-State Sciences, vol. 100, 3rd edn. (Springer, Berlin, 2002)
62. H. Keiter, N. Grewe, in *Valence Fluctuations in Solids*. ed. by L.M. Falicov, W. Hanke, M.B. Maple (North Holland, Amsterdam, 1981), p.451
63. M.Yu. Kagan, A.G. Aronov, Czech. J. Phys. **46**(5), 2661 (1996)
64. M.Yu. Kagan, V.V. Val'kov, J. Exp. Theor. Phys. **113**(1), 156 (2011)
65. M.Yu. Kagan, V.V. Valkov, Low Temp. Phys. **37**(1), 69 (2011)
66. M.Yu. Kagan, V.V. Val'kov, in *A Lifetime in Magnetism and Superconductivity: A Tribute to Professor David Shoenberg*, ed. by G. Lonzarich, S.S. Saxena, V.A. Sirenko, V.V. Eremenko (Cambridge Scientific Publishers, Cambridge, 2012)
67. M.Yu. Kagan, V.V. Val'kov, J. Supercond. Novel Magn. **25**(5), 1379 (2012)
68. J. Kondo, Prog. Theor. Phys. **32**(1), 37 (1964)
69. G. Iche, P. Noziéres, Physica A **91**(3), 485 (1978)
70. P. Nozières, L.T. de Dominicis, Phys. Rev. **178**, 1097 (1969)
71. Yu. Kagan, N.V. Prokof'ev, J. Exp. Theor. Phys. **63**(6), 1276 (1986)
72. Yu. Kagan, N.V. Prokof'ev, J. Exp. Theor. Phys. **66**(1), 211 (1987)
73. P.W. Anderson, Phys. Rev. Lett. **18**, 1049 (1967)
74. P.W. Anderson, Phys. Rev. **164**, 352 (1967)
75. A.O. Sboychakov, K.I. Kugel, A.L. Rakhmanov, Phys. Rev. B **76**, 195113 (2007)
76. S.V. Demishev, A.N. Samarin, J. Huang, V.V. Glushkov, I.I. Lobanova, N.E. Sluchanko, N.M. Chubova, V.A. Dyadkin, S.V. Grigoriev, M.Yu. Kagan, J. Vanacken, V.V. Moshchalkov, JETP Lett. **104**(2), 116 (2016)
77. M.Yu. Kagan, V.V. Val'kov, S.V. Aksenov, Phys. Rev. B **95**, 035411 (2017)
78. M.Yu. Kagan, V.V. Val'kov, S.V. Aksenov, J. Magn. Magn. Mater. **440**(Supplement C), 15 (2017)
79. M.Yu. Kagan, S.V. Aksenov, JETP Lett. **107**(8), 493 (2018)
80. L.V. Keldysh, Sov. Phys. JETP **20**, 1018 (1965)
81. G.M. Zhao, K. Conder, H. Keller, K.A. Müller, Nature **381**, 676 (1996)
82. I. Isaac, J.P. Franck, Phys. Rev. B **57**, R5602 (1998)
83. N.A. Babushkina, A.N. Taldenkov, L.M. Belova, E.A. Chistotina, O.Y. Gorbenko, A.R. Kaul, K.I. Kugel, D.I. Khomskii, Phys. Rev. B **62**, R6081 (2000)
84. A.N. Taldenkov, V.V. Snegirev, N.A. Babushkina, V.S. Kalitka, A.R. Kaul', J. Exp. Theor. Phys. **126**(3), 383 (2018)
85. A.V. Kalinov, O.Y. Gorbenko, A.N. Taldenkov, J. Rohrkamp, O. Heyer, S. Jodlauk, N.A. Babushkina, L.M. Fisher, A.R. Kaul, A.A. Kamenev, T.G. Kuzmova, D.I. Khomskii, K.I. Kugel, T. Lorenz, Phys. Rev. B **81**, 134427 (2010)
86. N.A. Babushkina, L.M. Belova, O.Y. Gorbenko, A.R. Kaul, A.A. Bosak, V.I. Ozhogin, K.I. Kugel, Nature **391**, 159 (1998)
87. G.M. Zhao, H. Keller, J. Hofer, A. Shengelaya, K.A. Müller, Solid State Commun. **104**(2), 57 (1997)
88. N.A. Babushkina, L.M. Belova, V.I. Ozhogin, O.Y. Gorbenko, A.R. Kaul, A.A. Bosak, D.I. Khomskii, K.I. Kugel, J. Appl. Phys. **83**(11), 7369 (1998)
89. D.I. Khomskii, K.I. Kugel, EPL (Europhys. Lett.) **55**(2), 208 (2001)
90. D.I. Khomskii, K.I. Kugel, Phys. Rev. B **67**, 134401 (2003)
91. A.O. Sboychakov, K.I. Kugel, A.L. Rakhmanov, D.I. Khomskii, Phys. Rev. B **83**, 205123 (2011)
92. E. Dagotto, T.M. Rice, Science **271**(5249), 618 (1996)
93. M.Yu. Kagan, S. Haas, T. M. Rice, Physica C **317-318**(Supplement C), 185 (1999)
94. M.Yu. Kagan, K.I. Kugel, A.V. Mikheyenkov, A.F. Barabanov, JETP Lett. **100**(3), 187 (2014)
95. K.I. Kugel, A.L. Rakhmanov, A.O. Sboychakov, Phys. Rev. Lett. **95**(26), 267210 (2005)
96. M.I. Katsnelson, K.S. Novoselov, A.K. Geim, Nat. Phys. **2**(9), 620 (2006)

97. A.V. Rozhkov, A.O. Sboychakov, A.L. Rakhmanov, F. Nori, Phys. Rep. **648**(1), 1 (2016)
98. A.L. Rakhmanov, A.V. Rozhkov, A.O. Sboychakov, F. Nori, Phys. Rev. B **87**, 075128 (2013)
99. A.O. Sboychakov, A.V. Rozhkov, K.I. Kugel, A.L. Rakhmanov, F. Nori, Phys. Rev. B **88**, 195142 (2013)
100. A.O. Sboychakov, A.L. Rakhmanov, A.V. Rozhkov, F. Nori, Phys. Rev. B **87**, 121401 (2013)
101. T.M. Rice, Phys. Rev. B **2**, 3619 (1970)
102. W. Kohn, J.M. Luttinger, Phys. Rev. Lett. **15**, 524 (1965)
103. M.Yu. Kagan, V.A. Mitskan, M.M. Korovushkin, Phys. Usp. **58**(8), 733 (2015)
104. M.Yu. Kagan, JETP Lett. **103**(11), 728 (2016)
105. E.J.W. Verwey, Nature **144**(3642), 327 (1939)
106. E.J.W. Verwey, P.W. Haayman, Physica **8**(9), 979 (1941)
107. S. Schubin, S. Wonsowsky, Proc. Roy. Soc. (London) A **145**(A854), 159 (1934)
108. M.Yu. Kagan, D.V. Efremov, M.S. Marienko, V.V. Val'kov, JETP Lett. **93**(12), 725 (2011)
109. B. Valenzuela, M.A.H. Vozmediano, New J. Phys. **10**(11), 113009 (2008)
110. E.G. Mishchenko, Phys. Rev. Lett. **98**, 216801 (2007)
111. Ya.I. Rodionov, K.I. Kugel, F. Nori, Phys. Rev. B **92**, 195117 (2015)
112. E.L. Nagaev, *Colossal Magnetoresistance and Phase Separation in Magnetic Semiconductors* (World Scientific, Singapore, 2002)
113. E. Dagotto, *Nanoscale Phase Separation and Colossal Magnetoresistance: The Physics of Manganites and Related Compounds, Springer Series in Solid-State Sciences*, vol. 136 (Springer Science & Business Media, Berlin, 2003)
114. E. Dagotto, Science **309**(5732), 257 (2005)
115. Y. Aharonov, D. Bohm, Phys. Rev. **115**, 485 (1959)
116. V.V. Val'kov, M.Yu. Kagan, S.V. Aksenov, J. Phys.: Condens. Matter **31**(22), 225301 (2019)
117. S.V. Aksenov, M.Yu. Kagan, JETP Lett. **111**(5), 286 (2020)
118. V.V. Val'kov, M.S. Shustin, S.V. Aksenov, A.O. Zlotnikov, A.D. Fedoseev, V.A. Mitskan, M.Yu. Kagan, Phys. Usp. **65**(1), 2 (2022)
119. E. Majorana, Nuovo Cim. (1924-1942) **14**, 171 (2008)
120. A.Y. Kitaev, Phys. Usp. **44**(10S), 131 (2001)
121. A. Kitaev, Ann. Phys. **303**(1), 2 (2003)
122. U. Fano, Phys. Rev. **124**, 1866 (1961)
123. M.Yu. Kagan, R. Frésard, M. Capezzali, H. Beck, Physica B **284–288**, 447 (2000)
124. M.Yu. Kagan, E.A. Mazur, J. Exp. Theor. Phys. **132**, 596 (2021)
125. M.Yu. Kagan, E.A. Mazur, J. Phys. Conf. Ser. **1740**, 012045 (2021)
126. E.A. Mazur, RSh. Ikhsanov, M.Yu. Kagan, J. Phys. Conf. Ser. **2036**(1), 012019 (2021)
127. E. Burovski, R.Sh. Ikhsanov, A. Kuznetsov, M.Yu. Kagan, J. Phys. Conf. Ser. **1163**, 012046 (2019)
128. M.A. Baranov, Y. Kagan, M.Yu. Kagan, JETP Lett. **64**(4), 301 (1996)
129. A.V. Turlapov, M.Yu. Kagan, J. Phys.: Condens. Matter **29**(38), 383004 (2017)
130. M.Yu. Kagan, A.V. Turlapov, Phys. Usp. **62**(3), 215 (2019)
131. M.Yu. Kagan, I.V. Brodsky, D.V. Efremov, A.V. Klaptsov, Phys. Rev. A **70**, 023607 (2004)
132. G. Modugno, G. Roati, F. Riboli, F. Ferlaino, R.J. Brecha, M. Inguscio, Science **297**(5590), 2240 (2002)
133. M.Yu. Kagan, A. Bianconi, Condens. Matter **4**(2) (2019)
134. M.Yu. Kagan, S.V. Aksenov, A.V. Turlapov, R.S. Ikhsanov, K.I. Kugel, E.A. Mazur, E.A. Kuznetsov, V.M. Silkin, E.A. Burovski, JETP Lett. **117**(10), 755 (2023)
135. L.V. Keldysh, Yu.V. Kopaev, Sov. Phys. Solid State **6**, 2219 (1965)
136. B.I. Halperin, T.M. Rice, Rev. Mod. Phys. **40**, 755 (1968)
137. W. Kohn, Phys. Rev. Lett. **19**, 439 (1967)
138. A.N. Kozlov, L.A. Maksimov, Sov. Phys. JETP **21**, 790 (1965)
139. A.N. Kozlov, L.A. Maksimov, Sov. Phys. JETP **22**, 889 (1966)
140. L.V. Keldysh, A.N. Kozlov, Sov. Phys. JETP **27**, 521 (1968)
141. D. Snoke, Science **298**(5597), 1368 (2002)
142. J.P. Eisenstein, A.H. MacDonald, Nature **432**(7018), 691 (2004)

143. D. Pines, P. Nozières, *Theory of Quantum Liquids*, 1st edn. (Advanced Book Classics (CRC Press, Boca Raton, 1999)
144. D. Pines, Can. J. Phys. **34**(12A), 1379 (1956)
145. V.M. Silkin, A. Balassis, P.M. Echenique, E.V. Chulkov, Phys. Rev. B **80**, 054521 (2009)
146. A. Balassis, E.V. Chulkov, P.M. Echenique, V.M. Silkin, Phys. Rev. B **78**, 224502 (2008)
147. V.P. Zhukov, V.M. Silkin, E.V. Chulkov, P.M. Echenique, Phys. Rev. B **64**, 180507 (2001)
148. V.E. Zakharov, E.A. Kuznetsov, Sov. Phys. JETP **64**, 773 (1986)
149. V.E. Zakharov, E.A. Kuznetsov, Phys. Usp. **55**(6), 535 (2012)
150. L.V. Keldysh, Phys. Usp. **60**(11), 1180 (2017)
151. V.M. Pudalov, JETP Lett. **116**(8), 286 (2022)
152. A.F. Andreev, M.Yu. Kagan, Sov. Phys. JETP **66**(3), 504 (1987)
153. M.Yu. Kagan, D.V. Efremov, J. Low Temp. Phys. **158**, 749 (2010)
154. A. Balatskii, G. Volovik, V. Konyshev, Sov. Phys. JETP **63**, 1194 (1986)
155. D.T. Son, B.Z. Spivak, Phys. Rev. B **88**, 104412 (2013)
156. B.Z. Spivak, A.V. Andreev, Phys. Rev. B **93**, 085107 (2016)
157. A.V. Andreev, B.Z. Spivak, Phys. Rev. Lett. **120**, 026601 (2018)

Chapter 2
Magnetic Polarons and Related Transport Phenomena

Abstract We give simple estimates for the radius of small spherical FM metallic droplets in insulating AFM matrix of CMR manganites in the framework of the double exchange model. We discuss also the droplet contribution to different transport properties of phase-separated magnetic oxides, such as resistivity, magnetoresistance, magnetic susceptibility, and the spectrum of $1/f$ noise.

2.1 Magnetic Polarons in Magnetic Semiconductors and Manganites

Historically, nanoscale phase separation with the formation of small FM droplets inside AFM insulating matrices was first predicted at the end of 1960s in seminal papers of Mott, Nagaev, and Kasuya on magnetic semiconductors [1–5]. In this context, it is important to mention also the papers of Krivoglaz [6] on heterophase inhomogeneities (so-called fluctuons) arising, e.g., in paramagnetic insulating matrices at finite temperatures, as well as a very important paper by de Gennes [7] on the double exchange model, canted spin states, and other type of (bound) magnetic polarons (or magnetic impurities), which can arise in such model. Later on, analogous predictions for bound magnetic polarons on regular and frustrated lattices were made in [8–10], in the framework of the specific model, which effectively combines exchange interaction, magnetic anisotropy, and attractive electron–ion Coulomb interaction for manganites in the limit of very small doping (see Chap. 3 for more details). The nanoscale electronic phase separation in manganites and other systems exhibiting the phenomena of the colossal magnetoresistance (CMR) [11–15] was experimentally confirmed by several groups [16–22], which utilized different experimental techniques including elastic and inelastic neutron scattering, NMR, STM, electron diffraction, dark image spectroscopy, and so on. Let us recall that the CMR phenomenon represents a colossal decrease of resistivity (often by a factor of 100–10000) in moderately strong magnetic fields $H = 2 - 4$ T. Physically it is often related to the increase in the size of magnetic inhomogeneities (magnetic polarons or ferrons) in the applied magnetic

M. Yu. Kagan et al., *Electronic Phase Separation in Magnetic and Superconducting Materials*, Springer Series in Solid-State Sciences 201,
https://doi.org/10.1007/978-3-031-55467-4_2

field, which, in its turn, increases the metallic fraction in the sample. As a result, we have a strong decrease in the resistivity. The magnetoresistance is negative in this case

$$MR = \frac{\rho(H) - \rho(0)}{\rho(H)}\,. \tag{2.1}$$

Let us remind that according to the Mott–Nagaev–Kasuya approach, the radius of a 3D spherical droplet in the absence of magnetic field is (see the details below)

$$R_{pol} = a\left(\frac{\pi t}{2zJ_{ff}S^2}\right)^{1/5}, \tag{2.2}$$

where a is the period of the lattice, z is the number of nearest spins in the given lattice, S is the value of a local spin, J_{ff} is the Heisenberg (AFM) exchange interaction between local spins at the neighboring sites, and t is the hopping matrix element between the neighboring sites for conduction electrons. In the 2D case for FM droplets of the circular shape, the power 1/5 in Eq. (2.2) is replaced by the power 1/4. The droplet radius in Eq. (2.2) can be found from the minimization of the energy of the nanoscale phase-separated state, which reads [23]

$$E_{pol} = -tna\left(z - \frac{\pi^2 a^2}{R^2}\right) + \frac{4}{3}\pi\frac{zJ_{ff}S^2}{2}\left(\frac{R}{a}\right)^3 n - \frac{1}{2}zJ_{ff}S^2\left[1 - \frac{4}{3}\pi\left(\frac{R}{a}\right)^3\right]n\,. \tag{2.3}$$

From the minimization condition $dE_{pol}/dR = 0$, we find the optimum droplet radius given by Eq. (2.2). The structure of a spherical FM droplet within the AFM matrix is illustrated in Fig. 2.1

Let us now show that the homogeneous canted state for the typical values of parameter $\alpha = t/J_{ff}S^2$ of the order of 80–100 (which is typical of manganites) is less favorable in energy than the (inhomogeneous) nanoscale phase-separated state. Moreover, in a wide range of parameters, the homogeneous canted state has a negative compressibility $\kappa = d^2E/dn^2$, which signals the instability of the ground state toward the electronic phase separation. This instability can be easily demonstrated for the classical canted state proposed by de Gennes. In the classical picture, an electron cannot move being surrounded by AFM-ordered local spins, and that is why, it is favorable for the electron to cant one sublattice relative to another for having a possibility of hopping to a neighboring site. As a result, the homogeneous canted state has the energy

$$E = -tn\cos\frac{\theta}{2} + \frac{1}{2}zJ_{ff}S^2\cos\theta\,, \tag{2.4}$$

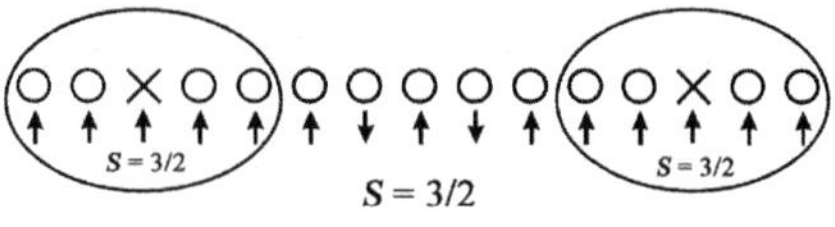

Fig. 2.1 FM polarons with one conduction electron inside the insulating AFM matrix [23]

where θ is the canting angle between two AFM sublattices and the arising effective hopping integral $t_{eff} = t\cos(\theta/2)$ is related to the spinor nature of the electron wave function [24]. Minimizing the energy in Eq. (2.4) with respect to $\cos(\theta/2)$, we get

$$\cos\frac{\theta}{2} = \frac{tn}{2J_{ff}S^2}, \tag{2.5}$$

and, consequently,

$$E = -\frac{zt^2n^2}{4J_{ff}S^2} - \frac{zJ_{ff}S^2}{2}. \tag{2.6}$$

As a result, the compressibility of the classical canted state reads

$$\kappa = -\frac{zt^2}{2J_{ff}S^2} < 0. \tag{2.7}$$

It is obviously negative, thus suggesting a tendency toward the phase separation .

It is important to stress that the energies of both the nanoscale phase-separated state (2.3) and the homogeneous canted state (2.5) can be considered as different approximations to the exact energy of the ground state in the double exchange model, which is specified by the Hamiltonian

$$\hat{H} = -J_H\sum_i \mathbf{S}_i\boldsymbol{\sigma}_i - t\sum_{\langle i,j\rangle} Pc^{\dagger}_{i\sigma}c_{i\sigma}P + J_{ff}\sum_{\langle i,j\rangle}\mathbf{S}_i\mathbf{S}_j\,, \tag{2.8}$$

where J_H is a strong onsite Hund's rule coupling, $\boldsymbol{\sigma} = c^{\dagger}\boldsymbol{\tau}c/2$ is the spin operator of conduction electrons, $\boldsymbol{\tau}$ are Pauli matrices, J_{ff} is the AFM Heisenberg exchange between local spins at the neighboring sites, $c\dagger$ and c are creation and annihilation operators, and $W = 2zt$ is the electron bandwidth. In manganites, projection operator P is responsible for the single occupation of e_g orbitals, for which a strong onsite Coulomb repulsion U prevents two conduction electrons from occupying the same site. In actual manganites, we have $J_H \simeq 1$ eV, $t \simeq 0.3$ eV, $J_{ff} \simeq 0.001$ eV, $z = 6$ for the simple cubic lattice, $S = 3/2$ and thus we are in the strong-coupling limit of the FM Kondo lattice model (FM KLM) $J_H \gg t \gg J_{ff}S^2$, which in the literature is just called the double exchange model of de Gennes. In this limit, a conduction electron and local spin form together a coupled state with a total spin $S_{tot} = S + 1/2$ at one site. Both in the energy of the phase-separated state (2.3) and the energy of the classical canted state (2.5), we omit the large constant term $J_H S/2$ (related to the parallel alignment of the local and conduction electron spins) since this term does not change the difference between the energies of these states. The result for t_{eff} in the classical canted state was derived in [24] just in the strong-coupling limit of the FM Kondo lattice model.

In the quantum canted state introduced by Nagaev [2], there are two bands t_+ and t_- for the motion of conduction electrons corresponding to two different projections of the total spin $S^z_{tot} = S + 1/2$ and $S^z_{tot} = S - 1/2$. The de Gennes classical description of the canted state corresponds to the limit $t_- \to 0$, whereas, in a quantum treatment for the canting angle $\theta = \pi$ (which corresponds to the collinear AFM structure), both hopping integrals are equal to each other $t_- = t_+ = t/\sqrt{2S+1}$. In this point lies the main difference between quantum and classical canted state. In the quantum picture, the hopping integral for large values of local spin $S \gg 1$ is small only as $1/\sqrt{S}$, while in the de Gennes classical limit, it is small as $1/S$. In quantum description, the physical mechanism allowing the electron hopping in the AFM environment (even for the angle $\theta = \pi$, where it is classically forbidden) is related to the possibility of coherent string-type motion (see [25]). This type of motion occurs when an electron hops from the site with a total spin projection $S^z_{tot} = S + 1/2$ to the neighboring site (where it forms the state with the total spin projection $S^z_{tot} = S - 1/2$), then hops back to the site where it forms the state with $S^z_{tot} = S + 1/2$, and so on. As a result of this motion, the coherent band of the width $W = 2zt/\sqrt{2S+1}$ is formed. However, even in the quantum case, it is possible to show that, at least for substantially large values of the parameter $\alpha = t/J_{ff}S^2$, the nanoscale phase-separated state is more favorable in energy [26]. The homogeneous string-like motion in collinear AFM environment in exact quantum treatment (which includes the spatial variations of the canting angle $\theta(x)$ and its gradients) can be stabilized only when $\alpha < \alpha_c$, where α_c is the threshold value of the order of 75 (see [27] and Fig. 2.2). For cuprates, we have $\alpha = t/J_{ff}S^2 \approx 2 - 3$ (for $S = 1/2$) and instead of FM polarons, the AFM strings discussed in [28, 29] can be stabilized in the system at low hole doping.

In the applied magnetic field, an effective Heisenberg exchange decreases due to the Zeeman splitting and reads

$$J^{eff}_{ff} S^2 = J_{ff} S^2 - g\mu_{BHS}, \tag{2.9}$$

where g is the gyromagnetic ratio and μ_B is the Bohr magneton. Correspondingly, the ferron radius in Eq. (2.2) and its volume

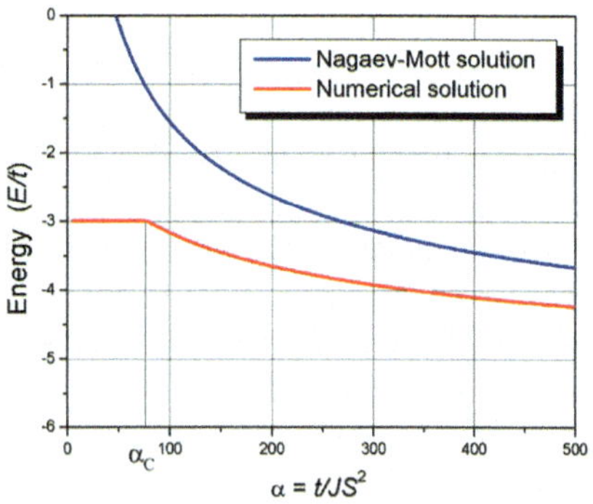

Fig. 2.2 Comparison of the Nagaev–Mott solution (taking into account the quantum effects) and the exact numerical solution for the energy of a FM polaron [27]

$$\Omega = \frac{4\pi}{3}\left(\frac{R_{pol}}{a}\right)^3 \tag{2.10}$$

also increase due to the decrease in the effective exchange.

In the case of electron-doped system, the central spin $\sigma = 1/2$ of the conduction electron in a FM droplet is surrounded by local spins $S = 3/2$, whereas in case of hole doping (typical for manganites), the central unpaired spin $S = 3/2$ is surrounded by paired spins $S_{tot} = 2$.

The similar effect of the FM polaron growth in magnetic field takes place for the so-called temperature ferrons, which arise in the paramagnetic (PM) phase. In 3D systems, their radius in the absence of magnetic field is given by

$$R_T \approx a\left(\frac{\pi t}{2T\ln(2S+1)}\right)^{1/5}, \tag{2.11}$$

and can be obtained from the minimization of the corresponding change in the free energy ΔF. This change takes place due to the formation of the FM droplets above the Curie temperature T_C in the PM matrices and can be written as [23]

$$\Delta F = -tn\left(z - \frac{\pi^2 a^2}{R^2}\right) + \frac{4}{3}\pi T\left(\frac{R}{a}\right)^3 n\ln(2S+1)\,. \tag{2.12}$$

In the denominator of Eq. (2.11), instead of the Heisenberg exchange term $J_{eff}S^2$, we have the term $T\ln(2S+1)$, which determines the contribution of the spin entropy to ΔF. This term also decreases under the effect of the applied field due to the Zeeman effect, which reduces the spin degeneracy, thus increasing again the radius and the volume of the 3D spherical droplet.

It should be emphasized again that in the case of the nanoscale phase separation, electron transport is described by the spin-dependent tunneling of conduction electron between the neighboring metallic FM droplets separated by AFM insulating barriers. When we increase the concentration of droplets, we get closer to the percolation threshold. In this case, the average distance between the droplets becomes strongly reduced and we can effectively consider the system as a network of the tunneling contacts (or bridges). This limiting case is described by the physics of the tunneling magnetoresistance (TMR) [30]. If we apply an external magnetic field, we can control the contact resistance. In this way, we can control the electric current flowing across contact, and hence can read and write the information. This is a main goal and the main challenge for the applications in spintronics [31].

In the system of 3D spherical droplets, the percolation threshold is achieved at the critical density of charge carriers

$$n_c = 1/\Omega = \frac{3}{4\pi}\left(\frac{a}{R_{pol}}\right)^3. \tag{2.13}$$

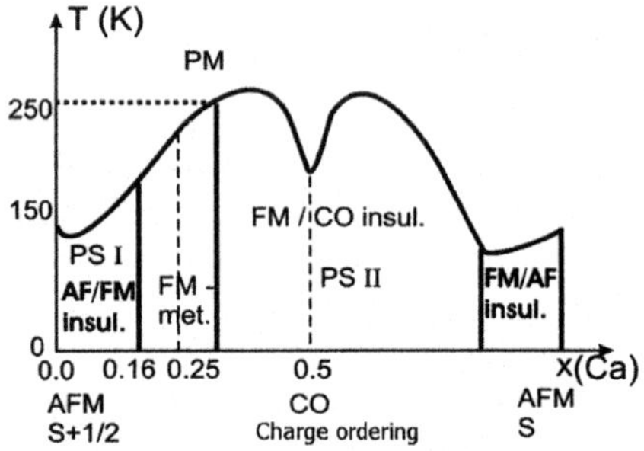

Fig. 2.3 Schematic illustration of the global phase diagram of manganites [32]. PSI and PSII are different kinds of the phase-separated state: FM metallic droplets in the AFM and charge-ordered (CO) insulating matrices, respectively

In real 3D manganites, the percolation threshold corresponds to the critical density of about 0.16. For larger densities, the ferrons start to overlap forming the infinite FM metallic cluster. The CMR phenomenon is especially pronounced for the charge carrier density within the $n = 0.2 - 0.4$ range (see Fig. 2.3 illustrating the global phase diagram of manganites). Here, we have two phase transitions, which occur simultaneously at the Curie temperature T_C, namely, the metal–insulator and FM–PM transitions. That is why, even within this density range, for temperatures above T_C, the electrical resistivity is mainly determined by the presence of temperature ferrons inside the PM matrix. At the same time, the temperature ferrons overlap at the critical density

$$\delta_c = \frac{3}{4\pi}\left(\frac{a}{R_T}\right)^3. \tag{2.14}$$

Substitution of Eqs. (2.2) and (2.11) to Eqs. (2.13) and (2.14) yields for the ratio

$$\frac{n_c}{\delta_c} \propto \left[\frac{T\ln(2S+1)}{zJ_{ff}S^2}\right]^{3/5} \propto \left[\frac{T_C\ln(2S+1)}{T_N}\right]^{3/5} \tag{2.15}$$

near the triple point between FM, AFM, and PM phases. Here, T_C and T_N are the Curie and Néel temperatures, respectively. For the hole-doped manganites, $T_C \backsim T_N \approx 120 - 150$ K and $\ln(2S+1) \approx 1.6$ for $S = 2$. Thus, $\delta_c < n_c$ in agreement with the experimental data.

2.2 Electron Transport in Inhomogeneous Phase-Separated Manganites

2.2.1 Electrical Conductivity in the Phase-Separated Systems

Let us consider in more detail the transport properties of the phase-separated manganites. As we will show below, the electrical resistivity in the nanoscale phase-separated state reads

$$\rho(T) = BT \exp \frac{A}{2T}, \tag{2.16}$$

where A is the energy corresponding to the Coulomb repulsion of two conduction electrons occupying the same FM metallic droplet. In the model of spherical droplets, A is given by

$$A \approx \frac{e^2}{\varepsilon R_{pol}}, \tag{2.17}$$

where ε is the static dielectric constant. In manganites, it can be of the order of 10–20, so the typical values of A are of the order of $(3–4)\times 10^3$ K. In Eq. (2.17), the droplet radius R_{pol} is given by Eq. (2.2).

The temperature-independent factor B in Eq. (2.16) in the spherical droplet model can be written as [33]

$$B = \frac{1}{128\pi e^2 \omega_0 l^5 n^2 k}, \tag{2.18}$$

where e is the electron charge, n is the concentration of the FM droplets, l is the characteristic tunneling length, ω_0 is a frequency corresponding to the characteristic energy of an electron in a droplet, and k is the number of itinerant charge carriers in the droplet. Let us emphasize that in small droplets with the radius given by Eq. (2.2), we have $k = 1$, whereas in large droplets usually observed in transport measurements [34–38], we are dealing with $k \gg 1$. A typical size of a large droplet can be $(7–8)a$, and hence the number of charge carriers can reach such values as $k = 50 - 100$. Thus, effectively we can visualize a large droplet as an object which can be divided into many Wigner–Seitz spheres of small radius, each sphere containing one charge carrier to minimize the Coulomb energy.

Let us try to understand qualitatively the results of Eqs. (2.16)–(2.18). For this purpose, we should follow [33, 39] considering a sample of volume V_S in external electric field $\mathbf{E}$. We assume that the sample is in the insulating state and there are N metallic droplets (polarons) embedded in the sample. Then, for the density of the droplets, we find $n = N/V_S$. In the simplest case, upon doping, the number of droplets is equal to the number of electrons (or holes). We neglect the conductivity of the insulating phase and thus consider the case, when we have charge carriers only inside the droplets. In this case, a charge transfer is governed by the motion of the (heavy) droplets as a whole or (if the droplets are considered to be immobile) by the electron tunneling between the droplets. The first mechanism is usually ignored because the motion of a droplet as a whole is related to a strong reconstruction of the local spins surrounding the droplet. Hence a droplet acquires a large effective mass. Moreover, quite often the droplets are localized at dislocations and at other lattice defects. As a result, we can consider only the tunneling of electrons between almost immobile droplets.

If the size of the droplet is small enough, the polaron typically possess only one electron (or hole) in the ground state of the droplet. However, if we take the tunneling processes into account, then we should consider also the droplets with more than one electron (e.g., with two electrons) as well as empty droplets (assuming that their lifetimes are quite large). Let the energy of the empty droplet be zero, $E(0) = 0$. In this case, the energy of a one-electron droplet is $E(1) \approx t(a/R_{pol})^2$. This estimate corresponds to the kinetic energy of electron localization inside a spherical droplet of radius R_{pol}. Accordingly, for the energy of a two-electron droplets, we can give an estimate $E(2) \approx 2E(1) + U$, where U is the energy of interaction between two electrons. Here, we neglect the surface energy. By doing this, we assume that the surface energy for the FM droplets (or ferrons) is small according to the estimates given in [26, 27, 40]. Therefore, we obtain $E(2) + E(0) > 2E(1)$. This means that to form the two-electron droplets, we need to overcome a potential barrier with the energy $A = E(2) - 2E(1) \approx U$. It is easy to understand that U is governed by the Coulomb repulsion between two electrons localized at one droplet. In this way, we come to the estimate for A given by Eq. (2.17). We assume that we are working in the low-doping limit, where the mean distance between the droplets is $n^{-1/3} \gg R_{pol}$ (the polarons do not overlap). Thus, in accordance with our estimates and transport measurements (see below), the typical values of $A \simeq 3500$–3700 K are larger than the average Coulomb energy $e^2 n^{1/3}/\varepsilon_0$. Moreover, $A \gg T$ for all realistic temperatures. That is why, we can neglect the creation of the droplets with the number of electrons larger than two. It is possible to demonstrate that even if the many-electron droplets are stable, their contribution to conductivity is strongly reduced by the Coulomb interaction, and hence it is very small for low droplet densities (when the system is far from the percolation threshold).

Let us evaluate factor B in Eq. (2.18), which is essentially temperature independent [41]). It is convenient to introduce parameters N_1, N_2, and N_3 for the numbers of droplets with one electron, two electrons, and of empty droplets, respectively. According to our model, we assume that the total number of the droplets in the system is constant. Thus, $N_2 = N_3$ and $N_1 + 2N_2 = N$, and the partition function of the system is given by

$$Z = \sum_{m=0}^{N} C_N^m C_{N-m}^m \exp(-m\beta), \quad \beta = A/T\,, \tag{2.19}$$

where C_N^m are the binomial coefficients. Using the Stirling formula, replacing the summation by the integration, and calculating the corresponding integral by the steepest descent method, we obtain the following expressions for the average values of N_1, N_2, and N_3

$$\overline{N_2} = \overline{N_3} = N \exp\left(-\frac{A}{2T}\right),$$
$$\overline{N_1} = N - 2\overline{N_2} = N\left[1 - 2\exp\left(-\frac{A}{2T}\right)\right]. \tag{2.20}$$

The electron tunneling implies one of the four possible events (see Fig. 2.4) (i) two droplets with one electron are converted into the droplet with two electrons and

Fig. 2.4 Elementary tunneling processes [33]. The arrows and σ_i denote spin projections

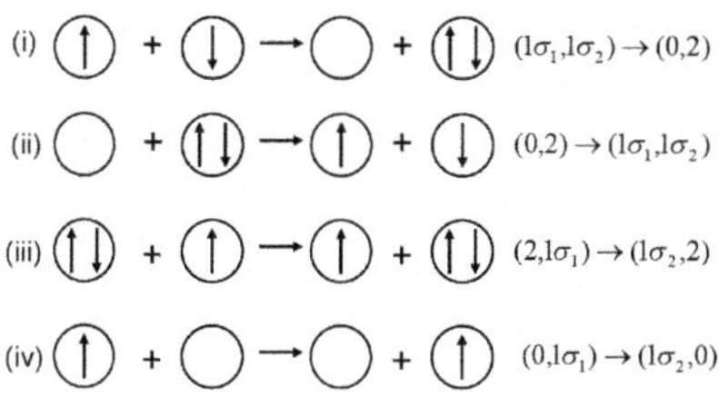

an empty droplet; (ii) process, which is inverse to the previous one; (iii) droplets with one and two electrons exchange their positions; and (iv) the same for the droplet with one electron and an empty droplet. The corresponding total current density is given by the sum of the contributions of all these processes $j = j_1 + j_2 + j_3 + j_4$. It was shown in [33] that all these processes equally contribute to the total electric current.

Let us consider, for example, the contribution of the first two processes

$$j_{1,2} = en_{1,2}\langle\sum_i v^i_{1,2}\rangle = en_{1,2}\langle\sum_i \frac{r^i\cos\theta^i}{\tau_{1,2}(r^i,\theta^i)}\rangle\,. \tag{2.21}$$

In Eq. (2.21), $n_{1,2} = N_{1,2}/V_S$ denote the densities of one-electron and two-electron droplets, $\langle ... \rangle$ corresponds to the statistical and time averages, $\langle v^i\rangle$ is the average velocity of electrons along the electric field $\mathbf{E}$, r^i is the distance between the droplets involved in these processes, θ^i is the angle between vector $\mathbf{E}$ and the direction of the electron motion, and $\tau_{1,2}(r^i,\theta^i)$ are characteristic times associated with these processes. The conventional expressions for $\tau_{1,2}$ (which are derived in [33] using the detailed balance equation) have the following form:

$$\tau_{1,2} = \omega_0^{-1}\exp\left(\frac{r}{l} \pm \frac{A}{2T} - \frac{eEr\cos\theta}{T}\right), \tag{2.22}$$

where l is the tunneling length and ω_0 is a characteristic magnon (or depolarization) frequency. Let us stress that when we are far from the percolation threshold, the average in (2.21) is reduced to the space average for velocity v_i multiplied by the number of droplets, to which the hopping is allowed. These numbers are equal to N_1 for process (i) and, respectively, to N_2 for process (ii). If we assume that the external electric field is small, $eEl/T \ll 1$, and moreover the repulsion energy is large, $A \gg T$, then in the first order in E, we have

$$\langle\sum_i v^i_{1,2}\rangle = \frac{eE\omega_0}{T}N_{1,2}\exp\left(\mp\frac{A}{2T}\right)\langle r^2\cos^2\theta\exp\left(-\frac{r}{l}\right)\rangle_V\,, \tag{2.23}$$

where $\langle ... \rangle_V$ denotes the average over the sample volume

$$\langle ... \rangle_V = V_S^{-1}\int ...d^3\mathbf{r}\,. \tag{2.24}$$

Performing the integration, we get

$$j_{1,2} = \frac{32\pi e^2 E\omega_0 l^5 n_{1,2}^2}{T} \exp\left(\mp\frac{A}{2T}\right). \tag{2.25}$$

Analogously, it is possible to demonstrate rather straightforwardly that for other tunneling processes, the corresponding current densities read

$$j_{3,2} = \frac{32\pi e^2 E\omega_0 l^5 n_1 n_2}{T}. \tag{2.26}$$

Comparing Eqs. (2.25) and (2.26), we can see that in the last one, the exponential factor is missing and the factors $n_{1,2}^2$ are replaced by $n_1 n_2$.

From Eqs. (2.20), (2.25), and (2.26), we can obtain the expression for dc conductivity

$$\sigma = 1/\rho = \frac{128\pi e^2 n^2 \omega_0 l^5}{T} \exp\left(-\frac{A}{2T}\right)k. \tag{2.27}$$

For small droplets described by Eq. (2.2), we can show that only one-electron droplet is stable, while the empty droplets and two-electron droplets decay. The decay time of the empty droplets is governed by the inverse depolarization frequency. At the same time, an empty droplet can get an electron from the one-electron or two-electron droplets located nearby. This process involves a time scale, which can be estimated as (see [37]) in agreement with Eqs. (2.16) and (2.18). For small droplets described by Eq. (2.2), we can show that, in fact, only one-electron droplets are stable, whereas empty droplets and two-electron droplets decay. The decay time of empty droplets is governed by the inverse depolarization frequency $1/\omega_0$. At the same time, an empty droplet can get an electron from one-electron or two-electron droplets located nearby. This process involves a time scale τ_0, which can be estimated as (see [33])

$$\tau_0 = \frac{\exp\left(-\frac{A}{2T}\right)}{8\pi\omega_0 l^5 n}. \tag{2.28}$$

The similar estimate is valid for an electron, which leaves a two-electron droplet. The validity of this picture is determined by the inequality $\tau_0 \ll \omega_0^{-1}$. Thus, our approach requires inequality $A \gg T$ and not too small droplet density n.

2.2.2 Comparison with Experimental Data

In Fig. 2.5, we present the plots for resistivity $\rho(T)$ in six families of manganites using the experimental data reported in [34–38]. As can be seen from these data, the samples differ in chemical composition, the type of the crystal structure, the magnitude and the low-temperature behavior of the resistivity (some of the samples

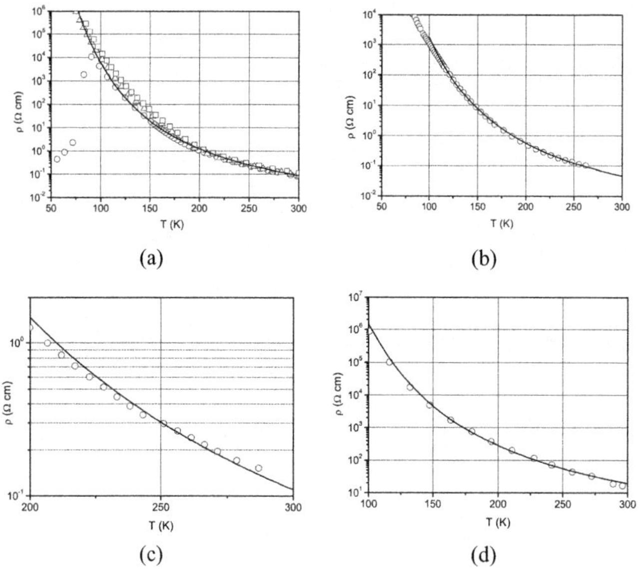

Fig. 2.5 Temperature dependence of electrical resistivity for **a** $(La_{1-y}Pr_y)_{0.7}$ $Ca_{0.3}MnO_3$ with $^{16}O \rightarrow {}^{18}O$ oxygen isotope substitution for $y = 1$ (squares) and $y = 0.75$ (triangles); **b** $Pr_{0.71}Ca_{0.29}MnO_3$; **c** a layered manganite with the chemical composition $(La_{0.4}Pr_{0.6})_{1.2}$ $Sr_{1.8}Mn_2O_7$; and **d** $La_{0.8}Mg_{0.2}MnO_3$ [41]

behave as metals, some as the insulators, whereas the resistivity at a given temperature varies more than in two orders of magnitude). However, in the high-temperature range above the Curie temperature, the resistivity exhibits an identical behavior for all the samples governed by the universal relation (2.27). This relation is typical of many tunneling systems [1] and is represented by solid curve in Fig. 2.5.

Using the experimental data and Eq. (2.27), we can determine the value of the Coulomb barrier energy A and make a reasonable estimate for the characteristic depolarization frequency $\omega_0 \simeq 10^{13} - 10^{14}$ Hz. We can also determine the important factor $l^5 n^2 k$ which enters Eq. (2.18) for the temperature-independent coefficient B (see Table 2.1).

2.2.3 *Magnetoresistance of Phase-Separated Manganites*

Now we can proceed to the analysis of the magnetoresistance in phase-separated manganites. Using Eq. (2.9) for the effective Heisenberg exchange, we can obtain the following expression for the droplet radius in experimentally accessible magnetic fields

$$R_{pol}(H) = R_{pol}(0)(1 + bH) , \tag{2.29}$$

Table 2.1 Coulomb energy A, resistivity ρ at 200 K, and product l^5n^2k for several families of manganites [41]

Samples	A (K)	ρ(200 K) Ω·cm	l^5n^2k cm^{-1}	Data source
$(La_{1-y}Pr_y)_{0.7}Ca_{0.3}MnO_3$	3650	1.25	2×10^5	Figure 2.5a (Babushkina et al. [34])
$Pr_{0.71}Ca_{0.29}MnO_3$	3500	0.57	3×10^5	Figure 2.5b (Fisher et al. [35])
$(La_{0.4}Pr_{0.6})_{1.2}Sr_{1.8}Mn_2O_7$	3600	1.5	1.5×10^5	Figure 2.5c (Wagner et al. [36])
$La_{0.8}Mg_{0.2}MnO_3$	3700	283	1×10^3	Figure 2.5d (Zhao et al. [37])

where in the 3D case, the coefficient b is

$$b = \frac{g\mu_B}{5J_{ff}S}. \tag{2.30}$$

For the corresponding decrease in the Coulomb barrier height A, we find

$$A(H) = A(0)(1 - bH). \tag{2.31}$$

Thus, using Eq. (2.16), we find that the magnetoresistance in the nanoscale phase-separated state is negative and can be written as

$$MR = 1 - \exp\left(\frac{AbH}{2T}\right). \tag{2.32}$$

It is shown in [41] that such exponential behavior really takes place in strong magnetic fields $H > 10$ Tesla. Since the gyromagnetic ratio $g \sim 10$ is typical of manganites, this behavior corresponds just to the magnetic field range where the argument of exponential in Eq. (2.32) becomes larger than unity. At the same time, at low and moderate magnetic fields, the magnetoresistance is quadratic in the field (in agreement with the textbooks on solid-state physics [42]) and is given by the relation

$$MR \propto -H^2/T^n, \tag{2.33}$$

where the exponent n in Eq. (2.33) is either 1, 2, or 5 depending on the range of magnetic fields under consideration, the value of magnetic anisotropy, and so on [41, 43, 44]. The magnetoresistance in this case is determined by the effects of the spin-dependent tunneling. The value $n = 5$ is typical of the magnetoresistance in manganites in the range of low magnetic fields and high temperatures. In this case, we have [41]

$$|MR| = 5 \times 10^{-3} \frac{\mu_B^3 S^5 J_{FM}^2 N_{eff}^3 Z^2 g^3 H_a}{T^5} H^2 = 5 \times 10^{-3} \frac{M_0^3 J_{FM}^2 Z^2 S^2 H_a}{T^5} H^2 , \tag{2.34}$$

where we introduced the magnetic moment of a droplet

$$M_0 = g\mu_B S N_{eff} . \tag{2.35}$$

In Eq. (2.34), N_{eff} is the number of Mn atoms inside the droplet, $S = 2$ is the total spin of Mn^{3+} ions in manganites under the hole doping, and Z is the number of nearest neighbors of a Mn ion. Let us stress that for 3D cubic manganites with $Z = 6$, H_a is the effective magnetic anisotropy field of a droplet. At the same time, an effective FM exchange integral J_{FM} is specified by the well-known relationship of the molecular field theory [45]

$$T_C = S(S+1)ZJ_{FM}/3 . \tag{2.36}$$

The behavior of the magnetoresistance given by Eq. (2.34) was also experimentally observed in different families of manganites [34–38] (see Fig. 2.6).

The solid curves in Fig. 2.6 were calculated for the value of $T_C = 120$ K. At the same time, the fitting parameter N_{eff} was determined from the relation $k = N_{eff}x$,

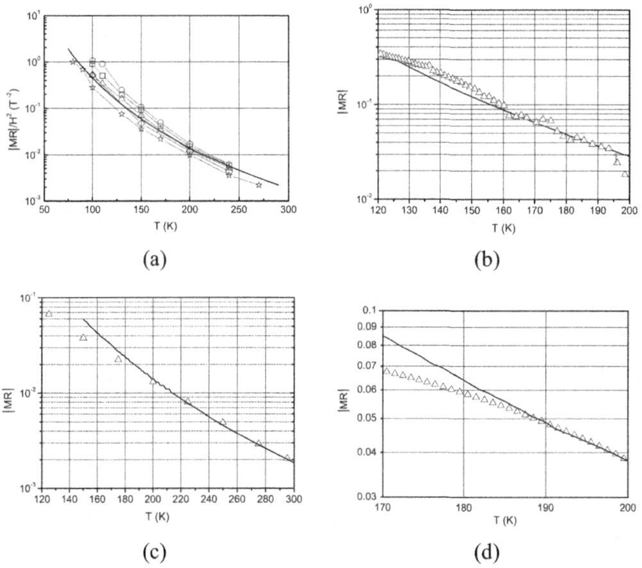

Fig. 2.6 Temperature dependence of the absolute value of magnetoresistance. Solid curve shows the results of the theoretical calculations using relation (2.34), $|MR| \propto 1/T^5$. **a** $|MR|/H^2$ ratio for $(La_{1-y}Pr_y)_{0.7}$ $Ca_{0.3}MnO_3$ at $y = 1$ and $y = 0.75$; **b** $Pr_{0.71}Ca_{0.29}MnO_3$ at $H = 2$ T; **c** layered manganite $(La_{0.4}Pr_{0.6})_{1.2}$ $Sr_{1.8}Mn_2O_7$ at $H = 1$ T; and **d** $La_{0.8}Mg_{0.2}MnO_3$ for $H = 1.5$ T [41]

Table 2.2 Effective number N_{eff} of manganese atom and the number of electrons k in a droplet, and the dopant fraction x for several different families of manganites, estimated using Eq. (2.34) and experimental data presented in Fig. 2.6

Samples	N_{eff}	x	k	Data source
$(La_{1-y}Pr_y)_{0.7}Ca_{0.3}MnO_3$	250	0.35	75	Figure 2.6a (Babushkina et al. [34])
$Pr_{0.71}Ca_{0.29}MnO_3$	200	0.29	58	Figure 2.6b (Fisher et al. [35])
$(La_{0.4}Pr_{0.6})_{1.2}Sr_{1.8}Mn_2O_7$	250	0.4	100	Figure 2.6c (Wagner et al. [36])
$La_{0.8}Mg_{0.2}MnO_3$	265	0.2	53	Figure 2.6d (Zhao et al. [37])

where x is the atomic fraction of the dopant. The values of N_{eff}, x, and k are presented in Table 2.2.

In brief, the derivation of Eq. (2.34) is as follows. The conductivity of the system is controlled by the "spin" contribution $\Sigma(H)$ to σ, that is, $\sigma(H) = \sigma_0\langle\Sigma(H)\rangle$). Therefore, we have $|MR| = \langle\Sigma(H)\rangle - 1$. Assuming the interaction between the droplets to be negligibly small and using an expression (2.35) for the effective magnetic moment, we can represent the free energy of the droplet in a magnetic field in the following form:

$$U(H) = U(0) - M(H\cos\theta + H_a\cos^2\Psi)\,. \tag{2.37}$$

In Eq. (2.37), θ is the angle between the external magnetic field **H** and the magnetic moment **M**, and H_a stands for the field of magnetic anisotropy. Finally, Ψ corresponds to the angle between the axis of the uniaxial anisotropy and **M**. Let **H** be parallel to the z-axis and let the anisotropy axis lie in the (x, z)-plane and have the angle β with vector **H**. In this configuration, we find

$$\cos\Psi = \sin\theta\sin\beta\cos\phi + \cos\theta\cos\beta\,. \tag{2.38}$$

Equation (2.38), ϕ denotes the angle between the x-axis and the projection of magnetic moment **M** onto the (x, y)-plane.

In the classical limit, a given orientation of vector **M** corresponds to the probability

$$P(H,\theta,\varphi) = C(H)\exp\frac{M(H\cos\theta + H_a\cos^2\Psi(\theta,\varphi))}{T}\,, \tag{2.39}$$

where $C(H)$ is the normalization factor. The eigenstates for a conduction electron correspond to the conservation of the spin projection $s = \pm 1/2$ onto the effective field direction in FM correlated region. Let the electron interact with Z magnetic moments of the droplet. The energy of this interaction is $E_s = -J_{FM}SZs$. Since the value of $J_{FM}SZ$ product according to Eq. (2.34) is of the order of the Curie temperature, E_s is much larger than the energy of interaction between the electron spin and magnetic field for realistic magnetic fields $H < 100$ T. In this situation, the direction of the effective field is parallel to the direction of the magnetic moment **M**. Therefore, the probability for the projection of the electron spin to be equal to s can be written as

$$P_s = \frac{\exp(-E_s/T)}{2\cosh(E_s/T)}. \tag{2.40}$$

In the process of the transfer from droplet 1 to droplet 2, an electron moves in the effective field, which forms an angle ν with the effective field in the initial state. This angle is naturally defined via the obvious relation $\cos\nu = \cos\theta_1\cos\theta_2 + \sin\theta_1\sin\theta_2\cos(\varphi_1 - \varphi_2)$, where subscripts 1 and 2 refer to the droplet numbers. Then, the work performed in the process of electron transfer from droplet 1 to droplet 2 is $\Delta E_s(1 - \cos\nu)$. Accordingly, the probability of this transfer is proportional to $\exp(-\Delta E_s/T)$. Taking into account all the probability factors introduced above, we obtain the final expression for the "spin contribution"

$$\langle\Sigma(H)\rangle = \int_0^{2\pi} d\varphi_1 \int_0^{2\pi} d\varphi_2 \int_0^{\pi} \sin\theta_1 d\theta_1 \int_0^{\pi} \sin\theta_2 d\theta_2 P(\theta_1,\phi_1)P(\theta_2,\phi_2)$$
$$\sum_{s=\pm 1/2} P_s \exp\left(\frac{-\Delta E_s}{T}\right). \tag{2.41}$$

The high-temperature expansion of the expression (2.41) for $T \gg \{\mu_B g S N_{eff} H, \mu_B g S N_{eff} H_a\}$ yields formula (2.34).

2.2.4 *Specific Features of the Magnetic Susceptibility in the Phase-Separated State*

At the high-temperature limit, the susceptibility of nanoscale phase-separated state is given by the Curie–Weiss law [41]

$$\chi(T) = \frac{n(\mu_B g S N_{eff})^2}{3(T-\Theta)} = \frac{nM_0^2}{3(T-\Theta)}, \tag{2.42}$$

where Θ is the Curie–Weiss constant.

The linear temperature dependence for the inverse susceptibility is also confirmed in experiments [34–38] for several different families of manganites (see Fig. 2.7, where solid curves illustrate the fitting procedure based on Eq. (2.42)). Using these experimental results [34–38] and theoretical results of Eq. (2.42), we can evaluate the Curie–Weiss constant Θ, the density of ferrons n, the FM phase fraction $p = nN_{eff}a^3$

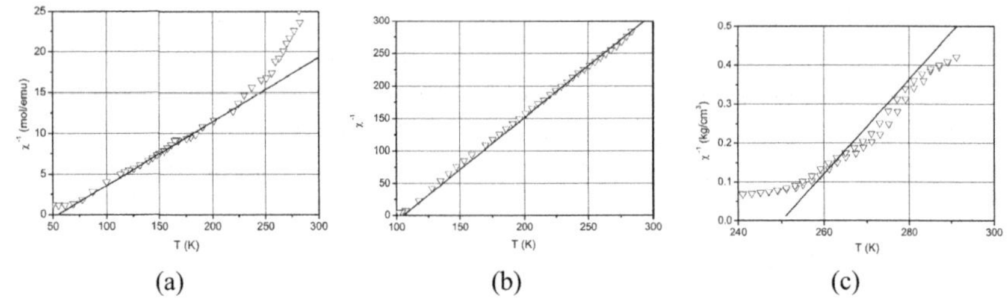

Fig. 2.7 Temperature dependence of the inverse magnetic susceptibility of **a** $(La_{1-y}Pr_y)_{0.7}Ca_{0.3}MnO_3$ for $y = 1$; **b** $Pr_{0.71}Ca_{0.29}MnO_3$; and **c** $(La_{0.4}Pr_{0.6})_{1.2}Sr_{1.8}Mn_2O_7$ [41]

Table 2.3 Curie–Weiss constant Θ, density of ferrons n, FM phase fraction p, and effective tunneling length l in different families of manganites

Samples	Θ, K	n, cm^{-3}	p	l, Å	Data source
$(La_{1-y}Pr_y)_{0.7}Ca_{0.3}MnO_3$	55	1.8×10^{18}	0.03	24	Figure 2.7a (Babushkina et al. [34])
$Pr_{0.71}Ca_{0.29}MnO_3$	105	6.0×10^{18}	0.07	17	Figure 2.7b (Fisher et al. [35])
$(La_{0.4}Pr_{0.6})_{1.2}Sr_{1.8}Mn_2O_7$	255	2.5×10^{18}	0.04	19	Figure 2.7c (Wagner et al. [36])
$La_{0.8}Mg_{0.2}MnO_3$	150	0.6×10^{18}	0.01	14	Zhao et al. [37]

(where for all the samples, we used the value $a = 3.9$Å as a lattice constant), as well as the effective tunneling length l (see Table 2.3).

Let us emphasize that effectively we can regard a small FM droplet as a qubit since the direction of the magnetic moment of the droplet can be parallel or antiparallel to the easy axis. If we apply a high magnetic field, we can strongly polarize FM droplets and thus can write down the information. In our case, an entangled two-qubit state (which is very important for quantum calculations) is organized by a pair of two droplets located nearby. The droplets in the pair are connected with the tunneling bridges. Their interaction has a magneto-dipole origin. Thinking about the applications in the spintronics, we can say that the network of FM droplets or nanocomposites (multilayers) Fe/Cr, Co/Cu, which is connected by the tunneling bridges, can serve as the basic element for the memory device [31].

Returning back to qubits, note that a popular nowadays magnetic principle of their realization is based on the quantum spin tunneling in the system of large magnetic molecules [46] and thus is very similar to the spin-dependent tunneling of conduction electron between the neighboring FM droplets in our case.

In addition, there are interesting possibilities for the applications of CMR manganites in the nanoscale phase-separated state for optoelectronics and nanoplasmonics. Possible application for the fabrication of the new optical devices can be related to a strong change in the infrared behavior of the transmission coefficient in the phase-separated state, when we apply an external magnetic field to our system. This change in the behavior gives us a possibility to construct an infrared radiation modulator based on such manganites [47].

Another possible application of the phase-separated manganites in the nanoplasmonics [48] was first proposed in [49]. The main idea here is to construct on the surface of a bulk metal an artificial layer with small metallic droplets embedded in the insulating charge-ordered (CO) matrix. Let us consider an incident electromagnetic wave coming at the surface from the vacuum. It is very interesting then to study the angular and frequency dependence of the transmission (T) and reflection (R) coefficients in this geometry assuming that metallic droplets form a periodic superstructure within the surface layer. The reflection and transmission coefficients will have the resonance peaks of different shapes for some critical angles including the angles of total internal reflection. It is important to stress that the critical angles themselves will correspond to the resonance excitation of the surface plasmon modes here. Thus, we can calculate the spectrum of surface plasmons and find the resonance frequencies. The experimental realization of this idea can lead to the development of a new generation of the devices for Raman spectroscopy, which will have a strongly enhanced sensitivity.

2.2.5 $1/f$ Voltage Noise in the Systems with Phase Separation

The difficulties, which we have on the way of the creation of magnetic devices, are mostly associated with a very large amplitude of the $1/f$ noise in the phase-separated state. In [50], we find the following expression for the intensity of the $1/f$ noise spectrum (or the Hooge constant) in the framework of the droplet model [33, 39, 43, 44])

$$\alpha_H = \frac{\langle \delta U^2 \rangle_\omega V_s \omega}{U_{dc}^2} = 2\pi^2 l^3 \ln^2\left(\frac{\bar{\omega}}{\omega}\right). \tag{2.43}$$

In Eq. (2.43), $\langle \delta U^2 \rangle_\omega$ is the spectral density of voltage fluctuations, U_{dc} is the applied voltage, $\langle \delta U^2 \rangle_\omega / U_{dc}^2 = \langle \delta \sigma^2 \rangle_\omega / \sigma^2$, and $\bar{\omega} = \omega_0 \exp(A/2T)$. To derive Eq. (2.43), we should use the important relation [33]

$$\delta\sigma = \sigma \frac{\delta n_2}{\bar{n}_2} [1 - 2\exp(-A/2T)] \tag{2.44}$$

between conductivity fluctuations and the fluctuations of the density of two-electron droplets. We use then the general prescription from the textbooks on statistical physics (see, e.g., [51, 52]), which relates fluctuation spectrum to the thermal average

$$\langle \delta n_2^2 \rangle_\omega = \langle \delta n_2^2 \rangle_T \langle \sum_i \frac{2\tau(r^i)}{1+\omega^2\tau^2} \rangle , \tag{2.45}$$

where neglecting the effect of electric field on the relaxation time, we have $\tau(r) = \omega_0^{-1} \exp(r/l - A/2T)$. In Eq. (2.45), we calculate the sum over the pairs which include empty and two-electron droplets. Correspondingly, in Eq. (2.45), r^i is the intersite distance within the pair. In our case, all the pairs contribute to the noise spectrum. That is why, in Eq. (2.45), we can safely proceed from the average to the spatial integration, in which the dominant contribution is governed by small distances

$$\langle \delta n_2^2 \rangle_\omega = 8\pi \bar{n}_2 \langle \delta n_2^2 \rangle_T \int_0^\infty \frac{\tau(r)}{1+\omega^2\tau^2(r)} r^2 dr . \tag{2.46}$$

As a result, we obtain the final expression (2.43) for the noise spectrum, which in the broad range of frequencies has almost the $1/f$ form.

Substituting in Eq. (2.43) the parameters typical of manganites, we obtain $\alpha_H \simeq 10^{-16}\,\text{cm}^3$ at temperatures 100–200 K and frequencies in the 1–1000 s^{-1} range. The value of Hooge constant α_H in phase-separated state is by 3–5 orders of magnitude higher than that in homogeneous semiconductors. The application of external magnetic field or the proximity to the percolation threshold can lead to an additional increase in the Hooge constant [53].

References

1. N.F. Mott, E.A. Davis, *Electronic processes in non-crystalline materials*. Monographs on Physics (Oxford University Press, 1971)
2. E.L. Nagaev, JETP Lett. **6**, 18 (1967)
3. E.L. Nagaev, *Physics of magnetic semiconductors* (Mir Publishers, 1983)
4. T. Kasuya, A. Yanase, T. Takeda, Solid State Commun. **8**(19), 1543 (1970)
5. T. Kasuya, A. Yanase, T. Takeda, Solid State Commun. **8**(19), 1551 (1970)
6. M.A. Krivoglaz, Sov. Phys. Uspekhi **16**(6), 856 (1974)
7. P.G. de Gennes, Phys. Rev. **118**(1), 141 (1960)
8. S.L. Ogarkov, M.Yu. Kagan, A.O. Sboychakov, A.L. Rakhmanov, K.I. Kugel, Phys. Rev. B **74**, 014436 (2006)
9. K.I. Kugel, A.L. Rakhmanov, A.O. Sboychakov, M.Yu. Kagan, S.L. Ogarkov, Physica B **403**(5), 1353 (2008)
10. M.Yu. Kagan, S.L. Ogarkov, K.I. Kugel, A.O. Sboychakov, A.L. Rakhmanov, J. Phys.: Condens. Matter **20**(42), 425214 (2008)
11. G.H. Jonker, J.H.V. Santen, Physica **16**(3), 337 (1950)
12. G.H. Jonker, Physica **22**(6), 707 (1956)
13. E.O. Wollan, W.C. Koehler, Phys. Rev. **100**, 545 (1955)
14. S. Jin, T.H. Tiefel, M. McCormack, R.A. Fastnacht, R. Ramesh, L.H. Chen, Science **264**(5157), 413 (1994)
15. E. Dagotto, *Nanoscale phase separation and colossal magnetoresistance:The physics of manganites and related compounds*, *Springer Series in Solid-State Sciences*, vol. 136 (Springer Science & Business Media, 2003)

16. M. Hennion, F. Moussa, G. Biotteau, J. Rodríguez-Carvajal, L. Pinsard, A. Revcolevschi, Phys. Rev. Lett. **81**, 1957 (1998)
17. G. Allodi, R. De Renzi, G. Guidi, F. Licci, M.W. Pieper, Phys. Rev. B **56**, 6036 (1997)
18. N.A. Babushkina, L.M. Belova, D.I. Khomskii, K.I. Kugel, O.Yu. Gorbenko, A.R. Kaul, Phys. Rev. B **59**, 6994 (1999)
19. I.F. Voloshin, A.V. Kalinov, S.E. Savel'ev, L.M. Fisher, N.A. Babushkina, L.M. Belova, D.I. Khomskii, K.I. Kugel, JETP Lett. **71**(3), 106 (2000)
20. A. Yakubovskii, K. Kumagai, Y. Furukawa, N. Babushkina, A. Taldenkov, A. Kaul', O. Gorbenko, Phys. Rev. B **62**, 5337 (2000)
21. M. Fäth, S. Freisem, A.A. Menovsky, Y. Tomioka, J. Aarts, J.A. Mydosh, Science **285**(5433), 1540 (1999)
22. Y. Moritomo, A. Machida, S. Mori, N. Yamamoto, A. Nakamura, Phys. Rev. B **60**, 9220 (1999)
23. M.Yu. Kagan, K.I. Kugel, Phys. Usp. **44**(6), 553 (2001)
24. P.W. Anderson, H. Hasegawa, Phys. Rev. **100**, 675 (1955)
25. J. Zaanen, A.M. Oleś, P. Horsch, Phys. Rev. B **46**, 5798 (1992)
26. M.Yu. Kagan, D.I. Khomskii, M.V. Mostovoy, Eur. Phys. J. B **12**(2), 217 (1999)
27. M.Yu. Kagan, A.V. Klaptsov, I.V. Brodskii, K.I. Kugel, A.O. Sboichakov, A.L. Rakhmanov, Phys. Usp. **46**(8), 851 (2003)
28. L.N. Bulaevskii, E.L. Nagaev, D.I. Khomskii, Sov. Phys. JETP **27**, 836 (1968)
29. W.F. Brinkman, T.M. Rice, Phys. Rev. B **2**, 1324 (1970)
30. M. Julliere, Physics Letters A **54**(3), 225 (1975)
31. M.Yu. Kagan, A. E. Abrameshin, Physical grounds of the information storage and recording in magnetic nanoclusters, in *Proceedings of the International Conference on Innovative Information Technologies, Prague, Czech Republic* (2014), Part II, pp. 119 – 121
32. M.Yu. Kagan, *Modern trends in superconductivity and superfluidity*, Lecture Notes in Physics, vol. 874. (Springer, Netherlands, 2014)
33. A.L. Rakhmanov, K.I. Kugel, Ya.M. Blanter, M.Yu. Kagan, Phys. Rev. B **63**(17), 174424 (2001)
34. N.A. Babushkina, E.A. Chistotina, K.I. Kugel, A.L. Rakhmanov, O.Yu. Gorbenko, A.R. Kaul, Phys. Solid State **45**(3), 508 (2003)
35. L.M. Fisher, A.V. Kalinov, I.F. Voloshin, N.A. Babushkina, K.I. Kugel, D.I. Khomskii, Phys. Rev. B **68**, 174403 (2003)
36. P. Wagner, I. Gordon, V.V. Moshchalkov, Y. Bruynseraede, M. Apostu, R. Suryanarayanan, A. Revcolevschi, EPL (Europhysics Letters) **58**(2), 285 (2002)
37. J.H. Zhao, H.P. Kunkel, X.Z. Zhou, G. Williams, J. Phys.: Condens. Matter **13**(41), 9349 (2001)
38. J.H. Zhao, H.P. Kunkel, X.Z. Zhou, G. Williams, Phys. Rev. B **66**, 184428 (2002)
39. A.L. Rakhmanov, K.I. Kugel, Ya.M. Blanter, M.Yu. Kagan, Physics of Metals and Metallography **918**(Suppl. 1), 207 (2001)
40. M.Yu. Kagan, M.V. Mostovoy, D.I. Khomskii, Physica B **284-288**(Part 2), 1209 (2000)
41. K.I. Kugel, A.L. Rakhmanov, A.O. Sboychakov, M.Yu. Kagan, I.V. Brodsky, A.V. Klaptsov, J. Exp. Theor. Phys. **98**(3), 572 (2004)
42. A.A. Abrikosov, *Fundamentals of the theory of metals* (North Holland, Amsterdam, 1988)
43. A.O. Sboychakov, A.L. Rakhmanov, K.I. Kugel, M.Yu. Kagan, I.V. Brodsky, J. Exp. Theor. Phys. **95**(4), 753 (2002)
44. A.O. Sboychakov, A.L. Rakhmanov, K.I. Kugel, M.Yu. Kagan, I.V. Brodsky, J. Magn. Magn. Mater. **258**(Supplement C), 296 (2003)
45. J.S. Smart, *Effective field theories of magnetism* (W.B. Saunders Company, Philadelphia, PA, 1966)
46. B. Barbara, L. Gunther, Physics World **12**(3), 35 (1999)
47. Y.P. Sukhorukov, N.N. Loshkareva, A.V. Telegin, E.V. Mostovshchikova, V.L. Kuznetsov, A.R. Kaul', O.Y. Gorbenko, E.A. Gan'shina, A.N. Vinogradov, Tech. Phys. Lett. **29**(11), 904 (2003)
48. V.M. Shalaev, A.K. Sarychev, *Electrodynamics of metamaterials* (World Scientific, Singapore, 2007)
49. A.K. Sarychev, S.O. Boyarintsev, A.L. Rakhmanov, K.I. Kugel, Y.P. Sukhorukov, Phys. Rev. Lett. **107**, 267401 (2011)

50. V. Podzorov, M. Uehara, M.E. Gershenson, T.Y. Koo, S.W. Cheong, Phys. Rev. B **61**, R3784 (2000)
51. P. Dutta, P.M. Horn, Rev. Mod. Phys. **53**, 497 (1981)
52. L.D. Landau, E.M. Lifshitz, *Statistical Physics, part I* (Pergamon, Oxford, 1980)
53. A.O. Sboychakov, A.L. Rakhmanov, K.I. Kugel, M.Yu. Kagan, I.V. Brodsky, J. Phys.: Condens. Matter **15**(10), 1705 (2003)

Chapter 3
Some Model Systems. Internal Structure of Magnetic Polarons

Abstract We consider different anisotropic doped antiferromagnets on regular and frustrated (triangular) lattices and determine the optimum shape and volume of the magnetic droplets (mainly of the ellipsoidal type) in them. We also consider magnetic polarons bound by Coulomb interaction with the impurity (Sr ion) and a long-range coat of spin distortions produced by them.

3.1 FM Polarons in Layered Manganites

In the previous section, we mostly considered isotropic 3D manganites (such as $La_{1-x}Ca_xMnO_3$ or $La_{1-x}Sr_xMnO_3$), which typically have a cubic perovskite structure. In addition to this family of manganites, a serious attention was also focused on the family of layered manganites with a structure specified by $La_{2-x}Ca_xMnO_4$, $La_{3-x}Ca_xMn_2O_7$, and so on (see, e.g., [1, 2]). In the literature, these compounds are usually referred to as the Ruddlesden–Popper phases. The first member of this family $La_{2-x}Ca_xMnO_4$ structurally is close to the well-known high-T_c cuprate $La_{2-x}Sr_xCuO_4$, whereas the second one $La_{3-x}Ca_xMn_2O_7$ contains two neighboring MnO planes.

In the absence of doping, at low concentration of Ca or Sr, the layered manganites usually exhibit the so-called A-type AFM structure. In this structure (see Fig. 3.1), spins in the conducting MnO layer are aligned ferromagnetically but the FM moments of neighboring layers are antiparallel to one another. At finite doping, the physics of layered manganites is similar to the physics of high-T_c compounds. Indeed, the basic magnetic and transport properties of layered manganites are determined by 2D metallic MnO layers, while the LaO and SrO layers play a role of charge reservoir similar to that in cuprates.

At low doping levels, the main properties of layered manganites are captured by the anisotropic double exchange model. The Hamiltonian of this model can be written as

M. Yu. Kagan et al., *Electronic Phase Separation in Magnetic and Superconducting Materials*, Springer Series in Solid-State Sciences 201,
https://doi.org/10.1007/978-3-031-55467-4_3

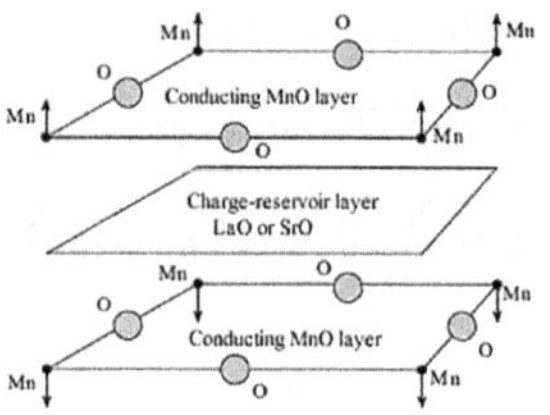

Fig. 3.1 Spin and charge arrangement in layered manganites [2]

$$\hat{H} = -J_H \sum_{ia} \mathbf{S}_{ia}\boldsymbol{\sigma}_{ia} - t_{\|} \sum_{\langle i,j\rangle a\sigma} c^{\dagger}_{ia\sigma}c_{ja\sigma} - t_{\perp} \sum_{i,a\sigma} c^{\dagger}_{ia\sigma}c_{ia+1\sigma} - J_{FM} \sum_{\langle i,j\rangle a} \mathbf{S}_{ia}\mathbf{S}_{ja}$$
$$+ J_{ff} \sum_{ia} \mathbf{S}_{ia}\mathbf{S}_{ia+1}\,, \tag{3.1}$$

where subscript a denotes the layer number, i and j refer to a lattice site within the layer, and $\langle i, j\rangle$ means the summation over neighboring sites. Accordingly, $t_{\|}$ and $t_{\perp}$ are the intralayer and interlayer hopping integrals, respectively. Finally, J_{FM} describes the FM exchange between local spins inside the layer, whereas J_{ff} takes into account the AFM exchange between local spins of the neighboring layers. We again consider the case of a large local spin $S \gg 1$ and work in the strong-coupling limit, where the following set of inequalities is satisfied

$$J_H S \gg \{t_{\|}, t_{\perp}\} \gg \{J_{FM}S^2, J_{ff}S^2\}\,. \tag{3.2}$$

We start the analysis of the different possible states in the layered manganites with Hamiltonian (3.1) in the strong-coupling limit (3.2) with the homogeneous classical canted state. In this state, a conduction electron traveling parallel to the layer should occupy the bottom ofthe conduction band. At the same time, its motion perpendicular to the layer should be an 1D analogue of the de Gennes classical canted state. Then, it is reasonable to assume that the spectrum of conduction electrons in the layered case has the form

$$\varepsilon = -4t_{\|} - 2t_{\perp}\cos\frac{\theta}{2}\,, \tag{3.3}$$

where θ is the canting angle between the FM moments of the neighboring metallic MnO layers and $z = 4$ is the number of the nearest neighbors in a MnO layer on a square lattice. The spectrum in Eq. (3.3) corresponds to the total spin at one site equal to $S_{tot} = S + 1/2$ and the total spin projection $S^z_{tot} = S + 1/2$. At the same time, it is possible to show (see [1–5]) that unlike the isotropic situation, the second conduction band, which corresponds to the total spin projection $S^z_{tot} = S - 1/2$, has the spectrum

$$\varepsilon_- = -\frac{-4t_{\|} - 2t_{\perp}\cos\frac{\theta}{2}}{2S+1}\,. \tag{3.4}$$

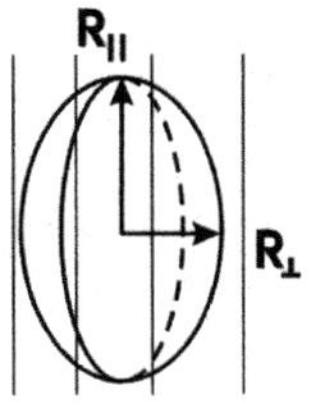

Fig. 3.2 FM droplet of the ellipsoidal shape in layered manganites [2]

Thus, $\varepsilon_{-}/\varepsilon \propto 1/(2S+1) \ll 1$, so we can ignore the second band and consider the classical de Gennes canted state. Minimization of the classical canting energy with respect to the parameter $\cos(\theta/2)$ leads for the layered case to the following result:

$$\cos\frac{\theta}{2} = \frac{t_{\perp}n}{2J_{ff}S^2}, \tag{3.5}$$

where n is the number of electrons per site. It is possible to demonstrate that the compressibility of the canted state ($\kappa^{-1} = d^2E/dn^2$) is negative. Moreover, the nanoscale phase-separated state with small FM droplets inside AFM matrix is more favorable in energy.

We will assume that ferrons have an ellipsoidal shape (Fig. 3.2). Similar to the isotropic case, the energy of a FM polaron can be written as

$$E_{pol} = E_0 - 4t_{\parallel}n - t_{\perp}n\left(2 - \frac{\pi^2a^2}{R_{\perp}^2}\right) + 2J_{ff}S^2n\Omega\,, \tag{3.6}$$

where Ω is the volume of FM droplet, $R_{\perp}$ is the droplet radius in the direction perpendicular to the layers, and E_0 is given by

$$E_0 = -J_{ff}S^2 - 2J_{FM}S^2\,. \tag{3.7}$$

The volume of ellipsoidal droplet is

$$\Omega_{ell} = \frac{4}{3}\pi\frac{R_{\perp}^2R_{\parallel}}{a^3}\,. \tag{3.8}$$

It is possible to show that in the anisotropic double exchange model, the optimum energy of the system corresponds to

$$R_{\parallel} = R_{\perp}\sqrt{\frac{t_{\parallel}}{t_{\perp}}} \gg R_{\perp}\,. \tag{3.9}$$

Relation (3.9) is in agreement with the neutron scattering experiments of Hennion et al. [6]. Using this relation, we can write the expression for the droplet volume in the following form:

$$\Omega_{ell} = \frac{4}{3}\pi \left(\frac{R_\perp}{a}\right)^3 \frac{t_\parallel}{t_\perp} . \tag{3.10}$$

The minimization of the droplet energy (3.6) with respect to parameter $R_\perp$ yields

$$R_\perp = a \left(\frac{\pi t_\perp^2}{4J_{ff}S^2 t_\parallel}\right)^{1/5} . \tag{3.11}$$

Correspondingly, the optimum volume of an ellipsoidal droplet is

$$\Omega_{ell} = \frac{4}{3}\pi \left(\frac{\pi t_\perp^2}{4J_{ff}S^2 t_\parallel}\right)^{3/5} \frac{t_\parallel}{t_\perp} . \tag{3.12}$$

It is interesting to compare the energies of the droplets of different shapes and thus to get the feeling what shape of the droplet is the optimum one for layered manganites. Let us illustrate this point considering also the droplets of the cylindrical shape [2, 4].

The volume of a cylindrical droplet is (see Fig. 3.3)

$$\Omega_{cyl} = \frac{\pi \rho^2 L}{a^3} . \tag{3.13}$$

The energy of cylindrical droplet is given by

$$E_{pol} = E_0 - t_\parallel n \left(4 - \frac{9\pi^2 a^2}{16\rho^2}\right) - t_\perp n \left(2 - \frac{\pi^2 a^2}{L^2}\right) + 2J_{ff}S^2 n\Omega , \tag{3.14}$$

where E_0 is defined in (3.7).

The coefficient $9\pi^2/16$ in the expression for the energy shift from the bottom of the band (which is just $-4t_\parallel n$) is determined by a zero of the Bessel function $J_0 = 0$ for $x \approx 3\pi/4$, where we have introduced the dimensionless variable $x = \rho/a$. It corresponds to the energy of the lowest bound state of an itinerant electron in the 2D cylindrically symmetric potential well.

The minimization of the droplet energy in Eq. (3.14) with respect to both ρ and L yields

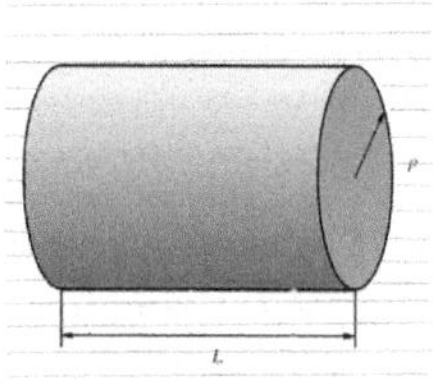

Fig. 3.3 FM droplet of the cylindrical shape in layered manganites [2, 4]

$$\rho \approx 0.53L\sqrt{\frac{t_{\|}}{t_{\perp}}},$$
$$L \approx 0.98a\left(\frac{4\pi t_{\perp}^2}{t_{\|}J_{ff}S^2}\right)^{1/5}. \tag{3.15}$$

As a result, the optimum volume of the cylindrical droplet reads

$$\Omega_{cyl} \approx \pi\frac{L^3}{a^3}\frac{t_{\|}}{4t_{\perp}} \approx 0.26\pi\left(\frac{4\pi t_{\perp}^2}{t_{\|}J_{ff}S^2}\right)^{3/5}\frac{t_{\|}}{t_{\perp}}. \tag{3.16}$$

The direct comparison of the optimum energies of ellipsoidal and cylindrical droplets shows that

$$\frac{E_{ell.pol} - \tilde{E}_0}{E_{cyl.pol} - \tilde{E}_0} \approx \frac{\Omega_{ell}}{\Omega_{cyl}} \approx 0.96, \tag{3.17}$$

where $\tilde{E}_0 = E_0 - 2t_{\perp}n - 4t_{\|}n$.

Hence, we come to the conclusion that the optimum shape of the droplet for layered manganites is an ellipsoidal one. Physically, this is quite reasonable since the optimum shape of FM polaron (its "ground state" to some extent) repeats the form of the electronic spectrum in the layered case, which is elliptically symmetrical and is given by $\varepsilon(p) \propto t_{\|}p_{\|}^2 + t_{\perp}p_{\perp}^2$. Of course, after the dilatation transformation of the momenta $p_{\|} = \tilde{p}_{perp}\sqrt{\frac{t_{\|}}{t_{\perp}}}$ to the new reference frame, an electronic spectrum becomes spherical. In the layered case, the percolation threshold, which corresponds to the overlap of the droplets, takes place at the critical density

$$\tilde{n}_c = \frac{1}{\Omega_{ell}} = \frac{3}{4\pi}\left(\frac{4J_{ff}S^2t_{\|}}{\pi t_{\perp}^2}\right)^{3/5}\frac{t_{\perp}}{t_{\|}}. \tag{3.18}$$

For all densities $n < \tilde{n}_c$, the energy of the droplet state is lower than all the optimum energies of the homogeneous state.

3.2 FM Polarons in Doped Anisotropic AFM Systems

We are now ready to face a more general situation and to consider the shape and the size of FM droplets in anisotropic two-dimensional (2D) or 3D cases when the electron hopping integrals along x-, y-, and z-directions, t_x, t_y, and t_z, as well as the AFM exchange integrals J_x, J_y, and J_z are different. We find that in analogy with the situation in layered manganites, the most favorable shape of the FM droplet is again an ellipsoidal one. The optimum energy and the optimum volume of the droplet are expressed in terms of the universal averaged parameters $\overline{J} = (J_x + J_y + J_z)S^2$ and $t_{eff} = (t_xt_yt_z)^{1/3}$ in 3D or, respectively, $\overline{J} = (J_x + J_y)S^2$ and $t_{eff} = (t_xt_y)^{1/2}$ in 2D anisotropic AFM systems [7]. These results are, in particular, interesting in relation to neutron scattering experiments [6, 8–10], which prove the existence of FM droplets of different shapes in perovskite and layered manganites. Additional experimental evidence of the existence of nanoscale FM droplets in layered manganites comes from scanning tunneling microscopy with atomic resolution [11]. The ferrons in low-doped layered manganites often proved to be "platelets" characteristic of 2D rather than spherical droplets characteristic of the 3D case [9, 10].

3.2.1 Shape of FM Droplets in Anisotropic 2D Case

In the anisotropic 2D case, the system is described by the anisotropic FM Kondo lattice model in the double exchange limit

$$\hat{H} = -J_H \sum_i \mathbf{S}_i \boldsymbol{\sigma}_i + \sum_{\langle i,j \rangle_a} J_a \mathbf{S}_{ia} \mathbf{S}_{ja} - \sum_{\langle i,j \rangle_a \sigma} t_a c^\dagger_{ia\sigma} c_{ja\sigma} \,, \tag{3.19}$$

where $\langle i, j \rangle_a$ denotes the neighboring sites in the square lattice along the a-direction, $a = \{x, y\}$, and $J_H \gg \{t_x, t_y\} \gg \{J_x, J_y\}$.

The analysis of the volume and energy of the FM droplets of different shapes in anisotropic 2D case is very similar to the analysis of the energy and volume of the ellipsoidal and cylindrical droplets in the layered manganites. It uses again the dilatation transformation of the coordinates (which restores the cylindrical symmetry in the new reference frame in 2D) alongside with the properties of the zeroth-order Bessel function J_0 (the value of its first zero) for the lowest binding state of an electron in the cylindrical potential well. As a result, the most energetically favorable shape of the droplet in the anisotropic 2D case is an elliptical one with the volume

$$\Omega = \frac{j_{0,1}}{2} \left(\frac{\pi t_{eff}}{\overline{J}} \right)^{1/2} , \tag{3.20}$$

where $j_{0,1} \approx 3\pi/4$ is the first zero of the zeroth-order Bessel function $J_0(x)$ corresponding to the solution of the Schrödinger equation in polar coordinates and t_{eff}, $\overline{J}$ are the universal averaged parameters mentioned above.

Similarly, in the 3D anisotropic case the most favorable shape of the FM droplet is an ellipsoidal one. Its optimum volume is

$$\Omega = \frac{\pi^{8/5} 2^{1/5}}{3} \left(\frac{t_{eff}}{\overline{J}} \right)^{3/5} , \tag{3.21}$$

and is again determined by the ratio $t_{eff}/\overline{J}$ of the universal averaged parameters.

3.2.2 *FM Droplets in the Case of Frustrated Triangular Lattice*

Just in the same manner, we can find the volume and the most favorable shape of the FM droplet on the 2D frustrated triangular lattice (with the same magnitude of AFM exchanges and hopping integrals along the sides of triangle) for the planar spin configuration . Similar to the situation described by Eq. (3.20), the most favorable shape of the ferron corresponds to the circle with an area [12]

$$\Omega = j_{0,1} \left(\frac{\pi t}{3JS^2} \right)^{1/2} . \tag{3.22}$$

3.3 Bound Magnetic Polarons with Extended Spin Distortions

In Chap. 2, we have mentioned an old idea, which belongs to de Gennes about the magnetic polarons of a different type, which resemble the magnetic impurity and can produce a long-range tail of spin distortions due to the magneto-dipole interaction. This type of bound magnetic polarons was introduced in [12–14], based on the model for an electron, which is bound at a donor impurity by the Coulomb attractive potential. The Hamiltonian of such a model has the following form:

$$\hat{H} = \hat{H}_{el} + J_{ff} \sum_{\langle i,j \rangle} [\mathbf{S}_i \mathbf{S}_j + S^2] - K \sum_i [(S_i^x)^2 - S^2] , \tag{3.23}$$

$$\hat{H}_{el} = -t \sum_{\langle i,j \rangle \sigma} c_{i\sigma}^{\dagger} c_{j\sigma} - -J_H \sum_i \mathbf{S}_i \boldsymbol{\sigma}_i - V \sum_{i,\sigma} \frac{c_{i\sigma}^{\dagger} c_{i\sigma}}{|\mathbf{n} - \mathbf{n}_0|} , \tag{3.24}$$

where K is the constant of magnetic anisotropy. The last term in Eq. (3.24) describes the Coulomb interaction between conduction electrons and the impurity ion located in the middle of the main diagonal of the unit cell (e.g., in the 3D case between the lattice sites $\mathbf{n} = (0, 0, 0)$ and $\mathbf{n} = (1, 1, 1)$, i.e., $\mathbf{n}_0 = (1/2, 1/2, 1/2)$ and similarly, $\mathbf{n}_0 = (1/2, 1/2)$ in the 2D case, see Fig. 3.4.

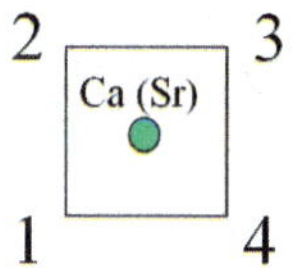

Fig. 3.4 Nonmagnetic Ca(Sr) donor impurity is located in the middle of the quadratic unit cell in 2D [15]

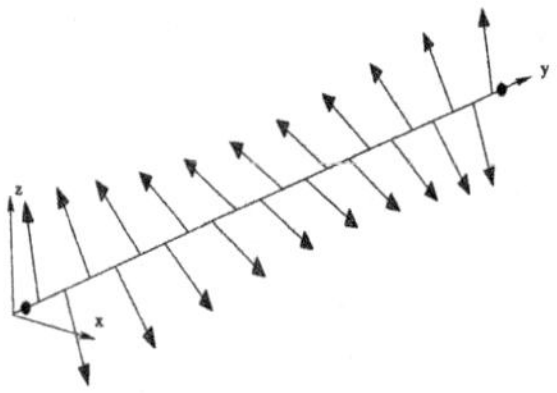

Fig. 3.5 Magnetic structure with slowly decaying spin distortions between two donor impurities corresponding to the coated magnetic polaron in the 1D AFM chain [18]

We work in the limit of strong Coulomb interaction and strong Hund's rule coupling

$$V \gg J_H S \gg t \gg J_{ff} S^2, K S^2 . \tag{3.25}$$

In this range of parameters, the radius of electron localization is of the order of interatomic distance a. The magnetic distortions outside the electron localization region (the ferron core), as it was shown in [12–14], decrease with the distance r as $1/r^4$ and $1/r^2$ in the 3D cubic and 2D square lattices, respectively. At the same time, they decay even slower, as $1/r$ in the 2D frustrated triangular lattice (for the planar spin configuration).

In both 3D and 2D cases, these "coated" magnetic polarons correspond to the ground state of the system at very small doping levels. At the same time, in the 1D case for the AFM chain of interacting neighboring spins with donor impurities located just in the middle (between the nearest neighbor spins) the situation is quite different. Here, as it was shown in [16–18], the "coated" magnetic polarons with slowly decaying tails of spin distortions (shown in Fig. 3.5) correspond usually to the excited state of the system, while the ground state corresponds to the more standard ("bare" or rigid) magnetic polarons with very rapidly decreasing tails of spin distortions. In other words, we can say that in rigid ferrons, the intermediate region where the canting angle ν changes from zero (FM domain) to π (AFM domain) is narrow (of the order of the interatomic distance a), so the radius of magnetic polaron R_{pol} is a well-defined quantity. Moreover, the tails of magnetic distortions exponentially decrease for the distances $r > R_{pol}$ (outside the ferron).

Let us now focus on the 3D and 2D cases. At low electron or hole doping ($n \ll 1$ or $x \ll 1$), we can consider a one-particle problem. In the limit of strong Hund's rule coupling $J_H \to \infty$, the spin $\boldsymbol{\sigma}_n$ of a conduction electron at site n should be parallel to the local spin $\mathbf{S}_n$. If $S \gg 1$, it is sufficient to consider the local spin as a classical vector, i.e., to represent it in the 3D case as [12–14]

$$\mathbf{S}_n = S(\sin\theta_n \cos\varphi_n, \sin\theta_n \sin\varphi_n, \cos\theta_n) = S\mathbf{e}_n , \tag{3.26}$$

where $\mathbf{e}_n$ is the unit vector, θ_n and φ_n are polar and azimuthal angles of $\mathbf{e}_n$.

For the planar structure, when all local spins lie in (x, y)-plane, we have $\theta_n = \pi/2$ and

$$\mathbf{S}_n = S(\cos\varphi_n, \sin\varphi_n, 0), \;\; (-\pi < \varphi_n \leq \pi) . \tag{3.27}$$

In the limit of strong electron–impurity coupling $V \to \infty$, the electron wave function Ψ_n will be nonzero only at sites nearest to the impurity. Let us focus on the 2D case (the 3D case is considered in a similar way). In this case, we can assume that $\Psi_n \neq 0$ only at four sites nearest to the impurity: $\mathbf{n}_1 = (1, 1)$, $\mathbf{n}_2 = (1, 0)$, $\mathbf{n}_3 = (0, 1)$, and $\mathbf{n}_4 = (0, 0)$ in the square lattice (see Fig. 3.4), and similarly Ψ_n is nonzero at eight nearest sites for the 3D cubic lattice, and three nearest sites at the frustrated triangular lattice. The wave function of the conduction electron in the explicit form can be written as [12–14]

$$|\Phi\rangle = \frac{1}{\sqrt{2}} \sum_n \Psi_n \left(\hat{c}^\dagger_{n\uparrow} + \exp(-i\varphi_n) \hat{c}^\dagger_{n\downarrow} \right) |0\rangle , \tag{3.28}$$

where $|0\rangle$ is a vacuum state. The energy of a conduction electron in the state $|\Phi\rangle$ is

$$E_{el}[\Phi] = \langle \Phi | \hat{H}_{el} | \Phi \rangle = -t \sum_{\langle n,m \rangle} [T_{mn} \Psi_n^* \Psi_m + T_{mn}^* \Psi_n \Psi_m^*] - \frac{1}{2} J_H S - V \sum_n \frac{|\Psi_n|^2}{|\mathbf{n} - \mathbf{n}_0|} , \tag{3.29}$$

where $\sum_n |\Psi_n|^2 = 1$,

$$T_{mn} = \frac{1}{2} \left[1 + \exp(i(\varphi_n - \varphi_m)) \right] = \cos \frac{\nu_{nm}}{2} \exp(i\omega_{nm}) ,$$
$$\nu_{nm} = \varphi_n - \varphi_m , \quad \omega_{nm} = \frac{\nu_{nm}}{2} . \tag{3.30}$$

In Eq. (3.30), ν_{nm} is the angle between the local spins $\mathbf{S}_n$ and $\mathbf{S}_m$ (canting angle) and ω_{nm} is the Berry phase [19, 20].

The electron energy in the ground state for fixed planar spin configuration (at given angles φ_n) can be found from the minimization of Eq. (3.29) with respect to Ψ_n. As a result, we have

$$E_{el}(\varphi_n) = \frac{1}{2} J_H S - \frac{V}{b} - t\varepsilon(a_{nm}) , \tag{3.31}$$

where $b = 1/\sqrt{2}$ for the square lattice, $a_{nm} = \cos \frac{\nu_{nm}}{2}$, and

$$\varepsilon(a_{nm}) = \frac{1}{\sqrt{2}} \{ a_{12}^2 + a_{23}^2 + a_{34}^2 + a_{41}^2 + [((a_{12} - a_{34})^2 + (a_{23} + a_{41})^2)((a_{12} + a_{34})^2 + (a_{23} - a_{41})^2)]^{1/2} \}^{1/2} . \tag{3.32}$$

Correspondingly, for the triangular lattice, we have $b = 1/\sqrt{32}$ and

$$\varepsilon(a_{nm}) = g(a_{nm}) + \frac{a_{12}^2 + a_{23}^2 + a_{31}^2}{3g(a_{nm})} , \tag{3.33}$$

where

$$g(a_{nm}) = \{a_{12}a_{23}a_{31} + \frac{1}{3\sqrt{3}}[(a_{12}^2 + a_{23}^2 + a_{31}^2)^3 - 27a_{12}^2a_{23}^2a_{31}^2]^{1/2}\}^{1/3}. \quad (3.34)$$

The electron energy E_{el} has a minimum when all neighboring local spins surrounding an impurity are parallel (or as we will see later, almost parallel) to each other. Hence, at this level of approximation, we have a bound magnetic polaron state, which can be described by a ferromagnetic core of radius b (with four local spins at the square lattice and three local spins at the triangular one) embedded into the antiferromagnetic host. When in the next section, we include magnetic degrees of freedom into account, then after the minimization of the total energy, we will observe that for the most energetically favorable "coated" magnetic polaron, its magnetic core will not be fully polarized and the AFM background around it will be substantially distorted on rather large distances.

3.4 Magnetic Structure of a Coated Ferron

Magnetic contribution to the total energy of a coated ferron can be found in the general case of 3D spins by the minimization of the total energy with respect to the angles θ_n and φ_n. To do this, it is convenient, first, to perform the following transformation of angles: $\varphi_n \to \varphi_n + \pi$, $\theta_n \to \pi - \theta_n$ for one of the sublattices of the cubic (or square) lattice. For planar spin configuration in the (x, y)-plane, after the transformation, θ_n is still equal to $\pi/2$. For the 2D triangular lattice and planar spin configuration, the corresponding transformation reads $\varphi_n \to \varphi_n \pm 2\pi/3$ for two of the three magnetic sublattices. As a result of this transformation, the AFM order becomes FM and *vice versa*. Such a transformation allows us to work with continuously changing orientation of spins outside the ferron core. Moreover, all the angles belong to the range from 0 to $\pi/2$ for cubic and square lattices and to the range from 0 to $2\pi/3$ for the triangular lattice. The total energy is

$$E = E_{el} + E_{mag}, \quad (3.35)$$

where E_{el} is given by Eq. (3.31) and

$$E_{mag} = J_{ff}S^2 \sum_{\langle n,m\rangle\sigma} (1 - \cos\nu_{nm}) - KS^2 \sum_n (\sin^2\theta_n \cos^2\varphi_n - 1). \quad (3.36)$$

After this transformation, we should replace a_{nm} in Eqs. (3.32)–(3.34) by $\alpha_{nm} = \sin\frac{\nu_{nm}}{2}$.

Minimization of the total energy with respect to the angles θ_n and φ_n yields two possible types of the solutions for the bound magnetic polaron. The first one corresponds to the magnetic polaron with completely polarized spins inside its core, which is embedded in the purely AFM background. The total magnetic moment of the polaron is parallel to the easy axis. We refer this trivial solution as a "bare" magnetic polaron. Its energy for the 2D square lattice is

$$E^0_{pol} = -2t + 16J_{ff}S^2 , \tag{3.37}$$

and

$$E^0_{pol} = -3t + 48J_{ff}S^2 , \tag{3.38}$$

for the 3D cubic lattice [12–14]. The corresponding expression for the energy of a "bare" magnetic polaron on the 2D triangular lattice is more cumbersome. Its explicit form is given in [12].

There exists another solution corresponding to the magnetic polaron state with magnetic moment perpendicular to the easy axis. We call this solution a "coated" magnetic polaron. Let us consider the planar spin configuration in the (x, y)-plane on the 2D square lattice. Then, from the symmetry requirements, it is clear that the angles $\varphi_n,\ n = 1, ..., 4$ for the local spins inside the core of magnetic polaron should satisfy the conditions

$$\varphi_1 = \varphi_3 = \varphi_0,\ \ \varphi_2 = \varphi_4 = -\varphi_0 , \tag{3.39}$$

where $0 < \varphi_0 \leq \pi/2$ (see Fig. 3.6).

For the 3D cubic lattice and the 2D triangular lattice, the angles φ_n satisfy, respectively, the conditions

$$\begin{aligned} 1 = \varphi_3 = \varphi_5 = \varphi_7 = \varphi_0,\ \ \varphi_2 = \varphi_4 = \varphi_6 = \varphi_8 = -\varphi_0 , \text{ for the 3D cubic lattice,} \\ \varphi_1 = 0,\ \varphi_2 = -\varphi_3 = \varphi_0,\ 0 < \varphi_0 \leq 2\pi/3 , \text{ for the 2D triangular lattice} . \end{aligned} \tag{3.40}$$

Minimization of the total energy of a coated magnetic polaron with respect to φ_n yields the following set of nonlinear equations:

$$\sum_{\Delta} \sin\left(\varphi_{\mathbf{n}+\Delta} - \varphi_{\mathbf{n}}\right) - \frac{\kappa}{2}\sin 2\varphi_0 = \frac{t}{2J_{ff}S^2}\sum_i \delta_{\mathbf{n}\mathbf{n}_i}(-1)^i \cos\varphi_i , \tag{3.41}$$

where $\mathbf{n} = (n_x, n_y)$, $\kappa = 2K/J_{ff}$, $\delta_{\mathbf{nm}}$ is the Kronecker symbol, and Δ takes the values $(\pm 1, 0)$ and $(0, \pm 1)$.

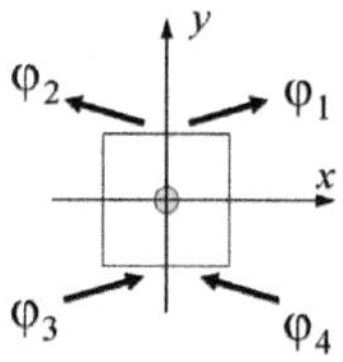

Fig. 3.6 Spin symmetry inside the ferron after the lattice transformation [15]

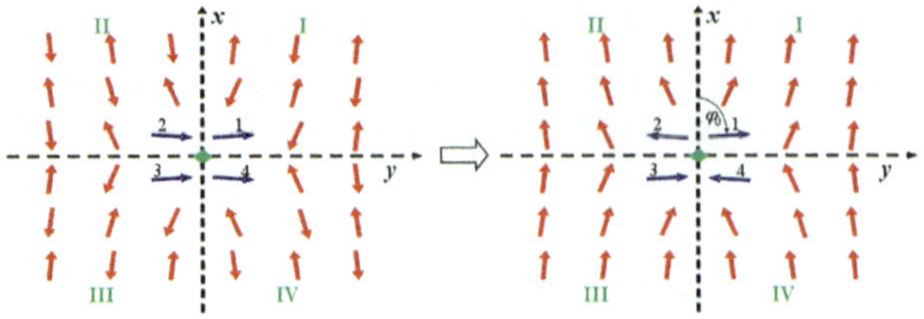

Fig. 3.7 "Coated" magnetic polaron on the square lattice before (left panel) and after (right panel) transformation of angles in one of sublattices. The magnetic structure is calculated by solving Eq. (3.41) at $t/J_{ff}S^2 = 50$ and $\kappa = 5 \times 10^{-3}$. At these values of parameters, we have $\varphi_0 = 85°$ [13]

Equation (3.41) with condition (3.39) was solved numerically for the cluster containing 40×40 sites. The further growth in the number of sites in the cluster does not change the obtained results. The initial angle φ_0 was also found. The calculated magnetic structure is shown in Fig. 3.7.

We see from this figure that in contrast to the "bare" ferron, the "coated" magnetic polaron produces spin distortions of the AFM background outside the region of electron localization (the core region). The "coated" magnetic polaron has a magnetization lower than its saturation value ($\varphi_0 < \pi/2$). Moreover the "coat" has a magnetic moment opposite to that of the core (sites 1–4 in Fig. 3.7 for the square lattice). To obtain analytical estimations for the spatial distribution of the spin distortions, it is important to find an approximate solution of Eq. (3.41) in the continuum limit. To do this, we should treat the angles φ_n as the values of the continuous function $\varphi(\mathbf{r})$ at points $\mathbf{r} = \mathbf{n} - \mathbf{n}_0$. Assuming that outside the magnetic polaron, the following condition for spatial variations of the angles $|\varphi_{\mathbf{n}+\Delta} - \varphi_{\mathbf{n}}| \ll 1$ is met, we can expand $\varphi(\mathbf{r} + \Delta)$ in the Taylor series up to the second order in Δ

$$\varphi(\mathbf{r} + \Delta) \approx \varphi(\mathbf{r}) + \Delta^\alpha d_\alpha \varphi(\mathbf{r}) + \frac{1}{2}\Delta^\alpha \Delta^\beta d_\alpha d_\beta \varphi(\mathbf{r})\,. \tag{3.42}$$

Substituting this expansion in Eq. (3.41), we find that the function $\varphi(\mathbf{r})$ outside the magnetic polaron should satisfy the 2D sine-Gordon equation

$$\Delta\varphi - \frac{\kappa}{2}\sin 2\varphi = 0\,. \tag{3.43}$$

In the range of the parameters under study, $K < J_{ff}$, that is, $\kappa < 1$, we can linearize this equation. As a result we obtain

$$\Delta\varphi - \kappa\varphi = 0\,. \tag{3.44}$$

This equation should be solved with the boundary conditions at infinity $\varphi(\mathbf{r})_{\mathbf{r}\to\infty} \to 0$ and with some boundary conditions at the surface of the magnetic

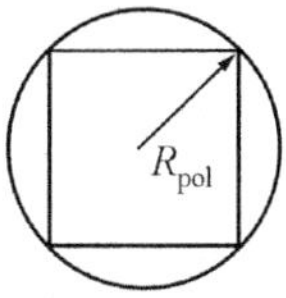

Fig. 3.8 Dirichlet boundary conditions at the surface of magnetic polaron for the 2D square lattice. We model the core of magnetic polaron by the circle of radius $R_{pol} = a/\sqrt{2}$ [15]

polaron. We model the magnetic polaron by a circle of radius $R_{pol} = a/\sqrt{2}$, where a is the intersite distance, and choose the Dirichlet boundary conditions (see Fig. 3.8)

$$\varphi(\mathbf{r})|_{r=b} = \tilde{\varphi}(\zeta)\,, \tag{3.45}$$

where we introduced the polar coordinates (r, ζ) in the (x, y)-plane.

The function $\tilde{\varphi}(\zeta)$ can be found in the following way. For the square lattice, $\tilde{\varphi}(\zeta)$ should meet symmetry conditions (3.39) at the points $\zeta_i = \pi(2i-1)/4$

$$\tilde{\varphi}(\zeta) = \varphi_i\,, \quad i = 1, ..., 4\,. \tag{3.46}$$

Since the function $\tilde{\varphi}(\zeta)$ is periodic, it can be expanded into Fourier series

$$\tilde{\varphi}(\zeta) = \sum_{m=0}^{\infty} [a_m \cos m\zeta + b_m \sin m\zeta]\,. \tag{3.47}$$

In Eq. (3.47), we neglect the terms with $m > 2$, which allows us to keep the minimum number of terms to satisfy condition (3.39). It follows from Eq. (3.46) that $a_0 = a_1 = a_2 = b_1 = 0$. Finally, we obtain

$$\tilde{\varphi}(\zeta) = \varphi_0 \sin 2\zeta\,. \tag{3.48}$$

The solution of Eq.(3.44) with boundary condition (3.45) yields

$$\varphi(\mathbf{r}) = \frac{\varphi_0}{K_2(R_{pol}/r_0)} K_2\left(\frac{r}{r_0}\right) \sin 2\zeta\,, \quad r_0 = \frac{a}{\sqrt{\kappa}}\,, \tag{3.49}$$

where $K_2(x)$ is the Macdonald function of the second order. Within the $|\mathbf{r}| < r_0$ range, the function $\varphi(\mathbf{r})$ behaves as R_{pol}^2/r^2, whereas at large distances it decreases exponentially, $\varphi(\mathbf{r}) \propto \exp(-r/r_0)$. Thus, R_{pol} and r_0 play role of the size of the core and the radius of the "coat", respectively. Moreover, in the case of small anisotropy ($\kappa \ll 1$), the distance, where function $\varphi(\mathbf{r})$ has power-law behavior, can be large enough in agreement with the de Gennes conjecture [21]. The function $\varphi(\mathbf{r})$ at $\zeta = \pi/4$ and the numerical results for $\varphi_{n,n}$ are plotted in Fig. 3.9.

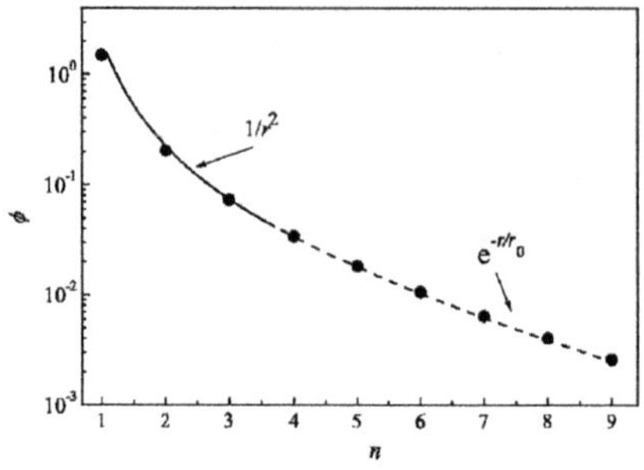

Fig. 3.9 Decrease with the distance of an angle $\varphi(\mathbf{r})$ between misaligned spins and easy axis for a coated magnetic polaron on 2D square lattice [13]. The circles correspond to the numerical solution of the discrete problem on a square cluster at $t/J_{eff}S^2 = 50$ and $\kappa = 5 \times 10^{-2}$ ($r_0 = 4.5a$). The solid curve corresponds to analytical solution (3.49) in the continuum limit. At small distances, $\varphi(\mathbf{r})$ decreases proportionally to $1/r^2$, while at large distances $r \gg r_0$, it decreases exponentially

For the 3D cubic lattice, the similar analytical solution of Eq. (3.44) with symmetry conditions (3.40), the Dirichlet boundary conditions, and a spherical core of magnetic polaron with the radius $R_{pol} = a\sqrt{3}/2$ yields

$$\varphi(x, y, z) = \varphi_0 \frac{R_3(r/r_0)}{R_3(R_{pol}/r_0)} \frac{3\sqrt{3}xyz}{r^3}, \tag{3.50}$$

where $r_0 = a/\sqrt{\kappa}$ and

$$R_3(\rho) = \left(1 + \rho + \frac{2\rho^2}{5} + \frac{\rho^3}{15}\right)\frac{e^{-\rho}}{\rho^4}, \quad \rho = \frac{r}{r_0}. \tag{3.51}$$

It follows from Eqs. (3.50) and (3.51) that $\varphi(\mathbf{r})$ decreases as $(R_{pol}/r)^4$ at $r < r_0$. At distances r exceeding the characteristic radius r_0, the spin distortions decay exponentially $\varphi(\mathbf{r}) \propto \exp(-r/r_0)$ as in the 2D case.

Finally, for the 2D triangular lattice an analytical solution of Eq. (3.44) with symmetry conditions (3.40) and the Dirichlet boundary conditions for polar coordinates and a circular core of magnetic polaron with the radius $R_{pol} = a/\sqrt{3}$ is given by [12]

$$\varphi(\mathbf{r}) = \frac{2\varphi_0}{\sqrt{3}} \frac{K_1(r/r_0^*)}{K_1(R_{pol}/r_0^*)} \sin\zeta, \tag{3.52}$$

where $r^* = a\sqrt{4\kappa/3} = r_0\sqrt{3}/2$ and $K_1(x)$ is the Macdonald function of the first order. Within the $r < r_0^*$ range, the function $\varphi(\mathbf{r})$ slowly decreases as R_{pol}/r. At the same time, at large distances, it decreases exponentially again, $\varphi(\mathbf{r}) \propto \exp(-r/r_0^*)$ [12].

Let us compare now the energies of the "coated" magnetic polarons with the "bare" ones. For the 2D square lattice, the energy of the "bare" magnetic polaron is given by Eq. (3.37) and the energy of the "coated" magnetic polaron is given by Eqs. (3.35) and (3.36), where $\theta_n = \pi/2$ and φ_n are the solutions of Eqs. (3.41). In the continuum approximation, we expand $\cos \nu_{nm}$ and $\cos^2 \varphi_n$ in the Taylor series up to the second order outside the ferron core. Changing the summation by the integration over r, we can represent the total energy of the "coated" magnetic polaron as

$$E = E_{el} + E_M = \left[-\frac{1}{2}J_H S - V\sqrt{2} - 2t \sin \varphi_0\right] + 8J_{ff}S^2\left(1 + \frac{\kappa}{4}\right)\sin^2 \varphi_0 + \frac{J_{ff}S^2}{2}\int_{r \geq r_0} d^2r[(\nabla\varphi)^2 + \kappa\varphi^2], \tag{3.53}$$

In Eq. (3.53), the last two terms correspond to the magnetic part of the total energy. The first one of them comes from the summation over spins in the core of magnetic polaron. Substituting the solution for the function $\varphi(\mathbf{r})$ from Eq. (3.49) into Eq. (3.53) and performing the integration, we get

$$E_{pol} = 8J_{ff}S^2\left(1 + \frac{\kappa}{4}\right)\sin^2 \varphi_0 + \frac{J_{ff}S^2\varphi_0^2}{2}I(R_{pol}/r_0) - 2t \sin \varphi_0 , \tag{3.54}$$

where

$$I(\rho_0) = 2\pi\left[1 + \frac{\rho_0 K_1(\rho_0)}{2K_2(\rho_0)}\right] \tag{3.55}$$

and $\rho_0 = R_{pol}/r_0$.

Since $R_{pol} < r_0$, the value of the function $I(\rho_0)$ is close to 2π. The optimum angle φ_0 is found by the minimization of the energy in Eq. (3.54) and is given by

$$\cos \varphi_0 - \frac{4J_{ff}S^2}{t}\left(1 + \frac{\kappa}{4}\right)\sin 2\varphi_0 - \frac{J_{ff}S^2\varphi_0^2}{2t}I(R_{pol}/r_0) = 0 . \tag{3.56}$$

Hence,

$$\varphi_0 \approx \frac{\pi}{2}\left[1 - \frac{\pi J_{ff}S^2}{t}\right] . \tag{3.57}$$

At $t/J_{ff}S^2 = 50$ corresponding to Fig. 3.9, estimate (3.57) gives us just the required value $\varphi_0 \approx 85°$. The dependence of energy difference between "coated" and "bare" magnetic polarons $\Delta E = E_{pol} - E^0_{pol}$ at the 2D square lattice on the parameter of anisotropy κ is shown in Fig. 3.10. As it follows from this figure, "coated" polarons are more favorable if the anisotropy parameter κ is not too large. Otherwise, "bare" magnetic polarons are more favorable.

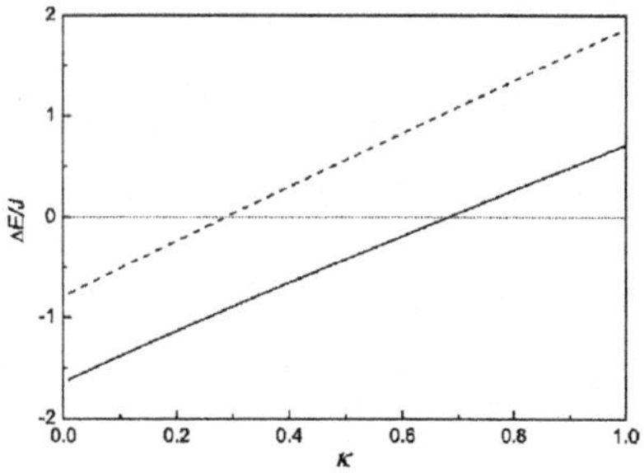

Fig. 3.10 Energy difference between "coated" and "bare" magnetic polarons *vs* $\kappa = 2K/J_{ff}$ calculated at $t/J_{ff}S^2 = 50$. Solid curve corresponds to the numerical calculations, whereas dashed curve is calculated using Eq. (3.54) [13]

It can be shown in the same manner that for 3D cubic lattice and 2D triangular lattice, the energies of the "coated" magnetic polarons for not very high values of the magnetic anisotropy can be also lower than the energies of the corresponding "bare" magnetic polarons [12–14].

References

1. E.L. Nagaev, Phys. Rep. **346**(6), 387 (2001)
2. M.Yu. Kagan, K.I. Kugel, Phys. Usp. **44**(6), 553 (2001)
3. E.L. Nagaev, JETP Lett. **6**, 18 (1967)
4. E.L. Nagaev, Phys. Usp. **39**(8), 781 (1996)
5. E.L. Nagaev, *Physics of Magnetic Semiconductors* (Mir Publishers, 1983)
6. M. Hennion, F. Moussa, G. Biotteau, J. Rodríguez-Carvajal, L. Pinsard, A. Revcolevschi, Phys. Rev. Lett. **81**, 1957 (1998)
7. M.Yu. Kagan, K.I. Kugel, A.L. Rakhmanov, K.S. Pazhitnykh, J. Phys.: Condens. Matter **18**(48), 10905 (2006)
8. M. Hennion, F. Moussa, G. Biotteau, J. Rodríguez-Carvajal, L. Pinsard, A. Revcolevschi, Phys. Rev. B **61**, 9513 (2000)
9. M. Hennion, F. Moussa, New J. Phys. **7**(1), 84 (2005)
10. M. Hennion, F. Moussa, P. Lehouelleur, F. Wang, A. Ivanov, Y.M. Mukovskii, D. Shulyatev, Phys. Rev. Lett. **94**, 057006 (2005)
11. H.M. Rønnow, C. Renner, G. Aeppli, T. Kimura, Y. Tokura, Nature **440**(7087), 1025 (2006)
12. M.Yu. Kagan, S.L. Ogarkov, K.I. Kugel, A.O. Sboychakov, A.L. Rakhmanov, J. Phys.: Condens. Matter **20**(42), 425214 (2008)
13. S.L. Ogarkov, M.Yu. Kagan, A.O. Sboychakov, A.L. Rakhmanov, K.I. Kugel, Phys. Rev. B **74**, 014436 (2006)
14. K.I. Kugel, A.L. Rakhmanov, A.O. Sboychakov, M.Yu. Kagan, S.L. Ogarkov, Physica B **403**(5), 1353 (2008)
15. M.Yu. Kagan, *Modern Trends in Superconductivity and Superfluidity*, Lecture Notes in Physics, vol. 874. (Springer, Netherlands, 2014)
16. I. González, J. Castro, D. Baldomir, A.O. Sboychakov, A.L. Rakhmanov, K.I. Kugel, Phys. Rev. B **69**, 224409 (2004)
17. A.O. Sboychakov, A.L. Rakhmanov, K.I. Kugel, I. González, J. Castro, D. Baldomir, Phys. Rev. B **72**, 014438 (2005)
18. K.I. Kugel, A.L. Rakhmanov, A.O. Sboychakov, M.Yu. Kagan, A.V. Klaptsov, I. González, *Modern Problems of the Physics of Phase Transitions and Critical Phenomena* (Institute of Physics, Makhachkala, 2005), pp.210–234

19. M.V. Berry, Proc. Roy. Soc. (London) A **392**(1802), 45 (1984)
20. E. Müller-Hartmann, E. Dagotto, Phys. Rev. B **54**, R6819 (1996)
21. P.G. de Gennes, Phys. Rev. **118**(1), 141 (1960)

Chapter 4
Charge Ordering and Phase Separation

Abstract We consider an instability of the checkerboard charged ordered state toward phase separation in the Verwey and Shubin–Vonsovsky models when we are away from the quarter filling ($x \neq 1/2$). We find the polaronic radius in the extended model with AFM interaction and Coulomb repulsion at the neighboring sites.

4.1 Introduction

In this section, we present the results of our group [1–5], as well as of other groups [6–9] on the charge ordering (CO) and formation of nanoscale metallic droplets in CO insulating matrices in framework of the Verwey [10, 11] and Shubin–Vonsovsky models [12]. We will also consider ferromagnetic metallic droplets, which arise in CO and antiferromagnetic insulating matrices in more extended models [1–4] combining the double exchange magnetic interactions with the strong intersite Coulomb interactions. Finally, we will comment on the large-scale phase separation and formation of stripes and zig-zag structures typical of these models and confirmed by different experimental techniques including the electron diffraction and dark-image electron microscopy [13–15].

Historically, the problem of CO in magnetic oxides has attracted attention of theorists since the discovery of the Verwey localization transition in magnetite Fe_3O_4 at the end of the 1930s [10]. The early theoretical description of this phenomenon is given, for example, in [16]. At the end of 1990s, this problem was reexamined in a number of papers in relation to the colossal magnetoresistance (CMR) phenomena in manganites and related compounds [17–19]. Mechanisms stabilizing the CO state can be different: the Coulomb repulsion of charge carriers (the energy minimization requires keeping the charge carriers as far away as possible from each other, similarly to the Wigner crystallization) or the electron–lattice interaction leading to the effective repulsion of electrons at the nearest-neighbor sites [20–23] if the electron bandwidth is sufficiently small. In the opposite case, large electron kinetic energy stabilizes the homogeneous metallic state. In actual materials, in contrast to the Wigner

M. Yu. Kagan et al., *Electronic Phase Separation in Magnetic and Superconducting Materials*, Springer Series in Solid-State Sciences 201,
https://doi.org/10.1007/978-3-031-55467-4_4

× ○ × ○ × ○ × ○

○ × ○ × ○ × ○ ×

× ○ × ○ × ○ × ○

Fig. 4.1 Verwey type of charge ordering at the doping level $x = 1/2$ with the checkerboard distribution of Mn^{3+} (crosses) and Mn^{4+} (open circles) ions, which here play the role of electrons and holes, respectively [2]

crystallization, the underlying lattice periodicity determines the preferential type of CO. Thus, in the simplest bipartite lattice, to which belong, for example, both cubic ($La_{1-x}Sr_xMnO_3$) and layered ($La_{2-x}Sr_xMnO_4$, $La_{2-2x}Sr_{1+2x}Mn_2O_7$) hole-doped manganites, the optimum conditions for the formation of CO state exist for $x = 1/2$. At this value of x, the concentrations of Mn^{3+} and Mn^{4+} ions are equal and the simple checkerboard arrangement is possible (see Fig. 4.1). A remarkable experimental fact is that even at $x \neq 1/2$ (in underdoped manganites with $x < 1/2$), only the simplest version of CO is experimentally observed with the alternating checkerboard structure of the occupied and empty sites in the basal plane. In other words, this structure corresponds to the doubling of the unit cell, whereas the more complicated structures with a longer period (or even incommensurate structures) do not actually appear in this case.

Then, a natural question arises, namely, how the extra or missing electrons can be redistributed for an arbitrary doping level such that the superstructure remains the same as for $x = 1/2$? To answer this question, the experimentalists introduced the concept of the incipient CO state corresponding to the distortion of a long-range charge ordering by microscopic metallic clusters [14, 15, 24]. In fact, the existence of this state implies a kind of phase separation. In this chapter, we show that the interplay between the CO and the tendency toward the phase separation plays a very important role in the physics of manganites and other related compounds (magnetite [10], cobaltites [25], and nickelates [26]).

4.2 The Simplest Model for Charge Ordering

We start our theoretical considerations of the CO and phase separation with the simplest Verwey model, which includes the Coulomb repulsion of electrons at the neighboring sites and the intersite electron hopping. The Hamiltonian of the Verwey model is

$$\hat{H}' = -t \sum_{<i,j>} c_i^\dagger c_j + V \sum_{<i,j>} n_i n_j - \mu \sum_i n_i \, , \tag{4.1}$$

where t is the hopping integral, V is the nearest-neighbor Coulomb repulsion (a similar nearest-neighbor repulsion can be obtained via the interaction with the breathing type optical phonons), μ is the chemical potential, $c_i^\dagger$ and c_j are respectively the electron creation and annihilation operators at site i, $n_i = c_i^\dagger c_i$. The symbol $\langle i, j \rangle$

denotes as usual the summation over the nearest-neighbor sites. In Eq. (4.1), we omit spin and orbital indices for simplicity, thus considering one orbital for spinless electrons and emphasizing the most robust effects related to the nearest-neighbor Coulomb repulsion. We also assume that the double occupancy does not occur in this model because of the strong on-site repulsion $U \to \infty$ between electrons.

The explicit inclusion of the arbitrary (either weak or strong) on-site Hubbard repulsion U in the model together with an account taken of spin degrees of freedom transforms the Verwey model [10] to the Shubin–Vonsovsky model [12] (or, using a more common term, the extended Hubbard model) with the Hamiltonian [5]

$$\hat{H}' = -t \sum_{<i,j>\sigma} c^{\dagger}_{i\sigma} c_{j\sigma} + U \sum_{i} n_{i\uparrow} n_{i\downarrow} + \frac{V}{2} \sum_{<i,j>} n_i n_j - \mu \sum_{i\sigma} n_{i\sigma}, \tag{4.2}$$

where $n_{i,\sigma} = c^{\dagger}_{i\sigma} c_{i\sigma}$ is the electron density at site i with spin projection $\sigma = |\uparrow, \downarrow\rangle$, and $n_i = \sum_{\sigma} n_{i\sigma}$. We will show later in this section (see also [2, 5]) that in the strong-coupling limit of the Shubin–Vonsovsky model ($U \gg V \gg W = 2zt$ and densities $n \neq 1/2$), we obtain the same results for the radius of the nanoscale metallic droplets in the CO matrix as in the strong-coupling limit of the Verwey model ($V \gg W$) after the evident substitution $V \to V/2$.

In the Shubin–Vonsovsky model, the Verwey localization transition at densities $n \to 1/2$ is an additional one to the more standard Mott–Hubbard localization transition [27, 28] at $U \gg W$ and $n \to 1$.

In the main part of this chapter (see Eqs. (4.1) and (4.2)), we usually speak about electrons. However, in application to manganites, for example, we mostly have in mind underdoped systems like $La_{1-x}Sr_xMnO_3$ with the hole doping level lower than 1/2 ($x < 1/2$). For them, we must therefore think in terms of holes. All the theoretical treatment definitely remains the same (from the very beginning, we could define the operators in the Hamiltonian of the Verwey model (4.1) as the operators of holes). We hope that this does not lead to any misunderstanding.

In what follows, we consider the simplest case of square (2D) and cubic (3D) lattices, where the simple two-sublattice ordering occurs (in manganites for $x = 1/2$). This is the case for layered manganites, whereas in 3D perovskite manganites, this ordering occurs as we already mentioned only in the basal plane: the ordering is not checkerboard "in-phase") along the c-axis direction. A more complicated model is apparently needed to account for this behavior.

If the bandwidth W is even smaller than the Coulomb repulsion between electrons located at more distant sites ($V \gg V_2 \gg V_3 \gg \cdots \gg V_m \gg W$, where $V_m \sum_i n_i n_{m+i}$ accounts for the Coulomb interaction between electrons at a distance ma apart), then the variation of the charge carrier density will give rise to the whole cascade of localization phase transitions, which differ in their ground state structure. For instance, if $V_2 \gg W$ is included to the model, the system will exhibit an additional phase transition to a localized state at electron densities $n \to 1/3$. A more general localization criterion for narrow-band systems ($\mathbf{a}\nabla V(\mathbf{r}) \gg W$ was originally proposed in [29, 30].

Let us return to the simplest model with only the nearest-neighbor Coulomb repulsion (Eq. (4.1)). For densities $n = 1/2$, the Verwey model was analyzed in many papers [16–18, 31, 32]). In this section, we follow the treatment of [16] (see also [2–4]). As mentioned above, the Coulomb repulsion stabilizes the CO in the form of a checkerboard arrangement of the occupied and the empty sites, whereas the first term (band energy) opposes this tendency. At arbitrary values of the electron density n, we first consider a homogeneous CO solution and use the same ansatz as in [16], namely,

$$n_i = n[1 + (-1)^i \tau]\,. \tag{4.3}$$

This expression implies doubling the lattice periodicity with the local densities

$$n_1 = n(1+\tau), \quad n_2 = n(1-\tau)\,. \tag{4.4}$$

at the neighboring sites. At $n = 1/2$ for a general form of the electron dispersion without nesting, the CO state exists only for a sufficiently strong repulsion $V > 2t$ [16]. The order parameter is $\tau < 1$ for finite $V/2t$, so the average electron density differs from zero or unity even at $T = 0$.

We use the coupled Green's function approach [16]), which yields [3, 4]

$$\begin{aligned} (E+\mu)G_1 - t_k G_2 - zVn(1-\tau)G_1 &= 1/2\pi\,, \\ (E+\mu)G_2 - t_k G_1 - zVn(1+\tau)G_2 &= 0\,. \end{aligned} \tag{4.5}$$

In Eq. (4.5), G_1 and G_2 are the Fourier transforms of the normal lattice Green functions $G_{ij} = \langle\langle c_i c_j^\dagger \rangle\rangle$ for the sites i and j belonging respectively to the same or different sublattices, z is the number of nearest neighbors, and t_k is the Fourier transform of the hopping matrix element. In deriving Eq. (4.5), we performed a mean-field decoupling and replaced the averages $\langle c_i^\dagger c_i \rangle$ by the on-site densities from Eq. (4.3). The solution of Eq. (4.5) gives the following spectrum

$$E + \mu = Vnz \pm \sqrt{(Vn\tau z)^2 + t_k^2} = Vnz \pm \omega_k\,. \tag{4.6}$$

The spectrum defined by Eq. (4.6) resembles the spectrum of superconductor and, hence, the first term under the square root is analogous to the superconducting gap squared. In other words, we can introduce the CO gap by the formula

$$\Delta = Vn\tau z\,. \tag{4.7}$$

It depends on the density not only explicitly, but also via the density dependence of τ. We thus obtain

$$\omega_k = \sqrt{\Delta^2 + t_k^2}\,. \tag{4.8}$$

Using Eqs. (4.5), (4.7), and (4.8), we can find Green's functions

$$G_1 = \frac{A_k}{E+\mu - Vnz - \omega_k + i0} + \frac{B_k}{E+\mu - Vnz + \omega_k + i0},$$
$$G_2 = \frac{t_k}{4\pi\omega_k}\left[\frac{1}{E+\mu - Vnz - \omega_k + i0} + \frac{1}{E+\mu - Vnz + \omega_k + i0}\right], \quad (4.9)$$

where

$$A_k = \frac{1}{4\pi}\left(1 - \frac{\Delta}{\omega_k}\right), \quad B_k = \frac{1}{4\pi}\left(1 + \frac{\Delta}{\omega_k}\right). \quad (4.10)$$

After a standard Wick transformation $E + i0 \to iE$ in the expression for G_i, we find the densities in the following form

$$n_1 = n(1+\tau) = \int \left[\left(1 - \frac{\Delta}{\omega_k}\right) f_F(\omega_k - \mu + Vnz) + \left(1 + \frac{\Delta}{\omega_k}\right) f_F(-\omega_k - \mu + Vnz)\right] \frac{d^D\mathbf{k}}{2\Omega_{BZ}},$$
$$n_2 = n(1-\tau) = \int \left[\left(1 + \frac{\Delta}{\omega_k}\right) f_F(\omega_k - \mu + Vnz) + \left(1 - \frac{\Delta}{\omega_k}\right) f_F(-\omega_k - \mu + Vnz)\right] \frac{d^D\mathbf{k}}{2\Omega_{BZ}}, \quad (4.11)$$

where

$$f_F(y) = \frac{1}{e^{y/T} + 1}$$

is the Fermi distribution function and Ω_{BZ} is the volume of the first Brillouin zone. Adding and subtracting the equations for n_1 and n_2, we obtain the final system of equations for n and μ

$$n = \int [f_F(\omega_k - \mu + Vnz) + f_F(-\omega_k - \mu + Vnz)] \frac{d^D\mathbf{k}}{2\Omega_{BZ}},$$
$$1 = Vz \int \frac{1}{\omega_k} [f_F(-\omega_k - \mu + Vnz) - f_F(\omega_k - \mu + Vnz)] \frac{d^D\mathbf{k}}{2\Omega_{BZ}}. \quad (4.12)$$

For low temperatures ($T \to 0$) and $n \le 1/2$, it is reasonable to assume the filling of only the lowest levels (negative $\mu - Vnz$). Therefore,

$$f_F(\omega_k - \mu + Vnz) = 0$$

and

$$f_F(-\omega_k - \mu + Vnz) = \theta(-\omega_k - \mu + Vnz)$$

is the step function.

It can be shown that for $n = 1/2$, the system of equations (4.12) yields identical results for all $-\Delta \le \mu - Vnz \le \Delta$. This means that $n = 1/2$ is a point of indifferent equilibrium. For infinitely small deviations from $n = 1/2$, i.e., for densities $n = 1/2 - 0$, the chemical potential should be defined as

$$\mu = -\Delta + \frac{Vz}{2} = \frac{Vz}{2}(1-\tau). \quad (4.13)$$

If we consider the strong-coupling $V \gg 2t$ and assume a constant density of states inside the band, then for $n = 1/2$, we can find [3, 4]

$$\tau = 1 - \frac{2W^2}{3V^2z^2} \tag{4.14}$$

with $z = 6$ for the simple cubic lattice and, therefore,

$$\mu = \frac{W^2}{3Vz}. \tag{4.15}$$

For the density $n = 1/2$, the CO gap Δ appears for an arbitrary interaction strength V. This is due to the existence of nesting in our simple model. In the weak coupling case $V \ll 2t$ and with perfect nesting, we have

$$\Delta \propto W \exp\left(-\frac{W}{Vz}\right), \tag{4.16}$$

and τ is exponentially small. For $zV \gg W = 2tz$, or accordingly for $V \gg 2t$, it follows that $\Delta \approx Vz/2$ and $\tau \to 1$ (Eq. (4.7)). In the general case, when there is no nesting, the CO exists only if the interaction strength V exceeds a certain critical value of the order of the bandwidth W [16]. In what follows, we restrict ourselves to the physically more instructive strong-coupling case $V \gg 2t$.

For the constant density of states, the integrals in Eq. (4.12) can be taken explicitly and the system of equations (4.12) can be easily solved for arbitrary n. However, in the strong-coupling case $V \gg 2t$ and for small density deviations from 1/2 ($\delta \ll 1$), the results are not very sensitive to the form of the electron dispersion.

When $n = 1/2 - \delta$, the chemical potential is $\mu = \mu(\delta, \tau)$, and we have two coupled equations for μ and τ. As a result, we obtain [3]

$$\mu(\delta) \approx Vnz(1-\tau) - \frac{4W^2}{Vz}\delta^2 \approx \frac{W^2}{3Vz} + \frac{4W^2}{3Vz}\delta + O(\delta^2). \tag{4.17}$$

The energy of CO state is therefore given by

$$E_{CO}(\delta) = E_{CO}(0) - \frac{W^2}{3Vz}\delta - \frac{2W^2}{3Vz}\delta^2 + O(\delta^2), \tag{4.18}$$

where

$$E_{CO}(0) = -\frac{W^2}{6Vz} \tag{4.19}$$

is the energy precisely corresponding to the density $n = 1/2$. Moreover, $|E_{CO}(0)| \ll W$. At the same time, the CO gap Δ is given by

$$\Delta \approx \frac{Vz}{2}\left[1 - 2\delta - \frac{2W^2}{3V^2z^2}(1 + 4\delta)\right]. \tag{4.20}$$

The dependence of the chemical potential μ and the total energy E on δ in Eqs. (4.17) and (4.18) is in fact related to the linear decrease of the CO gap δ with the deviation from half-filling.

For $n > 1/2$, the energy of the CO state starts to increase rapidly due to a large contribution of the Coulomb repulsion (the upper Verwey band is partially filled at $n > 1/2$). For $n > 1/2$, in contrast to the case of $n < 1/2$, each extra electron placed into the checkerboard CO state necessarily has the occupied nearest-neighbor sites, increasing the total energy by $Vz|\delta|$. For $|\delta| = n - 1/2 > 0$, we then have

$$E_{CO}(\delta) = E_{CO}(0) + \left(Vz - \frac{W^2}{3Vz}\right)|\delta| - \frac{2W^2}{3Vz}\delta^2 + O(\delta^3). \tag{4.21}$$

Accordingly, the chemical potential is given by

$$\mu(\delta) = Vz - \frac{W^2}{3Vz} - \frac{4W^2}{3Vz}|\delta| + O(\delta^2). \tag{4.22}$$

It undergoes a stepwise change equal to Vz at $\tau \to 1$. The gap Δ is symmetric for $n > 1/2$ and is given by

$$\Delta \approx \frac{Vz}{2}\left[1 - 2|\delta| - \frac{2W^2}{3V^2z^2}(1 + 4|\delta|)\right]. \tag{4.23}$$

We could make the entire picture symmetric with respect to $n = 1/2$ by shifting all the one-electron energy levels and the chemical potential by $Vz/2$, i.e., defining $\mu' = \mu - Vz/2$. In terms of μ', Eqs. (4.17) and (4.22) can be written as

$$\begin{aligned} \mu' &= -\frac{Vz}{2} + \frac{W^2}{3Vz} + \frac{4W^2}{3Vz}\delta, \quad n < 1/2, \\ \mu' &= \frac{Vz}{2} - \frac{W^2}{3Vz} - \frac{4W^2}{3Vz}|\delta|, \quad n > 1/2. \end{aligned} \tag{4.24}$$

Similarly to the situation in semiconductors, we have $\mu' = 0$ precisely at the point $n = 1/2$, which means that the chemical potential lies in the middle of the band gap (see Fig. 4.2). At densities $n = 1/2 - 0$, the chemical potential $\mu' = -Vz/2$ coincides with the upper edge of the filled Verwey band. We get just the same band structure in the strong-coupling limit of the Shubin–Vonsovsky model [2].

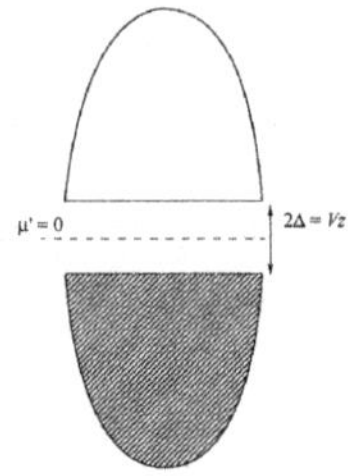

Fig. 4.2 Band structure of the Verwey and Shubin–Vonsovsky models at $n = 1/2$ [3, 5]. The lower Verwey band is completely filled. The upper Verwey band is empty. The chemical potential $\mu' = 0$ lies in the middle of the gap with the width of 2Δ

4.3 Phase Separation in the Verwey Model

We now check the stability of the CO state in the Verwey model. At the densities close to $n = 1/2$, the dependence of the energy on the charge density has the form illustrated in Fig. 4.3.

Figure 4.3 clearly demonstrates an instability of the CO state. Indeed, the most remarkable implication of Eqs. (4.17)–(4.24) is that the compressibility κ of the homogeneous CO state is negative for the densities different from 1/2,

$$\kappa^{-1} \propto \frac{d\mu}{dn} = \frac{d^2E}{d\delta^2} = -\frac{4W^2}{3Vz} < 0\,, \tag{4.25}$$

where $\delta = 1/2 - n$. As we already mentioned, this is a manifestation of the tendency toward the phase separation characteristic of the CO system with $\delta \neq 0$. The presence of a kink in the $E_{CO}(\delta)$ curve determined by Eqs. (4.18) and (4.21) implies that one of the states. into which the system might separate, would correspond to the checkerboard CO state with $n = 1/2$, whereas the other would have a certain density n' smaller or larger than n. This conclusion (based on very general considerations [33]) resembles that discussed in [18, 34, 35], although the detailed physical mechanism is different. The possibility of the phase separation in model (4.1) away from half-filling was also reported earlier in [53] for the infinite-dimensional case. In what follows, we focus on the situation with $n < 1/2$. The case when $n > 1/2$ apparently has certain special properties—the existence of stripe phases, etc. [14, 15], the detailed origin of which should be analyzed separately.

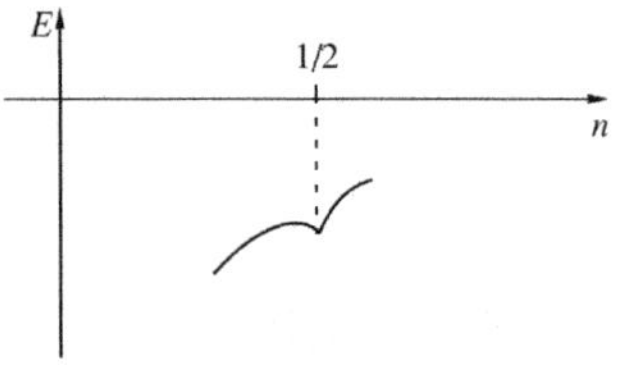

Fig. 4.3 Energy of the CO state versus charge density for $n \to 1/2$. The energy has a kink at $n = 1/2$ [3]

It is easy to understand the physics of phase separation in our case. As follows from Eq. (4.20), the CO gap decreases linearly with the deviation from the half-filling. Correspondingly, the energy of the homogeneous CO state rapidly increases, and it is more favorable to "extract" extra holes from the CO state, putting them into one part of the sample, while creating the "pure" checkerboard CO state in the other part. The energy loss due to this redistribution of holes is overcompensated by the gain provided by the better CO.

However, the hole-rich regions would not be completely "empty" similarly to pores or voids (clusters of vacancies) in crystals: we can gain extra energy by "dissolving" a certain amount of electrons there. In doing this, we decrease the band energy of the electrons due to their delocalization. Thus, the second phase would be a metallic one. The simplest state of this kind is a homogeneous metal with the electron density n_{met}. This density, as well as the relative volumes of the metallic and CO phases, can be easily calculated by minimizing the total energy of the system. The energy of the metallic part of the sample, E_{met}, in the 3D case is given by

$$E_{met} = -tzn_{met} + ct(n_{met})^{5/3} + \frac{Vz}{2}(n_{met})^2 , \tag{4.26}$$

where c is a constant.

Minimizing Eq. (4.26) with respect to n_{met}, we find the equilibrium electron density in the metallic phase. For the strong-coupling case $V \gg 2t$, we obtain the following simple expression (neglecting the relatively small correction provided by the term which contains $n_{met}^{5/3}$)

$$n_{met}^{(0)} = t/V . \tag{4.27}$$

The optimum energy of the metallic phase is given by

$$\left(E_{met} \approx -\frac{t^2 z}{2V} = -\frac{W^2}{8Vz}\right) > \left(E_{CO} = -\frac{W^2}{6Vz}\right) , \tag{4.28}$$

and it is larger than the energy of CO state.

4.4 Phase Separation into Two Large Clusters (Complete Phase Separation)

In accordance with the general treatment in statistical physics, the system with the density $n_{met}^{(0)} < n < 1/2$ separates fully on two large clusters (Fig. 4.4): a metallic one with the density $n_{met}^{(0)}$ and a CO one with the density 1/2. For arbitrary n, the relative volumes Ω_{met} and Ω_{CO} of these clusters can be found from the Maxwell construction

$$\frac{\Omega_{met}}{\Omega_{CO}} = \frac{1/2 - n}{n - n_{met}^{(0)}} , \tag{4.29}$$

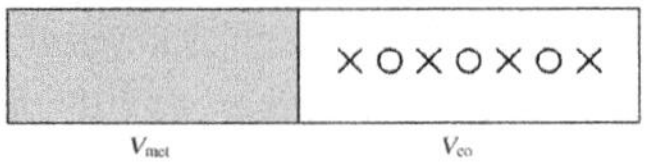

Fig. 4.4 Complete phase separation into two large clusters (metallic and CO ones) [1]

which implies that the metallic cluster occupies the part Ω_{met} of the total volume $\Omega = \Omega_{met} + \Omega_{CO}$ given by

$$\frac{\Omega_{met}}{\Omega} = \frac{1/2 - n}{1/2 - n_{met}^{(0)}} = \frac{\delta}{\delta_0}, \tag{4.30}$$

where we introduced deviations from the half-filling, $\delta = 1/2 - n$ and $\delta_0 = 1/2 - n_{met}^{(0)}$.

Accordingly, the energy of the fully separated state $E_{comp.sep.} = E_{met}(\Omega_{met}/\Omega) + E_{CO}(\Omega_{CO}/\Omega)$ takes the form

$$E_{comp.sep.} \approx -\frac{W^2}{8Vz}\frac{\delta}{\delta_0} - \frac{W^2}{6Vz}(1 - \frac{\delta}{\delta_0}). \tag{4.31}$$

The metallic cluster will occupy the entire sample volume when the total electron density n is less than $n_{met}^{(0)}$, while CO cluster occupies all the volume for $n = 1/2$.

4.5 Nanoscale Phase-Separated State with Small Metallic Droplets Within the Charge-Ordered Insulating Host Material

We now consider a different possibility for the phase-separated state, namely the possibility, which is closer to the nanoscale phase separation in manganites and magnetic semiconductors, described in Chap. 2. To be more specific, we will analyze metallic droplets of the small radius in the CO insulating host. The expression for the energy of this state in the 3D cubic case can be written in the form [1, 3]

$$E_{drop} = -tn_{drop}\left(z - \frac{\pi^2 a^2}{R^2}\right) - \frac{W^2}{6Vz}\left[1 - \frac{4}{3}\pi\left(\frac{R}{a}\right)^3 n_{drop}\right]. \tag{4.32}$$

It corresponds to the situation illustrated in Fig. 4.5 and describes the process, in which small metallic droplets with one conduction electron are formed in the CO matrix.

In expression (4.32), similarly to the situation considered in Chap. 2, R is the radius of a droplet, and a is the interatomic spacing. The first term in Eq. (4.32) describes the kinetic energy gain due to the localization of an electron inside the

Fig. 4.5 Simplest nanoscale phase separation scenario involving the formation of small metallic droplets with one conduction electron embedded into a CO host [1, 5]

droplet. The second term is the energy of CO in the insulating region of the sample. The parameter n_{drop} in Eq. (4.32) is the concentration of metallic droplets. Similarly to the situation of FM metallic droplets in AFM matrix, a surface energy term can be incorporated into the total energy of the system. This term, however, is of the order of W^2R^2/V for a metallic droplet in a CO matrix, so that for $R \gg a$, it is always small compared to the bulk energy term (which is proportional to R^3 in the 3D case) and it will be neglected in the following discussion. The minimization of energy in (4.32) with respect to the droplet radius R results in the following condition

$$\frac{R}{a} \propto \left(\frac{V}{t}\right)^{1/5}. \tag{4.33}$$

Therefore, at the critical concentration of droplets

$$n_{drop}^{(c)} = \frac{3}{4\pi}\left(\frac{a}{R}\right)^3 \propto \left(\frac{t}{V}\right)^{3/5} \tag{4.34}$$

the metallic droplets begin to overlap and the sample as a whole undergoes a transition to the metallic state. Expression (4.32) is in fact the analogue of the Nagaoka theorem [36] for the Verwey model. Similarly, in the 2D case, the exponent 1/5 in Eq. (4.33) for the droplet radius is replaced by 1/4 and the exponent 3/5 in the expression for the critical density in Eq. (4.34) is replaced by 1/2 [2, 5].

Just the same expression for the critical density corresponding to the phase separation was obtained in [5] for the strong-coupling limit of the Shubin–Vonsovsky model, whereas for low electron densities $n \ll 1$ the ground state of the model corresponds to the triplet p-wave superconductivity, which is governed by the Kohn–Luttinger mechanism [5, 37, 38]. The qualitative phase diagram for the phase separation regions in the strong-coupling limit of the 2D Shubin–Vonsovsky model is presented in Fig. 4.6.

Fig. 4.6 Qualitative phase diagram of the nanoscale phase-separated state in the strong-coupling limit of the 2D Shubin–Vonsovsky model

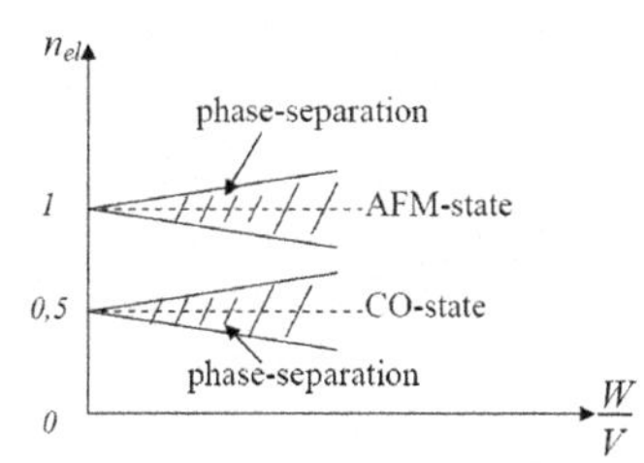

If we use Eq. (4.34) for the critical droplet concentration, the total energy (4.32) of the system can be rewritten as

$$E_{drop} = -tn_{drop}\left[z - \tilde{c}(n^{(c)}_{drop})^{2/3}\right] - \frac{W^2}{6Vz}\left[1 - \frac{n_{drop}}{n^{(c)}_{drop}}\right], \tag{4.35}$$

where $\tilde{c}$ is a numerical coefficient of the order of unity.

The ratio $n_{drop}/n^{(c)}_{drop}$ in Eq. (4.35) describes the relative volume Ω_{drop}/Ω occupied by droplets. This volume can be found from the condition for the conservation of the total number of particles in the system, which reads

$$N = \frac{1}{2}(\Omega - \Omega_{drop}) + n_{drop}\Omega_{drop} = n\Omega\,, \tag{4.36}$$

where n is the average density of the conduction electrons. From condition (4.36), the required relation (which resembles relation (4.30) for complete phase separation) follows immediately and has the form

$$\frac{\Omega_{drop}}{\Omega} = \frac{1/2 - n}{1/2 - n^{(c)}_{drop}} = \frac{\delta}{\delta_c}\,, \tag{4.37}$$

where again $\delta = 1/2 - n$ and $\delta_c = 1/2 - n^{(c)}_{drop}$. From Eq. (4.36), it follows that $n_{drop} = 0$ at $n = 1/2$. Accordingly, at the critical density $n_{drop} = n^{(c)}_{drop}$, we have $\delta = \delta_c = 1/2 - n^{(c)}_{drop}$. Finally, the energy of the phase-separated state can be written as

$$E_{drop} = -t\frac{\delta}{\delta_c}n^{(c)}_{drop}[z - \tilde{c}(n^{(c)}_{drop})^{2/3}] - \frac{W^2}{6Vz}\left(1 - \frac{\delta}{\delta_c}\right) \approx -\frac{W}{2}\frac{\delta}{\delta_c}n^{(c)}_{drop} - \frac{W^2}{6Vz}\left(1 - \frac{\delta}{\delta_c}\right). \tag{4.38}$$

Comparison of energy (4.38) with that of the CO state given by (4.18) shows that for the densities $n^{(c)}_{drop} < n < 1/2$, we have

$$E_{drop} - E_{CO} \approx -\frac{W}{2}\frac{\delta}{\delta_c}n^{(c)}_{drop} < 0\,. \tag{4.39}$$

Thus, a nanoscale phase-separated state with small metallic droplets inside a CO matrix is more favorable than a homogeneous CO state. The direct comparison of the energy for nanoscale phase-separated state (4.38) and the energy for the complete separation into two large clusters (4.31) shows that for $n^{(0)}_{met} < n^{(c)}_{drop}$, we have $E_{drop} < E_{comp.sep}$ and the nanoscale phase separation is more favorable than complete one, as well. Physically, it is related to a kind of Nagaoka theorem for the Verwey model. The point is that even when metallic droplets overlap, each conduction electron

Fig. 4.7 Nanoscale phase separation scenario with an undermelted CO state inside a metallic droplet embedded into the CO matrix [1]

occupies a sphere of the radius $R/a \sim (V/t)^{1/5}$ and thus effectively it is located at the larger distance from the other electrons than it would be in the case of complete phase separation. Therefore, at the complete phase separation, the energy of Coulomb interactions between electrons turns out to be higher than for metallic droplets of a small radius.

In addition to the one-electron metallic droplets, a nanoscale phase separation scenario of another type can be organized [1, 2] (see Fig. 4.7). In this scenario, a metallic droplet can be formed by replacing one electron with a hole in the center of the droplet. Note, however, that the energy of such an undermelted CO state (a resonance valence bond or RVB state for the Verwey model) is much more difficult to evaluate than that for a one-electron droplet, and this problem is not considered in this book.

Concluding this section, we can summarize the main results as follows. In the strong-coupling limit, for $V \gg 2t$ and $(t/V)^{3/5} < n < 1/2$, the 3D Verwey model is unstable toward the nanoscale phase separation. Moreover, in the simplest case, the most energetically beneficial scenario of the phase separation corresponds to small one-electron metallic droplets inside the CO insulating host. Similar results for the nanoscale phase separation, we get in the strong-coupling limits of the 2D Verwey model [2] and the 2D Shubin–Vonsovsky model [12] for the densities $(t/V)^{1/2} < n < 1/2$.

4.6 Phase Separation in the Extended Double Exchange Model with the Nearest-Neighbor Coulomb Interaction

Let us now return to the basic double exchange model (or the ferromagnetic Kondo-lattice model) considered in Chap. 2 (see Eq. (2.8) there) and add the term with nearest-neighbor Coulomb interaction to this model. The Hamiltonian of this extended model reads

$$\hat{H}' = -J_H \sum_i \mathbf{S}_i \sigma_i - t \sum_{<i,j>\sigma} P c^{\dagger}_{i\sigma} c_{j\sigma} P + J_{ff} \sum_{<i,j>} \mathbf{S}_i \mathbf{S}_j + V \sum_{<i,j>} n_i n_j - \mu \sum_{i\sigma} n_{i\sigma} . \tag{4.40}$$

It is physically reasonable to consider this model in the strong-coupling limit

$$J_H S > V > W > J_{ff} S^2 . \tag{4.41}$$

Fig. 4.8 Formation of metallic FM droplets of small radius embedded into the AFM CO insulating host material] [1]

In the absence of the Coulomb term it is exactly the conventional double exchange model.

Let us consider the possibility of the nanoscale phase separation in the extended model (4.40). The simplest scenario of the phase separation in this model corresponds in analogy with Chap. 2 and previous subsection to small FM metallic droplets (with one conduction electron inside) embedded into the AFM CO insulating host [1, 34, 39] (see Fig. 4.8).

The energy of this state in the 3D case is given by the expression

$$E = -tn_{drop}\left(z - \frac{\pi^2 a^2}{R^2}\right) + \frac{1}{2}zJ_{ff}S^2\frac{4}{3}\pi\left(\frac{R}{a}\right)^3 n_{drop}$$
$$-\frac{1}{2}zJ_{ff}S^2\left[1 - \frac{4}{3}\pi\left(\frac{R}{a}\right)^3 n_{drop}\right] - \frac{W^2}{6Vz}\left[1 - \frac{4}{3}\pi\left(\frac{R}{a}\right)^3 n_{drop}\right]. \quad (4.42)$$

The first three terms in Eq. (4.42) are in fact identical to the FM polaron energy in the double exchange model (see Eq. (2.3) in Chap. 2) with the electron density replaced by the droplet density n_{drop}. At the same time, the last term in Eq. (4.42) is identical to the second term in Eq. (4.32) corresponding to the energy of a homogeneous CO Verwey state. Minimization of the droplet energy with respect to radius R yields

$$\frac{R}{a} \propto \left(\frac{t}{V} + \frac{J_{ff}S^2}{t}\right)^{-1/5}. \quad (4.43)$$

For $t/V \ll J_{ff}S^2/t$, we obtain $R/a \propto (t/J_{ff}S^2)^{1/5}$ and we reproduce the double exchange result for the metallic droplet radius. In the opposite limit $t/V \gg J_{ff}S^2/t$, we have $R/a \propto (V/t)^{1/5}$, and we arrive at the result of the Verwey model (4.33). Then, we have

$$n_{drop}^{(c)} \propto \left(\frac{t}{V} + \frac{J_{ff}S^2}{t}\right)^{3/5}. \quad (4.44)$$

Naturally, the droplet concentration in the expression for the energy of the phase-separated state (4.42) is again given by

$$n_{drop} = n_{drop}^{(c)}\frac{\delta}{\delta_c} \quad (4.45)$$

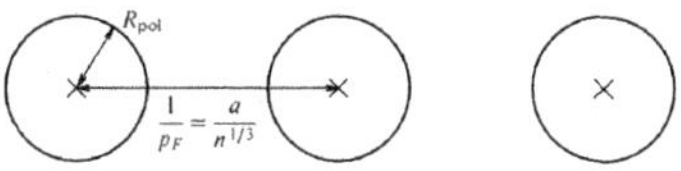

Fig. 4.9 Distribution of conduction electrons in a nanoscale phase-separated state with FM polarons embedded into the AFM host [1]

with $n^{(c)}_{drop}$ from Eq. (4.44). Physically, the minimization of the energy in (4.42) with respect to the droplet radius implies that there is only one conduction electron inside a metallic droplet and that this electron is surrounded by FM ordered local spins. At the same time, outside the droplets, we have a checkerboard CO arrangement of conduction electrons as well as AFM arrangement of the local spins.

The latter result illustrates the main difference between the phase-separated states that are obtained in the extended model (4.40) at densities $n \to 0$ and $n \to 1/2$. The point is that in the low-density regime for $n \ll n_c \propto \left(\frac{J_{ff}S^2}{t}\right)^{3/5}$, the mean distance between conduction electrons $a/n^{1/3}$ in the 3D case far exceeds the polaronic radius R, which in its turn is larger than the intersite distance a $(a/n^{1/3} \gg R \gg a)$ (see Fig. 4.9).

Thus, for FM polarons (ferrons) with one conduction electron inside the ferron the inclusion to the model of even strong Coulomb repulsion between electrons at the neighboring sites does not lead to a charge redistribution. Therefore, upon the inclusion of the Coulomb interaction, both the energy of the phase-separated state with FM polarons inside AFM matrix and the energy of the homogeneous canted state (see Eq. ((4.6)) in Chap. 2) acquires only a Hartree–Fock correction term proportional to $zVn^2/2$, so that the energy difference $E_{pol} - E_{hom}$ between the state with magnetic polarons and the homogeneous state remains negative and does not change. Hence, the global minimum for the energy of the system in the extended model (4.40) corresponds at low density to the phase-separated state with FM polarons embedded into the AFM host just as in the conventional double exchange model. Thus, at low average densities ($n \ll 1$), the local conduction electron density outside FM polarons is zero and the entire electric charge is contained within metallic droplets. At the same time, at densities close to 1/2, most conduction electrons are localized within the CO regions outside metallic droplets.

4.7 Concluding Remarks

Eventually, the sequence of phases with the increase in the doping level for the double exchange model with the Coulomb interaction can be described as follows.

1. For $0 < n < (J_{ff}S^2/t)^{3/5}$, the system separates into nanoscale FM metallic droplets inside the AFM insulating matrix.
2. For $(J_{ff}S^2/t)^{3/5} < n < (t/V + J_{ff}S^2/t)^{3/5} < 1/2$, the system is a FM metal. Of course, we should have a "window" of parameters to satisfy the inequality in the

right-hand side of this expression. In manganites, for example, $t/V \sim 1/2 - 1/3$, and at the same time, $J_{ff}S^2/t \sim 0.01$. Therefore, the inequality $n < (t/V + J_{ff}S^2/t)^{3/5} < 1/2$ is not necessarily met. Experimental evidence indicates that the desired parameter range exists for $La_{1-x}Ca_xMnO_3$, but definitely not for $Pr_{1-x}Ca_xMnO_3$.

3. Finally, for $(t/V + J_{ff}S^2/t)^{3/5} < n < 1/2$, the phase separation into nanoscale FM metallic droplets inside an AFM CO matrix occurs. It should be emphasized that an ideal AFM structure in the CO matrix is obtained only accounting for the Heisenberg interaction between local spins **S**. A more thorough account of the large Hund's rule exchange between a local spin **S** and spin of a conduction electron $\boldsymbol{\sigma}$ can change this naive expectation and lead to the FM order in the CO matrix. At $n = 1/2$, a homogeneous CO state is realized.

Here, we limit ourselves to the case of $n < 1/2$, since at higher doping levels, the model under study should take into account some additional factors, such as interactions with the lattice distortions.

We can get analogous results for the phase diagram in the 2D extended model containing the double exchange magnetic interactions and the nearest-neighbor Coulomb interaction in the strong-coupling limit.

Concluding this chapter, let us emphasize that the phase diagram discussed above is in a good qualitative agreement with many available experiments [40–44] and numerical calculations [45] concerning the nanoscale phase separation in manganites and, in particular, with the experimental observation of the nanoscale phase separation close to the density $n = 1/2$ [13] (or to the doping $x = 1/2$ in hole-doped manganites). We also note that our phase diagram has certain similarities with the phase diagram obtained in [46, 47] for the problem of spontaneous ferromagnetism in doped excitonic insulators.

The coexistence of FM reflections and the reflections of the CE type of the magnetic structure (typical of the CO state at $x = 1/2$) were observed by neutron scattering in [48]. This coexistence provides additional arguments in favor of the proposed phase diagram for the 3D underdoped ($x \leq 1/2$) manganites.

The long-range Coulomb interaction may modify our results, but we do not expect any qualitative changes. For realistic values of the parameters, we can conclude that the radius of FM polarons inside CO host is of the order of 10 Å, and the excess charge contained in them is rather small for the doping close to 1/2.

Our treatment is also applicable to other systems with the charge ordering such as cobaltites [25] and nickelates [26]. It would be interesting to study them for charge carrier densities different from the commensurate "checkerboard" one. Among the important questions, which deserve further investigations, remain also the origin of the "in-phase" ordering along the c direction in perovskite manganites, as well as the behavior of the CO state at finite temperatures.

Let us now draw attention to interesting experimental results reported in Ref. [49] on the electronic phase separation in $CaFe_3O_5$ compound exhibiting the charge ordering, which manifests itself as the alternation of Fe^{3+} and Fe^{2+} sites similar to that in magnetite, Fe_3O_4. It is demonstrated that below the magnetic transition temperature at 302 K, this compound undergoes separation into two phases with different electronic and spin orders. One of them has the charge order; however, it involves not a simple alternation of charges in a checkerboard manner, but also a kind of orbital ordering leading to the formation of electronic molecules (trimerons). Another phase is characterized by the Fe^{3+}/ Fe^{2+} charge averaging including, nevertheless, ferromagnetic chains. Lattice symmetry is unchanged but different strains from the electronic orders probably drive the phase separation. Indeed, as it is shown in [50], the trimeron–phonon interaction is rather strong, thus determining the specific features of the ground state. However, for the phase separation in charge ordered materials, a deviation from the stoichiometry may be needed [51]. Thus, we clearly see here that the electronic phase separation reveals a complex interplay of charge, spin, orbital, and lattice degrees of freedom, which will be discussed in the following sections. Let us mention that magnetite itself exhibits an inhomogeneous state characterized by structural, spin, and charge fluctuations even above the Verwey transition [52].

References

1. M.Yu. Kagan, K.I. Kugel, Phys. Usp. **44**(6), 553 (2001)
2. M.Yu. Kagan, *Modern Trends in Superconductivity and Superfluidity*, Lecture Notes in Physics, vol. 874. (Springer, Netherlands, 2014)
3. M.Yu. Kagan, K.I. Kugel, D.I. Khomskii, J. Exp. Theor. Phys. **93**(2), 415 (2001)
4. M.Yu. Kagan, K.I. Kugel, A.L. Rakhmanov, D.I. Khomskii, Low Temp. Phys. **27**(8), 601 (2001)
5. M.Yu. Kagan, D.V. Efremov, M.S. Marienko, V.V. Val'kov, JETP Lett. **93**(12), 725 (2011)
6. C. Castellani, C.Di Castro., M. Grilli, Phys. Rev. Lett. **75**(7), 4650 (1995)
7. C. Castellani, C. Di Castro., M. Grilli, J. Phys. Chem. Solids **59**(10–12), 1694 (1998)
8. A. Perali, C. Castellani, C. Di Castro, M. Grilli, Phys. Rev. B **54**, 16216 (1996)
9. J. Lorenzana, C. Castellani, C. Di Castro, Phys. Rev. B **64**, 235128 (2001)
10. E.J.W. Verwey, Nature **144**(3642), 327 (1939)
11. E.J.W. Verwey, P.W. Haayman, Physica **8**(9), 979 (1941)
12. S. Schubin, S. Wonsowsky, Proc. Roy. Soc. (London) A **145**(A854), 159 (1934)
13. Y. Moritomo, A. Machida, S. Mori, N. Yamamoto, A. Nakamura, Phys. Rev. B **60**, 9220 (1999)
14. S. Mori, C.H. Chen, S.W. Cheong, Nature **392**(6675), 473 (1998)
15. M. Uehara, S. Mori, C.H. Chen, S.W. Cheong, Nature **399**(6736S), 560 (1999)
16. D.I. Khomskii, *About Electronic Ordering in Crystals 1969*. Preprint No. 105, Lebedev Physical Institute of Physics (USSR Academy of Sciences, Moscow, 1969)
17. T. Mutou, H. Kontani, Phys. Rev. Lett. **83**, 3685 (1999)
18. J. van den Brink, G. Khaliullin, D. Khomskii, Phys. Rev. Lett. **83**, 5118 (1999)
19. G. Jackeli, N.B. Perkins, N.M. Plakida, Phys. Rev. B **62**, 372 (2000)
20. D.M. Newns, N. Read, Adv. Phys. **36**(6), 799 (1987)
21. P. Coleman, Phys. Rev. B **35**, 5072 (1987)
22. A.M. Tsvelik, *Quantum Field Theory in Condensed Matter Physics*, 2nd edn. (Cambridge University, 2007)
23. C.M. Varma, Physica B **378–380**(Supplement C), 17 (2006)
24. A. Arulraj, A. Biswas, A.K. Raychaudhuri, C.N.R. Rao, P.M. Woodward, T. Vogt, D.E. Cox, A.K. Cheetham, Phys. Rev. B **57**, R8115 (1998)
25. Y. Moritomo, M. Takeo, X.J. Liu, T. Akimoto, A. Nakamura, Phys. Rev. B **58**, R13334 (1998)
26. J.A. Alonso, J.L. García-Muñoz, M.T. Fernández-Díaz, M.A.G. Aranda, M.J. Martínez-Lope, M.T. Casais, Phys. Rev. Lett. **82**, 3871 (1999)

27. J. Hubbard, Proc. Roy. Soc. A **205**, 238 (1963)
28. N.F. Mott, *Metal-Insulator Transitions*, 2nd edn. (Taylor and Francis, London, 1990)
29. Yu. Kagan, L.A. Maksimov, J. Exp. Theor. Phys. **60**(1), 201 (1984)
30. Yu. Kagan, L.A. Maksimov, J. Exp. Theor. Phys. **61**(3), 583 (1985)
31. Z. Jirák, S. Krupička, Z. Šimša, M. Dlouhá, S. Vratislav, J. Magn. Magn. Mater. **53**(1), 153 (1985)
32. R. Pietig, R. Bulla, S. Blawid, Phys. Rev. Lett. **82**, 4046 (1999)
33. L.D. Landau, E.M. Lifshitz, *Statistical Physics, Part I* (Pergamon, Oxford, 1980)
34. M.Yu. Kagan, D.I. Khomskii, M.V. Mostovoy, Eur. Phys. J. B **12**(2), 217 (1999)
35. D.P. Arovas, F. Guinea, Phys. Rev. B **58**, 9150 (1998)
36. Y. Nagaoka, Phys. Rev. **147**, 392 (1966)
37. W. Kohn, J.M. Luttinger, Phys. Rev. Lett. **15**, 524 (1965)
38. M.Yu. Kagan, V.A. Mitskan, M.M. Korovushkin, Phys. Usp. **58**(8), 733 (2015)
39. M.Yu. Kagan, M.V. Mostovoy, D.I. Khomskii, Physica B **284-288**(Part 2), 1209 (2000)
40. M. Hennion, F. Moussa, G. Biotteau, J. Rodríguez-Carvajal, L. Pinsard, A. Revcolevschi, Phys. Rev. Lett. **81**, 1957 (1998)
41. G. Allodi, R. De Renzi, G. Guidi, F. Licci, M.W. Pieper, Phys. Rev. B **56**, 6036 (1997)
42. N.A. Babushkina, L.M. Belova, D.I. Khomskii, K.I. Kugel, O.Yu. Gorbenko, A.R. Kaul, Phys. Rev. B **59**, 6994 (1999)
43. I.F. Voloshin, A.V. Kalinov, S.E. Savel'ev, L.M. Fisher, N.A. Babushkina, L.M. Belova, D.I. Khomskii, K.I. Kugel, JETP Lett. **71**(3), 106 (2000)
44. A. Yakubovskii, K. Kumagai, Y. Furukawa, N. Babushkina, A. Taldenkov, A. Kaul', O. Gorbenko, Phys. Rev. B **62**, 5337 (2000)
45. S. Yunoki, T. Hotta, E. Dagotto, Phys. Rev. Lett. **84**, 3714 (2000)
46. L. Balents, C.M. Varma, Phys. Rev. Lett. **84**, 1264 (2000)
47. V. Barzykin, L.P. Gor'kov, Phys. Rev. Lett. **84**, 2207 (2000)
48. R. Kajimoto, H. Yoshizawa, H. Kawano, H. Kuwahara, Y. Tokura, K. Ohoyama, M. Ohashi, Phys. Rev. B **60**, 9506 (1999)
49. K.H. Hong, A.M. Arevalo-Lopez, J. Cumby, C. Ritter, J.P. Attfield, Nat. Commun. **9**(1), 2975 (2018)
50. P. Piekarz, D. Legut, E. Baldini, C.A. Belvin, T. Kolodziej, W. Tabiś, A. Kozlowski, Z. Kakol, Z. Tarnawski, J. Lorenzana, N. Gedik, A.M. Oleś, J.M. Honig, K. Parlinski, arXiv:2010.11564 (2020)
51. S.J. Cassidy, F. Orlandi, P. Manuel, S.J. Clarke, Nat. Commun. **10**(1), 5475 (2019)
52. G. Perversi, E. Pachoud, J. Cumby, J.M. Hudspeth, J.P. Wright, S.A.J. Kimber, J.P. Attfield, Nat. Commun. **10**(1), 2857 (2019)
53. G.S. Uhrig and R. Vlaming, Phys. Rev. Lett. **71**, 271 (1993)

Chapter 5
Phase Separation in Strongly Correlated Systems with Two Electron Bands

Abstract We discuss the tendency toward phase separation in the two-band model with intraband and interband Coulomb interaction at a large mismatch in the electron densities of the bands.

As it was stated above, the phase separation in strongly correlated electron systems arises due to the competition between the contributions of kinetic and potential energies to the total energy of charge carriers. The interplay between the effects related to magnetic ordering and kinetic energy considered above is one of such possibilities. The second rather general case of this competition can be observed in multiband materials [1]. Actually, the bandwidth in a strongly correlated multiband material depends on the number of electrons in the band. The competition between the phases with more localized and more itinerant electrons can give rise to the phase separation. The phase separation in the classical Falicov–Kimball model [2] is the most evident and simple illustration for this. This model is often used as a toy model for heavy-fermion compounds. It describes the system of itinerant and localized electrons with a strong on-site Coulomb repulsion. The numerical simulations for the Falicov–Kimball model demonstrated an inhomogeneous charge density distribution at some relation between the itinerant electron bandwidth and the distance between the localized level and the bottom of conduction band [3–5]. The competition between the metallicity and localization in a similar system with magnetic interactions was studied in Refs. [6–8] with an emphasis on the phase diagram of magnetic oxides such as manganites. The system with a band and localized level is a limiting case of a more common case of two bands having different widths.

M. Yu. Kagan et al., *Electronic Phase Separation in Magnetic and Superconducting Materials*, Springer Series in Solid-State Sciences 201,
https://doi.org/10.1007/978-3-031-55467-4_5

5.1 Phase Separation in the Two-Band Hubbard Model

5.1.1 General Equations

Here, we use a two-band Hubbard model for the description of a strongly correlated electron system with two types of charge carriers. We demonstrate that the phase separation in this system arises even without any ordering if the ratio of the bandwidths exceeds some threshold value.

We consider the system with two bands a and b. Let the first band, a, be wider than the second one, b. In the Hubbard approach, the Hamiltonian of the system can be written in the form [1]

$$\hat{H} = -\sum_{\langle \mathbf{ij}\rangle\alpha,\sigma} t^{\alpha} a^{\dagger}_{\mathbf{i}\alpha\sigma} a_{\mathbf{j}\alpha\sigma} - \epsilon \sum_{\mathbf{i}\sigma} n_{\mathbf{i}b\sigma} - \mu \sum_{\mathbf{i}\alpha\sigma} n_{\mathbf{i}\alpha\sigma} + \frac{1}{2}\sum_{\mathbf{i}\alpha\sigma} U^{\alpha} n_{\mathbf{i}\alpha\sigma} n_{\mathbf{i}\alpha\bar{\sigma}} + \frac{U'}{2}\sum_{\mathbf{i}\alpha\sigma'} n_{\mathbf{i}\alpha\sigma} n_{\mathbf{i}\bar{\alpha}\sigma'}. \quad (5.1)$$

Here, $a^{\dagger}_{\mathbf{i}\alpha\sigma}$ and $a_{\mathbf{i}\alpha\sigma}$ are the creation and annihilation operators for electrons corresponding to bands $\alpha = a, b$ at site $\mathbf{i}$ with spin projection σ, and $n_{\mathbf{i}\alpha\sigma} = a^{\dagger}_{\mathbf{i}\alpha\sigma} a_{\mathbf{i}\alpha\sigma}$. The first term in the right-hand side of Eq. (5.1) corresponds to the kinetic energy of the conduction electrons in bands a (b) with the hopping integral t_a (t_b). We ignore the interband hopping. The second term describes the shift ϵ of the center of band b with respect to the center of band a. The last two terms describe the on-site Coulomb repulsion of two electrons either in the same state α with the Coulomb energy U^{α} or in the different states U'. The assumption of the strong electron correlations means that $U^{\alpha}, U' \gg t^{\alpha}, \epsilon$. The total number n of electrons per site is a sum of electrons in the a and b states, $n = n_a + n_b$. Below, for definiteness sake, we consider the case $n \leq 1$.

We define the single-particle and two-particle Green's functions as

$$G_{\alpha\sigma}(\mathbf{j}-\mathbf{j}_0, t-t_0) = -i\langle \hat{T} a_{\mathbf{j}\alpha\sigma}(t) a^{\dagger}_{\mathbf{j}_0\alpha\sigma}(t_0)\rangle,$$
$$G_{\alpha\sigma\,\beta\sigma'}(\mathbf{j}-\mathbf{j}_0, t-t_0) = -i\langle \hat{T} a_{\mathbf{j}\alpha\sigma}(t) n_{\mathbf{j}\beta\sigma'}(t) a^{\dagger}_{\mathbf{j}_0\alpha\sigma}(t_0)\rangle, \quad (5.2)$$

where $\hat{T}$ is the time-ordering operator. The equation of motion for the single-particle Green's function with Hamiltonian (5.1) can be written as ($\hbar = 1$)

$$\left(i\frac{\partial}{\partial t} + \mu + \epsilon^{\alpha}\right) G_{\alpha\sigma}(\mathbf{j}-\mathbf{j}_0, t-t_0) = \delta_{\mathbf{ij}}\delta(t-t_0) - t^{\alpha}\sum_{\Delta} G_{\alpha\sigma}(\mathbf{j}-\mathbf{j}_0+\Delta, t-t_0)$$
$$+ U^{\alpha} G_{\alpha\sigma\,\alpha\bar{\sigma}} + U'\sum_{\sigma'} G_{\alpha\sigma\,\bar{\alpha}\sigma'}, \quad (5.3)$$

where $\epsilon^{\alpha} = 0$ for $\alpha = a$ and $\epsilon^{\alpha} = \epsilon$ for $\alpha = b$, the summation in the second term in the right-hand side of Eq. (5.3) is performed over the sites nearest to $\mathbf{j}$, and Δ are the vectors connecting site $\mathbf{j}$ with its nearest neighbors. Equation (5.3) includes the two-particle Green's functions. Then, we should write the equations of motion for these functions, which will include the next-order Green's functions, etc.

Then, it is necessary to cut such an infinite chain of equations; this problem was treated in Ref. [1] within a standard mean-field approach. In the limit of strong Coulomb repulsion, the presence of two electrons at the same site is unfavorable, and the two-particle Green's function is of the order of $1/U$, where $U \sim U^{\alpha}, U'$. In the equations of motion for the two-particle Green's functions, the three-particle terms are of the order of $1/U^2$ and were neglected. Following the Hubbard I approximation [9], in the equations of motion for the two-particle Green's functions, the term $\langle \hat{T} a_{\mathbf{j}+\Delta\alpha\sigma}(t) n_{\mathbf{j}\beta\sigma'}(t) a^{\dagger}_{\mathbf{j}_0\alpha\sigma}(t_0)\rangle$ [coming from the commutator of $a_{\mathbf{j}\alpha\sigma}(t)$ with the kinetic energy terms of Hamiltonian (5.1)] is replaced by $\langle n_{\mathbf{j}\beta\sigma'}\rangle\langle \hat{T} a_{\mathbf{j}+\Delta\alpha\sigma}(t) a^{\dagger}_{\mathbf{j}_0\alpha\sigma}(t_0)\rangle$. The similar decoupling in the terms coming from the commutator of $n_{\mathbf{j}\alpha\sigma}$ with the same kinetic energy operator yields zero. As a result, the equations for the two-particle Green's functions can be written as [1]

$$\left(i\frac{\partial}{\partial t} + \mu + \epsilon^{\alpha} - U\right)\begin{bmatrix} G_{\alpha\sigma\,\alpha\bar{\sigma}}(\mathbf{j}-\mathbf{j}_0, t-t_0) \\ G_{\alpha\sigma\,\bar{\alpha}\sigma}(\mathbf{j}-\mathbf{j}_0, t-t_0) \end{bmatrix}$$
$$= \begin{pmatrix} n_{\alpha\bar{\sigma}} \\ n_{\bar{\alpha}\sigma} \end{pmatrix}\left[\delta_{\mathbf{j}\mathbf{j}_0}\delta(t-t_0) - t^{\alpha}\sum_{\Delta} G_{\alpha\sigma}(\mathbf{j}-\mathbf{j}_0+\Delta, t-t_0)\right], \tag{5.4}$$

where $n_{\alpha\sigma} = \langle n_{\mathbf{j}\alpha\sigma}\rangle$ is the average number of electrons per site in the state (α, σ) and we put for brevity $U^{\alpha}, U' = U$. The Hubbard I approximation is an appropriate method to find out the main features of the electron band structure, which should be confirmed by comparison with experiments and numerical results [10, 11]. The detailed analysis of the applicability of the Hubbard I approximation to the problems of phase separation is given in Ref. [12].

Equations (5.3) and (5.4) can be solved in a conventional manner [9] by passing from the time–space to the frequency–momentum representation. Eliminating the two-particle Green's functions, we can calculate the single-particle Green's functions, which have poles corresponding to the Hubbard subbands for bands a and b. The splitting between these subbands is determined by the on-site Coulomb repulsion U of electrons belonging to the same bands. The on-site Coulomb repulsion of electrons from different bands gives rise to the correlation between the fillings of a and b bands [1, 6, 7]. In the case when the total number of the electrons per site is $n < 1$, the upper Hubbard subbands are empty, and one can proceed to the limit $U \to \infty$. As a result, we derive

$$G_{\alpha\sigma}(\mathbf{k}, \omega) = \frac{g_{\alpha\sigma}}{\omega + \mu + \epsilon^{\alpha} - g_{\alpha\sigma} w_{\alpha} \zeta(\mathbf{k})}, \tag{5.5}$$

where $w_\alpha = zt^\alpha$, z is the number of nearest neighbors

$$g_{\alpha\sigma} = 1 - \sum_{\sigma'} n_{\bar{\alpha}\sigma'} - n_{\alpha\bar{\sigma}}, \quad \varsigma(\mathbf{k}) = \frac{1}{z}\sum_{\Delta} e^{i\mathbf{k}\Delta}. \tag{5.6}$$

The spectral function $\varsigma(\mathbf{k})$ depends on the lattice symmetry. For example, in the case of a simple cubic lattice, we have $\varsigma(\mathbf{k}) = \left[\cos k_x d + \cos k_y d + \cos k_z d\right]/3$, where d is the lattice constant.

In the main approximation in $1/U$ considered here, the magnetic ordering does not appear. Then, one can put $n_{\alpha\downarrow} = n_{\alpha\uparrow} \equiv n_\alpha/2$ and the filling of each band $g_\alpha \equiv (g_{\alpha\downarrow} + g_{\alpha\uparrow})/2$. It follows from the expression for the density of states $\rho_\alpha(E) = -\pi^{-1}\mathrm{Im}\int G_\alpha(\mathbf{k}, E + i0)d^3\mathbf{k}/(2\pi)^3$ that [1]

$$n_\alpha = 2g_\alpha n_0\left(\frac{\mu + \epsilon^\alpha}{g_\alpha w_\alpha}\right), \tag{5.7}$$

where

$$n_0(\varepsilon) = \int_{-1}^{\varepsilon} dE \int \frac{d^3\mathbf{k}}{(2\pi)^3}\delta[E - \varsigma(\mathbf{k})]. \tag{5.8}$$

The chemical potential μ in Eq. (5.7) can be determined from the equality $n = n_a + n_b$.

Let the energy difference ϵ between centers of a and b bands be of the order of the characteristic width of the b band, w_b. In this case, there exist only a electrons at low doping until the chemical potential reaches the bottom of the b band, $\mu = -\epsilon - w_b$, at some electron density n_c. If $n > n_c$, the b electrons appear in the system, and the effective width of the a band, $W_a = 2w_a g_a(n_a, n_b)$, starts to decrease. In other words, at $n = n_c$ the topology of the Fermi surface transforms since the number of the electronic bands at the Fermi surface changes from one to two. Such a singularity is also referred to as the topological Lifshitz transition or so-called 2.5 order transition [13]. The plots of n_a, n_b, and the effective bandwidths as functions of n are shown in Figs. 5.1 and 5.2 for the simple cubic lattice.

The energy of the system in the homogeneous state, E_{hom}, is the sum of electron energies in all filled bands. Similar to Eq. (5.7), we get [1]

$$E_{\mathrm{hom}} = 2\sum_\alpha g_\alpha^2 w_\alpha \varepsilon_0\left(\frac{\mu + \epsilon^\alpha}{g_\alpha w_\alpha}\right), \tag{5.9}$$

where

$$\varepsilon_0(\varepsilon) = \int_{-1}^{\varepsilon} E\, dE \int \frac{d^3\mathbf{k}}{(2\pi)^3}\delta[E - \varsigma(\mathbf{k})]. \tag{5.10}$$

The $E_{\mathrm{hom}}(n)$ function is plotted in Fig. 5.3 by the solid line.

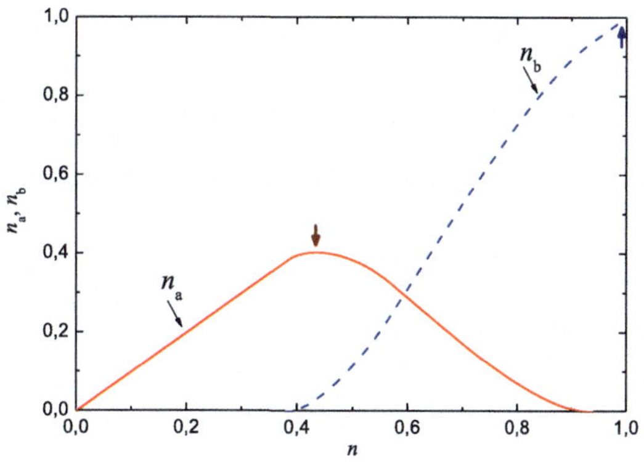

Fig. 5.1 Electron densities n_a (solid line) and n_b (dashed line) vs the total number of charge carriers n; $w_b/w_a = 0.2$, $\epsilon/w_a = 0.12$. Vertical arrows show the values of n_a and n_b corresponding to the inhomogeneous state [1]

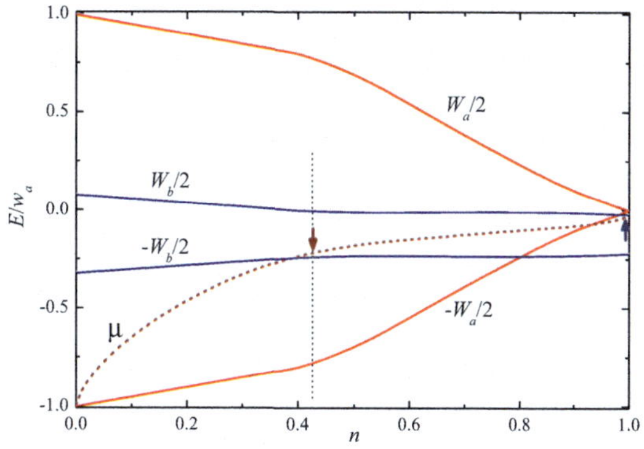

Fig. 5.2 Effective bandwidths $W_\alpha = 2w_\alpha g_\alpha$ versus n. The dashed curve is the chemical potential μ. The values of the parameters are the same as in Fig. 5.1. Vertical arrows show the values of n_a and n_b in the inhomogeneous state [1]

5.1.2 Phase Separation

As one can see in Fig. 5.3, the energy of the homogeneous state, $E_{\text{hom}}(n)$, has two minima. Thus, we can expect that a system forms two phases with different electron densities. However, the phase separation may be hindered by the increase of the energy due to surface effects and a charge redistribution. However, at first, we ignore these. We consider two phases, I (low charge carrier density per site n_1 and volume fraction p) and II (high charge carrier density n_2 and volume fraction $1 - p$). We seek a minimum of the system energy

$$E^0_{\text{ps}} = pE_{\text{hom}}(n_1) + (1 - p)E_{\text{hom}}(n_2) \qquad (5.11)$$

under the condition of the conservation of the number of charge carriers

$$n = pn_1 + (1 - p)n_2. \qquad (5.12)$$

The calculated energy of the system with simple cubic lattice in the phase-separated state is shown in Fig. 5.3 by the dashed curve [7]. The phase separation

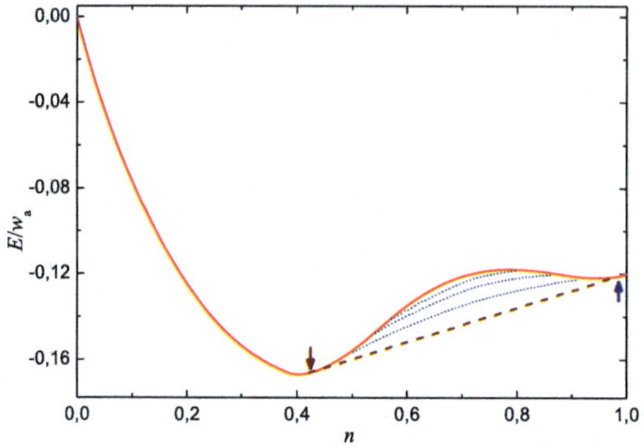

Fig. 5.3 Energy of the system vs doping level n. The solid curve corresponds to the homogeneous state, whereas the dashed curve is the energy of the phase-separated state without taking into account electrostatic and surface contributions to the total energy. Dotted curves are the energies of the inhomogeneous state at $V_0/w_a = 0.1$, 0.05, and 0.01 from top to bottom, respectively, (see the text below). Here, $w_b/w_a = 0.2$ and $\epsilon/w_a = 0.12$. Vertical arrows show the values of n_a and n_b in the inhomogeneous state [1]

exists within the range of n where both types of charge carriers coexist in the homogeneous state. The densities of charge carriers in each phase vary slowly with n, remaining close to certain optimal values for each phase: $n_1 \approx n_a \approx 0.5$ and $n_2 \approx n_b \approx 1$. Phase II can be considered as a Mott–Hubbard insulator since the corresponding lower Hubbard subband is almost completely filled. Therefore, we can conclude that the system may become separated into metallic and insulating phases in a certain parameter range. The above discussion demonstrates that the width and filling of one band depend on the width and filling of the other band.

The phase separation can be favorable only if the bands are appreciably different. As we know, the negative compressibility $\chi = \partial^2 E/\partial n^2 < 0$ is a signature of the possible phase separation. The numerical analysis of the considered particular model in Ref. [1] demonstrates that $\partial^2 E/\partial n^2 < 0$ in a certain range of n if $w_b/w_a < 0.4$. Naturally, this threshold value of the bandwidth ratio may be different for different systems. The region of the phase separation in the $(\epsilon/w_a, 1-n)$ plane for the model under discussion is shown in Fig. 5.4.

5.2 Role of the Interband Hybridization

In the previous discussion of the two-band Hubbard model, we did not take into account a possibility of the interband charge transfer or the hybridization between bands a and b. Such a hybridization may be especially relevant in the case of cuprate superconductors. Cuprate superconductors can be considered as typical examples of the strongly correlated electron systems. Effects occurring in them are usually treated in the framework of the multiband Hubbard model or its generalizations. In

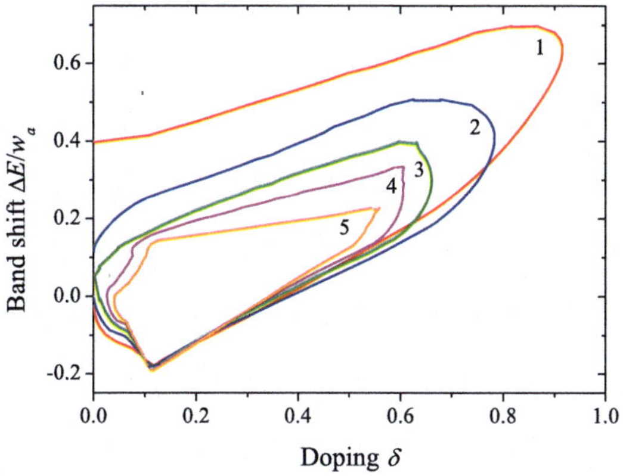

Fig. 5.4 Phase separation regions (value ΔE in the figure corresponds to ϵ in the text and $\delta = 1 - n$) for different values of the bandwidth ratio: $w_b/w_a = 0.1$ (curve 1), 0.2 (2), 0.3 (3), 0.35 (4), and 0.4 (5). If $w_b/w_a > 0.4$, the phase separation is unfavorable (the long-range Coulomb interaction is neglected, $V = 0$) [14]

such situations, we are usually dealing with the hybridization of oxygen p orbitals and copper d orbitals. Then, the a and b bands under discussion can be treated as p and d bands of cuprates. Thus, we can generalize Hamiltonian (5.1) taking the kinetic energy term in the form

$$-\sum_{\langle \mathbf{ij}\rangle\alpha\beta\sigma} t^{\alpha\beta}\left(a^{\dagger}_{\mathbf{i}\alpha\sigma}a_{\mathbf{j}\beta\sigma} + h.c.\right). \tag{5.13}$$

Here, the notation is the same as in Eq. (5.1), and α and β mean indices of a and b (or p and d) bands. Hence, we have three different hopping integrals t_{pp}, t_{dd}, and t_{pd} (later on we also use notation $w_{pp} = zt_{pp}$, $w_{dd} = zt_{dd}$, and $w_{pd} = zt_{pd}$). The detailed analysis based on such two-band Hubbard Hamiltonian with the interband hybridization is given in Ref. [15]. The problem is considered following the same lines as those for the two-band Hubbard model without hybridization studied in the previous chapters. The corresponding results can be summarized as follows. First, we can diagonalize the kinetic energy part of the Hamiltonian. Solving the corresponding eigenvalue problem, we obtain the energy spectrum of charge carriers in two new unhybridized bands (labeled by $j = \pm 1$). In contrast to p and d holes with short lifetime due to interband hoping, the new quasiparticles have a longer lifetime due to scattering by, e.g., phonons and impurities. We denote the lower band as $j = 1$ and the upper one as $j = -1$. The densities of states (DOS) for these bands are plotted in Fig. 5.5 in the case of the simple cubic lattice. Again, each band is split into two Hubbard subbands due to the on-site Coulomb repulsion. So, in the two-band Hubbard model, we have four bands, two lower and two upper bands separated by the gap of the order of U. As above, we can consider here the doping range $n < 1$, for

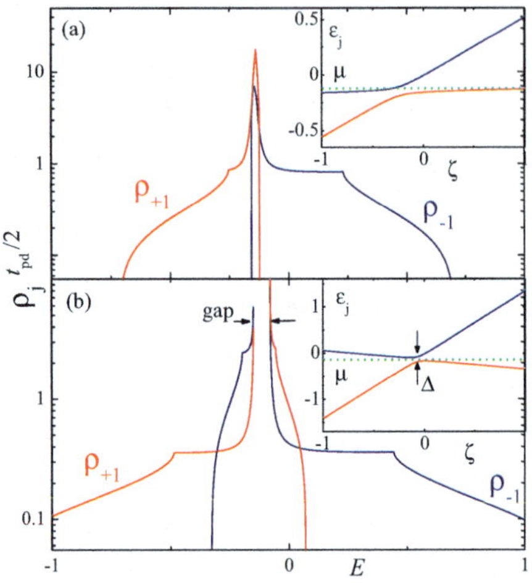

Fig. 5.5 Density of states $\rho_{\pm 1}$ versus energy E for quasiparticles (holes) located in bands $j = \pm 1$ for two different types of spectra, shown in the corresponding insets. The gapless spectrum shown in inset (**a**) was calculated at $n = 0.6$, $\varepsilon_d = 0.15$ eV, $w_{pp} = 1$ eV, $w_{dd} = 0.1$ eV, and relatively small interband hybridization $w_{pd} = 0.25$ eV. The gapped spectrum in inset of panel (**b**) was calculated for the same parameters but for stronger interband hybridization $w_{pd} = 1$ eV. In case (**b**), the transition to the insulating state occurs at a certain doping level for which the chemical potential μ (shown by the green dotted line) is located inside the gap. In both cases, a large peak in the DOS is observed at energies corresponding to the anticrossing of the bands, where a significant flattening of the Fermi surface (see insets) takes place. The quasiparticle energy spectra are shown in the insets as a function of the variable ζ, since ε_j depends on the crystal momentum **k** only via $\zeta(\mathbf{k})$ [15]

which we can limit ourselves to the case $U \to \infty$, when the upper Hubbard subbands are empty and do not contribute to the total electron energy.

Two qualitatively different quasiparticle spectra are shown in the insets of Fig. 5.5a and b. If the hybridization is small (see inset of Fig. 5.5a), the anticrossing of two bands corresponds to a metallic behavior for any doping. When doping increases, the chemical potential μ [shown by the dotted (green) line] shifts upward: at low doping, we have one metallic band and one empty band; then two metallic bands; with further increase of the doping we have one metallic and one filled band. For larger interband hybridization, when $t^{pd} > (t^{pp}t^{dd})^{1/2}$ (see inset of Figs. 5.5b), there exists a transition to an insulator at some doping level. Indeed, for a certain doping, μ is located in the gap between the bands. Nearby the anticrossing point of the two bands, the quasiparticle spectrum becomes flattened. This could be related to the nesting of the Fermi surface. As a result, the DOS exhibits peaks at the energies corresponding to the anticrossing points (Fig. 5.5). The optimum doping for superconductivity corresponds to the case when μ is close to the energy, where these peaks are observed.

The energy of the homogeneous state versus doping n is shown in the inset of Fig. 5.6a. The curvature of $E_{\text{hom}}(n)$ is negative between the two marked points n_1 and n_2. This indicates the instability of the homogeneous state with respect to the

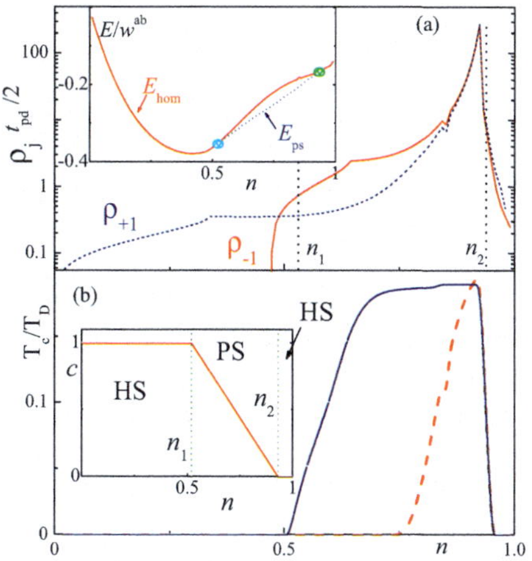

Fig. 5.6 **a** Density of states at the Fermi level versus doping in the two-band Hubbard model. Inset of panel (**a**): energy of the homogeneous (red solid line) and the phase-separated (blue dashed line) states. The dependence of the energy in the homogeneous state, E_{hom}, on n has a negative curvature if $n_1 < n < n_2$. In this range of doping, the phase-separated state becomes more favorable since its energy, E_{ps}, is lower than E_{hom}. The values $n_{1,2}$ are indicated in the main panel of (**a**) by black vertical dotted lines. The point n_2 is near the peak in the DOS. **b** Critical temperature, T_c, of the superconducting transition versus doping level n for the homogeneous (red dashed line) and phase-separated (blue solid line) states calculated using Eq. (5.14). The $T_c(n)$ of the homogeneous state decreases fast when the doping n deviates from its optimum value about n_2. In contrast to this, $T_c(n)$ in the phase-separated state is a broad function and exhibits a plateau within the $n_1 < n < n_2$ range. The inset of panel (**b**) demonstrates the dependence of concentration of the phase with lower hole content on n (the notation c in the figure corresponds to the notation p in the text). The regions of homogeneous (HS) and the phase-separated (PS) states are indicated in the inset. Here, we use the following parameters: $\varepsilon_d = 0.2$ eV, $w_{pp} = 1$ eV, $w_{dd} = 0.3$ eV, and $w_{pd} = 0.7$ eV [15]

separation into two phases with the hole densities n_1 and n_2, when $n_1 < n < n_2$. The energy $E_{\text{ps}}(n) = pE_{\text{hom}}(n_1) + (1 - p)E_{\text{hom}}(n_2)$ of the phase-separated state is lower than the energy $E_{\text{hom}}(n)$ of the homogeneous phase in this doping range (see dashed line in the inset of Fig. 5.6a). In a wide parameter range, the hole density n_2 turns out to be near the optimum value for superconductivity. A typical dependence of the phase concentration p versus doping is shown in the inset of Fig. 5.6b. At a low doping level, the system is in a homogeneous state ($p = 1$). If $n > n_1$, the sample becomes segregated into droplets with two different hole densities, n_1 and n_2. With the further increasing in n, the relative concentration p of the phase with lower hole content n_1 decreases almost linearly, and when $n > n_2$, this phase disappears and the system becomes homogeneous again. Further analysis shows that the phase-separated state occurs in the parameter range, where $w_{pd} < w_{pp}$. If $w_{pd} \gg w_{pp}$, the two-band Hubbard Hamiltonian reduces to an effective single-band model and the discussed mechanism for the phase separation in the system does not work.

The phase separation leads to a redistribution of the charge carriers and charge neutrality breaking. The structure of the phase-separated state can be either ordered (e.g., checkerboard structure) or random (i.e., randomly distributed droplets of one phase within a matrix of the other phase). The charge redistribution and finite droplet size give rise to the additional size-dependent terms in Eq. (5.11). For the random phase-separated state, the droplet size D is determined by the competition between the Coulomb and surface energies. This problem will be discussed in detail below (in Sect. 5.5). Following the approach used in that section, we can estimate D in the range from three to six lattice constants d when $w_{pd} \sim w_{pp} \sim 1$ eV, in agreement with the experimental data reported for the cuprates, e.g., in Ref. [16], where spatial variations of the DOS and superconducting gap were measured using STM.

Let us discuss the possible effect of the considered type of the phase separation on the critical temperature T_c of the superconducting transition. We do not focus on any particular mechanism of superconductivity in cuprates. We should only relate somehow the DOS to T_c. Then, to illustrate the possible effects, we use the simplest BCS formula for T_c accounting for the Coulomb repulsion [17]

$$T_c = T_D \exp\left[-\,1/\left(\rho_j V_p - \nu_c\right)\right], \tag{5.14}$$

where T_D is the Debye temperature, V_p is the BCS-type electron–phonon coupling constant,

$$\nu_c = \rho_j V_c / \left[1 + \rho_j V_c \ln\left(\mu/k_B T_D\right)\right] \tag{5.15}$$

is the Coulomb pseudopotential, and $V_c \sim U^p \approx 5$ eV is the Coulomb matrix element. The possible applicability of Eq. (5.14) in the case of high-T_c cuprates is discussed in Ref. [18].

In the approach under discussion, the homogeneous superconducting state can appear in different bands depending on doping, see Fig. 5.6: in the band $j = 1$ for low doping and in the band $j = -1$ for higher doping. The dependence of T_c on doping n for the homogeneous state is shown by the dashed line in Fig. 5.6b. For the chosen range of parameters, T_c is nonzero only for high doping level and achieves a maximum at the optimum doping, which corresponds to the value of chemical potential near the anticrossing point (see insets of Fig. 5.5). Away from the optimum doping, $T_c(n)$ decreases fast with n. In the phase-separated state, one of the phases retains the optimum hole density (which is about n_2) within a wide interval of doping. This results in an plateau of $T_c(n)$ (see blue solid line in Fig. 5.6b). This can provide a possible explanation for the observed dependence of T_c on the hole doping in cuprates (see Refs. [18, 19] and references therein).

The above picture relates T_c to the variation of the density of states and doping. It can be qualitatively applicable for other models of the superconductivity. The relation between DOS and T_c is widely discussed in literature in connection with many other models of superconductivity (see, e.g., Refs. [20, 21]).

5.3 Several Remarks: Beyond the Hubbard I Approximation

In our analysis of the electronic phase separation, we were mostly dealing with an approach based on the Hubbard I approximation. As we have already mentioned, Hubbard I is an appropriate method to find out the main features of the electron band structure capturing the essential physics related to electron correlations. Moreover, it is a good starting point for revealing the tendency to electronic phase separation in various systems and finding out the structure of corresponding phase diagrams, especially in the limit of strong correlations.

However, Hubbard I misses some important effects such as interband quasiparticle peak clearly seen if we use more refined approaches such as the dynamic mean-field theory (DMFT). The detailed comparison of the results obtained within the Hubbard I and DMFT approaches have been undertaken in Ref. [12]. The spinless Falicov–Kimball model for the simple cubic lattice and a more general model, in which heavier charge carriers correspond to a band with finite width W, were analyzed. Green's function for itinerant charge carriers, the evolution of the system with doping, and the possibility of the phase separation were studied. It was shown that both approximations lead to nearly identical results for the Falicov–Kimball model, i.e., in the narrow-band limit, $W \rightarrow 0$. For finite W and in the strong correlation limit, $U \gg W$, both approximations do not contradict each other giving qualitatively similar results in the spectral properties of itinerant charge carriers and the behavior of the system with doping. Both approximations predict an instability of the homogeneous state toward the phase separation in a wide range of model parameters. It was also shown that the system remains unstable toward the phase separation even for finite bandwidth W for heavy charge carriers. This suggests that the phase separation phenomenon is the inherent feature of Hubbard-like models with wide and narrow bands, and not an artifact of this or that approximation scheme.

Moreover, the Hubbard I approximation was used for the direct evaluation of the equal-time anomalous two-electron propagator for the Hubbard model on a two-dimensional square lattice [22]. The results were compared with the quantum Monte Carlo calculations reported in Ref. [23] in the limit of strong electron–electron interaction. The Hubbard I predictions appear to be in a good qualitative agreement with the Monte Carlo results. In particular, d-wave type correlations decay as cr^{-3} ("free-electron" behavior), if the separation r exceeds 2–3 lattice constants. However, the Hubbard I approximation underestimates the coefficient c by a factor of about 3. This demonstrates that the Hubbard I approximation, despite its simplicity and artifacts, captures the qualitative behavior of the two-particle propagator for the Hubbard model, at least, for moderate values of r.

Note, however, that using the Hubbard I approximation, we can miss additional mechanisms, which can also lead to the phase separation. Quite a vivid example is the so-called charge transfer instability, which was widely discussed in the context of cuprate superconductors [24–27]. The charge transfer instability manifests itself in the vanishing with doping of the lowest excitonic particle-hole excitation. This leads to the diverging compressibility giving rise to a phase separation. A thorough analysis of this effect undertaken in Ref. [27] shows that it has a rather fundamental nature and is related to the violation of the Landau stability criterion for the Fermi liquid (so-called Pomeranchuk instability). In this analysis, the used slave-boson technique [28] appears to be very fruitful since it reveals nontrivial electron correlation effects beyond the Hubbard I approximation. In this connection, let us mention a series of papers of Irkhin et al. [29–32], where the slave-boson technique has been successfully implemented for finding out the phase separation regions for different magnetic transitions in strongly correlated electron systems.

5.4 Electron–Lattice Interaction and the Phase Separation

Electron–lattice coupling is frequently incorporated in the multiband Hubbard model to describe electronic structure of cuprates and manganites. Indeed, such compounds contain Jahn–Teller ions (copper or manganese) and this implies the crucial role of this coupling. Even at the initial stages of the studies of the colossal magnetoresistance effect, it was shown that both the strong electron correlations and electron–lattice interactions are fundamental ingredients for any adequate interpretation of this effect [33–36]. Moreover, the electron–lattice coupling can lead to quite unusual forms of the phase separation (see, e.g., Ref. [37]).

In Ref. [38], the effect of this coupling on the phase separation in the two-band Hubbard model, Eq. (5.1), was considered. The interaction of electrons with the lattice distortions results, first, in the change of the hopping probability and, second, in the relative shifts of the electron bands. If the electron–lattice interaction is strong enough, then there appears a competition between states with different values of strains, and the transition between these states can occur in a jump-like manner. It was also demonstrated that the electron–lattice interaction produces a pronounced effect on the conditions of the electronic phase separation since it affects the value of the bandwidth ratio and the relative positions of the bands.

The electron–lattice interaction can be chosen in the following form [38]

$$H_{el-latt} = \sum_{\langle \mathbf{ij}\rangle,\sigma} \left(\lambda\, a^{\dagger}_{\mathbf{i}\sigma} a_{\mathbf{j}\sigma} u_{\mathbf{j}} + \lambda' b^{\dagger}_{\mathbf{i}\sigma} b_{\mathbf{j}\sigma} u_{\mathbf{j}} + h.c.\right) + \sum_{\mathbf{i},\sigma} \left(\lambda_a u_{\mathbf{i}} n^{a}_{\mathbf{i}\sigma} + \lambda_b u_{\mathbf{i}} n^{b}_{\mathbf{i}\sigma}\right), \quad (5.16)$$

where λ, λ' and $\lambda_{a,b}$ are corresponding constants of the electron–lattice interaction and $u_{\mathbf{i}}$ are the local distortions of the unit cell containing site $\mathbf{i}$. The first sum in this equation describes the change of the hopping probabilities due to lattice distortions affecting both intersite distance and the form of electron wave functions. The second sum corresponds to the change in the electron–ion interaction induced by strain. Thus, we take into account the effect of lattice strains both on the on-site electron energy and intersite charge transfer.

The elastic energy of the system depending on distortions at different sites can be written as

$$\mathcal{F}_{\text{elast}} = \frac{K}{2}\sum_{\mathbf{i}} u_{\mathbf{i}}^2, \tag{5.17}$$

where K is the elastic modulus.

To find a self-consistent solution to the problem in the adiabatic approximation, we first perform averaging of the Hamiltonian over the electronic degrees of freedom. From the condition of energy minimum with respect to strains, $\partial\langle H(u_{\mathbf{i}})\rangle/\partial u_{\mathbf{i}} = 0$, we obtain for the lattice local distortions

$$\bar{u}_{\mathbf{i}} = -\frac{\lambda z\langle a_{\mathbf{i}\sigma}^{\dagger}a_{\mathbf{j}\sigma}\rangle + \lambda' z\langle b_{\mathbf{i}\sigma}^{\dagger}b_{\mathbf{j}\sigma}\rangle + \lambda_a\langle n_{\mathbf{i}}^a\rangle + \lambda_b\langle n_{\mathbf{i}}^b\rangle}{K}, \tag{5.18}$$

where $\mathbf{i}$ and $\mathbf{j}$ are the nearest-neighbor sites. Using Eq. (5.18), we can present the effective electron Hamiltonian as

$$\begin{aligned} H_{\text{eff}} = &-\sum_{\langle\mathbf{ij}\rangle,\sigma}\left[\left(t^a - \lambda\bar{u}_{\mathbf{i}}\right)a_{\mathbf{i}\sigma}^{\dagger}a_{\mathbf{j}\sigma} + \left(t^b - \lambda'\bar{u}_{\mathbf{i}}\right)b_{\mathbf{i}\sigma}^{\dagger}b_{\mathbf{j}\sigma} + h.c.\right] - \sum_{\mathbf{i},\sigma}\left[\Delta E + (\lambda_a - \lambda_b)\bar{u}_{\mathbf{i}}\right]n_{\mathbf{i}\sigma}^b \\ &+ H_U - \sum_{\mathbf{i},\sigma}(\mu - \lambda_a\bar{u}_{\mathbf{i}})\left(n_{\mathbf{i}\sigma}^a + n_{\mathbf{i}\sigma}^b\right) + \frac{K}{2}\sum_{\mathbf{i}}\bar{u}_{\mathbf{i}}^2, \end{aligned} \tag{5.19}$$

where H_U includes all the terms responsible for the on-site electron–electron interaction [see Eq. (5.1)].

Hamiltonian (5.19) clearly demonstrates that the effect of electron–lattice interaction is twofold. This interaction in its nondiagonal terms changes the effective bandwidth, whereas in diagonal terms, it shifts the positions of the bands and the chemical potential. In the previous sections, we have demonstrated that the qualitative features of the phase diagram for the two-band Hubbard model are mainly determined by two dimensionless parameters: the ratio of the bandwidths and the relative positions of the bands. Thus, to construct a minimal model capturing the main physical effects of electron–lattice interaction, it is sufficient to keep only λ and λ_b. In addition, we can put $\Delta E = 0$ to emphasize the effect of band shift related only to the electron–lattice interaction. As a result, we get

$$\bar{u}_{\mathbf{i}} = -\frac{\lambda z\langle a_{\mathbf{i}\sigma}^{\dagger}a_{\mathbf{j}\sigma}\rangle + \lambda_b\langle n_{\mathbf{i}}^b\rangle}{K} \tag{5.20}$$

and

$$\begin{aligned} H_{\text{eff}} = &-\sum_{\langle\mathbf{ij}\rangle,\sigma}\left[\left(t^a - \lambda\bar{u}_{\mathbf{i}}\right)a_{\mathbf{i}\sigma}^{\dagger}a_{\mathbf{j}\sigma} + t^b b_{\mathbf{i}\sigma}^{\dagger}b_{\mathbf{j}\sigma} + h.c.\right] + H_U \\ &+ \sum_{\mathbf{i},\sigma}\lambda_b\bar{u}_{\mathbf{i}}n_{\mathbf{i}\sigma}^b - \sum_{\mathbf{i},\sigma}\mu\left(n_{\mathbf{i}\sigma}^a + n_{\mathbf{i}\sigma}^b\right) + \frac{K}{2}\sum_{\mathbf{i}}\bar{u}_{\mathbf{i}}^2. \end{aligned} \tag{5.21}$$

Starting from Hamiltonian (5.21), we can point out the main qualitative effects of the electron–lattice interaction in the two-band model. In the absence of doping, $n = 0$, bands a, and b are empty and their centers coincide. With the growth of n, the wider band a begins to be filled up from the bottom. The band filling is accompanied by strain $\bar{u}_\mathbf{i} = -\lambda z \langle a^\dagger_{\mathbf{i}\sigma} a_{\mathbf{j}\sigma}\rangle / K$. The average $\langle a^\dagger_{\mathbf{i}\sigma} a_{\mathbf{j}\sigma}\rangle$ is proportional to the hopping probability and thus is positive. The strain $\bar{u}_\mathbf{i}$ is positive if $\lambda < 0$ and negative if $\lambda > 0$. From Eq. (5.21), it is easy to see that at any sign of λ the bandwidth increases with the strain. At a certain doping level, the chemical potential attains the bottom of the narrower band b and this second band starts to be filled up. In this range of doping, we have two types of the electrons and the energy of the system depends on n in a rather complicated manner due to electron–electron correlations. Such a situation is favorable for the phase separation even in the absence of the electron–lattice interaction. However, the characteristic feature of the system under study is the dependence of the effective band shift $\Delta E_{\text{eff}} = -\lambda_b \bar{u}_\mathbf{i}$ on the strain and, hence, on the doping. The sign of the shift depends on the signs of λ and λ_b. A simple analysis of Eqs. (5.20) and (5.21) shows that for the same signs of λ and λ_b the value of ΔE_{eff} is negative and the sign of $\bar{u}_\mathbf{i}$ remains the same at any n. If the signs of λ and λ_b are different, the strain can change its sign at some doping level. The change of the sign of the strain results in the change of the sign of the effective band shift ΔE_{eff}.

The dependence of ΔE_{eff} on doping can give rise to a more sophisticated behavior of the system. If at some doping level n^*, the narrower band crosses the bottom of the wider band, that is, $\lambda_b^2 n^*/K > z t_a$, then, it could be favorable to have almost all electrons in the b band. So, there appears a competition between two states with different values of the strain. It suggests the possibility of a transition between these two states, which can have a jump-like form. The corresponding results are illustrated in Figs. 5.7 and 5.8.

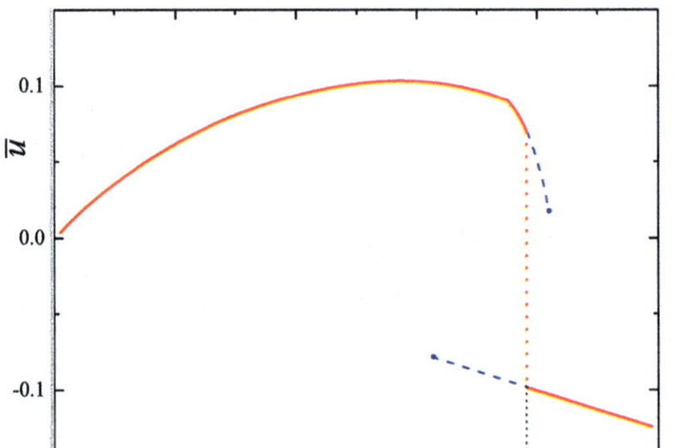

Fig. 5.7 Lattice strain as a function of doping. The parameters are $\lambda/w_a = -1.2$, $\lambda_b/w_a = 2$, $K = 16w_a$ and $w_b/w_a = 0.25$. Jump-like transition between two states with different values of lattice distortions occurs at $n = n^*$. Solid (red) lines correspond to the states with minimum energy, whereas dashed (blue) lines correspond to metastable states [38]

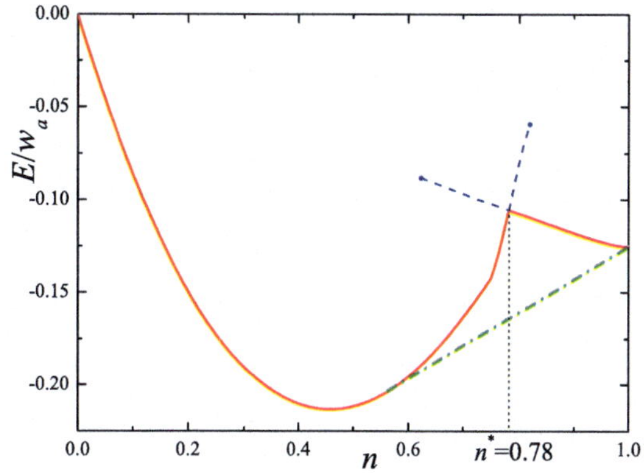

Fig. 5.8 Energy of the system versus doping level n. Solid (red) curve corresponds to the homogeneous state, whereas the dot-dash (green) curve is the energy of the phase-separated state. The dashed (blue) curves correspond to the energies of metastable states. The parameters are the same as in Fig. 5.7. The kink in the curve $E_{\text{hom}}(n)$ corresponds to the jump-like change in the lattice distortion [38]

5.5 Characteristic Size and Geometry of Inhomogeneities

The phase separation leads to the violation of charge neutrality ($n_1 \neq n_2$). Therefore, we should take into account the electrostatic contribution, E_C, to the total energy of the phase-separated state. The value E_C evidently depends on the size and shape of inhomogeneities. The second contribution E_S to the total energy depending on the size of inhomogeneities is related to the interface between two phases. Therefore, the interplay between E_C and E_S determines the equilibrium sizes and geometry of the inhomogeneities. The simplest is a droplet-like geometry of the inhomogeneities. However, the long-range Coulomb interaction can give rise to a more sophisticated geometry of phase separation (stripes, layers, rods, etc.) [14, 39–42].

In Refs. [14, 40, 41], the Coulomb contribution to the total energy was calculated for different shapes of the inhomogeneities: spherical ($D = 3$), cylindrical ($D = 2$), and layered ($D = 1$). The calculations were performed using the Wigner–Seitz approximation, that is, the system is divided into cells, where an internal region (core) containing phase a is surrounded by a shell of phase b, if the concentration p of phase a is lower than 0.5. When $p > 0.5$, the core contains the phase b and the shell contains the phase a. The relative volumes of the core and shell are determined by the charge conservation condition Eq. (5.12). It was assumed that the characteristic sizes of inhomogeneities are larger than the lattice constant d and the macroscopic approach can be applied to calculate the electrostatic energy. The interaction between neutral cells is neglected. In so doing, the electrostatic energy, E_C, for the inhomogeneities with dimensionality D can be expressed as [14, 40, 41]

$$E_C = V(n_1 - n_2)^2 \left(\frac{R_S}{d}\right)^2 u_D(p), \tag{5.22}$$

where $V = e^2/d\varepsilon$ is the characteristic Coulomb interaction, ε is the average permittivity, R_s is the size of the core (which is the radius of the internal sphere or cylinder in $3D$ or $2D$ cases, respectively, and the half-width of the internal layer in the $1D$ case), and $u_D(p)$ is a dimensionless function. When $p < 0.5$, we have

$$u_3 = \frac{4\pi p}{5}\left(2 - 3p^{1/3} + p\right), \quad D = 3 \text{ (droplets)},$$

$$u_2 = \pi p\,(-\ln p + p - 1)\,, \quad D = 2 \text{ (rods)}, \tag{5.23}$$

$$u_1 = \frac{4\pi}{3}(1 - p)^2\,, \quad D = 1 \text{ (planes or layers)}. \tag{5.24}$$

In the case $p > 0.5$, we should replace $n_1 \Leftrightarrow n_2$ and $p \Leftrightarrow 1 - p$ in all equations.

The second term in the total energy depending on the size of the inhomogeneities, E_S, is related to the boundary between two phases. It comes from the size quantization and depends on the electron densities in both phases. In Ref. [1], the energy E_S was calculated using the perturbative approach proposed by R. Balian and C. Bloch [43]. If $R_s > d$, the surface tension $\sigma(n_1, n_2)$ does not depend on the shape of inhomogeneity. Then, we can write the surface energy per unit cell as $E_S = \sigma(n_1, n_2)S_D d^3/V_D$, where S_D is the surface area of the core and V_D is the total volume of the Wigner–Seitz cell. It is convenient to write

$$E_S = \sigma(n_1, n_2)pDd^2\frac{d}{R_s}. \tag{5.25}$$

Minimization of the sum $E_{CS} = E_S + E_C$ with respect to R_s gives

$$E_{CS} = 3\left[\frac{d^2\sigma(n_1, n_2)|n_1 - n_2|}{2}pD\sqrt{Vu_D}\right]^{2/3}. \tag{5.26}$$

The equations used for the numerical calculation of $\sigma(n_1, n_2)$ can be found in the Appendix of Ref. [1]. We do not rewrite them here. Note only, that the function $\sigma(n_1, n_2)$ varies rather slowly with the change of its arguments.

The behavior of energy E_{CS} as a function of p is illustrated in Fig. 5.9. We see that the droplet geometry is the most favorable at small p ($p < p_1 = 0.215$). At intermediate values of p ($p_1 < p < p_2 = 0.355$), we have rods, which are transformed to layers at $p > p_2$. However, the different geometries are close in energy. Thus, we can use E_{CS} calculated, e.g., for droplets in different estimates as it was done in Fig. 5.3 for calculation of the energy of the phase-separated phase. Note also, that the presented derivation of E_{CS} does not depend on the particular electron Hamiltonian and can be applied in different cases.

The electroneutrality condition (5.12) determines the relationship between the phase content and the number of electrons per site $p = (n - n_2)/(n_1 - n_2)$. In the

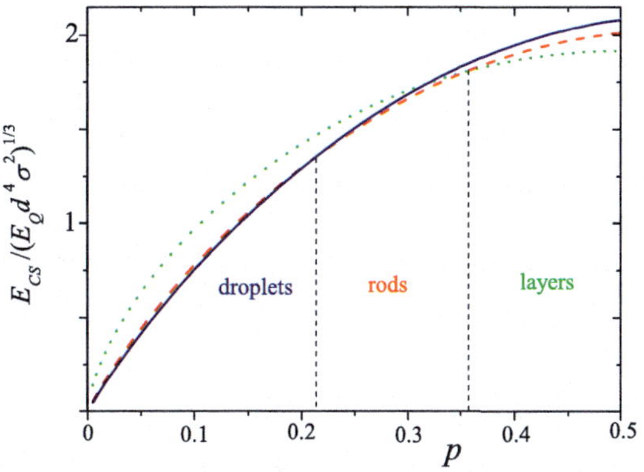

Fig. 5.9 Normalized energy of an inhomogeneity as a function of the phase content p at $p < 0.5$. The picture for $p > 0.5$ is a mirror reflection of this figure with respect to the vertical line $p = 0.5$, $E_Q = V(n_1 - n_2)^2$. Solid (blue), dashed (red), and dotted (green) lines correspond to droplets, rods, and planes or layers, respectively [14]

main approximation, the values $n_{1,2}$ are independent of n. So, the phase content is almost a linear function of n or doping $\delta = 1 - n$. Then, the inhomogeneities evolve with doping as follows. Phase separation arises if the doping exceeds some value $\delta = \delta^*(w_a/w_b, \epsilon) > 1 - n_2$. If $\delta^* < \delta_1 = 1 - n_1 p_1 - n_2(1 - p_1)$, the phase separation appears in the form of the nanoscale droplets of the phase b in a host. With the increase of doping, the geometry changes to the rod-like [$\delta_1 < \delta < \delta_2 = 1 - n_1 p_2 - n_2(1 - p_2)$], and, finally, to the stripe-like layered configuration when $\delta > \delta_2$. If the value of δ^* is larger than δ_1 and/or δ_2, droplet-like or even rod-like structure can be missed.

5.6 Electronic Phase Separation in High-T_c Superconducting Cuprates and Pnictides

The solution of the problem of high-T_c superconductivity in cuprates and pnictides requires a correct description of the normal state where spin, charge, orbital, and lattice degrees of freedoms compete, with the formation of nanoscale puddles of spin density wave stripes, puddles of charge density wave stripes, and/or puddles of ordered mobile oxygen interstitials. In this way, it was suggested that the minimal model to capture the essential physics of high-temperature superconductors needs to take into account both the presence of two electronic components with different orbital symmetry [44–46] and a nanoscale phase separation [47–50]. There is now an experimental evidence that the high-temperature superconductivity in cuprates and pnictides emerges in the proximity of a topological Lifshitz transition [51–53]. In Ref. [54], a theory for the high-temperature superconductivity was suggested based on the shape resonance near the Lifshitz transition between a BCS-like superconducting gap and a second gap in the Bose–Einstein condensate.

In Refs. [14, 15, 50, 55], a model with two-band Hubbard Hamiltonian (5.1) was proposed as a version of minimal model for the description of the normal state in high-T_c cuprate superconductors. This approach naturally provides a reasonable theoretical background for the phase diagram region where the nanoscale phase separation emerges near the topological Lifshitz transition. The detailed analysis of experimental evidence on the electronic phase separation in cuprate superconductors is given in review article [56].

The phase separation in pnictides will be discussed in Sect. 8.5. In pnictides, the electron–electron interaction is not so strong as in cuprates and an important feature of these systems is an imperfect nesting of the Fermi surface.

5.7 Electronic Phase Separation in Magnetic Oxides with Jahn–Teller Ions

As we have already mentioned above, the phase separation phenomena play a fundamental role in the physics of magnetic oxides, especially manganites and related compounds. The most frequently discussed type of the phase separation in them is the formation of nanoscale inhomogeneities such as ferromagnetic (FM) metallic droplets in an insulating antiferromagnetic (AF) matrix arising due to self-trapping of charge carriers [57, 58]. The analysis of the experimental data, especially on the magnetotransport [59] and voltage noise [60], shows that such type of the phase separation is indeed a common feature of these materials.

The model for the phase separation in manganites and related compounds was suggested in Refs. [6–8]. First, we should take into account the Jahn–Teller (JT) nature of the magnetic ions, which could give rise to the localization of charge carriers at the lattice distortions. This situation is typical of manganites and could be described in terms of the Kondo-lattice model in the double exchange limit with account taken for the JT distortions and the superexchange interaction between the localized electrons. The intraatomic exchange coupling is assumed to be large enough to align the spins of e_g electrons in a magnetic ion parallel to spin $\mathbf{S}$ of core (t_{2g}) electrons. That is, we assume Hund's rule coupling constant to be large compared to hopping integrals. The effective electron hopping integral for neighboring sites depends on an angle ν between the directions of spins $\mathbf{S}$ at these sites: $t_{\text{eff}} = t\cos(\nu/2)$ [61]. The JT effect leads to the splitting of the double-degenerate e_g level. Then, following Ref. [37], we can divide e_g electrons into two groups: "localized" (l, with hoping integral $t^l \ll t$) producing the maximum splitting of the e_g level and "itinerant" (b) electrons with nonzero hopping integrals t, leading to smaller distortions of MnO_6 octahedra. Thus, we come to the two-band picture, which reduces to the Falicov–Kimbal model in the limit $t^l \to 0$. Really, this is only some "minimal model" describing the magnetic phases and the effect of the phase separation in the manganites and related systems.

The corresponding effective Hamiltonian can be written as [6–8]

$$\hat{H} = -t\sum_{\langle \mathbf{ij}\rangle} b_{\mathbf{i}}^{\dagger} b_{\mathbf{j}} \cos\left(\frac{\nu_{\mathbf{ij}}}{2}\right) - \epsilon_{\mathrm{JT}}\sum_{\mathbf{i}} n_{\mathbf{i}l} - \mu\sum_{\mathbf{i}} (n_{\mathbf{i}b} + n_{\mathbf{i}l}) + U\sum_{\mathbf{i}} n_{\mathbf{i}b} n_{\mathbf{i}l}$$
$$+ J\sum_{\langle \mathbf{ij}\rangle} \cos\left(\nu_{\mathbf{ij}}\right). \tag{5.27}$$

Here $n_{\mathbf{i}b} = b_{\mathbf{i}}^{\dagger} b_{\mathbf{i}}$, $n_{\mathbf{i}l} = l_{\mathbf{i}}^{\dagger} l_{\mathbf{i}}$, are the number operators of b and l electrons, $b_{\mathbf{i}}^{\dagger}$, $b_{\mathbf{i}}$, $l_{\mathbf{i}}^{\dagger}$, and $l_{\mathbf{i}}$ are creation and annihilation operators of the itinerant and localized electrons at site **i**, respectively, $\nu_{\mathbf{ij}}$ is the angle between the localized spins at sites **i** and **j**, ϵ_{JT} is the energy gain due to electron localization at the Jahn–Teller distortions. The last term in the Hamiltonian (5.27) is the AFM ($J > 0$) exchange interaction between local spins. The number of localized, n_l, and band, n_b, electrons per lattice site obeys an obvious relation $n_b + n_l = 1 - x$, where x is the number of doped charge carriers per site. The half-filling, $n_b + n_l = 1$, corresponds to the undoped system. In the considered approximation, the charge carriers are "spinless", since their spins are aligned along the direction of the local spins **S**.

Hamiltonian (5.27) was studied in Refs. [6, 7] in the framework of the mean-field approximation within entire range of doping, $0 \leq x \leq 1$. A similar model has been studied numerically in Refs. [62, 63]. Depending on parameters of the model, doping, and temperature, it can describe either homogeneous (FM, AFM, or canted) or inhomogeneous (phase-separated) states. The phase-separated state turns out to be favorable in a large part of the phase diagram. The relative number of different kinds of charge carriers depends on the doping level, temperature, and magnetic field. In Ref. [7] it was taken into account that the hopping integral t in Eq. (5.27) is temperature dependent due to the polaron band narrowing [64]: $t(T) = t(0)\exp\left[-2\lambda^2/\left(e^{\hbar\Omega/T} - 1\right)\right]$, where λ is the dimensionless electron–phonon coupling constant and Ω is the characteristic phonon frequency. The energy spectrum of band electrons was calculated based Green's function technique using the Hubbard I decoupling similar to that described in Sect. 5.1 of this book.

The number of band and localized electrons depends on the relative positions of μ and $\varepsilon_{\mathrm{JT}}$. If $\mu < \varepsilon_{\mathrm{JT}}$, then $n_l = 0$ and $n_b = 1 - x$. With the growth of n_b, the chemical potential attains the value of $-\varepsilon_{\mathrm{JT}}$ and the further growth of the number of band electrons becomes unfavorable in energy. The chemical potential μ becomes pinned at the level $-\varepsilon_{\mathrm{JT}}$ and the localized electrons arise in the system. As a result, there appear two critical values of x (each depends on temperature), x_1 and x_2. When $x < x_1$, we have $n_b = 0$, while $n_l = 0$ if $x > x_2$. The behavior of n_l and n_b as functions of x is illustrated in Fig. 5.10. Depending on the doping level and temperature, the system may be in the homogeneous ferromagnetic (FM), AFM ($n_b = 0$), and canted ($0 < \nu < \pi$) states, as well as in two different kinds of paramagnetic (PM) states (PM insulating phase, $n_b = 0$, and PM metallic phase, $n_b \neq 0$). The AFM state is favorable at small x, whereas the canted state appears close to $x = 1$.

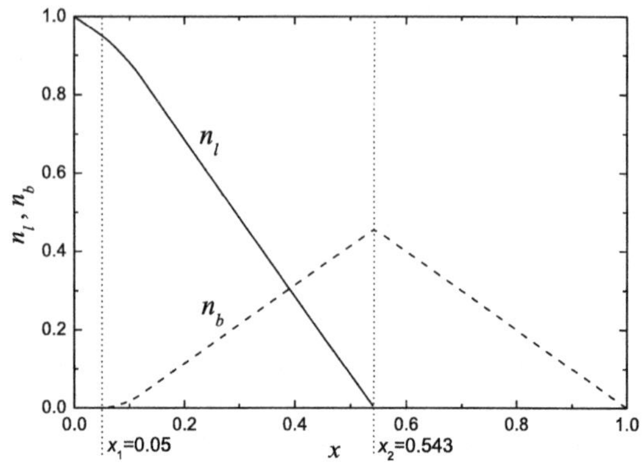

Fig. 5.10 Dependence of the number of localized n_l (solid line) and band n_b (dashed line) electrons on doping x at $\epsilon_{\rm JT}/t(0) = 1.8$ [7]

The phase-separated states now can be considered based on the above analysis of the homogeneous phases of the model. According to Refs. [6, 7], in the phase-separated state, the system consists of the metallic ($n_b \neq 0$) and insulating ($n_b = 0$) phases. The metallic phase can be either FM or PM while insulating phase is AFM or PM. As it was done in the previous analysis, when calculating the free energy of the phase-separated state, we should take into account the Coulomb contribution due to the redistribution of the charge density, Eq. (5.22), and the correction to the electron density of states related to the finite size of inhomogeneities, Eq. (5.25), and find the minimum of their sum, Eq. (5.26). As it was discussed above, the geometry of the inhomogeneous regions is different, however, the energies of these configurations are close to each other, Fig. 5.9. Thus, to derive the phase diagram of the model, we can consider only the inhomogeneities of one type, e.g., in the form of spherical droplets.

In Fig. 5.11, we present the phase diagram of the model in the (x, T) plane from Ref. [7]. The phase diagram is very rich. The details of this diagram are described in the figure caption. Some phases in the diagram can disappear if we choose different model parameters. The obtained results demonstrate that in the considered model the phase separation exists in a wide range of intermediate doping values and disappears at low and high doping levels. These predictions are in agreement with the general features of the experimentally found phase diagrams of manganites [57, 65, 66]. For the characteristic values of parameters for manganites, the FM droplet includes 10–30 unit cells.

The applied magnetic field **H** can affect significantly the phase diagram of the systems under study. This effect was considered in Ref. [7] with taking into account the interaction of the field **H** with the local spins **S** in the classical limit. Thus, in the presence of the external magnetic field, we should add the term $\mu_{\rm B} g \sum_{\rm i} \mathbf{S}_{\rm i}\mathbf{H}$ in the Hamiltonian, where $\mu_{\rm B}$ is the Bohr magneton and g is the Landé factor. The applied magnetic field evidently favors the FM ordering and, consequently, the number of itinerant electrons increases.

As it is seen from the phase diagram in Fig. 5.11, the metal–insulator transition can take place at some characteristic values of the doping x corresponding to the

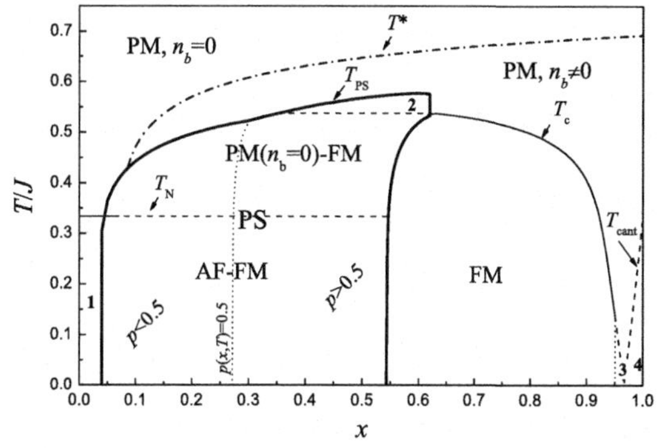

Fig. 5.11 Phase diagram of the model at $\epsilon_{JT}/t(0) = 1.8$, $J/t(0) = 0.36$, $\lambda = 10$, $\Omega/t(0) = 0.18$, and $V/t(0) = 3$. AF means AFM state, FM—ferromagnetic state, PM—paramagnetic state, PS—phase-separated state. The numbers denote: 1—homogeneous AFM phase; 2— the mixture of two PM states with $n_b \neq 0$ and $n_b = 0$; 3 and 4—homogeneous canted states with $\cos\nu > 0$ and $\cos\nu < 0$, respectively. In the picture, p is the volume content of the metallic phase, T_N is the Néel temperature, T_C is the Curie temperature, T_{PS} is the transition temperature between the PS and homogeneous states, T^* is the transition temperature between the PM metallic and insulating phases, and T_{cant} is the transition temperature between the canted and PM phases [7]

crossover between different types of the phase separation. Such a transition can be induced by changing temperature or magnetic field and is of a percolation type. This transition can be related to the colossal magnetoresistance effect [6, 7].

References

1. A.O. Sboychakov, K.I. Kugel, A.L. Rakhmanov, Phys. Rev. B **76**, 195113 (2007)
2. L.M. Falicov, J.C. Kimball, Phys. Rev. Lett. **22**(19), 997 (1969)
3. J.K. Freericks, C. Gruber, N. Macris, Phys. Rev. B **60**(3), 1617 (1999)
4. J.K. Freericks, E.H. Lieb, D. Ueltschi, Phys. Rev. Lett. **88**(10), 106401 (2002)
5. M.M. Maśka, K. Czajka, Phys. Status Solidi B **242**(2), 479 (2005)
6. K.I. Kugel, A.L. Rakhmanov, A.O. Sboychakov, Phys. Rev. Lett. **95**(26), 267210 (2005)
7. A.O. Sboychakov, K.I. Kugel, A.L. Rakhmanov, Phys. Rev. B **74**, 014401 (2006)
8. K.I. Kugel, A.L. Rakhmanov, A.O. Sboychakov, J. Magn. Magn. Mater. **310**(2), 1024 (2007)
9. J. Hubbard, Proc. Roy. Soc. A **205**, 238 (1963)
10. P. Fulde, *Electron Correlations in Molecules and Solids*. Springer Series in Solid-State Sciences, vol. 100, 3rd edn. (Springer, Berlin, Heidelberg, 2002)
11. S.G. Ovchinnikov, V.V. Val'kov, *Hubbard Operators in the Theory of Strongly Correlated Electrons* (World Scientific, 2004)
12. A.O. Sboychakov, Phys. B **417**, 49 (2013)
13. I.M. Lifshitz, Sov. Phys. JETP **11**(5), 1130 (1960)
14. K.I. Kugel, A.L. Rakhmanov, A.O. Sboychakov, F. Kusmartsev, N. Poccia, A. Bianconi, Supercond. Sci. Technol. **22**(1), 014007 (2008)
15. A.O. Sboychakov, S. Savel'ev, A.L. Rakhmanov, K.I. Kugel, F. Nori, Phys. Rev. B **77**, 224504 (2008)

16. K.M. Lang, V. Madhavan, J.E. Hoffman, E.W. Hudson, H. Eisaki, S. Uchida, J.C. Davis, Nature **415**(6870), 412 (2002)
17. P.G. de Gennes, *Superconductivity of Metals and Alloys* (Perseus Books, New York, 1999)
18. E.G. Maksimov, O.V. Dolgov, Phys. Usp. **50**(9), 933 (2007)
19. E.G. Maksimov, Phys. Usp. **47**(9), 957 (2004)
20. E. Dagotto, A. Nazarenko, A. Moreo, Phys. Rev. Lett. **74**(2), 310 (1995)
21. L.F. Feiner, J.H. Jefferson, R. Raimondi, Phys. Rev. Lett. **76**(26), 4939 (1996)
22. A.V. Rozhkov, A.L. Rakhmanov, J. Phys.: Condens. Matter **23**(6), 065601 (2011)
23. T. Aimi, M. Imada, J. Phys. Soc. Jpn. **76**(11), 113708 (2007)
24. M. Grilli, R. Raimondi, C. Castellani, C. Di Castro, G. Kotliar, Phys. Rev. Lett. **67**, 259 (1991)
25. M. Grilli, R. Raimondi, C. Castellani, C. Di Castro, G. Kotliar, Int. J. Mod. Phys. B **5**, 309 (1991)
26. Y. Bang, G. Kotliar, C. Castellani, M. Grilli, R. Raimondi, Phys. Rev. B **43**, 13724 (1991)
27. R. Raimondi, C. Castellani, M. Grilli, Y. Bang, G. Kotliar, Phys. Rev. B **47**, 3331 (1993)
28. G. Kotliar, A.E. Ruckenstein, Phys. Rev. Lett. **57**, 1362 (1986)
29. P.A. Igoshev, M.A. Timirgazin, V.F. Gilmutdinov, A.K. Arzhnikov, V.Yu. Irkhin, J. Magn. Magn. Mater. **383**, 2 (2015)
30. P.A. Igoshev, M.A. Timirgazin, V.F. Gilmutdinov, A.K. Arzhnikov, V.Yu. Irkhin, J. Phys.: Condens. Matter **27**(44), 446002 (2015)
31. M.A. Timirgazin, P.A. Igoshev, A.K. Arzhnikov, V.Y. Irkhin, J. Phys.: Condens. Matter **28**(50), 505601 (2016)
32. P. Igoshev, M. Timirgazin, A. Arzhnikov, V. Irkhin, J. Magn. Magn. Mater. **459**, 311 (2018)
33. A.J. Millis, P.B. Littlewood, B.I. Shraiman, Phys. Rev. Lett. **74**, 5144 (1995)
34. A.J. Millis, B.I. Shraiman, R. Mueller, Phys. Rev. Lett. **77**, 175 (1996)
35. A.J. Millis, Nature **392**(6672), 147 (1998)
36. A.J. Millis, T. Darling, A. Migliori, J. Appl. Phys. **83**(3), 1588 (1998)
37. T.V. Ramakrishnan, H.R. Krishnamurthy, S.R. Hassan, G.V. Pai, Phys. Rev. Lett. **92**(15), 157203 (2004)
38. A.O. Sboychakov, A.L. Rakhmanov, K.I. Kugel, J. Phys.: Condens. Matter **22**(41), 415601 (2010)
39. U. Löw, V. Emery, K. Fabricius, S. Kivelson, Phys. Rev. Lett. **72**(12), 1918 (1994)
40. J. Lorenzana, C. Castellani, C. Di Castro, Phys. Rev. B **64**(23), 235127 (2001)
41. J. Lorenzana, C. Castellani, C. Di Castro, EPL (Euro Phys. Lett.) **57**(5), 704 (2002)
42. C. Ortix, J. Lorenzana, C. Di Castro, Phys. Rev. Lett. **100**(24), 246402 (2008)
43. R. Balian, C. Bloch, Ann. Phys. **60**(2), 401 (1970)
44. J.G. Bednorz, K.A. Müller, Rev. Mod. Phys. **60**(3), 585 (1988)
45. A. Bianconi, A.C. Castellano, M. De Santis, P. Rudolf, P. Lagarde, A.M. Flank, A. Marcelli, Solid State Commun. **63**(11), 1009 (1987)
46. A. Bianconi, M. De Santis, A. Di Cicco, A.M. Flank, A. Fontaine, P. Lagarde, H. Katayama-Yoshida, A. Kotani, A. Marcelli, Phys. Rev. B **38**(10), 7196 (1988)
47. K.A. Müller, A. Bussmann-Holder, *Superconductivity in Complex Systems*, vol. 114 (Springer Science & Business Media, 2005)
48. V.Z. Kresin, Yu.N. Ovchinnikov, S.A. Wolf, Phys. Rep. **431**(5), 231 (2006)
49. A. Bianconi, Nat. Phys. **9**(9), 536 (2013)
50. A. Bianconi, N. Poccia, A.O. Sboychakov, A.L. Rakhmanov, K.I. Kugel, Supercond. Sci. Technol. **28**(2), 024005 (2015)
51. C. Liu, A.D. Palczewski, R.S. Dhaka, T. Kondo, R.M. Fernandes, E.D. Mun, H. Hodovanets, A.N. Thaler, J. Schmalian, S.L. Bud'ko, P.C. Canfield, A. Kaminski, Phys. Rev. B **84**(2), 020509 (2011)
52. D. LeBoeuf, N. Doiron-Leyraud, B. Vignolle, M. Sutherland, B.J. Ramshaw, J. Levallois, R. Daou, F. Laliberté, O. Cyr-Choiniere, J. Chang, Y.J. Jo, L. Balicas, R. Liang, D.A. Bonn, W.N. Hardy, C. Proust, L. Taillefer, Phys. Rev. B **83**(5), 054506 (2011)
53. F. Laliberté, J.J. Chang, N. Doiron-Leyraud, E. Hassinger, R. Daou, M. Rondeau, B.J. Ramshaw, R. Liang, D.A. Bonn, W.N. Hardy, et al., Nat. Commun. **2**, 432 (2011)

54. D. Innocenti, N. Poccia, A. Ricci, A. Valletta, S. Caprara, A. Perali, A. Bianconi, Phys. Rev. B **82**(18), 184528 (2010)
55. K.I. Kugel, A.L. Rakhmanov, A.O. Sboychakov, N. Poccia, A. Bianconi, Phys. Rev. B **78**(16), 165124 (2008)
56. E. Liarokapis, Condens. Matter **4**, 87 (2019)
57. E. Dagotto, *Nanoscale Phase Separation and Colossal Magnetoresistance: The Physics of Manganites and Related Compounds*. Springer Series in Solid-State Sciences, vol. 136 (Springer Science & Business Media, 2003)
58. M.Yu. Kagan, K.I. Kugel, Phys. Usp. **44**(6), 553 (2001)
59. K.I. Kugel, A.L. Rakhmanov, A.O. Sboychakov, M.Yu. Kagan, I.V. Brodsky, A.V. Klaptsov, J. Exp. Theor. Phys. **98**(3), 572 (2004)
60. A.L. Rakhmanov, K.I. Kugel, Ya. M. Blanter, M.Yu. Kagan, Phys. Rev. B **63**(17), 174424 (2001)
61. P.G. de Gennes, Phys. Rev. **118**(1), 141 (1960)
62. V.B. Shenoy, T. Gupta, H.R. Krishnamurthy, T.V. Ramakrishnan, Phys. Rev. Lett. **98**, 097201 (2007)
63. V.B. Shenoy, T. Gupta, H.R. Krishnamurthy, T.V. Ramakrishnan, Phys. Rev. B **80**, 125121 (2009)
64. A.S. Alexandrov, N.F. Mott, *Polarons and Bipolarons* (World Scientific Singapore, 1995)
65. E. Dagotto, T. Hotta, A. Moreo, Phys. Rep. **344**(1), 1 (2001)
66. E.L. Nagaev, Phys. Rep. **346**(6), 387 (2001)

Chapter 6
Spin-State Transitions and Magnetic Instabilities

Abstract We analyze the effects related to the low-, intermediate-, and high-spin states in different materials such as cobaltites. In these materials, the interior of magnetic polaron is formed by low-spin state in the surrounding of high-spin states (or vice versa).

6.1 Spin-State Transitions

In this chapter, we consider specific features of the electronic phase separation, which can be observed in the systems with transition-metal ions exhibiting different multiplet states. The main emphasis will be put here on oxides with Co^{3+} ions. This ion (as well as Fe^{2+}) can be characterized either by a low-spin (LS) state with $S = 0$ (t_{2g}^6), intermediate-spin (IS) state, $S = 1$ ($t_{2g}^5 e_g^1$), or high-spin (HS) state ($t_{2g}^4 e_g^2$) with $S = 2$, see, e.g., Ref. [1]. The characteristic energies of these states usually do not differ much, thus the crossover between such states is quite possible, and it is usually referred to as the spin-state transition (SST). A popular compound with SST is $LaCoO_3$ [1–4]. The ions with different spin states can form a regular array [5, 6]. Therefore, spin (or, in general, multiplet) ionic configuration can be treated as another degree of freedom in addition to charge, orbital, and spin ones typical of transition-metal compounds. In the previous chapters, we have demonstrated that under doping, the electronic phase separation can arise owing to the competition between the delocalization of charge carriers related to their kinetic energy and the possible formation of some localized magnetic or orbital structure. Therefore, in the materials exhibiting SST, the tendency to form some spin-state array can limit the motion of charge carriers appearing due to the doping. Here, again, we can locally modify the regular background to have an additional gain in kinetic energy. We know that in the framework of both the two-band double exchange model [7, 8] and single-band Hubbard model [9, 10], an antiferromagnetic background hinders the motion of doped holes, whereas they can freely move on the ferromagnetic background. As it was discussed above, this favors the formation of an inhomogeneous state of the spin-polaron type [10–12].

M. Yu. Kagan et al., *Electronic Phase Separation in Magnetic and Superconducting Materials*, Springer Series in Solid-State Sciences 201,
https://doi.org/10.1007/978-3-031-55467-4_6

It is also well known and is discussed in detail in the next chapter that the orbital ordering can also suppress the motion of charge carriers and this favors a local modification of orbitals giving rise to orbital polarons [13–16]. In the case of SST, such interplay between the localization and delocalization can be related to the so-called spin blockade [17]. Indeed, if we add some electrons to the material with Co^{3+} in the low-spin state ($S = 0$), we could get the high-spin ($S = 3/2$) Co^{2+}. However, if one electron move between LS Co^{3+} and HS Co^{2+}, the eventual configuration will include Co^{3+} and Co^{2+} in IS states rather than in the initial states. Indeed, the one-electron hopping leads to the $\pm 1/2$ spin change, while here the starting spin states differ by 3/2. Hence, the situation becomes similar to that discussed for the Hubbard model [10], namely, an electron can only move creating wrong spin states. This leads to the energy loss and electron localization.

Such energy loss can be minimized by local changes in the underlying spin states, i.e., by the phase separation, which has been actually reported for such cobaltites as $La_{1-x}Sr_xCoO_3$ [18–22]. In particular, the magnetic measurements [2] at low hole doping ($< 1\%$ of Sr) that the magnetic moment per doped hole (per Sr) far exceeds that of LS Co^{4+} with $S = 1/2$. Actually, we have magnetic impurities with unusually large spin $S = 5 - 10$, implying the formation of magnetic polaron owing to a transfer of some of Co^{3+} ions to a magnetic state. It turns out to be possible to find out the size and shape of such magnetic inhomogeneities through the use of neutron scattering, supplemented by the ESR and NMR [23].

Below, we formulate a model allowing us to demonstrate an instability of inhomogeneous state in the systems with SST like doped cobaltites. Our discussion is based mostly on the results of Ref. [24].

6.2 Spin States of Cobalt Ions

As a typical example of materials with the spin-state transitions, we will first study hole-doped perovskites of $LaCoO_3$ type. A detailed analysis of the characteristic features of such materials with the reference to the corresponding experimental data can be found in review articles [25, 26]. Cobalt ions in these compounds are located within oxygen octahedra. In the octahedral coordination, the d levels of transition-metal ions are split into a lower t_{2g} level and the higher e_g one. In the case of hole doping, we are dealing with Co^{3+} and Co^{4+} ions, which have $3d^6$ and $3d^6$ configurations, respectively. As we have mentioned above, three low-energy states are possible for such ions, namely, low-spin (LS), intermediate-spin (IS), and high-spin (HS) states.

The spin states of Co^{3+} and Co^{4+} ions and their energies are illustrated in Table 6.1. For Co^{3+}, the lowest energy E_0 corresponds to the LS state ($S = 0$), with fully occupied t_{2g} level and empty e_g one. In the IS state ($S = 1$), one electron is promoted to the e_g level leaving five electrons at the t_{2g} level. In the HS state ($S = 2$), one more electron appears at the e_g level, and so there are four t_{2g} and two e_g electrons. We denote energy gap between t_{2g} and e_g as Δ and Hund's rule coupling constant as J_H. Thus,

Table 6.1 Energy levels and spin states of Co ions [24]

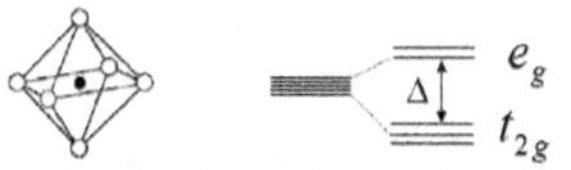

Co^{3+}_{LS} $S = 0$ $E^{(3+)}_{LS} = E_0$	Co^{4+}_{LS} $S = 1/2$ $E^{(4+)}_{LS} = E_1$
Co^{3+}_{IS} $S = 1$ $E^{(3+)}_{IS} = E_0 + (\Delta - J_H)$	Co^{4+}_{IS} $S = 3/2$ $E^{(4+)}_{IS} = E_1 + (\Delta - 2J_H)$
Co^{3+}_{HS} $S = 2$ $E^{(3+)}_{HS} = E_0 + 2(\Delta - 2J_H)$	Co^{4+}_{HS} $S = 5/2$ $E^{(4+)}_{HS} = E_1 + 2(\Delta - 3J_H)$

we have $E_{IS}^{(3+)} = E_0 + \Delta - J_H$ and $E_{HS}^{(3+)} = E_0 + 2\Delta - 4J_H$. For the Co^{4+} ion with five $3d$ electrons, these electrons can be distributed between the corresponding states in a similar way, with the only difference that the number of t_{2g} electrons is smaller (one electron is missing), see Table 6.1. Therefore, the spins for LS, IS, and HS are equal to 1/2, 3/2, and 5/2, respectively, and the corresponding energies in the case of Co^{4+} ion are $E_{LS}^{(4+)} = E_1$, $E_{IS}^{(4+)} = E_1 + \Delta - 2J_H$, and $E_{HS}^{(4+)} = E_1 + 2\Delta - 6J_H$. As we demonstrate below, the specific values E_0 and E_1 do not produce much effect on the obtained results.

As we can see in Table 6.1, the ground state of isolated cobalt ions is determined by the ratio of Δ and J_H. In particular, at $\Delta > 3J_H$, the LS state corresponds to the lowest energy for both Co^{3+} and Co^{4+} ions. Within the $2J_H < \Delta < 3J_H$ range, we have LS Co^{3+} and HS Co^{4+}. At large Hund's rule coupling constant, $2J_H > \Delta$, both ions have the HS ground state. This means that for noninteracting cobalt ions, the IS ground state is not favorable.

However, the possibility of charge transfer between cobalt ions can qualitatively change the aforementioned set of ground states. To make the further treatment clearer, let us emphasize the main effect, namely, the electron hopping between the e_g states. Indeed, in cobaltites, the probability of electron hopping between the t_{2g} states is usually quite low and can be neglected without any loss in generality. We also do not take into account the states with more than six electrons per site since the strong on-site Coulomb repulsion of electrons suppresses their existence. Finally, the spin blockade mentioned above prevents the electron transfer corresponding to the spin changes exceeding one half.

Therefore, we are dealing below with two main charge transfer channels typical of doped cobaltites, namely, the electron hoppings between the IS Co^{3+} and LS Co^{4+} and those between the HS Co^{3+} and IS Co^{4+}. Actually, to attain the maximum gain in the kinetic energy, the ground state should involve the IS cobalt ions. In particular, if the ground state for isolated cobalt ions contains LS Co^{4+} and LS Co^{3+} ($\Delta > 3J_H$), we can excite some LS Co^{3+} ions to the IS state. Then the energy loss related to Δ could be compensated by the gain in kinetic energy due to the charge transfer between IS Co^{3+} and LS Co^{4+} states. In the opposite limit $\Delta < 2J_H$, favoring the high-spin states of cobalt ions, another charge transfer channel becomes operational, namely the hopping between HS Co^{3+} and IS Co^{4+}. Within the intermediate parameter range, $2J_H < \Delta < 3J_H$, the charge transfer electron hopping can occur if we simultaneously excite both Co^{3+} and Co^{4+}. However the excitation of two ions is clearly less probable. In the next subsection, we start from the $\Delta > 3J_H$ case at different doping levels, which is more often corresponds to the situation in actual cobaltites. After that, we perform the same analysis for the case of $\Delta < 2J_H$. In both ranges of parameters, we draw a special attention to the possible electronic phase separation and eventually determine the phase diagram in the Δ/J_H−doping plane. Δ is closely related to the ionic radius r_A of rare-earth R in $RCoO_3$ perovskite cobaltites and increases when r_A becomes smaller.

6.3 Effects of Electron Delocalization and Spin-State Transitions: LS Ground State for Isolated Ions

We start with the case of $\Delta > 3J_H$. As it was mentioned above, it is characteristic of hole-doped $La_{1-x}Sr_xCoO_3$ perovskite cobaltites. Since Δ is growing with the rare-earth ionic radius, this case in also typical for some other doped $RCoO_3$ compounds [27]. In such cobaltites, Co^{3+} and Co^{4+} ions occupy the sites of simple cubic lattice. If these ions do not interact, they have the LS ground state and the doping level x corresponds to the relative number of Co^{4+}. If some LS Co^{3+} ions are transferred to the IS state, this results to the kinetic energy gain owing to the charge transfer from IS Co^{3+} to LS Co^{4+}.

Let us formulate a mathematical approach allowing us to justify the semiqualitative arguments put forward above. As a vacuum state, we choose the LS state of Co^{3+} on and introduce creation operators $a^\dagger_{\mathbf{n}}$ and $c^\dagger_{\mathbf{n}}$ for an electron at the e_g level and a hole at the t_{2g} level, respectively, at site $\mathbf{n}$. Their action in the vacuum state can be represented as

$$\begin{aligned} |0\rangle &= |\mathrm{Co}^{3+}_{LS}\rangle, \quad E^{(vac)} = E_0, \\ a^\dagger_{\mathbf{n}}|0\rangle &= |\mathrm{Co}^{2+}\rangle, \quad E^{(2+)} = U', \\ c^\dagger_{\mathbf{n}}|0\rangle &= |\mathrm{Co}^{4+}_{LS}\rangle, \quad E^{(h)} = E_1. \end{aligned} \tag{6.1}$$

The IS state of Co^{3+} ions is written in the form

$$|\mathrm{Co}^{3+}_{IS}\rangle = c^\dagger_{\mathbf{n}} a^\dagger_{\mathbf{n}}|0\rangle, \quad E^{(3+)}_{IS} = E_0 + \Delta - J_H = E_2. \tag{6.2}$$

Then, the single-site Hamiltonian involving all low-energy states has the form

$$H_{\mathbf{n}} = E_0 + (E_1 - E_0)n^h_{\mathbf{n}} + (U' - E_0)n^e_{\mathbf{n}} + \left[(E_2 - E_0) - (E_1 - E_0) - (U' - E_0)\right] n^h_{\mathbf{n}} n^e_{\mathbf{n}}\,, \tag{6.3}$$

where $n^e_{\mathbf{n}} = a^{\dagger}_{\mathbf{n}} a_{\mathbf{n}}$ and $n^h_{\mathbf{n}} = c^{\dagger}_{\mathbf{n}} c_{\mathbf{n}}$ are the operators corresponding to the numbers of electrons at e_g levels and holes at t_{2g} levels, respectively. Hamiltonian can be rewritten as (6.3) in a more compact form, we have

$$H_{\mathbf{n}} = [E_0 + (E_1 - E_0)(n^h_{\mathbf{n}} - n^e_{\mathbf{n}})] + (\Delta - J_H)n^e_{\mathbf{n}} + U n^e_{\mathbf{n}}(1 - n^h_{\mathbf{n}})\,, \tag{6.4}$$

where $U = U' + E_1 - \Delta + J_H - 2E_0$. After that, we can write the Hamiltonian for the lattice as a whole taking also into account the intersite electron transfer

$$H = \sum_{\mathbf{n}} [E_0 + (E_1 - E_0 - \mu)(n^h_{\mathbf{n}} - n^e_{\mathbf{n}})] + \Delta_1 \sum_{\mathbf{n}} n^e_{\mathbf{n}} + U \sum_{\mathbf{n}} n^e_{\mathbf{n}}(1 - n^h_{\mathbf{n}})$$
$$- t \sum_{\langle \mathbf{nm} \rangle} \left(a^{\dagger}_{\mathbf{n}} a_{\mathbf{m}} + h.c.\right)\,, \tag{6.5}$$

where $\Delta_1 = \Delta - J_H$ and $\langle \mathbf{nm} \rangle$ denotes the summation over the nearest-neighbor sites.

In the t term of Hamiltonian (6.5), we have left only the most important electron transfer process, namely, the hopping between the occupied e_g level of IS Co^{3+} to the empty e_g level of LS Co^{4+}. In addition, we assume that due to the Hund's rule coupling, the spin of an itinerant e_g electron has the same direction as the total spin of core t_{2g} electrons. Thus, we are dealing with a kind of ferromagnetic ground state and may not use the spin index of electron operators. Here, we do not take into account additional effects related to the degeneracy of e_g states, which can lead only to unnecessary complications. Some of such effects in cobaltites are addressed in [4, 28–30] using *ab initio* band structure calculations and the analysis of X-ray photoemission spectra.

Hamiltonian (6.5) turns out to be similar to that of the Falicov–Kimball model [31], which we have discussed in the previous chapter. Here, we have again the usual competition between the electron localization and itineracy: the LS cobalt ions play the role of localized states, whereas electrons promoted to the IS state become itinerant. As it was shown above, such competition in the framework of the Falicov–Kimball model leads to the phase separation. Note also that the atomic-scale charge and spin inhomogeneities related to electronic phase separation were revealed by the numerical solutions for small clusters [32]. Below, using the techniques reported in Refs. [33, 34], we analyze the specific features of systems with the spin-state transitions.

We introduce the average numbers of e_g electrons and t_{2g} holes per site $\langle n^e_{\mathbf{n}} \rangle = n^e$ and $\langle n^h_{\mathbf{n}} \rangle = n^h$ ($n^h - n^e = x$). We have in mind the average number of electrons appearing at initially empty e_g levels owing to the LS–IS transitions.

The corresponding energies are

$$E^{(1)} = E_0(1 - x) + E_1 x + \langle H_1 \rangle / N\,, \tag{6.6}$$

where

$$H_1 = \Delta_1 \sum_{\mathbf{n}} n^e_{\mathbf{n}} + U \sum_{\mathbf{n}} n^e_{\mathbf{n}}(1 - n^h_{\mathbf{n}}) - t \sum_{\langle \mathbf{nm} \rangle} \left(a^\dagger_{\mathbf{n}} a_{\mathbf{m}} + h.c. \right) . \tag{6.7}$$

To each value of n^e, we can put into the correspondence some homogeneous state and calculate the energy spectrum using the Hubbard I approximation [35] for the one-electron Green's function $G^e(\mathbf{n}, \mathbf{n}_0;\ t - t_0) = -i \langle T a_{\mathbf{n}}(t) a^\dagger_{\mathbf{n}_0}(t_0) \rangle$ for the e_g electrons appearing in the IS state. Here, we have a clear analogy to the band b electrons discussed in the previous section (see also Refs. [33, 36]). This Green's function has the following form in the frequency–momentum representation

$$G^e(\mathbf{k}, \omega) = -\frac{\omega + \mu - \Delta_1 - U n^h}{(\omega + \mu - \Delta_1 - E_1(\mathbf{k}))\,(\omega + \mu - \Delta_1 - E_2(\mathbf{k}))} , \tag{6.8}$$

where

$$E_{1,2}(\mathbf{k}) = \frac{U + \varepsilon(\mathbf{k})}{2} \mp \sqrt{\left(\frac{U - \varepsilon(\mathbf{k})}{2} \right)^2 + U \varepsilon(\mathbf{k})(1 - n^h)} , \tag{6.9}$$

In these expressions, $\varepsilon(\mathbf{k})$ is the dispersion law in the absence of electron correlations, $U = 0$. We use the simplest tight-binding form of $\varepsilon(\mathbf{k})$ without involving orbital effects and the characteristic form of the hopping integrals for e_g electrons. Thus, we use $\varepsilon(\mathbf{k}) = -2t(\cos k_x + \cos k_y + \cos k_z)$ for the simple cubic lattice. Below, we consider the case of strongly correlated electrons, $U/t \gg 1$. In this case, the Hubbard I approximation is an adequate one capturing the essential physics related to electron correlations and agrees well both with experiment and numerical calculations [37, 38].

Using Eq. (6.8) for Green's function, we calculate the densities of e_g charge carriers. The electron and hole densities, n^e and n^h, are calculated based on Eq. (6.8) in the $U \to \infty$ limit. As a result, we find the total energy as function of doping x. These calculations are similar to those performed in Ref. [33] and discussed in detail in Sect. 6.5. The b electrons in Ref. [33] correspond to our e_g electrons at IS Co^{3+} ions, whereas the number of localized l electrons in Ref. [33], n_l, corresponds to the number of LS Co^{3+} ions, $1 - n^h$. Using this similarity, we could make a direct mapping between the two systems. However, the specific feature of the systems with spin-state transitions is the existence of another homogeneous state in addition to that considered in Ref. [33]. In this state, we have all IS Co^{3+} ions ($n^h = 1, n^e = 1 - x$). In the usual conduction band picture, such state corresponds to an empty localized level located under the Fermi level of the conduction band. Such a situation is rather exotic and hardly possible. In our case, the state corresponding to the localized level is determined by the existence of LS Co^{3+} ions, and this analogy fails in the absence of such LS Co^{3+} ions (the localized level simply disappears).

We denote the state similar to that discussed in Ref. [33] (the state with coexisting LS and IS Co^{3+} ions) as a type 1 state, and the state without LS Co^{3+} as type 2 state. In Fig. 6.1, we plot the energies of such states versus the doping level x at $\Delta_1/zt = 0.2$

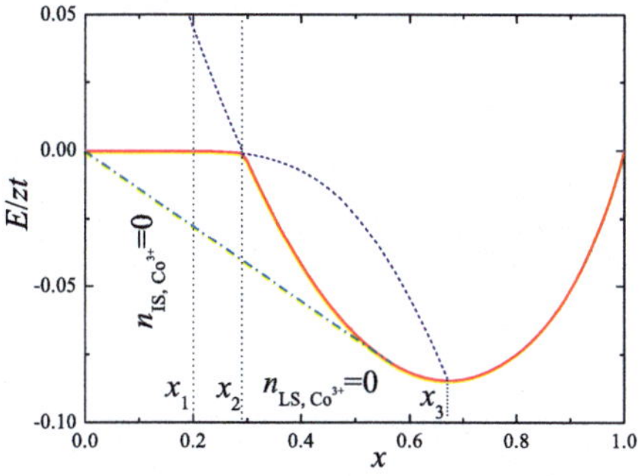

Fig. 6.1 Energies of type 1 and 2 states versus doping x. The red solid curve indicates a homogeneous state with the lowest energy. The blue dashed lines denote the states with higher energies (see the text). The green dot-and-dash line shows the energy of an inhomogeneous phase-separated state. $\Delta_1/zt = 0.2$ [24]

($z = 6$ is the number of nearest neighbors). The type 2 state corresponds to the ground state at $x > x_2$. At $x > x_3$ both states, 1 and 2, are equivalent. At $x < x_1$, the number of electrons appearing at the e_g level vanishes ($n^e = 0$), see Fig. 6.2.

At $x > x_1$, we observe a gradual increase in the number of e_g electrons. In the usual Falicov–Kimball model, this increase persists up to $x = x_3$ when all Co^{3+} ions transfer to the IS state. In our case, there appears the type 2 state, and at $x = x_2$, the system undergoes a jump-like transition to this state. The ground state energy E for the homogeneous system as function of x is shown in Fig. 6.1 by the red solid curve. At the same time, this figure clearly demonstrates that within the $0 < x \lesssim x_3$ doping range, we should expect the formation of inhomogeneous state, involving the domains with $n^h = 1$ and $n^e = 0$. In Fig. 6.1, the energy of such phase-separated state is shown by the green dot-and-dash line.

In Fig. 6.2, we illustrate the changes with doping x in the densities of IS Co^{3+} (n^e) and LS Co^{3+} $(1 - n^h)$ ions shown by blue dot-and-dash and red solid curves, respectively. At x_2, n^e undergoes a jump related to the transition to the type 2 state. The plots of n^e and $1 - n^h$ at $x > x_2$ corresponding to the Falicov–Kimball model (without type 2 state) are drawn by thin dashed blue and red lines, respectively.

If the number of electrons promoted to the IS state is small, $n^e \ll 1$, the approximate expression for the total energy E in the case of the Fermi surface has the form

$$E \simeq \Delta_1 n^e - tzn^e n^h + \frac{3t}{5}\left(36\pi^4 n^h\right)^{1/3} (n^e)^{5/3}. \tag{6.10}$$

The actual value of n^e is determined by the minimization of energy (6.10) under condition $n^h = x + n^e$. A nonzero value of n^e can be attained only at $\Delta_1 < tzx$. Hence, at $\Delta_1/tz > 1$, IS Co^{3+} ions do not arise within the whole doping range.

Since the spins of e_g electrons are parallel to the total spin of t_{2g} electrons due to the Hund's rule coupling, the changes in n^e with doping x are closely related to those in the magnetic moment of Co ions. The LS Co^{3+} ions are characterized

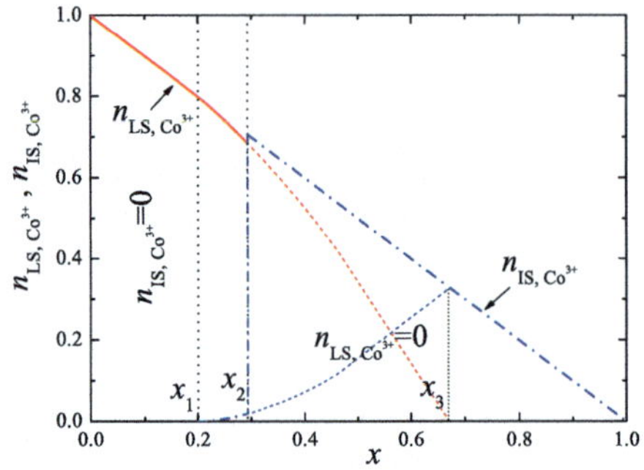

Fig. 6.2 The densities of IS Co^{3+} ions ($n^e = n_{IS, Co^{3+}}$) and LS Co^{3+} ions ($1 - n^h = n_{LS, Co^{3+}}$) versus doping x at $\Delta_1/zt = 0.2$ [24]. At $x_1 < x < x_2$, both IS an LS states of Co^{3+} exist simultaneously and at $x = x_2$, we have a jump to the type 2 state with only IS Co^{3+}. The plots of $n_{IS, Co^{3+}}$ and $n_{LS, Co^{3+}}$ corresponding to the Falicov–Kimball model (without type 2 state) and discussed in Ref. [33] are drawn by thin dashed lines

by zero magnetic moment, $S = 0$, and the doping gives rise to LS Co^{4+} ions with $S = 1/2$ and at the same time, some LS Co^{3+} ions becomes transferred to the IS state with $S = 1$. Therefore, the plots shown in Fig. 6.2 manifest themselves in the behavior of magnetic moments of cobalt ions illustrated in Fig. 6.3. In particular, the jump in n^e is associated in the jump of the magnetic moment, if we consider the homogeneous states. In the phase-separated state, the behavior of the magnetic moment is different: the magnetic moment per Co^{4+} ion does not change. Indeed, both the relative amount of the IS Co^{3+} phase and the density of Co^{4+} ions are proportional to x. In the case of phase separation, the magnetic moment is related to such x, at which the green dot-and-dash line in Fig. 6.1 touches the plot of energy corresponding to the homogeneous state. In Fig. 6.3, we can see the magnitude of the magnetic moment jump in the homogeneous state and the value of magnetic moment per Co^{4+} in the phase-separated state depend on the hopping integral t. Namely, the increase in t by a factor of two results in a much larger growth of two aforementioned parameters.

Let us emphasize here that in our analysis of magnetic moments, we imply that in the phase-separated state, the size of inhomogeneities exceeds the lattice constant. This is true for relatively high doping levels. At small x, Co^{4+} ion could act separately forming a kind of small droplets, or in other words, spin-state polarons with only one Co^{4+} ion surrounded by IS Co^{3+}. For such polaron, the effective spin value should be higher than that given by our quasimacroscopic approach. Such a situation has been indeed reported in Ref. [23] for $La_{1-x}Sr_xCoO_3$ at quite small values of x, where the existence of spin polarons with the effective spin of $13\mu_B$ has been suggested. We argue that the magnetic moment per Co^{4+} should become smaller at higher doping levels x. The exact calculations for small clusters [39] provide additional evidence that within a relevant range of parameters, the effective magnetic moment can achieve the saturation even for atomic-size clusters of different shapes.

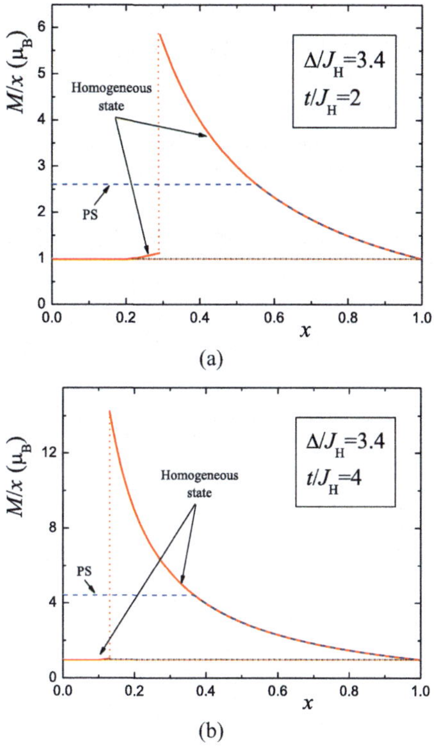

Fig. 6.3 The dependence of effective magnetic moment M per Co^{4+} ion on doping level x at two values of the hopping integral t: **a** $t/J_H = 2$; **b** $t/J_H = 4$. The magnetic moment for the homogeneous states is plotted by red curves. The blue dashed line illustrates the behavior of M corresponding to the phase-separated state [24]

Summarizing the above discussion, we can say that the spin-state transitions in cobaltites with the hole doping can be analyzed in the framework of the Falicov–Kimball type model treating an interplay of localized and itinerant electron states. In addition to the effects described by the similar model in the case of manganites [33, 34], we reveal the possibility of a jump-like transition to the state, in which all charge carriers become delocalized. In respect to cobaltites, this means than all Co^{3+} ions turn out to be in the intermediate-spin state. Moreover, before attaining such homogeneous state with itinerant electrons, the system has the doping range favorable for the phase separation, in which IS Co^{3+} ions are located within nanoscale inhomogeneities. This picture is in an agreement with experimental data on $La_{1-x}Sr_xCoO_3$ system [18–23].

6.4 The Case of HS Ground State for Isolated Ions

As we have mentioned in the beginning of this chapter, a considerable energy gain can be also achieved at $\Delta < 2J_H$, corresponding to the high-spin state for Co^{3+} ions.

In this case, the electron hopping between Co ions becomes possible if Co^{4+} passes to the IS state by promoting a hole to the e_g level.

Thus, we can have the electron transfer from HS Co^{3+} to IS Co^{4+}, or equivalently the transfer if a hole from IS Co^{4+} to HS Co^{3+}. Here, the relevant vacuum state is the HS state of Co^{4+}, and we can write the corresponding relations by the analogy with (6.1) and (6.2)

$$\begin{aligned} |0\rangle &= |\mathrm{Co}^{4+}_{HS}\rangle, \quad \tilde{E}^{(vac)} = E_1 + 2\Delta - 6J_H = \tilde{E}_0, \\ \tilde{c}^\dagger_{\mathbf{n}}|0\rangle &= |\mathrm{Co}^{5+}\rangle, \quad E^{(5+)} = \tilde{U}', \\ \tilde{a}^\dagger_{\mathbf{n}}|0\rangle &= |\mathrm{Co}^{3+}_{HS}\rangle, \quad E^{(e)} = E_0 + 2\Delta - 4J_H = \tilde{E}_1, \\ \tilde{a}^\dagger_{\mathbf{n}}\tilde{c}^\dagger_{\mathbf{n}}|0\rangle &= |\mathrm{Co}^{4+}_{IS}\rangle, \; E^{(4+)}_{IS} = E_0 + \Delta - 2J_H = \tilde{E}_2. \end{aligned} \tag{6.11}$$

By making the substitution $E_0, E_1, E_2, U \to \tilde{E}_0, \tilde{E}_1, \tilde{E}_2, \tilde{U}$ and also $n^e_{\mathbf{n}} \to \tilde{n}^h_{\mathbf{n}}, n^h_{\mathbf{n}} \to \tilde{n}^e_{\mathbf{n}}$ in (6.3) and (6.4), we obtain the single-site Hamiltonian. Then, instead of Hamiltonian (6.5), we get

$$\begin{aligned} H = &\sum_{\mathbf{n}}[\tilde{E}_0 + (\tilde{E}_0 - \tilde{E}_1 - \mu)(\tilde{n}^e_{\mathbf{n}} - \tilde{n}^h_{\mathbf{n}})] + + \Delta_2 \sum_{\mathbf{n}} \tilde{n}^h_{\mathbf{n}} + \tilde{U} \sum_{\mathbf{n}} \tilde{n}^h_{\mathbf{n}}(1 - \tilde{n}^n_{\mathbf{n}}) \\ &- t \sum_{\langle \mathbf{nm} \rangle} \left(\tilde{c}^\dagger_{\mathbf{n}}\tilde{c}_{\mathbf{m}} + h.c.\right) . \end{aligned} \tag{6.12}$$

In this Hamiltonian, $\Delta_2 = 4J_H - \Delta$ is the difference between energies of IS and HS of Co^{4+} ions, $\tilde{c}^\dagger_{\mathbf{n}}, \tilde{c}_{\mathbf{n}}$ are the creation and annihilation operators describing a hole exited to the e_g level of IS Co^{4+} ion located at site $\mathbf{n}$, $\tilde{n}^h_{\mathbf{n}} = \tilde{c}^\dagger_{\mathbf{n}}\tilde{c}_{\mathbf{n}}$, and $\tilde{n}^e_{\mathbf{n}} = \tilde{a}^\dagger_{\mathbf{n}}\tilde{a}_{\mathbf{n}}$ is the operator corresponding to the number (0 or 1) of additional t_{2g} electrons located at site $\mathbf{n}$ ($\tilde{a}^\dagger_{\mathbf{n}}$ and $\tilde{a}_{\mathbf{n}}$ are the creation and annihilation operators for such electrons). The relation between the averages have now the form $\tilde{n}^e - \tilde{n}^h = 1 - x$.

Instead of expression (6.6), we have now for the energy per site

$$E^{(2)} = E_0(1-x) + E_1 x + \langle H_2 \rangle / N , \tag{6.13}$$

where

$$\begin{aligned} H_2 = &\sum_{\mathbf{n}} \left(2\Delta - 6J_H + 2J_H(\tilde{n}^e_{\mathbf{n}} - \tilde{n}^h_{\mathbf{n}})\right) + \Delta_2 \sum_{\mathbf{n}} \tilde{n}^h_{\mathbf{n}} + \tilde{U} \sum_{\mathbf{n}} \tilde{n}^h_{\mathbf{n}}(1 - \tilde{n}^n_{\mathbf{n}}) \\ &- t \sum_{\langle \mathbf{nm} \rangle} \left(\tilde{a}^\dagger_{\mathbf{n}}\tilde{a}_{\mathbf{m}} + h.c.\right) . \end{aligned} \tag{6.14}$$

It is important that the energy difference $E^{(2)} - E^{(1)}$ is independent of the specific values of E_0 and E_1, and this will be useful below for the comparison of phase diagrams in different ranges of parameters.

As a result, we can see that in the case under study in this subsection, the energy of the system and charge carrier densities, $\tilde{n}^e$ and $\tilde{n}^h$, exhibit the behavior analogous to that depicted in Figs. 6.1 and 6.2. Note, however, that we should make the following replacement: $n^h \rightarrow \tilde{n}^e$, $n^e \rightarrow \tilde{n}^h$, and $x \rightarrow 1 - x$. This means that the densities of Co^{3+} ions in IS ($n^e = n_{IS, Co^{3+}}$) and LS ($1 - n^h = n_{LS, Co^{3+}}$) states are substituted by the densities of Co^{4+} ions in IS ($\tilde{n}^h = n_{IS, Co^{4+}}$) and HS ($1 - \tilde{n}^e = n_{HS, Co^{4+}}$) states. Such exact mapping of the LS–LS case to HS–HS one is a consequence of the parallel arrangement of the spins of charge carriers. This implies the electron–hole symmetry, that is, the similarity of an empty e_g level at LS Co^{3+} and the filled e_g level at HS Co^{3+}.

6.5 Phase Diagrams

The above analysis allows us to illustrate the behavior of the systems under study with the variation of doping at different values of the characteristic parameter Δ/J_H and the form of resulting phase diagrams. Such phase diagrams are quite different at different values of the hopping integral t, see in Fig. 6.4, where the possible homoge-

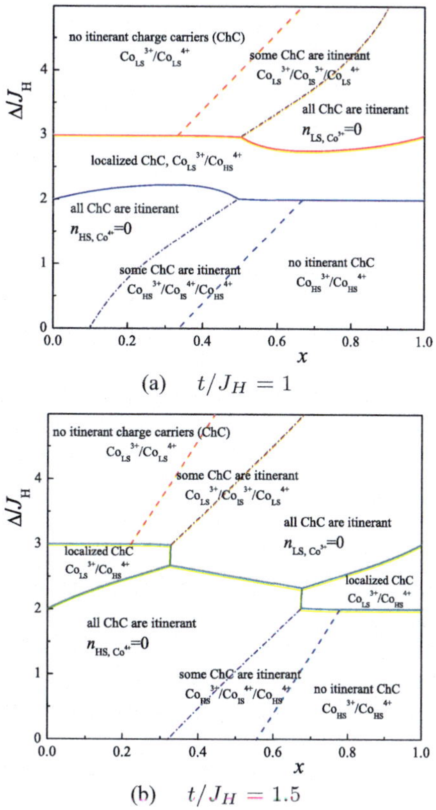

Fig. 6.4 Phase diagram illustrating the range of existence for the possible homogeneous states in the system with spin-state transitions at two characteristic values of the hopping integral t: **a** $t/J_H = 1$ and **b** $t/J_H = 1.5$. The phase boundaries for the similar phases in the upper and lower parts of both panels are shown denoted by the same lines (solid, dashed, or dot-and-dash) [24]. ChC is the abbreviation for charge carriers

neous states are depicted. In Fig. 6.4a corresponding to small values of t ($t/J_H \lesssim 1$), there are distinct regions of the phase diagram revealing the situation at $\Delta > 3J_H$ and $\Delta < 2J_H$ (see Sects. 6.3 and 6.4, respectively). Each of these regions exhibits the evolution of the system with doping, at which the phase involving only localized charge carries transforms to the phase, where the localized and itinerant charge carriers coexist, and finally, we have the phase with the complete delocalization of charge carriers. Between the regions corresponding to $\Delta > 3J_H$ and $\Delta < 2J_H$, there exists the phase with Co^{3+} in LS ($S = 0$) and Co^{4+} in HS ($S = 5/2$) states, with the charge carriers localized owing to the spin blockade [17].

With the growth of t (at $t/J_H \gtrsim 1$), see Fig. 6.4b, the intermediate phase shrinks and there appears the possibility of a direct spin-state transition between the phases with purely itinerant charge carriers.

If we take into account that the phase separation is possible, this leads to the drastic changes in the phase diagram. In Fig. 6.5, we present the resulting phase diagrams at different values of parameter t/J_H. In comparison to the phase diagram with only homogeneous states, here the range of phase separation occupies a considerable part of the phase diagram. Within this range, we have an inhomogeneous state with intermixed domains with either localized or delocalized charge carriers. For small t ($t/J_H \lesssim 1$), there is again intermediate region corresponding to charge carries localized at any doping level (that is, we have with LS Co^{3+} and HS Co^{4+}). An increase in t leads to the shrinking of this intermediate region. At $t/J_H \cong 1.24$, there appear two separated such regions (see Fig. 6.5b), and a direct boundary between PS I and PS II phases. With the further increase in t, these two regions of the localization are gradually vanishing and do not exist at $t/J_H > 1.44$.

Note again that the change in Δ axis can be implemented in cobaltites by varying the average ionic radius of rare-earth elements (see, e.g., Refs. [40, 41]).

Above, we have considered the $U \to \infty$ limit. This means that the spin–spin, spin–orbital, and orbital–orbital correlations between neighboring sites, which take place in two-band Hubbard Hamiltonians (6.5) and (6.12), are not taken into account. Such correlations can affect the actual types of magnetic and orbital orderings . However, the corresponding terms are of the order of $t^2/U \ll t,\ \Delta,\ J_H$. Hence, they are rather small and should not significantly change the phase diagrams. Using the mean-field approach, we also neglect the fluctuation effects, but it is known that such approach captures the main qualitative features of the phase diagrams.

The long-range Coulomb interaction related to the charge inhomogeneity in the phase-separated state can reduce the doping range corresponding to the phase separation and affect the size of inhomogeneities. The minimization of both the long-range Coulomb repulsion and the surface energy provides an opportunity to reveal the geometry of the phase-separated state (e.g., the inhomogeneities can take the shapes of droplets, alternating layers, or rods) and to determine the characteristic sizes of inhomogeneities and the percolation threshold for the doping, at which the insulator–metal transition occurs [33, 42, 43].

Fig. 6.5 Phase diagrams involving the phase-separated states in the system with spin-state transitions at three characteristic values of the hopping integral t: **a** $t/J_H = 1$; **b** $t/J_H = 1.3$; **c** $t/J_H = 1.5$. PS I is the state with the phase separation into the domains without itinerant charge carriers, corresponding (with LS Co^{3+}), and those containing purely itinerant charge carriers corresponding to IS Co^{3+}. PS II is the state with the similar type of the phase separated, state, where the domains with and without delocalized charge carriers correspond to IS Co^{4+} and HS Co^{4+}, respectively. Regions 1 in panel (**b**) correspond to charge carriers located at HS Co^{4+} and LS Co^{3+}, for which the electron hopping between Co ions is prohibited owing to the spin blockade [24]

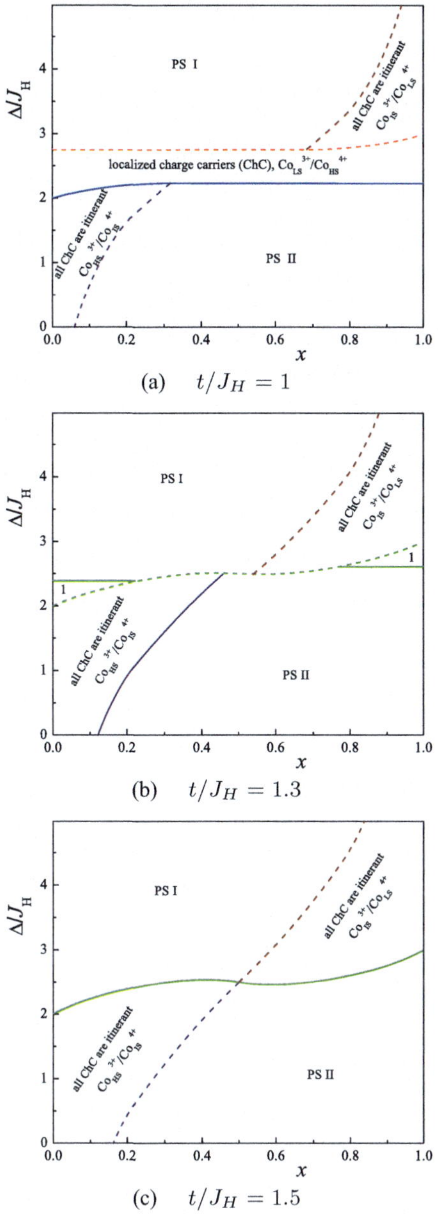

6.6 Conclusions

In this section, we have formulated a simple model capturing the main physics of strongly correlated electron systems with spin-state transitions. Based on this model, we predict the possibility of electronic phase separation for perovskite cobaltites under doping. We determined the corresponding phase diagrams exhibiting broad ranges of the phase separation. The specific shape of phase diagrams exhibits significant changes with the variation of the ratio of the electron hopping integral t and Hund's rule coupling constant J_H.

We did not analyze in detail the possible structure of the phase-separated state. However, as we have discussed in the previous chapters in the case of manganites, the calculations [33] and numerical simulations [36] taking into account the surface and long-range Coulomb contributions to the total energy lead to the characteristic size of nanoscale inhomogeneities of the order of several lattice constants. One can expect that the inhomogeneities caused by the phase separation in cobaltites with spin-state transitions would be of a similar scale.

Note that the existence of the electronic phase separation in the systems with the spin-state transitions is rather a robust phenomenon appearing in different models. For example, the numerical studies reported in [44] revealed a possibility of the phase separation in the two-band model with the bands differing in the spin state. In such a model, the phase separation appears upon doping if the widths of these bands are comparable to each other.

It is also worth noting rather active line research related to the photoinduced spin-state transitions and the formation of the corresponding inhomogeneous states, which can be revealed, for example, by the pump–probe techniques. In particular, a bound state of optically exited hole and the HS state can be created in cobaltites within the LS phase. This bound state brings about a characteristic peak structure in the optical pump–probe spectra which is distinct from the spectra in thermally excited states [45, 46]. This brings us a clue for the new optical spin control technique.

The nanoscale phase separation in materials with spin state transitions can involve some additional degrees of freedom. The detailed studies of $La_{2-x}Sr_xCoO_4$ layered cobaltites near $x = 1/3$ in the experiments including the resonant micro X-ray diffraction, neutron scattering, and muon spin rotation spectroscopy revealed nanosize undoped and hole-doped islands involving the charge order correlations in addition to the magnetic ones [47, 48].

References

1. P.M. Raccah, J.B. Goodenough, Phys. Rev. **155**(3), 932 (1967)
2. S. Yamaguchi, Y. Okimoto, H. Taniguchi, Y. Tokura, Phys. Rev. B **53**(6), R2926 (1996)
3. S. Yamaguchi, Y. Okimoto, Y. Tokura, Phys. Rev. B **55**(14), R8666 (1997)
4. M.A. Korotin, S.Yu. Ezhov, I.V. Solovyev, V.I. Anisimov, D.I. Khomskii, G.A. Sawatzky, Phys. Rev. B **54**(8), 5309 (1996)

5. M. Pouchard, A. Villesuzanne, J.P. Doumerc, J. Solid State Chem. **162**(2), 282 (2001)
6. D. Khomskii, U. Löw, Phys. Rev. B **69**(18), 184401 (2004)
7. P.W. Anderson, H. Hasegawa, Phys. Rev. **100**, 675 (1955)
8. P.G. de Gennes, Phys. Rev. **118**(1), 141 (1960)
9. L.N. Bulaevskii, D.I. Khomskii, Sov. Phys. JETP **25**, 1067 (1967)
10. L.N. Bulaevskii, E.L. Nagaev, D.I. Khomskii, Sov. Phys. JETP **27**, 836 (1968)
11. E.L. Nagaev, JETP Lett. **6**, 18 (1967)
12. T. Kasuya, A. Yanase, T. Takeda, Solid State Commun. **8**(19), 1543 (1970)
13. K.I. Kugel, A.L. Rakhmanov, A. Sboychakov, D.I. Khomskii, Phys. Rev. B **78**(15), 155113 (2008)
14. R. Kilian, G. Khaliullin, Phys. Rev. B **60**(19), 13458 (1999)
15. T. Mizokawa, D.I. Khomskii, G.A. Sawatzky, Phys. Rev. B **61**(6), R3776 (2000)
16. T. Mizokawa, D.I. Khomskii, G.A. Sawatzky, Phys. Rev. B **63**(2), 24403 (2001)
17. A. Maignan, V. Caignaert, B. Raveau, D. Khomskii, G. Sawatzky, Phys. Rev. Lett. **93**(2), 026401 (2004)
18. R. Caciuffo, D. Rinaldi, G.L. Barucca, J. Mira, J. Rivas, M.A. Señarís-Rodríguez, P.G. Radaelli, D. Fiorani, J.B. Goodenough, Phys. Rev. B **59**(2), 1068 (1999)
19. N.N. Loshkareva, E.A. Gan'shina, B.I. Belevtsev, Yu.P. Sukhorukov, E.V. Mostovshchikova, A.N. Vinogradov, V.B. Krasovitsky, I.N. Chukanova, Phys. Rev. B **68**(2), 024413 (2003)
20. D. Phelan, D. Louca, K. Kamazawa, S.H. Lee, S.N. Ancona, S. Rosenkranz, Y. Motome, M.F. Hundley, J.F. Mitchell, Y. Moritomo, Phys. Rev. Lett. **97**(23), 235501 (2006)
21. S. Giblin, I. Terry, D. Prabhakaran, A. Boothroyd, J. Wu, C. Leighton, Phys. Rev. B **74**(10), 104411 (2006)
22. C. He, M.A. Torija, J. Wu, J.W. Lynn, H. Zheng, J.F. Mitchell, C. Leighton, Phys. Rev. B **76**(1), 014401 (2007)
23. A. Podlesnyak, M. Russina, A. Furrer, A. Alfonsov, E. Vavilova, V. Kataev, B. Büchner, T. Strässle, E. Pomjakushina, K. Conder, D.I. Khomskii, Phys. Rev. Lett. **101**(24), 247603 (2008)
24. A.O. Sboychakov, K.I. Kugel, A.L. Rakhmanov, D.I. Khomskii, Phys. Rev. B **80**, 024423 (2009)
25. N.B. Ivanova, S.G. Ovchinnikov, M.M. Korshunov, I.M. Eremin, N.V. Kazak, Phys. Usp. **52**(8), 789 (2009)
26. V.A. Dudnikov, Y.S. Orlov, N.V. Kazak, M.S. Platunov, S.G. Ovchinnikov, JETP Lett. **104**(8), 588 (2016)
27. J. Baier, S. Jodlauk, M. Kriener, A. Reichl, C. Zobel, H. Kierspel, A. Freimuth, T. Lorenz, Phys. Rev. B **71**(1), 014443 (2005)
28. S.K. Pandey, A. Kumar, S. Patil, V.R.R. Medicherla, R.S. Singh, K. Maiti, D. Prabhakaran, A.T. Boothroyd, A.V. Pimpale, Phys. Rev. B **77**, 045123 (2008)
29. S.K. Pandey, S. Patil, V.R.R. Medicherla, R.S. Singh, K. Maiti, Phys. Rev. B **77**, 115137 (2008)
30. W. Luo, F. Wang, J. Solid State Chem. **182**(11), 3171 (2009)
31. L.M. Falicov, J.C. Kimball, Phys. Rev. Lett. **22**(19), 997 (1969)
32. A. Kocharian, G.W. Fernando, K. Palandage, J.W. Davenport, Phys. Lett. A **373**(11), 1074 (2009)
33. K.I. Kugel, A.L. Rakhmanov, A.O. Sboychakov, Phys. Rev. Lett. **95**(26), 267210 (2005)
34. A.O. Sboychakov, K.I. Kugel, A.L. Rakhmanov, Phys. Rev. B **74**, 014401 (2006)
35. J. Hubbard, Proc. Roy. Soc. A **205**, 238 (1963)
36. V.B. Shenoy, T. Gupta, H.R. Krishnamurthy, T.V. Ramakrishnan, Phys. Rev. Lett. **98**, 097201 (2007)
37. P. Fulde, *Electron Correlations in Molecules and Solids*. Springer Series in Solid-State Sciences, vol. 100, 3rd edn. (Springer, Berlin, Heidelberg, 2002)
38. S.G. Ovchinnikov, V.V. Val'kov, *Hubbard Operators in the Theory of Strongly Correlated Electrons* (World Scientific, 2004)
39. A. Kocharian, G. Fernando, K. Palandage, J. Davenport, Phys. Rev. B **78**(7), 075431 (2008)
40. T. Fujita, S. Kawabata, M. Sato, N. Kurita, M. Hedo, Y. Uwatoko, J. Phys. Soc. Jpn. **74**(8), 2294 (2005)

41. G.Y. Wang, X. Chen, T. Wu, G. Wu, X.G. Luo, C.H. Wang, Phys. Rev. B **74**(16), 165113 (2006)
42. J. Lorenzana, C. Castellani, C. Di Castro, EPL (Euro Phys. Lett.) **57**(5), 704 (2002)
43. K.I. Kugel, A.L. Rakhmanov, A.O. Sboychakov, F. Kusmartsev, N. Poccia, A. Bianconi, Supercond. Sci. Technol. **22**(1), 014007 (2008)
44. R. Suzuki, T. Watanabe, S. Ishihara, Phys. Rev. B **80**, 054410 (2009)
45. Y. Kanamori, H. Matsueda, S. Ishihara, Phys. Rev. Lett. **107**, 167403 (2011)
46. S. Ishihara, J. Ohara, Y. Kanamori, Y. Kanamori, Eur. Phys. J. Spec. Topics **22**, 1125 (2013)
47. Y. Drees, Z.W. Li, A. Ricci, M. Rotter, W. Schmidt, D. Lamago, O. Sobolev, U. Rütt, O. Gutowski, M. Sprung, A. Piovano, J.P. Castellan, A.C. Komarek, Nat. Commun. **5**(1), 5731 (2014)
48. H. Guo, Z.W. Li, S. Sakong, G. Ryu, L. Zhao, A. Piovano, W. Schmidt, M. Sprung, J. Strempfer, S. Francoual, D. Dzhigaev, S. Subakti, Z. Hu, H.J. Lin, C.T. Chen, H. Luetkens, O. Stockert, A.C. Komarek, Phys. Rev. B **100**, 014411 (2019)

Chapter 7
Effects of Orbital Degrees of Freedom, Different Shapes of Polarons

Abstract We consider the formation of orbital polarons (which contain only one type of d orbitals in the ferro-orbital state) in the matrix of the checkerboard (antiferro) d orbitals of two different types in the transition metals compounds. We analyze different shapes of orbital polarons in 2D systems.

7.1 Introduction

In this chapter, we study the effects related to one more degree of freedom, namely with orbitals and their ordering. The orbital ordering (OO) plays an important role in the physics of magnetic oxides, especially of those, which contain the ions with the orbitally degenerate ground state, so-called ion Jahn–Teller (JT) ions. The compounds with JT ions usually exhibit different kinds of superstructures (OO) [1, 2]. In fact, the OO is a specific feature of magnetic insulators. The additional charge carriers introduced by doping are able to transfer an OO insulator to the metallic state. Hence, here we are dealing again with the competition between the charge carrier localization and their itinerancy, which can give rise to the phase separation, especially at low doping levels [3–5]. In the analysis of phase separation in magnetic materials such as manganites, the OO is usually not addressed (except maybe the treatment of orbital and magnetic polarons [5–8]). In this chapter, we study the role of OO for the electronic phase separation employing minimal models including itinerant charge carriers interacting with the OO superstructure. We demonstrate that at a low doping level, the orbitally ordered systems can exhibit the phase separation, which not directly related to the possible magnetic structure.

Such phase separation is treated using two approaches. In Sect. 7.2, we base our analysis on a symmetrical model analogous to the considered above double exchange model where the orbital variables play a role in local spins. We assume that owing to the orbital degeneracy, localized electrons induce lattice distortions of JT type, thus giving rise to OO. Itinerant charge carriers appearing due to the doping move on OO background. Within the other approach (Sect. 7.3), we employ the e_g symmetry of the wave functions for the doped charge carriers. Using both approaches, we reveal

M. Yu. Kagan et al., *Electronic Phase Separation in Magnetic and Superconducting Materials*, Springer Series in Solid-State Sciences 201,
https://doi.org/10.1007/978-3-031-55467-4_7

the tendency to the phase separation related to the existence of the orbital degrees of freedom. In other words, itinerant charge carriers introduced by doping stimulate the formation of inhomogeneous states with nanosize domains having the orbital order differing from that taking place in the absence of doping. In Sect. 7.4, we analyze the possible shapes and sizes of such domains and demonstrate that at different relations between the electron hopping integral t and the parameter J characterizing the interaction between orbitals, such shapes can be quite different. For example, in the layered compounds, the two-dimensional case, below some critical value of t/J, nearly circular domains can undergo the transformation to the needle-like ones. This is a consequence of anisotropic charge distribution characteristic of specific orbitals, which brings forward additional possibilities to the physics of phase separation.

In Sect. 7.5, we are dealing with a picture more relevant to the actual JT compounds, in which the same charge carriers are involved both in OO and, after doping, in conduction. We assume that at zero doping, the orbitals are ordered due to the electron–lattice coupling. The doped charge carriers are moving on the OO background. In Sect. 7.6, we consider another model, where the same e_g electrons are responsible for the band structure and for the orbital ordering.

7.2 Symmetrical Model

Here, we analyze the situation typical of compounds with JT ions having a double-degenerate ground state. Different kinds of lattice distortions (e.g., compression or stretching) lift such degeneracy making favorable two possible ground states for each ion, say, either a or b. The a and b states at lattice site $\mathbf{n}$ describe, in fact, the occupation of electron orbitals at this site. Usually we deal with the linear combination of a and b orbitals, which can be expressed in terms of angle θ

$$|\theta\rangle = \cos\frac{\theta}{2}|a\rangle + \sin\frac{\theta}{2}|b\rangle\,. \tag{7.1}$$

The orbitals and the related local distortions at different sites can form regular arrays (orbital ordering). In the case of two basis states, the orbitals can be put into correspondence to some spin-1/2 variables, and the interaction between orbitals in its simplest version can be written in the form of Heisenberg Hamiltonian

$$H_{\mathrm{OO}} = J\sum_{\langle\mathbf{nm}\rangle} \boldsymbol{\tau}_{\mathbf{n}}\boldsymbol{\tau}_{\mathbf{m}}\,, \tag{7.2}$$

where $\boldsymbol{\tau}_n = \{\tau^x_{\mathbf{n}}, \tau^z_{\mathbf{n}}\}$ are the Pauli matrices, and a and b states of the ion $\mathbf{n}$ correspond to eigenvectors of operators $\tau^z_{\mathbf{n}}$, with eigenvalues 1 and -1, respectively. Hamiltonian (7.2), provides a possibility of two simplest types of orbital ordering: ferro-OO (similar orbitals at each site) and antiferro-OO (alternating orbitals at neighboring sites). At zero doping (no itinerant charge carriers), the system can have either the antiferro-OO ground state (at $J > 0$) or the ferro-OO one (at $J < 0$). In actual materials containing Jahn–Teller ions, orbital Hamiltonians are not so simple. Nevertheless, model (7.2) appears to be sufficient for capturing main physical effects related to the orbital ordering.

Let us now pass to the case of nonzero doping and hence of nonzero density of itinerant charge carriers in the system, $n \neq 0$. These itinerant charge carriers are moving on the background of orbitally ordered state formed by localized electrons (Below, we demonstrate that the main qualitative results remain the same, if itinerant charge carries such as e_g electrons, are responsible both for the orbital ordering and for the electrical conductivity). The electron hopping integrals are determined by the specific electron orbitals located at the neighboring lattice sites. The Hamiltonian describing only the electron states have the form

$$H_{\text{el}} = - \sum_{\langle \mathbf{nm} \rangle, \alpha, \beta, \sigma} t^{\alpha\beta} \left(P^{\dagger}_{\mathbf{n}\alpha\sigma} a^{\dagger}_{\mathbf{n}\alpha\sigma} a_{\mathbf{m}\beta\sigma} P_{\mathbf{m}\beta\sigma} + h.c. \right), \tag{7.3}$$

where, $a^{\dagger}_{\mathbf{n}\alpha\sigma}$, $a_{\mathbf{n}\alpha\sigma}$ are creation and annihilation operators corresponding to charge carriers at site $\mathbf{n}$ with spin projection σ at orbital α. Projection operators P in (7.3) take into account that the double occupancy of the lattice sites by electrons is highly unfavorable in energy due to the strong on-site Coulomb repulsion. Below, we are dealing mostly with the case $n < 1$. In this section, our analysis of the electron contribution to the total energy is based on the tight-binding approximation. To avoid unnecessary complications, we illustrate the essential physics of orbitally ordered systems considering the square lattice in the two-dimensional (2D) case and simple cubic lattice in the three-dimensional (3D) case. In the tight-binding approximation, the spectrum of charge carriers has the form

$$E^{\alpha\beta}(\mathbf{k}) = -t^{\alpha\beta} \frac{1}{D} \sum_{i=1}^{D} \cos k_i = t^{\alpha\beta} \xi(\mathbf{k}), \tag{7.4}$$

where D (2 or 3) is the number of spatial dimensions. Components k_i of the wave vector $\mathbf{k}$ are expressed in the units of inverse lattice constant $1/d$.

Within the approach adopted in this section, the charge carriers introduced by doping are built into the JT distorted background and occupy orbital states $|a\rangle$ or $|b\rangle$. Then, their intersite charge transfer is described by three hopping integrals: t^{aa}, t^{bb}, and $t^{ab} = t^{ba} = t'$. For simplicity, we can assume that $t^{aa} = t^{bb} = t$. Such a system is characterized by an interplay two contributions to the total energy: the possible gain in the kinetic energy can come from the increase in the electron bandwidth, whereas the potential energy can be reduced by the optimization of the orbital ordering.

In a more general case, when there is a superposition of $|a\rangle$ or $|b\rangle$ states at site $\mathbf{n}$ characterized by angle θ, (see, by Eq. (7.1)), we can write the expression for the hopping integral between the sites characterized by orbital states $|\theta_1\rangle$ and $|\theta_2\rangle$

$$t^{\theta_1\theta_2} = t\cos\frac{\theta_1-\theta_2}{2} + t'\sin\frac{\theta_1+\theta_2}{2} \tag{7.5}$$

This is in fact a generalization to the OO case of the semiclassical expression for the effective hopping integral in the double exchange model [9].

We shall start from the homogeneous phase, in which orbitals $|\theta_1\rangle$ and $|\theta_2\rangle$ alternate at different lattice sites ($J > 0$). Within the mean-field approach, the total energy per site is given by the expression

$$E_{tot}(\theta_1, \theta_2) = zt^{\theta_1\theta_2}\varepsilon_0(n) + \frac{zJ}{2}\cos(\theta_1-\theta_2)\,, \quad \varepsilon_0(n) < 0\,. \tag{7.6}$$

Here, z is the number of nearest-neighbor sites and $\varepsilon_0(n)$ is the dimensionless kinetic energy, which depends on the type of crystal lattice. In this section, we consider an isotropic case, for which t, t', and, hence, $t^{\theta_1\theta_2}$, are independent on the direction in space. For such isotropic system, $\varepsilon_0(n)$ does not depend on $\theta_{1,2}$, and we can easily calculate the orbital structure by minimization of total energy (7.6) with respect to angles θ_1 and θ_2. At high doping levels, namely for $t|\varepsilon_0(n)| > 2J$, the ferro-OO state with $\theta_1 = \theta_2 = \pi/2$ becomes favorable. In the opposite limiting case, $t|\varepsilon_0(n)| < 2J$, the orbital structure corresponds to $\theta_2 = \pi - \theta_1$, and

$$\theta_1 = \arcsin\left(\frac{t|\varepsilon_0(n)|}{2J}\right), \quad \frac{t|\varepsilon_0(n)|}{2J} < 1\,. \tag{7.7}$$

Such alternation of orbitals can be interpreted as a kind of canting for the directions of pseudospins τ in the orbital space. For the energy of such state, we have

$$E_{tot} = zt'\varepsilon_0(n) - \frac{zt^2}{4J}\varepsilon_0^2(n) - \frac{zJ}{2}\,. \tag{7.8}$$

If it is possible to represent the kinetic energy as $\varepsilon_0(n) = nf(n)$, where $f(n)$ a slowly changing function of n, then E_{tot} can exhibit a negative curvature, at least, at small n, which signals on an instability of a homogeneous orbitally ordered state (negative compressibility).

We can find a specific form of $\varepsilon_0(n)$ and determine the shape of the curve depicting the behavior of the total energy with doping. For the electron energy (7.4), the density of states $\rho_0(E)$, can be written as

$$\rho_0(E) = \int \frac{d\mathbf{k}}{(2\pi)^D}\delta(E - \xi(\mathbf{k})) = \int_0^\infty \frac{ds}{\pi}\cos(Es)J_0^D\left(\frac{s}{D}\right), \tag{7.9}$$

where J_0 is the Bessel function. The expression for $\varepsilon_0(n)$ n has the form

$$\varepsilon_0(n) = \int_{-1}^{\mu(n)} dE\, E\rho_0(E)\,, \tag{7.10}$$

where the chemical potential μ can be found based on the usual relation $n = \int_{-1}^{\mu} dE\rho_0(E)$.

At small density of itinerant charge carriers, $n \ll 1$, we can write an explicit expression for $\varepsilon_0(n)$. For example, in a 2D case, we have $\varepsilon_0(n) \approx -n + \pi n^2/2$, and the corresponding expression for the total energy is

$$E_{tot} \approx -zt'n - z\left(\frac{t^2}{4J} - \frac{\pi t'}{2}\right)n^2 - \frac{zJ}{2}\,. \tag{7.11}$$

Using Eq. (7.11), we can find the range of parameters corresponding to the negative curvature of the energy plot, i.e., $d^2E_{tot}/dn^2 < 0$. Such situation takes place if

$$\frac{t}{J} > \frac{2\pi t'}{t}, \tag{7.12}$$

This means that the homogeneous state with canted orbitals is unstable with respect to the phase separation into a inhomogeneous states with ferro- and antiferro-orbital ordered domains . Here, we have a close analogy to the situation typical of the double exchange (in the simplest single-band case) [10], which corresponds to $t' = 0$. In the opposite limiting case, that is, for $2\pi t'/t > t/J$, a homogeneous state is favorable within the whole doping range—in contrast to the usual double exchange.

A rough estimate for the phase separation range can be found if we neglect the quadratic term in Eq. (7.7), $\varepsilon_0(n) \approx -n$. Thus, we find

$$0 < n \lesssim \frac{2J}{t}. \tag{7.13}$$

Actually, the homogeneous orbitally canted state appears to be unfavorable in energy within almost the whole doping range, where the difference $\theta_2 - \theta_1 = \pi - 2\theta_1$ with θ_1 from Eq. (7.7) does not vanish. The situation does not qualitative change, if we use more accurate calculations for E_{tot} based on $\varepsilon_0(n)$ determined employing density of states found from (7.9). The corresponding $E_{tot}(n)$ plot in the 2D case is shown in Fig. 7.1.

In the 3D case, the situation is not so simple. At low doping levels, the corresponding expression for the kinetic energy has the form $\varepsilon_0(n) \approx -n + an^{5/3}$, where

$$a = \frac{3}{5}\left(\frac{\pi^2}{\sqrt{6}}\right)^{2/3},$$

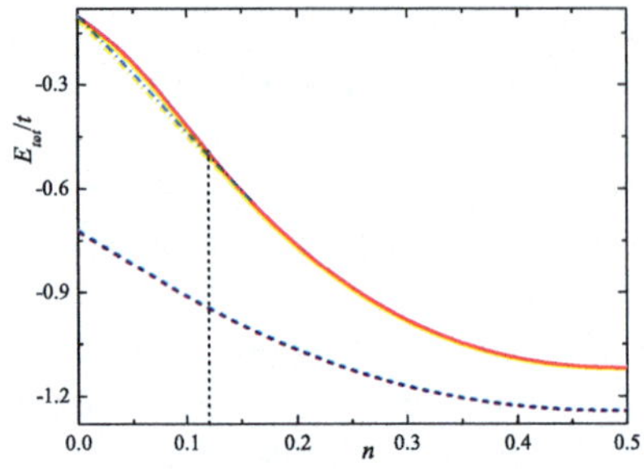

Fig. 7.1 Total energy of homogeneous orbitally ordered state (7.8) in 2D versus doping n. Two types of behavior are shown: with the range corresponding to the negative curvature (red solid line, $J/t = 0.05$), and without it (blue dashed line, $J/t = 0.35$); $t' = 0.5t$ for both curves. The range of instability for the homogeneous state (with the negative curvature) is located on the left-hand side of the vertical line. The energy of the phase-separated state is drawn by the dot-and-dash line. The blue dashed curve represents the energy of the homogeneous state, which at $J/t = 0.35$ is stable within the whole doping range [11]

and the total energy is

$$E_{tot} \approx -zt'n - z\left(\frac{t^2}{4J} - \frac{a}{n^{1/3}}\right)n^2 - \frac{zJ}{2}. \tag{7.14}$$

The second derivative of E_{tot} is positive at $n \to 0$ and becomes negative at

$$n_c \approx \left(\frac{5aJt'}{9t^2}\right)^3. \tag{7.15}$$

The same line of reasoning as that in the 2D case leads to the following estimate for the phase separation range

$$n_c \lesssim n \lesssim \frac{2J}{t}. \tag{7.16}$$

According to Eq. (7.15), a nonzero value of the nondiagonal hopping integral t' manifests itself in a nonzero critical concentration n_c for phase separation. The phase separation range obtained by the Maxwell construction is in fact broader and starts from some $n_0 < n_c$. Here, the orbital state is described by a classical vector, but nevertheless we have the lower critical concentration for the canted state and hence for the phase separation, which in the usual double exchange model appears only, if we take into account the quantum nature of core spins [10]. Inequalities (7.13) and (7.16) are approximate ones and are valid only at quite small values of parameter J/t.

7.3 Anisotropic Model

Let us now analyze the effects related to the specific form of e_g orbitals. To make the underlying physics clearer, we discuss in this subsection the two-dimensional case (square lattice). This case is relevant, e.g., to layered cuprates, like K_2CuF_4, or manganites (La_2MnO_4 or $La_2Mn_2O_7$). Another simplification is that we choose the Heisenberg-like form (7.2) for the orbital exchange. Such simplification do not lead to a loss in generality. At least, in the analysis based on the superexchange mechanism of the orbital ordering [1] in the case of e_g orbitals demonstrates the main physics of the phase separation discussed below remains qualitatively the same.

As we have mentioned above, any wave function can be written as a linear combination of basis orbitals. For e_g orbitals, we have $|a>=|x^2-y^2>$ and $|b>=|2z^2-x^2-y^2>$ and $|\theta>=\cos(\theta/2)|x^2-y^2>+\sin(\theta/2)|2z^2-x^2-y^2>$. Hopping integrals $t^{\alpha\beta}$ in Eq. (7.3) now become anisotropic and can be presented in the matrix form

$$(t_{x,y})^{\alpha\beta} = \frac{t_0}{4}\begin{pmatrix} 3 & \mp\sqrt{3} \\ \mp\sqrt{3} & 1 \end{pmatrix}, \tag{7.17}$$

where minus and plus signs correspond to the electron hopping along the axes x and y, respectively.

Again, we choose the orbital ground state in the form of alternating $|\theta_1\rangle$ and $|\theta_2\rangle$ orbitals. In this case, the spectrum of charge carriers takes the form

$$E(\mathbf{k}) = -t_0\left(A_x(\theta_1,\theta_2)\cos k_x + A_y(\theta_1,\theta_2)\cos k_y\right), \tag{7.18}$$

where

$$A_{x,y}(\theta_1,\theta_2) = \left|\cos\left(\frac{\theta_1-\theta_2}{2}\right) + \cos\left(\frac{\theta_1+\theta_2}{2} \pm \frac{\pi}{3}\right)\right|. \tag{7.19}$$

The corresponding expression for the total energy is

$$\begin{aligned} E_{tot}(\theta_1,\theta_2) = t_0\left(A_x(\theta_1,\theta_2) + A_y(\theta_1,\theta_2)\right) \times \\ \varepsilon(n;\theta_1,\theta_2) + 2J\cos(\theta_1-\theta_2), \end{aligned} \tag{7.20}$$

where $\varepsilon(n;\theta_1,\theta_2) = \int_{-1}^{\mu} dE\, E\rho(n;\theta_1,\theta_2)$, and the electronic density of states has the form

$$\rho(n;\theta_1,\theta_2) = \int_0^{\infty} \frac{ds}{\pi}\cos(Es) J_0\left(\frac{sA_x}{A_x+A_y}\right) J_0\left(\frac{sA_y}{A_x+A_y}\right). \tag{7.21}$$

Density of states (7.21) is the function of orbital angles θ_1 and θ_2. Therefore, to determine the optimum orbital configuration, it is necessary to minimize E_{tot}, Eq. (7.20), with respect to angles θ_1 and θ_2. As a result of such minimization, we find that at low doping levels (n less than some threshold value, n_1, which depends on parameter J/t_0), the ground state configuration corresponds to $\theta_1 = 0$, $\theta_2 = \pi$. This is the homogeneous phase with alternating $|x^2 - y^2>$ and $|2z^2 - x^2 - y^2>$ orbitals, i.e., the antiferro-OO. We do not take in to account here anharmonic terms and higher order interactions, which favor locally elongated octahedra corresponding in our notation to the angles equal to $\theta = \pi$, $\pm 2\pi/3$, (see Refs. [12, 13]). After the minimization, the total energy takes the form

$$E_{tot} = t_0\sqrt{3}\varepsilon_0(n) - 2J\,. \tag{7.22}$$

The plot of this energy has the positive curvature, $\partial^2 E_{tot}/\partial n^2 > 0$, corresponding to the stability of such state.

At $n = n_1$, the system passes in a jump-like manner to another state characterized by alternating orbital angles $\theta_2 = -\theta_1$, where

$$\theta_1 = \arccos\left(\frac{t_0|\varepsilon_0(n)|}{4J}\right)\,, \tag{7.23}$$

and $E_{tot}(n)$ exhibits a kink at $n = n_1$. This canted state arising at $n > n_1$ has the energy

$$E_{tot} = t_0\varepsilon_0(n) - \frac{t_0^2}{4J}\varepsilon_0^2(n) - 2J\,. \tag{7.24}$$

The angle θ_1 decreases with the increase in n and eventually vanishes at $n = n_2$. The value of $n = n_2$ can be found from the equation $t_0|\varepsilon_0(n_2)|/4J = 1$. The resulting state, $\theta_1 = \theta_2 = 0$, is the ferro-orbital one formed by $|x^2 - y^2\rangle$ orbitals. The plot total energy versus doping is presented in Fig. 7.2. In this figure, we see that the plot of energy (7.24) can exhibit both positive and negative curvature (see the inset of Fig. 7.2). For the positive curvature ensuring the stability, the phase separation can nevertheless arise due to the kink within the doping range close to $n = n_1$. At the negative curvature, the phase separation naturally appears. Thus, both the negative curvature of E_{tot} plot and the kink can give rise to different kinds of inhomogeneous states [14].

7.4 Inhomogeneities in the Orbitally Ordered Structures

In the previous sections, we have shown that itinerant charge carriers introduced by doping can favor the formation of charge inhomogeneities in the case of orbital ordering. Now, let us discuss possible shapes of such inhomogeneities. We base our discussion on the case of e_g electrons forming a regular array at the square lattice. We assume that the host lattice is characterized by the antiferro-OO structure with alternating $|x^2 - y^2>$ and $|2z^2 - x^2 - y^2>$ orbitals considered in the previous subsection. An additional charge carrier favors the formation of another type of

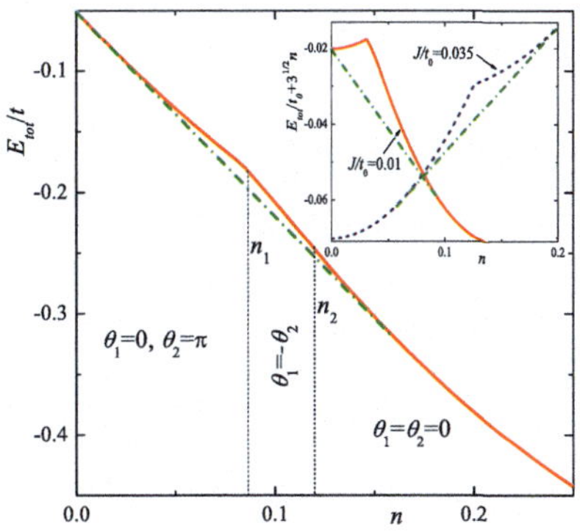

Fig. 7.2 Energy Eq. (7.20) versus doping at $J/t_0 = 0.025$ (red solid curve). Within the doping range close $n_1 \approx 0.08$, the homogeneous state is unstable, and a phase separation can arise. In the inset, we plot the function $E_{tot}(n) + t_0\sqrt{3}n$ (for the better visuality, we subtract the linear term in $E_{tot}(n)$) in the vicinity of $n < n_1$ at different values of parameter J/t. The red solid curve (blue dashed curve) corresponds to $J/t_0 = 0.01$ ($J/t_0 = 0.035$), and have a negative (positive) value of $\partial^2 E_{tot}/\partial n^2$ in the region $n > n_1$ close to n_1. The phase separation appears in both cases. The dot-and-dash line is the result of the Maxwell construction [11]

orbital ordering within a limited region close to it. The optimum structure can be found by minimization of the additional energy related to the doping. Without a loss in generality, we can consider alternating $|\theta_1\rangle$ and $|\theta_2\rangle$ orbitals.

In such a situation, the energy spectrum of charge carriers is specified by Eq. (7.18). Since we are dealing with a limited region, we can write an effective Hamiltonian by expanding Eq. (7.18) up to up to the second order in **k**. As a result, we have

$$\hat{H}_{eff} = -t_0\left(A_x + A_y\right) + \frac{t_0}{2}\left(A_x\frac{\partial^2}{\partial x^2} + A_y\frac{\partial^2}{\partial y^2}\right), \tag{7.25}$$

where A_x, A_y are defined in Eq. (7.19). Then, we assume that the region under study is an ellipse with semiaxes $\sqrt{A_x}\rho_0$ and $\sqrt{A_y}\rho_0$ and search for the solution of the Schrödinger equation within this ellipse employing Hamiltonian (7.25). In such elliptical droplet, the kinetic energy of a charge carrier can be written as

$$E_{kin} = -t_0\left(A_x + A_y\right) + \frac{t_0 j_{0,1}^2}{2\rho_0^2}. \tag{7.26}$$

Here, $j_{0,1} \cong 2.405$ is the first root of the Bessel function J_0.

For the orbitally ordered state, we can write the potential energy E_{pot} related to the orbital ordering, which is proportional to the volume v of the elliptical droplet ($v = \pi\sqrt{A_x A_y}\rho_0^2$). It includes the energy of the new canted-OO state $zJ\cos(\theta_1 - \theta_2)/2$ and the loss in energy of the antiferro-OO host structure occurring due to arising droplets with the modified OO structure ($zJ/2$, $z = 4$). For a set of identical droplets, the sum of these two contributions is

$$E_{pot} = 4\pi\rho_0^2 J\sqrt{A_x A_y}\cos^2\left(\frac{\theta_1 - \theta_2}{2}\right). \tag{7.27}$$

Minimizing the droplet energy $E_{kin} + E_{pot}$ with respect to ρ_0, we find

$$\rho_0 = \left(\frac{t_0 j_{0,1}^2}{8\pi J\sqrt{A_x A_y}\cos^2\left(\frac{\theta_1-\theta_2}{2}\right)}\right)^{1/4}. \tag{7.28}$$

The total energy (per lattice site) then reads

$$E_{tot} = -2J + E(\theta_1, \theta_2)\, n\,, \tag{7.29}$$

$$E(\theta_1, \theta_2) = -2t_0(A_x + A_y) \quad + j_{0,1}\left(8\pi t_0 J\sqrt{A_x A_y}\right)^{1/2}\left|\cos\left(\frac{\theta_1 - \theta_2}{2}\right)\right|. \tag{7.30}$$

To determine the optimum shapes of OO droplets, we perform the minimization of energy (7.29) with respect to orbital angles θ_1 and θ_2 using expressions (7.19) [for $A_{x,y}$]. The behavior of $E(\theta_1, \theta_2)$ at two different values of J/t_0 is illustrated in Fig. 7.3. In this figure, we can clearly see that the function $E(\theta_1, \theta_2)$ exhibits several minima, and the specific angles θ_1 and θ_2 corresponding to the lowest energy strongly depend on the value of J/t_0. At small J/t_0 (Fig. 7.3a), the lowest energy is achieved at to $\theta_1 = \theta_2 = 0$ (ferro-OO state with the filled $|x^2 - y^2>$ orbitals). For such state, the optimum shape of droplets is a circle (see the left panel of Fig. 7.4). At J/t_0 exceeding a certain critical value ($J_{cr}/t_0 \simeq 0.0075$), the minimum $\theta_1 = \theta_2 = 0$ corresponds to a metastable state, and the energy $E(\theta_1, \theta_2)$ attains the lowest value at four degenerate structures: with the alternation of either $\theta_1 = \pi/3$ and $\theta_2 = 2\pi/3$, or $\theta_1 = -\pi/3$ and $\theta_2 = -2\pi/3$, and the other two minima correspond to the similar structures

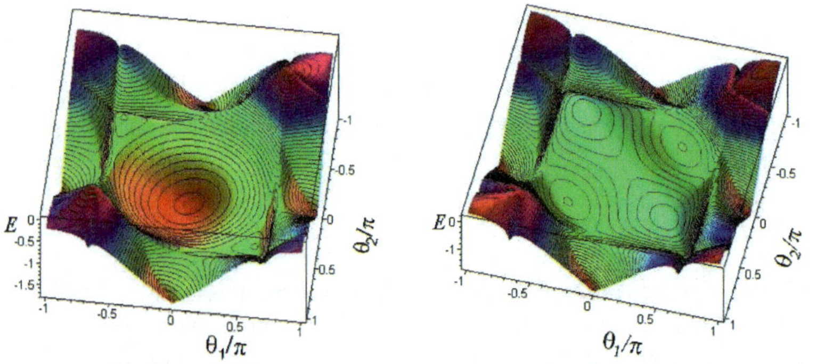

(a) $J/t_0 = 0.005$ (b) $J/t_0 = 0.02$

Fig. 7.3 Energy Eq. (7.29) of an orbital versus angles θ_1 and θ_2 at J/t_0 smaller (**a**) and larger (**b**) than the critical value $\simeq 0.0075$ [11]

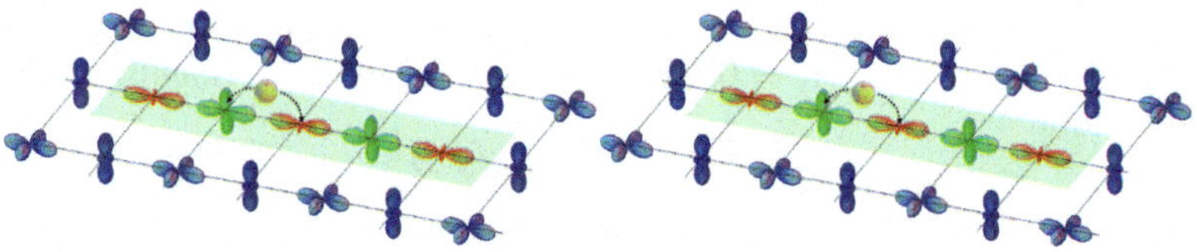

Fig. 7.4 Possible types of orbital inhomogeneities arising due to doping: circular (left panel) and needle-like (right panel) droplets. The motion of an itinerant charge carrier results in the formation of a ferro- or canted-OO structure within a limited region inside the host material with the antiferro-OO [11]

with the replacement $\theta_1 \leftrightarrow \theta_2$ (see Fig. 7.3b). Such four orbital arrays are the chains with alternating $|x^2 - z^2>$ and $|2y^2 - x^2 - z^2>$ (or $|y^2 - z^2>$ and $|2x^2 - y^2 - z^2>$) orbitals. This, in fact, leads to the formation of nearly one-dimensional (cigar-shaped) droplets directed along y or x axes (right panel of Fig. 7.4). At the further increase in J/t_0, the metastable $\theta_1 = \theta_2 = 0$ array splits into two more complicated structures with $\theta_1^* = -\theta_2^*$, in which θ_1^* can be either positive and (see Fig. 7.3b). The inhomogeneities related to the latter arrays should have the circular shape, but the OO structure inside them is canted.

The physical mechanism underlying the formation of two types of droplets with different shapes is quite clear. Indeed, the ferro-OO state with $|x^2 - y^2>$ orbitals provides the maximum gain in the kinetic energy. This gain prevails at small J/t_0 favoring circular droplets with such the ferro-OO. In the opposite case of dominant potential energy (large J/t_0), nearly one-dimensional structures turn out to be favorable. Therefore, the optimum droplets have now cigar- or needle-like shape. The orbital configuration in a needle-like droplet provides the maximum gain in the kinetic energy in the absence of hopping between neighboring chains.

It can be shown that in a certain doping range, $0 < n < n_c^*$, the inhomogeneous state including circular or needle-like OO droplets within the antiferro-OO host material corresponds to the energy lower than that of the homogeneous state. With the increase in the charge carrier density, the droplets can overlap; therefore, at some threshold value $n = n_c^*$, the considered droplet type disappears. However, the phase separation can exist in a broader doping range (see the previous subsection). For the threshold charge carrier density in the case of circular droplets, we obtain an estimate $n_c^* \sim 1/\pi\rho_0^2$. Thus, at $J/t_0 = 0.005$, we have $\rho_0 \approx 2$ (in units of lattice constant) and $n_c^* \sim 0.08$.

For the needle-like droplets, $A_x = 0$ (or $A_y = 0$), and Eq. (7.28) gives us the chains of infinite length (but with zero volume v), $\rho_0 = \infty$. This rather unphysical result stems from the assumption that the potential is proportional only to the droplet volume only. To estimate the characteristic length L of the needle, we should consider also the surface contribution to the energy (proportional to the length of droplet) to the potential energy E_{pot}. For such an estimate, we can consider, for example, an array of $|y^2 - z^2>$ and $|2x^2 - y^2 - z^2>$ orbitals stretched along the x axis. For this array, we have $A_x = \sqrt{3}$, $A_y = 0$. The corresponding effective Hamiltonian (7.25) takes the form

$$\hat{H}_{eff} = -t_0\sqrt{3} + \frac{t_0\sqrt{3}}{2}\frac{\partial^2}{\partial x^2},$$

and the kinetic energy of a charge carrier in the chain of length L can be written as $E_{kin} = -t_0\sqrt{3}(1 - \pi^2/2L^2)$. The surface energy results from the interaction between the orbitals in the chain and in the host material. The minimum surface energy corresponds to the configuration illustrated in Fig. 7.4, for which each $|y^2 - z^2>$ $(|2x^2 - y^2 - z^2>)$ orbital in the chain has its nearest-neighbor $|x^2 - y^2>$ $(|2z^2 - x^2 - y^2>)$ orbital in the host. Using the continuum approximation, we can write the potential energy as $E_{pot} = 9JL/4$. Then, we minimize the total energy $E_{kin} + E_{pot}$ with respect to L and obtain the expression for the characteristic length of a needle-like droplet

$$L_0 = \left(\frac{4\pi^2 t_0\sqrt{3}}{9J}\right)^{1/3}. \tag{7.31}$$

For $J/t_0 = 0.05 > J_{cr}/t_0$, we obtain $L_0 \approx 5.5$. If we assume a random distribution of the needle-like droplets in the host (with the needles directed both along x and y axes), the critical density is about $n_c^* \sim 1/L_0^2$. Note, however, that this density can be higher in a more regular array such as that corresponding to stripes.

Let us recall that for spin polarons discussed in the previous chapters, the existence of the gapless (Goldstone) excitations leads to slowly decaying spin structure distortions around magnetic defect (magnetic polaron) [9]. Therefore, it is natural to expect that orbital polarons should give rise to long-range strain fields will be created around them. In analogy to the case of spin polarons, these long-range fields should not affect the existence of orbital polarons themselves. They can only modify the parameters of an inhomogeneous state and lead to an additional long-range interaction between inhomogeneities.

In this and in the previous sections, we were dealing with a two-dimensional case of the anisotropic model. However, in contrast to the isotropic case treated in Sect. 7.2, the space dimensionality can drastically affect the obtained results. Indeed, the hopping integrals for e_g electrons are strongly anisotropic. This can produce a pronounced effect on the OO type in the homogeneous state as well as on the shape and orbital order of the inhomogeneities. That is why, the detailed study of the three-dimensional case is a separate and quite complicated task.

7.5 Simplified Two-Band Model with the Orbital Ordering

7.5.1 Formulation of the Problem

In the previous sections, we studied the effect of doping on already existing orbital ordering. Now, we shall treat a more natural situation when the same charge carriers are involved both in the orbital ordering and in charge transfer. We start with the analysis of the system with JT ions having double-degenerate ground state. As usual, such degeneracy is lifted by local lattice distortions. Then, each lattice site, can correspond to two different ground states, a or b (in perovskites, for example, a (b) state can be related to stretching (compression) of the surrounding anion octahedron). The a and b states at site $\mathbf{n}$ are, in fact, possible orbitals filled by a charge carrier. Of course, the local distortions are not independent, and their interaction can lead to certain regular structure. The simplest Hamiltonian describing the orbital interaction can be described in an Ising-like form [15]

$$H_{\mathrm{OO}} = J \sum_{\langle \mathbf{nm} \rangle} \tau_{\mathbf{n}}^{z} \tau_{\mathbf{m}}^{z} \,. \tag{7.32}$$

Here, the $\tau_{\mathbf{n}}^{z}$ operators have the eigenvalues 1 or -1 depending on the state of site $\mathbf{n}$ (a or b, respectively). In such system, we can consider two simplest types of the orbital ordering: either ferro-OO (with the same orbitals occupying each site) and antiferro-OO (corresponding to the alternation of orbitals).

The doping gives rise to nonzero density, $n \neq 0$, of itinerant charge carriers that appear in the system. As a result, the system includes localized charge carriers responsible for OO and itinerant ones. The electron motion is determined by hopping integrals depending on the states of the neighboring lattice sites. In the model under study, there are three possible hopping integrals, namely, t^{aa}, t^{bb}, and $t^{ab} = t^{ba}$. For simplicity, let us assume that $t^{aa} = t^{bb} > t^{ab}$. At both types of orbital ordering (ferro- and antiferro-OO) the charge carriers are described by the Hamiltonian

$$H_{\mathrm{el}} = -t \sum_{\langle \mathbf{nm} \rangle, \sigma} \left(a_{\mathbf{n}\sigma}^{\dagger} a_{\mathbf{m}\sigma} + h.c. \right) + \frac{U}{2} \sum_{\mathbf{n}\sigma} n_{\mathbf{n}\sigma} n_{\mathbf{n}\bar{\sigma}} \,. \tag{7.33}$$

The first term in H_{el} describes the kinetic energy of itinerant charge carriers. Symbol t denotes the hopping integral t^{aa} in for the ferro-OO and t^{ab} for antiferro-OO. The second (Hubbard) term in Eq. (7.33) takes into account the on-site Coulomb repulsion between electrons and the bar above σ means the spin component of the opposite sign. Angular brackets $\langle ... \rangle$ denote the summation over nearest-neighbor sites. We also assume that the parameters of Hamiltonian (7.33) are independent of the doping level n.

7.5.2 Homogeneous States and Phase Separation

Hamiltonian (7.33) is the usual Hubbard Hamiltonian for charge carriers moving on the ferro- or antiferro-OO background. We begin our analysis from the possible homogeneous states corresponding to such Hamiltonian. At $J > 0$ and in the absence of doping($n = 0$), we have the uniform antiferro-OO ground state, which is the ground state if $J > 0$ (the case of ferro-OO state can be treated in a similar way). In the antiferro-OO state, we can consider only the interorbital hopping $t = t^{ab}$ in Eq. (7.33). We determine the energy spectrum of charge carriers using the Hubbard I approximation [16] in equation of motion for one-electron Green's function $G_\sigma(\mathbf{n}, \mathbf{n}_0;\ t - t_0) = -i\langle T a_{\mathbf{n}\sigma}(t) a^\dagger_{\mathbf{n}_0\sigma}(t_0)\rangle$. In the frequency–momentum representation, this Green's function can be written as [16]

$$G_\sigma(\mathbf{k}, \omega) = \frac{\omega + \mu - U(1 - n/2)}{(\omega + \mu - E_1(\mathbf{k}))\,(\omega + \mu - E_2(\mathbf{k}))} = \frac{\omega+\mu-U(1-n/2)}{E_1(\mathbf{k}) - E_2(\mathbf{k})}\left\{\frac{1}{\omega+\mu-E_1(\mathbf{k})} - \frac{1}{\omega+\mu-E_2(\mathbf{k})}\right\}, \tag{7.34}$$

where

$$E_{1,2}(\mathbf{k}) = \frac{U + \varepsilon(\mathbf{k})}{2} \mp \sqrt{\left(\frac{U - \varepsilon(\mathbf{k})}{2}\right)^2 + U\varepsilon(\mathbf{k})\frac{n}{2}}\,, \tag{7.35}$$

and $\varepsilon(\mathbf{k})$ is the $\mathbf{k}$-dependence of electron energy in the absence of correlations ($U = 0$). We do not consider any spin ordering and assume that the numbers of electrons with each spin projections are equal, $n_\sigma = n_{-\sigma} = n/2$. In the case of simple cubic lattice, we have $\varepsilon(\mathbf{k}) = -2t^{ab}(\cos k_x + \cos k_y + \cos k_z)$, where for k_i, we use the units of inverse lattice constant d.

The charge carrier density can be determined using Green's functions (7.34)

$$Cn = -i\int \frac{d\omega}{\pi}\int \frac{d^3\mathbf{k}}{(2\pi)^3} G_\sigma(\mathbf{k}, \omega + i0\ \mathrm{sign}\omega)e^{i\omega 0}. \tag{7.36}$$

Changing in (7.34) ω by $\omega + \mu$ and using spectrum (7.35), we find

$$n = 2\int_{\tilde{w}}^{\mu} d\omega \int \frac{d^3\mathbf{k}}{(2\pi)^3} \frac{U\,(1 - n/2) - \omega}{\sqrt{(U - \varepsilon(\mathbf{k}))^2 + 2\varepsilon(\mathbf{k})nU}} \left\{\delta\left(\omega - E_1(\mathbf{k})\right) - \delta\left(\omega - E_2(\mathbf{k})\right)\right\}, \tag{7.37}$$

where $\tilde{w} = \tilde{w}(n)$ is the bottom of the lower band. When the lower band is not fully occupied filled, the second δ-function in (7.37) does not work. Contribute to the result. In fact, Eq. (7.37) allows us to find the dependence of chemical potential on the doping level n. The kinetic energy per lattice site has the form

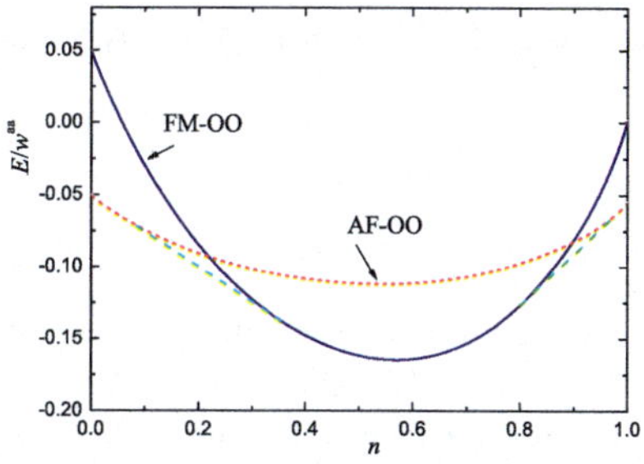

Fig. 7.5 Dependence of the energies of homogeneous ferro-OO (blue solid line) and antiferro-OO (red dashed line) states on the doping level$n$. Green dashed lines describe the energy of an inhomogeneous state. The plots correspond to the following choice of parameters: $t^{ab}/t^{aa} = 0.3$, $U/t^{aa} = 3$, and $J/t^{aa} = 0.1$ [15]

$$E_{\text{kin}} = 2\int_{\tilde{w}}^{\mu} \omega d\omega \int \frac{d^3\mathbf{k}}{(2\pi)^3} \frac{U(1-n/2)-\omega}{\sqrt{(U-\varepsilon(\mathbf{k}))^2 + 2\varepsilon(\mathbf{k})nU}} \{\delta(\omega - E_1(\mathbf{k})) - \delta(\omega - E_2(\mathbf{k}))\} . \tag{7.38}$$

After that, we can calculate the total energy of the homogeneous antiferro-OO state

$$E_{AFOO} = E_{kin}(t^{ab}) - Jz/2. \tag{7.39}$$

The plot of such energy as a function of doping n is shown in Fig. 7.5, where it is compared to the corresponding plot for the ferro-OO state (see below).

In the case of a homogeneous ferro-OO state (the charge carriers occupy similar orbitals), the energy loss per site related to the unfavorable orbital ordering is $zJ/2$. However, the gain in kinetic energy is larger because the hopping integral between similar orbitals exceeds that between different orbitals ($t^{aa} > t^{ab}$). Here, the kinetic energy is described by the same expression as (7.38) but with the different hopping integral, t^{aa}. Thus, instead of Eq. (7.39), we have

$$E_{FMOO} = E_{kin}(t^{aa}) + Jz/2. \tag{7.40}$$

In Fig. 7.5, we illustrate the behavior of energies for the system in ferro-OO and antiferro-OO states within the $0 < n < 1$ doping range. We can clearly see that at some dopings, antiferro-OO is unfavorable and the ferro-OO becomes the ground state owing to a larger kinetic energy gain. In the vicinity of the crossover range, we can see that the homogeneous state turns out to be unstable toward the phase separation into a mixture of ferro-OO and antiferro-OO domains characterized by different values of charge carrier density (see green dashed lines in Fig. 7.5).

7.6 A Model with Charge Carriers Occupying e_g Orbitals: 2D Case

7.6.1 A Model

Let us now generalize the previous analysis to the case of some specific orbitals such as e_g orbitals on the square 2D lattice [15]. This situation is relevant, e.g., to such compounds as layered cuprates, (K_2CuF_4,) or manganites (La_2MnO_4 and $La_2Mn_2O_7$). As it was discussed above, any e_g orbital can be represented as a linear combination of two basis states $|a>=|x^2-y^2>$ and $|b>=|2z^2-x^2-y^2>$

$$|\theta>=\cos\frac{\theta}{2}\,|x^2-y^2>+\sin\frac{\theta}{2}\,|2z^2-x^2-y^2>\,. \tag{7.41}$$

In the case under study, the same charge carries are responsible both for the formation of orbital order and for the charge transfer. This means that the Hamiltonian of the model should involve all possible types of the intersite hopping integrals. Of course, such hopping integrals $t^{\alpha\beta}$ $(\alpha,\beta=a,b)$ should depend on the direction of hopping. We can write the corresponding Hamiltonian in the form

$$H_{el}=-\sum_{\langle\mathbf{nm}\rangle\alpha\beta\sigma} t^{\alpha,\beta}_{\mathbf{n}-\mathbf{m}} d^{\dagger}_{\mathbf{n}\alpha\sigma}d_{\mathbf{m}\beta\sigma}+\frac{U}{2}\sum_{\mathbf{n}\alpha\sigma} n_{\mathbf{n}\alpha\sigma}n_{\mathbf{n}\alpha\bar{\sigma}}+\frac{U}{2}\sum_{\mathbf{n}\alpha\sigma\sigma'} n_{\mathbf{n}\alpha\sigma}n_{\mathbf{n}\bar{\alpha}\sigma'}\,, \tag{7.42}$$

where for the hopping integrals, we have $t^{\alpha\beta}_{\mathbf{n}-\mathbf{m}}=t^{\alpha\beta}_x$ ($t^{\alpha\beta}_{\mathbf{n}-\mathbf{m}}=t^{\alpha\beta}_y$) for $\mathbf{n}-\mathbf{m}$ directed along the x (y) axis. As usual, in Eq. (7.42), $\bar{\sigma}$ and $\bar{\alpha}$ denote not σ and not α, respectively. The hopping integrals $t^{\alpha\beta}_{x,y}$ can be represented in the matrix form

$$(t_{x,y})^{\alpha\beta}=\frac{t_0}{4}\begin{pmatrix}3 & \mp\sqrt{3}\\ \mp\sqrt{3} & 1\end{pmatrix}. \tag{7.43}$$

Here, signs minus and plus correspond to x and y hopping directions, respectively. Further on, we limit ourselves to the case of large U/t_0 in Hamiltonian (7.42).

7.6.2 Antiferro-Orbital Ordering Near Half Filling

The case of one electron per site (half-filling) is of special interest. Here, we study the situation near half-filling, $x=1-n\ll 1$. It can be demonstrated that at $n=1$ and large U, Hamiltonian (7.42) has the insulating the ground state with the antiferro-orbital ordering. Indeed, the second order of perturbation theory in terms of t_0/U gives rise to the effective Hamiltonian in the form [1]

$$H_{OO} = J \sum_{\langle \mathbf{nm} \rangle} \left\{ \mathbf{s_n s_m} \left[\tau^z_{\mathbf{n}} \tau^z_{\mathbf{m}} - 2\tau^z_{\mathbf{n}} + 1 \pm 2\sqrt{3}\tau^z_{\mathbf{n}}\tau^x_{\mathbf{m}} \mp 2\sqrt{3}\tau^x_{\mathbf{m}} + 3\tau^x_{\mathbf{n}}\tau^x_{\mathbf{m}} \right] \right. \tag{7.44}$$

$$\left. + \frac{1}{2} \left[\tau^z_{\mathbf{n}} \tau^z_{\mathbf{m}} + 2\tau^z_{\mathbf{n}} \pm \sqrt{3}\tau^x_{\mathbf{m}} 2\sqrt{3}\tau^z_{\mathbf{n}}\tau^x_{\mathbf{m}} \pm 2\sqrt{3}\tau^x_{\mathbf{m}} + 3\tau^x_{\mathbf{n}}\tau^x_{\mathbf{m}} \right] \right\}, \tag{7.45}$$

where $J = t_0^2/U$, $\mathbf{s_n} = \sum_{\alpha\sigma\sigma'} d^\dagger_{\mathbf{n}\alpha\sigma} (\boldsymbol{\sigma})_{\sigma\sigma'} d_{\mathbf{n}\alpha\sigma'}/2$ is the operator determining electron spin at site $\mathbf{n}$, and $\tau^{z,x}_{\mathbf{n}} = \sum_{\alpha\beta\sigma} d^\dagger_{\mathbf{n}\alpha\sigma} (\sigma^{z,x})_{\alpha\beta} d_{\mathbf{n}\beta\sigma}/2$ describe the occupation of orbitals, and $\boldsymbol{\sigma}$ are the Pauli matrices. Similarly to Eq. (7.43), signs minus and plus correspond to x and y directions of the $\mathbf{n} - \mathbf{m}$ bond, respectively. In the absence of spin ordering, $\mathbf{s_n s_m} = 0$, and the above Hamiltonian takes a simpler form

$$H_{OO} = \frac{J}{2} \sum_{\langle \mathbf{nm} \rangle} \left[\tau^z_{\mathbf{n}} \tau^z_{\mathbf{m}} + 2\tau^z_{\mathbf{n}} \pm 2\sqrt{3}\tau^z_{\mathbf{n}}\tau^x_{\mathbf{m}} \pm 2\sqrt{3}\tau^x_{\mathbf{m}} + 3\tau^x_{\mathbf{n}}\tau^x_{\mathbf{m}} \right]. \tag{7.46}$$

Hamiltonian Eq. (7.46) has the antiferro-OO ground state corresponding to the alternation of $|\theta_0>$ and $|-\theta_0>$ orbitals, where

$$\theta_0 = \arccos\left(-\frac{1}{4}\right). \tag{7.47}$$

The orbital ordering leads to the energy gain per lattice site $E_{OO} = -13J/4$.

At the hole doping, there appear some empty lattice sites. It is natural to assume that the long-range antiferro-orbital order is not destroyed at low doping levels, $x = 1 - n \ll 1$, but the energy gain related to the ordering becomes smaller. In the framework of the mean-field approximation, we can write $\langle \tau^{z,x}_{\mathbf{n}} \rangle \propto 1 - x$ and

$$E_{OO} = -\frac{13 + 11x^2}{4} J + 6J\,x. \tag{7.48}$$

To determine the kinetic energy of mobile holes in antiferro-OO structure, we can define the creation $b^\dagger_{\mathbf{n}\sigma}$ and annihilation $b_{\mathbf{n}\sigma}$ operators for such holes

$$b_{\mathbf{n}\sigma} = \cos\frac{\theta_0}{2} d^\dagger_{na\sigma} \pm \sin\frac{\theta_0}{2} d^\dagger_{nb\sigma}, \quad b^\dagger_{\mathbf{n}\sigma} = (b_{\mathbf{n}\sigma})^\dagger, \tag{7.49}$$

where signs plus and minus sign correspond to the sites with $|\theta_0>$ and $|-\theta_0>$ orbitals, respectively. Then, the kinetic energy described by the first term in Hamiltonian (7.42)) can be written as

$$H_{el} = \frac{t_0}{8} \sum_{\langle \mathbf{nm} \rangle \sigma} \left(P^\dagger_{\mathbf{n}\sigma} b^\dagger_{\mathbf{n}\sigma} b_{\mathbf{m}\sigma} P_{\mathbf{m}\sigma} + h.c. \right), \tag{7.50}$$

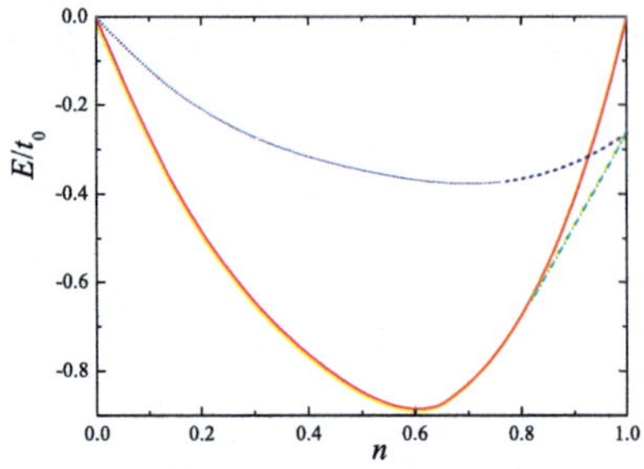

Fig. 7.6 Dependence of energies on the total number n of electrons for of orbitally disordered (red solid curve) and antiferro-OO (dashed blue curve) states. $J/t_0 = 0.08$. The green dot-and-dash line shows the energy of a possible phase-separated state [15]

where we projection operators P are introduced to exclude a double occupation of each sites. Let us analyze the situation with the 2D square lattice. Then, the energy spectrum of free holes is

$$\varepsilon(\mathbf{k}) = -\frac{t_0}{4}\left(\cos k_x + \cos k_y\right) . \tag{7.51}$$

In this case, the energy of the doped antiferro-OO array can be written as

$$E_{AFOO}(x) = -\frac{13 + 11x^2}{4}J + \int\limits_{-t_0/2}^{\mu(x)} dE\, E\rho_0(E) + 6Jx\,, \tag{7.52}$$

where the density of states $\rho_0(E)$ for holes in given by the expression

$$\rho_0(E) = \int \frac{d^2\mathbf{k}}{(2\pi)^2}\delta(E - \varepsilon(\mathbf{k})) = \int\limits_0^\infty \frac{ds}{\pi}\cos(Es)J_0^2\left(\frac{t_0 s}{4}\right). \tag{7.53}$$

The chemical potential μ is determined by the condition $x = \int_{-t_0/2}^{\mu(x)} dE\rho_0(E)$. The calculated energy (7.52) versus $n = 1 - x$ is plotted in Fig. 7.6, see the blue dashed curve and compared to the energy of the orbitally disordered state analyzed below.

7.6.3 *Orbitally Disordered State*

Now, let us discuss the situation characteristic of relatively high doping levels x, the orbital ordering discussed above disappears. We shall consider the state corresponding to Hamiltonian (7.42) in the absence of the long-range OO at the electron densities $0 < n < 1$ and compare the corresponding energy (7.52). The one-particle Green have the form similar to that considered above in Sect. 7.5.2 and analyze them using the Hubbard I approximation [16]. The corresponding equations of motion can be written as

$$(\omega + \mu)G_{\alpha\beta}(\mathbf{n}, \omega) = g_\alpha \left(1 - \sum_{\Delta\gamma} t_\Delta^{\alpha\gamma} G_{\gamma\beta}(\mathbf{n} + \Delta, \omega)\right), \tag{7.54}$$

where

$$g_\alpha = 1 - \frac{n_\alpha}{2} - n_{\bar{\alpha}}, \tag{7.55}$$

and the summation is taken over sites $\mathbf{m}$, which are the neighbors of the chosen site $\mathbf{n}$, ($\boldsymbol{\Delta} = \mathbf{n} - \mathbf{m}$). We assume the absence of magnetic ordering, $n_{\alpha\uparrow} = n_{\alpha\downarrow} = n_\alpha/2$ and work at in the limit of strong correlations, $U \to \infty$.

Passing to the momentum representation, we can write Eq. (7.54) as

$$(\omega + \mu)G_{\alpha\beta}(\mathbf{k}, \omega) = g_\alpha \left(1 - \sum_{\gamma} \varepsilon^{\alpha\gamma}(\mathbf{k}) G_{\gamma\beta}(\omega, \mathbf{k})\right), \tag{7.56}$$

where $\varepsilon^{\alpha,\beta}(\mathbf{k})$ is represented in the matrix form

$$\hat{\varepsilon}(\mathbf{k}) = \frac{-t_0}{2} \begin{pmatrix} 3(\cos k_x + \cos k_y) & \sqrt{3}(\cos k_y - \cos k_x) \\ \sqrt{3}(\cos k_y - \cos k_x) & \cos k_x + \cos k_y \end{pmatrix}. \tag{7.57}$$

Then, introducing the matrix

$$\hat{S} = \sqrt{\hat{g}} = \begin{pmatrix} \sqrt{g_a} & 0 \\ 0 & \sqrt{g_b} \end{pmatrix}, \tag{7.58}$$

it is possible to represent Green's function in the compact operator form

$$\hat{G} = \hat{S} \frac{1}{\omega + \mu - \hat{\varepsilon}_g} \hat{S}, \tag{7.59}$$

where

$$\hat{\varepsilon}_g = \begin{pmatrix} g_a \varepsilon_{aa} & \sqrt{g_a g_b} \varepsilon_{ab} \\ \sqrt{g_a g_b} \varepsilon_{ab} & g_b \varepsilon_{bb} \end{pmatrix}. \tag{7.60}$$

The energy spectrum is determined by the solution of equation

$$\det(\hat{\varepsilon}_g - E) = 0. \tag{7.61}$$

Eventually, we find

$$E_{\pm 1}(\mathbf{k}) = -\frac{t_0(3g_a + g_b)}{4}\Bigg[(\cos k_x + \cos k_y) \pm\sqrt{(\cos k_x + \cos k_y)^2 - \frac{48 g_a g_b}{(3g_a + g_b)^2}\cos k_x \cos k_y}\,\Bigg]. \quad (7.62)$$

Using the quasiparticle spectrum (7.62), we can write the expression for Green's function

$$G_{\alpha\beta}(\omega, \mathbf{k}) = \sum_{s=\pm 1} \frac{\sqrt{g_\alpha g_\beta}\, v_\alpha^{(s)}(\mathbf{k}) v_\beta^{(s)}(\mathbf{k})}{\omega + \mu - E_s(\mathbf{k})}. \quad (7.63)$$

In Eq. (7.63), $v_\alpha^{(s)}$ are the matrix elements of the transformation determined by the $\hat{v}^{(s)}$ obeying the relations

$$\left[\hat{\varepsilon}_g(\mathbf{k})\hat{v}^{(s)}\right]_\alpha = E_s(\mathbf{k}) v_\alpha^{(s)}. \quad (7.64)$$

As a result of straightforward but cumbersome calculations, we find an explicit form of matrix $\hat{v}^{(s)}$ corresponding to the basis orbitals $|b>=|2z^2 - x^2 - y^2>$ and $|a> = |x^2 - y^2>$

$$\hat{v}_\alpha^{(s)} = \frac{1}{\sqrt{(g_a\varepsilon_{aa} - E_s)^2 - g_a g_b \varepsilon_{ab}^2}} \begin{pmatrix} -\sqrt{g_b g_a}\varepsilon_{ab} \\ g_a\varepsilon_{aa} - E_s \end{pmatrix}, \quad (7.65)$$

where ε_{aa}, ε_{ab}, and ε_{ab} are matrix elements of matrix $-\hat{t}(\mathbf{k})$, see (7.57).

Using the above expressions, we can calculate the density of states corresponding to two new bands with indices $s = 1$ and -1

$$\rho^{(s)}(E) = \sum_\alpha \rho_{\alpha\alpha}^{(s)}(E), \quad (7.66)$$

where

$$\rho_{\alpha\alpha}^{(s)}(E) = \frac{1}{(2\pi)^2}\int d\mathbf{k}\left[v_\alpha^{(s)}(\mathbf{k})\right]^2 \delta(E - E_s(\mathbf{k})). \quad (7.67)$$

Then, we can determine the partial filling of orbital states

$$n_\alpha = 2g_\alpha \sum_{s=\pm 1} \int_{-\infty}^{\mu} \rho_{\alpha\alpha}^{(s)}(E)\, dE. \quad (7.68)$$

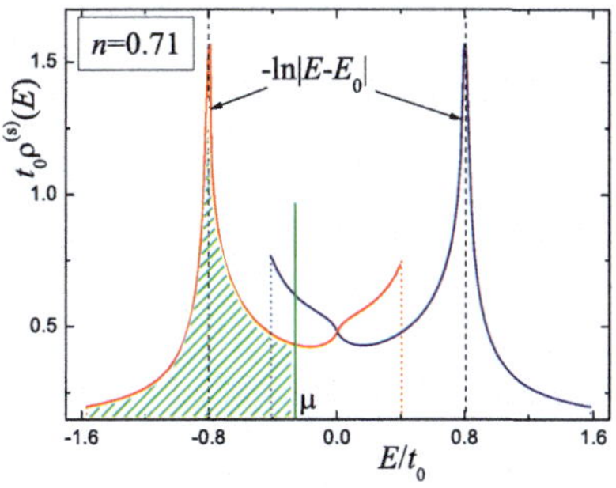

Fig. 7.7 Densities of states for bands $s = -1$ (red solid line) and $s = 1$ (blue solid line). The occupied states at the total number of electrons per site $n = 0.71$ are shown by the shaded area [15]

The chemical potential μ as can be obtained using the condition

$$n_a + n_b = 1 - x, \tag{7.69}$$

where x is the hole density (doping level). Substituting $\mu(n_a)$ to Eq. (7.68), we can determine the dependence of n_a and n_b on the doping level x. Then, the total energy corresponding to Hamiltonian (7.33) is given by the expression

$$E_{el} = \sum_{\alpha,s=\pm 1} (1 + g_\alpha) \int_{-\infty}^{\mu} \rho^{(s)}_{\alpha\alpha}(E) E dE. \tag{7.70}$$

In Fig. 7.7 we illustrate the behavior of the density of states for the new bands $s = \pm 1$, These densities of states exhibit the logarithmic van Hove singularity, which is related to the transition from the closed to open Fermi surface in each band (the evolution of the Fermi surface is represented in Fig. 7.9).

The behavior of occupation numbers n_a and n_b for $|a>$ and $|b>$ orbital states with the variation of the total number n of electrons per site is illustrated in Fig. 7.8. With the increase in n, first only $|a>=|x^2 - y^2>$ orbitals turn out to be occupied. This means that at high doping levels $1 - n$, we have the occupation of only one sort of orbitals (a kind of ferro-OO) and these $|x^2 - y^2>$ orbitals provide the largest gain in the kinetic energy of electrons. The further increase in n gives rise to electrons at $|b>=|2z^2 - x^2 - y^2>$ orbitals. Eventually, at $n = 1$, we get $n_a = 1/4 = \sin^2(\pi/6)$ and $n_b = 3/4 = \cos^2(\pi/6)$. This corresponds to the average occupation of orbitals at $n = 1$ equal to $|\theta = \pm\pi/3>$. However, the obtained results on the density of states and on the average occupation numbers n_a and n_b do not provide a conclusive evidence concerning the actual alteration of the $|\pm\theta>$ orbitals.

Evolution of the Fermi surface with the increase in the total number n of electrons is presented in Fig. 7.9.

At $n = 0.8$ (see Fig. 7.9e) the Fermi surface involves several flat portions. This is quite an interesting situation. In Fig. 7.8, we see that $n = 0.8$ corresponds to $n_a =$

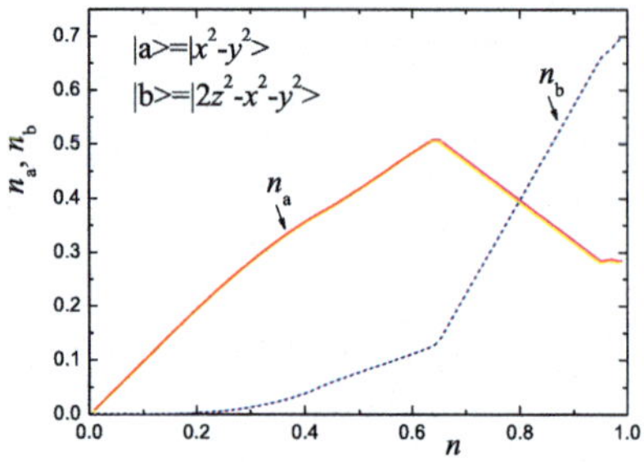

Fig. 7.8 Average occupation numbers of $|a>$ (red solid line) and $|b>$ (blue dashed line) orbitals versus total number n of electrons per site. A kink near $n \simeq 0.6$ corresponds to the doping level, at which the chemical potential crosses the bottom of the upper renormalized band with $s = 1$ [15]

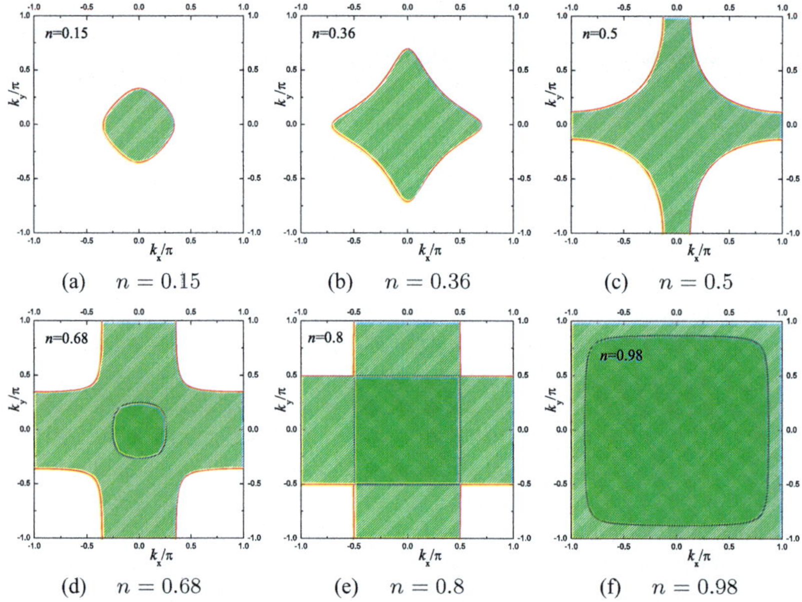

Fig. 7.9 Evolution of the Fermi surface with the increase in the total number of electrons n. The red solid curves indicate the Fermi surface sheet corresponding to the $s = -1$ band and the blue solid curves correspond to the $s = 1$ band [15]

$n_b = 0.4$, and hence, $g_a = g_b = 0.4$. Substituting $g_a = g_b$ into Eqs. (7.62) and (7.67), we can find out that in such a particular case, the quasiparticle spectrum and the corresponding density of states coincide with those for uncorrelated electrons up to a numerical factor. This means that for $n_a = n_b$, we are dealing with the compensation of two main correlation effects, namely, the changes in bandwidths and in number of states within the bands. Thus, we have a clear similarity to the case of half-filling ($n = 1$) for noninteracting electrons [17]. As well as in Ref. [17], at $n_a = n_b = 0.4$, there appears the nesting of Fermi surface sheets, which corresponds to the translation by vector $\mathbf{Q} = \{\pi, \pi\}$ implying a possibility of arising some orbital superstructure. Owing to the suppression of correlation effects at this specific doping level, the appearing superstructure could be the same as that discussed in Ref. [17], but, of course, this issue demands a special analysis.

For the case of the disordered state under study, dependence of energy on the total number n of electrons per site is shown in Fig. 7.6 (red solid line). It is interesting that the shape of this curve is similar to that obtained in Sect. 7.5.2 for the simplified model, see Fig. 7.5. In this case, the correlations themselves do not give rise to the phase separation. Nevertheless, the comparison of such orbitally disordered model with the model that discussed in Sect. 7.6.2 reveals that within a certain doping range (at small hole doping), the phase separation could provide a gain in energy. The energy of such a possible phase-separated state is shown by the green dot-and-dash line in Fig. 7.6.

7.7 Conclusions

In this chapter, we have analyzed a possibility of electronic phase separation in the systems with itinerant charge carriers interacting with an orbitally ordered array. It was shown that such itinerant charge carriers favor the formation of an inhomogeneous state with nanosize inhomogeneities having the orbital structure different from that in the host material. For the 2D lattice with e_g orbitals, we have found the shapes and sizes of such inhomogeneities. It is shown that the inhomogeneities have the shape strongly dependent on the ratio of the interorbital interaction constant and characteristic hopping integral for the charge carriers, J/t_0: there exists a threshold value of J/t_0, corresponding to the transition from the circular inhomogeneities to needle-like ones.

The orbital model under discussion has certain similarities with that corresponding to the double exchange. Here, the operators describing the orbital states are similar to the spin operator. In the previous chapters, we demonstrated that the double exchange model exhibits an instability toward phase separation into inhomogeneities with different types of magnetic structures. The inhomogeneous state with circular ferro-OO droplets is, in fact, similar to that involving magnetic polarons (ferromagnetic droplets within the antiferromagnetic host material. The latter state is inherent to the double exchange [4, 5, 10]. However, this similarity is not trivial: the orbital model has its own specific features owing to the contribution of nondiagonal hopping and the corresponding anisotropy. Such features give rise to the behavior specific of orbital systems, such as the kink in the energy of a homogeneous state and canted-OO needle-like droplets. Note also that the difference between spins and orbitals leads to many interesting effects, see, e.g., Refs. [18, 19].

Of course, in actual magnetic materials, we are often deal with the usual mechanism underlying the formation of magnetic polarons (in simple cases - ferromagnetic inhomogeneities in AF matrix), which ensures a gain in the kinetic energy of conduction electrons). However, even in this situation, the possible existence of orbital degeneracy and orbital ordering present, it is necessary to consider the orbital structure and its effect on the motion of charge carriers. For example, the magnetic ordering (parallel spins) could not impede the electron motion, but if it occurs in the presence of antiferro-OO (such as alternating $|2x^2 - y^2 - z^2\rangle$ and $|2y^2 - z^2 - x^2\rangle$ orbitals in the basal plane of manganites, which coexists with the ferromagnetic spin ordering in this plane), such antiferro-OO itself should affect the motion of itinerant charge carriers. Therefore, this favors the formation of orbital polarons even at the ferromagnetic spin background. This implies that in the compounds with JT ions, the ferromagnetism does not exclude the formation of orbital inhomogeneities. Such reasoning is in agreement with the Monte Carlo simulations for the two-orbital model reported in Ref. [20], which demonstrate the possibility of nanoscale phase separation even in the case of ferromagnetism. In this chapter, we have not considered any magnetic structure and spin ordering of charge carriers A due account taken for the spin degrees of freedom can give rise to a rich variety of inhomogeneities with different orbital and spin structures.

In the first part of this chapter, we considered the localized electrons forming an orbital order, whereas conduction electrons or holes were supposed to be two different groups of electrons. However, we have demonstrated that our main results are also valid for a model, where the same electrons take part both in the hopping and in the formation of orbitally ordered structure. For example, in magnetic oxides with Jahn–Teller ions, lattice distortions lift the orbital degeneracy, this leading to the formation of the ground state with an orbital ordering at $n = 1$. In this case, it is natural to assume the long-range orbital ordering does not disappear at low doping levels, $x = 1 - n \ll 1$. Thus, we are dealing with holes at an orbitally ordered background. Using the mean-field approximation for the analysis of orbitally ordered arrays, we should introduce the factor $1 - x$, thus obtaining the effective interorbital exchange interaction $J(1 - x)^2$. In the actual compounds with Jahn–Teller ions, the orbital Hamiltonians have more complicated forms than the Heisenberg-like one. Calculations involving the superexchange mechanism of the orbital ordering [1] show that the obtained results still remain valid.

In real substances, at high doping level, we can expect an orbitally disordered state or an orbital liquid. For manganites, for example, such a state was discussed in Ref. [21]. This problem is rather controversial, and other possibilities have been also addressed (e.g., complex orbitals [22]). In the usual spin systems, a disordered state is as a rule more favorable from the viewpoint of kinetic energy than the AF one (or in our case, the antiferro-OO state). However, the detailed analysis of spin systems undertaken in Refs. [3–5] clearly demonstrates that even in a disordered (i.e., paramagnetic) state, an additional energy gain can be achieved due to the formation a ferromagnetic polaron ("ferron") even at the paramagnetic background. It is natural to expect a similar possibility for the systems with orbital degrees of freedom, although this situation demands further analysis. To perform it accurately, it is necessary to take into account unusual specific features of the electron motion within disordered media. In particular, one should deal with the disorder, which adjusts itself in the course of electron motion. This is quite interesting and complicated separate task.

References

1. K.I. Kugel, D.I. Khomskii, Sov. Phys. Usp. **25**, 231 (1982)
2. M.D. Kaplan, B.G. Vekhter, *Cooperative Phenomena in Jahn-Teller Crystals* (Springer Science & Business Media, 2012)
3. E. Dagotto, *Nanoscale Phase Separation and Colossal Magnetoresistance: The Physics of Manganites and Related Compounds*. Springer Series in Solid-State Sciences, vol. 136 (Springer Science & Business Media, 2003)
4. E.L. Nagaev, *Colossal Magnetoresistance and Phase Separation in Magnetic Semiconductors* (World Scientific, 2002)
5. M.Yu. Kagan, K.I. Kugel, Phys. Usp. **44**(6), 553 (2001)
6. R. Kilian, G. Khaliullin, Phys. Rev. B **60**(19), 13458 (1999)
7. G. Khaliullin, S. Okamoto, Phys. Rev. Lett. **89**(16), 167201 (2002)
8. J. van den Brink, G. Khaliullin, D. Khomskii, in *Colossal Magnetoresistive Manganites*, ed. by T. Chatterji (Kluwer, 2004), Chap. 6, pp. 263–302
9. P.G. de Gennes, Phys. Rev. **118**(1), 141 (1960)
10. M.Yu. Kagan, D.I. Khomskii, M.V. Mostovoy, Eur. Phys. J. B **12**(2), 217 (1999)
11. K.I. Kugel, A.L. Rakhmanov, A.O. Sboychakov, D.I. Khomskii, Phys. Rev. B **78**(15), 155113 (2008)
12. J. Kanamori, J. Appl. Phys. **31**(5), S14 (1960)
13. D. Khomskii, J. van den Brink, Phys. Rev. Lett. **85**(15), 3329 (2000)
14. C. Ortix, J. Lorenzana, C. Di Castro, Phys. Rev. Lett. **100**(24), 246402 (2008)
15. K.I. Kugel, A.O. Sboychakov, D.I. Khomskii, J. Supercond. Novel Magn. **22**(2), 147 (2009)
16. J. Hubbard, Proc. Roy. Soc. A **205**, 238 (1963)
17. D. Efremov, D.I. Khomskii, Phys. Rev. B **72**(1), 012402 (2005)
18. M. Daghofer, K. Wohlfeld, A.M. Oleś, E. Arrigoni, P. Horsch, Phys. Rev. Lett. **100**(6), 066403 (2008)
19. Z. Nussinov, G. Ortiz, EPL (Euro Phys. Lett.) **84**(3), 36005 (2008)
20. C. Şen, G. Alvarez, H. Aliaga, E. Dagotto, Phys. Rev. B **73**(22), 224441 (2006)
21. S. Ishihara, M. Yamanaka, N. Nagaosa, Phys. Rev. B **56**(2), 686 (1997)
22. J. van den Brink, D. Khomskii, Phys. Rev. B **63**(14), 140416 (2001)

Chapter 8
Systems with Imperfect Nesting and Phase Separation

Abstract In the framework of the Rice model for chromium and its alloys, we analyze the instability toward the phase separation in the systems with imperfect nesting both in the absence and in the presence of the applied magnetic field. We consider also the phase separation in organic metals under pressure, in iron-based pnictide superconductors, and the competition between the inhomogeneous phases in the anisotropic Hubbard models.

8.1 Introduction

Electronic models having a band structure with the Fermi surface nesting are frequently used to analyze physical properties of a number of the multiband materials [1]. The nesting, that is, existence of two fragments of the Fermi surface, which can be matched upon translation by a certain reciprocal lattice vector (nesting vector $\mathbf{Q}_0$), entails an instability of a Fermi-liquid state. The instability leads to appearance of a superstructure or additional order parameter related to the nesting vector. The notion of nesting is often invoked for the analysis of charge density wave (CDW) states [2, 3], spin density waves (SDW) states [4, 5], mechanisms of high-T_c superconductivity [6–8], fluctuating charge/orbital modulation in magnetic oxides [9], chromium and its alloys [10–13], organic metals [14–17], etc. In general, various types of spatially inhomogeneous phases with free energies close to each other emerge in the system with nesting [18]. In this chapter we discuss why the multiband systems with the Fermi surface nesting exhibit the tendency to electronic phase separation.

In real materials, the Fermi surface fragments can match only approximately and in such case nesting may be referred to as imperfect. The concept of the imperfect nesting was introduced in the 60–70s of the last century and one of the earliest studies of imperfect nesting was performed by Rice [13] in the context of chromium and its alloys (see also review articles [19, 20]). In later works it was argued that the imperfect nesting can be responsible for the nanoscale phase separation in chromium alloys [21], iron-based superconductors [22], organic metals (Bechgaard salt) [23], and in doped bilayer graphene [24, 25]. The spin and charge inhomogeneities related

M. Yu. Kagan et al., *Electronic Phase Separation in Magnetic and Superconducting Materials*, Springer Series in Solid-State Sciences 201,
https://doi.org/10.1007/978-3-031-55467-4_8

to the mechanism of imperfect nesting are also considered in the physics of low-dimensional compounds [14, 26–28].

Inhomogeneities in the phase separated state can form different configurations, e.g., "stripes", domain walls, etc. Such structures were also discussed in the literature in the framework of similar models [29]. Moreover, it was shown that the possibility of SDW ordering in the systems with itinerant charge carriers gives rise to very rich and complicated phase diagrams involving phase-separation regions [18, 30–32].

Physical mechanisms underlying the nucleation of the inhomogeneous state in materials with nesting are, in general, common to other systems considered in this book. The electronic spectrum instability related to the nesting gives rise to an order parameter and, consequently, to lowering of the system free energy. The better is nesting, the larger is this energy gain. Thus, it may be favorable for the system with imperfect nesting to break up into two phases having different density of itinerant electrons with the better and worse (or even absent) nesting.

8.2 Phase Separation in the Rice Model for Chromium and Its Alloys

We start with the model proposed by Rice [13] to describe the incommensurate antiferromagnetism in chromium alloys (see also review article [20]). This model is rather simple and allows for a detailed examination of its phase diagram. Despite its simplicity, it can be applied to describe real important systems. In our consideration, we mostly follow Ref. [21].

The model band structure corresponds to one spherical electron pocket and one spherical hole pocket with different radii (imperfect nesting). It includes as well, another band or bands, which do not participate in the magnetic ordering and serve as a reservoir of electrons. All interactions are ignored except the repulsion between electrons and holes in the Fermi surface pockets giving rise to the ordering, since even a small coupling in this channel generates a band instability and opening a gap in the electron spectrum. The Hamiltonian of such model has the form

$$\hat{H} = \sum_{\mathbf{k},\sigma,\alpha} \epsilon^{\alpha}(\mathbf{k}) n^{\alpha}_{\mathbf{k}\sigma} + \frac{V}{\mathcal{V}} \sum_{\mathbf{k},\mathbf{k}',\mathbf{q},\sigma,\sigma'} a^{\dagger}_{\mathbf{k}+\mathbf{q}\sigma} a_{\mathbf{k}\sigma} b^{\dagger}_{\mathbf{k}'-\mathbf{q}\sigma'} b_{\mathbf{k}'\sigma'} + \text{H.c.}, \tag{8.1}$$

where α is equal to either a (electron pocket), b (hole pocket), or c (nonmagnetic bands), $a^{\dagger}$ ($b^{\dagger}$) are creation operators for an electron in the a (b) pocket, n is the number operator, V is the Coulomb interaction, and $\mathcal{V}$ is the volume. The nonmagnetic c band has a finite density of states N_r at the Fermi energy. The energy spectra for the electron and hole pockets measured relative to the Fermi energy are taken as

$$\begin{aligned} \epsilon^{a}(\mathbf{k}) &= \hbar v_F(k - k_{Fa}) = \hbar v_F(k - k_F) - \mu, \\ \epsilon^{b}(\mathbf{k} + \mathbf{Q}_0) &= -\hbar v_F(k - k_{Fb}) = -\hbar v_F(k - k_F) - \mu, \end{aligned} \tag{8.2}$$

where $k_F = (k_{Fa} + k_{Fb})/2$, $\mu = \hbar\, v_F(k_{Fa} - k_{Fb})/2$ is the chemical potential, and the wave vector $\mathbf{Q}_0$, which is commonly called the nesting vector, connects the centers of the electron and hole pockets in the reciprocal space. We consider the weak-coupling regime: $VN_m \ll 1$, where $N_m = k_F^2/2\pi^2\hbar v_F$. We treat Hamiltonian (8.1) using a mean-field approach, which gives an accurate result for weakly interacting electrons. In the case of perfect nesting, which corresponds to $\mu = 0$ in our notations, the radii of the electron and hole pockets are identical. If we translate the electron pocket by nesting vector $\mathbf{Q}_0$, its Fermi sphere coincides perfectly with the Fermi sphere of the hole pocket. If we perform the transformations, $b_{\mathbf{k}} \to b^\dagger_{\mathbf{k}}$, $b^\dagger_{\mathbf{k}} \to b_{\mathbf{k}}$, the interaction constant V and the hole pocket dispersion in Eq. (8.2) change sign. As a result, the Hamiltonian for a and b bands becomes identical to two copies of the BCS Hamiltonian. This mapping is useful since it allows us to use the familiar BCS mean-field approach to study the model.

Standard BCS-like calculations readily show that in the case of ideal nesting, $\mu = 0$, the system is unstable with respect to the ordering with the antiferromagnetic order parameter $\Delta_0 = \frac{V}{\mathcal{V}} \sum_{\mathbf{k}} \langle a^\dagger_{\mathbf{k}\sigma} b_{\mathbf{k}+\mathbf{Q}_0 -\sigma}\rangle$. In the weak-coupling limit under study, we have

$$\Delta_0 = \epsilon_F \exp\left(-1/N_m V\right) \ll \epsilon_F, \tag{8.3}$$

where $\epsilon_F = \hbar v_F k_F$. The order parameter Δ_0 couples electrons with unequal momenta $\mathbf{k}$ and $\mathbf{k} + \mathbf{Q}_0$. Consequently, in real space, the order parameter Δ_0 corresponds to the rotation of the magnetization axis with wave vector $\mathbf{Q}_0$. Usually, the a and b pockets are located at the high-symmetry points of the Brillouin zone and the vector $\mathbf{Q}_0$ is related to the underlying lattice structure. Thus, this order is usually called commensurate.

When $\mu \neq 0$, the electron and hole Fermi spheres have different radii, and do not coincide upon translation. However, the difference between the translated spheres remains small, if μ is small. In this case, the system free energy can be optimized by treating the translation vector $\mathbf{Q}_1 = \mathbf{Q}_0 + \mathbf{Q}$ as a variation parameter. The SDW order parameter has the form

$$\Delta = \frac{V}{\mathcal{V}} \sum_{\mathbf{k}} \langle a^\dagger_{\mathbf{k},\sigma} b_{\mathbf{k}+\mathbf{Q}_1,-\sigma}\rangle. \tag{8.4}$$

The vector $\mathbf{Q}$ is small: $|\mathbf{Q}| \ll |\Delta|/\hbar v_F \sim |\mathbf{Q}_0|$. Hence, the order parameter Δ describes an ordering with a slowly rotating antiferromagnetic magnetization axis. This rotation is unrelated to the underlying lattice, and such order is referred to as incommensurate.

If the transformation $b_{\mathbf{k}} \to b^\dagger_{\mathbf{k}}$, $b^\dagger_{\mathbf{k}} \to b_{\mathbf{k}}$ is applied to a system with nonzero μ, the Hamiltonian of our magnetic system becomes similar to the Fulde–Ferrel–Larkin–Ovchinnikov (FFLO) Hamiltonian [33, 34] of a superconductor in the Zeeman magnetic field. Note that we can introduce the order parameters of slightly different structure than in Eq. (8.4). For example, Larkin and Ovchinnikov in their original paper [34] considered the order parameter $(\Delta e^{i\mathbf{Q}_1\mathbf{r}} + \Delta^* e^{-i\mathbf{Q}_1\mathbf{r}})$, which periodically passes through zero in real space. However, the general result does not change significantly (for details see, e.g., Refs. [21, 34, 35]). In particular, the difference in the free energy and in the chemical potential calculated for different forms of the order parameter is small, and in actual systems, the balance may be shifted by factors unaccounted for by the present model (anisotropy, disorder, etc.) [21].

In the model under discussion, it is convenient to calculate equilibrium parameters of the system by minimization of the grand thermodynamic potential

$$\Omega = -T \ln \left[\mathrm{Tr}\, e^{-(\hat{H}-\mu\hat{N})} \right], \tag{8.5}$$

where $\hat{N}$ is the operator of the total number of particles. To evaluate Ω in the mean-field approximation, we need the eigenenergies of Hamiltonian (8.1) in the mean-field approach

$$\hat{H}_{\mathrm{mf}} = \sum_{\mathbf{k},\sigma,\alpha} \epsilon^{\alpha}(\mathbf{k}) n^{\alpha}_{\mathbf{k}\sigma} + \Delta \sum_{\mathbf{k},\sigma} a_{\mathbf{k}\sigma} b^{\dagger}_{\mathbf{k}+\mathbf{Q}_1-\sigma}.$$

These are

$$E_{1,2}(\mathbf{k}) = \frac{\epsilon^{a}(\mathbf{k}) + \epsilon^{b}(\mathbf{k}+\mathbf{Q}_1)}{2} \pm \sqrt{\Delta^2 + \left[\frac{\epsilon^{a}(\mathbf{k}) - \epsilon^{b}(\mathbf{k}+\mathbf{Q}_1)}{2}\right]^2}. \tag{8.6}$$

Then, the grand potential equals

$$\frac{\Omega}{\mathcal{V}} = \frac{2\Delta^2}{V} - 2T \sum_{s} \int \frac{d^3\mathbf{k}}{(2\pi)^3} \ln\left(1+e^{-E_s/T}\right) - 2T N_r \int \ln\left(1+e^{-(\epsilon-\mu)/T}\right) d\epsilon. \tag{8.7}$$

Here, the first and the second terms are the contributions of the bands responsible for the SDW ordering, while the third term corresponds to the nonmagnetic bands. To carry out the integration over $\mathbf{k}$, we expand the band energies in powers of $|\mathbf{Q}|$ and $\delta k = |\mathbf{k}| - k_F$

$$\epsilon^{b}(\mathbf{k}+\mathbf{Q}_1) + \epsilon^{a}(\mathbf{k}) = -2\mu - \hbar v_F Q\eta, \quad \epsilon^{b}(\mathbf{k}+\mathbf{Q}_1) - \epsilon^{a}(\mathbf{k}) = -2\hbar v_F \delta k - \hbar v_F Q\eta, \tag{8.8}$$

where η is the cosine of the angle between $\mathbf{k}$ and $\mathbf{Q}_1$. Performing the integration, one finds the following expression [21]

$$\delta\Omega = \Omega(\Delta, Q, \mu, T) - \Omega(0, Q, \mu, T) = \frac{k_F^3 \mathcal{V}}{\pi^2 \epsilon_F} \tag{8.9}$$

$$\left\{ \Delta^2 \left(\ln \frac{\Delta}{\Delta_0} - \frac{1}{2} \right) + \frac{Q^2}{3} + \mu^2 + \frac{\pi^2 T^2}{3} + T \int_0^{\infty} d\xi \int_{-1}^{1} d\eta \ln\left[f(Q\eta-\mu-\epsilon) f(\mu-Q\eta-\epsilon) \right] \right\},$$

where $f(\epsilon) = 1/(1+\exp \epsilon/T)$ is the Fermi function and $\epsilon = \sqrt{\Delta^2 + \xi^2}$.

The equations for the equilibrium values of the gap Δ and the magnitude of the structural vector Q are determined by minimizing $\delta\Omega$ [21]

$$\ln\frac{\Delta}{\Delta_0} = \int_0^{\infty}\frac{d\xi}{2\epsilon}\int_{-1}^{1}d\eta\,[f(\epsilon+\mu-Q\eta)+f(\epsilon-\mu+Q\eta)]\,,$$

$$\frac{2Q}{3} = -\int_0^{\infty}d\xi\int_{-1}^{1}\eta d\eta\,[f(\epsilon+\mu-Q\eta)+f(\epsilon-\mu+Q\eta)]\,. \quad (8.10)$$

The total number of electrons per unit volume $n(\mu)$ is the sum of the electrons from magnetic $n_m(\mu)$ and nonmagnetic bands $n_r(\mu)$. The doping x is defined as the difference $x = n(\mu) - n(0)$. Since $T, \mu \ll \epsilon_F$, we can write $n_r(\mu) - n_r(0) = N_r\mu$. For magnetic electrons, we have $n_m(\mu) = (2/\mathcal{V})\sum_{\mathbf{k},s} f(E_s(\mathbf{k}))$. After straightforward calculations, we derive

$$\frac{x}{x_0} = \frac{n\mu}{\Delta_0} + \int_0^{\infty}\frac{d\xi}{\Delta_0}\int_{-1}^{1}d\eta\,[f(\epsilon-\mu+Q\eta)-f(\epsilon+\mu-Q\eta)]\,, \quad (8.11)$$

where $x_0 = 4\Delta_0 N_m$ and $n = N_r/2N_m$.

The phase diagram of the model can be constructed by numerical solution of Eqs. (8.9), (8.10), and (8.11). First, one can calculate the Néel temperature $T_N(x)$ for homogeneous state of the system and the transition temperature T_Q between the commensurate and incommensurate antiferromagnetic phases, which corresponds to the highest doping, at which $Q = 0$. Then, we can determine the region of the phase-separated state and the phases in this state analyzing the dependence of the grand potential Ω or chemical potential μ versus doping x following the standard procedure described in the previous sections (see also for details Ref. [21]).

As an illustration, a typical dependence of the chemical potential μ on doping x for the considered model is shown in Fig. 8.1. It is seen from this figure that the function $\mu(x)$ has a descending part, which means a negative compressibility and, hence, instability of the homogeneous antiferromagnetic phase. In the phase-separated state, the system segregates into two phases with different doping values x_1 and x_2. The values x_1 and x_2 can be found using the Maxwell construction (see, e.g., Ref. [36]). Figure 8.1 illustrates the latter concept: the horizontal dashed line is drawn in such a manner that the areas of the shaded regions, S_1 and S_2, are equal to each other.

The calculated by the latter method boundary $T_{PS}(x)$ between the homogeneous and phase-separated states is shown by the dashed (red) curves in the (x, T) phase diagrams drawn in Fig. 8.2 for different values of n. The phase with lower doping, x_1 is the commensurate antiferromagnetic ($Q = 0$), whereas the phase with higher doping,

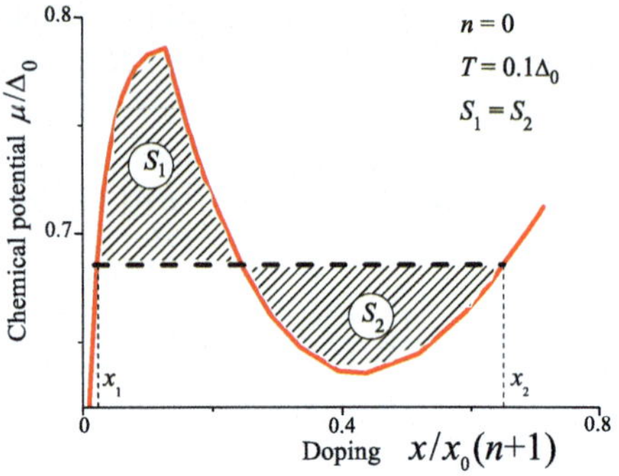

Fig. 8.1 Chemical potential μ versus doping x for the homogeneous phase, $T/\Delta_0 = 0.1$ and $n = 0$ [solid (red) line]. The horizontal dashed (black) line shows the Maxwell construction, with shaded areas $S_1 = S_2$ [21]

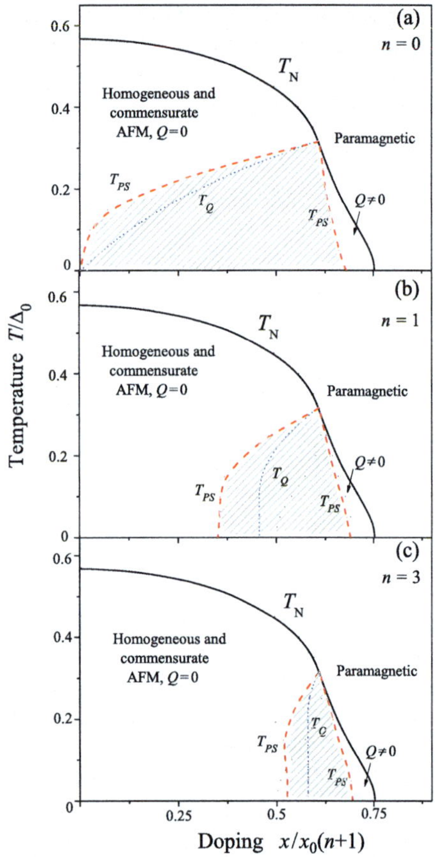

Fig. 8.2 Phase diagram of the Rice model for $n = 0$ (no electrons in nonmagnetic pockets) (**a**), $n = 1$ (**b**), and $n = 3$ (**c**). The solid (black) curves represent the Néel temperature $T_N(x)$, which separates the paramagnetic ($\Delta = 0$) and antiferromagnetic ($\Delta \neq 0$) phases. The dotted (blue) curves denote $T_Q(x)$, i.e. the boundary between the commensurate ($Q = 0$) and incommensurate ($Q \neq 0$) homogeneous antiferromagnetic phases. The dashed (red) curve, $T_{PS}(x)$, is the boundary between the uniform and phase-separated (shaded areas) phases [21]

x_2, is the incommensurate antiferromagnetic ($Q \neq 0$). Thus, the phase separation occurs due to the competition between two antiferromagnetic states with different structures.

Above, we have mentioned the mapping between the Rice model and the FFLO superconductor. However, the phase separation is absent from the phase diagram of the FFLO state in a superconductor. This has a very simple explanation: the Zeeman field in the FFLO superconductor corresponds to doping in the Rice model. The field, being an intensive thermodynamic quantity, does not allow for the phase separation. However, for cold atoms in an optical trap (where an analogue of the FFLO is predicted), it may be possible to control not the field, but the polarization, which is an extensive quantity. In this case, the phase separation occurs [37].

The FFLO state is very fragile with respect to the impurity scattering [38]. Thus, the incommensurate antiferromagnetic, which is the mathematical analog of the FFLO phase, is expected to be susceptible to microscopic imperfections. Therefore, the disorder-induced modifications to the phase diagram is an open question.

The study of characteristic scales and geometry of the phase-separated state could be performed by the method similar to that described in the previous sections of the book. It requires, as usual, to take into consideration an interplay between the long-range Coulomb interaction and the energy of the boundary between different phases.

8.3 The Effect of Magnetic Field on the Phase Separation in Systems with Imperfect Nesting

As we see, the electron–electron coupling gives rise to the nucleation of the magnetically ordered states and the phase separation in the system with nesting. Naturally, the magnetically ordered system is especially sensitive to the applied magnetic field. This problem was considered in detail in Ref. [39] within the framework of the presented above Rice model. We analyze here the effects of the applied magnetic field following the latter paper.

As usual, considering electrons or holes in the applied magnetic field $\mathbf{B}$, we should, first, replace the momentum operator $\hat{\mathbf{p}}$ by the gauge-invariant one $\hat{\mathbf{p}} + (e/c)\mathbf{A}$, where $\mathbf{A}$ is the vector potential of magnetic field. Then, we have to add the Zeeman term $g_\alpha \sigma \hbar\omega_\alpha$ in the single electron Hamiltonian, where g_α is the Landé g-factor and $\omega_\alpha = eB/cm_\alpha$ is the cyclotron frequency of a charge carrier with mass m_α. The energy spectrum of the electrons and holes is characterized now by two single-particle energy scales. The first scale is the Fermi energy ϵ_F, and the second one is $\hbar\omega_\alpha$, which is the distance between the Landau levels. The energy scale associated with the interactions will be characterized by the value of a spectral gap Δ_0 and in the case of the weak electron–hole coupling $\Delta_0 \ll \epsilon_F$. The Landau quantization is of importance in the range of high magnetic field, $\hbar\omega_\alpha > \Delta_0$, whereas at low fields, $\hbar\omega_\alpha < \Delta_0$, it can be neglected. Here, we limit ourselves by the regime of low magnetic fields, neglecting any corrections associated with the small ratio $\hbar\omega_\alpha/\epsilon_F \ll 1$. In the range

Fig. 8.3 Band structure of the model in an applied magnetic field. The applied magnetic field lifts the degeneracy of electron-like a and hole-like b bands with respect to the electron spin. The red arrows indicate the interband coupling giving rise to the order parameters [39]

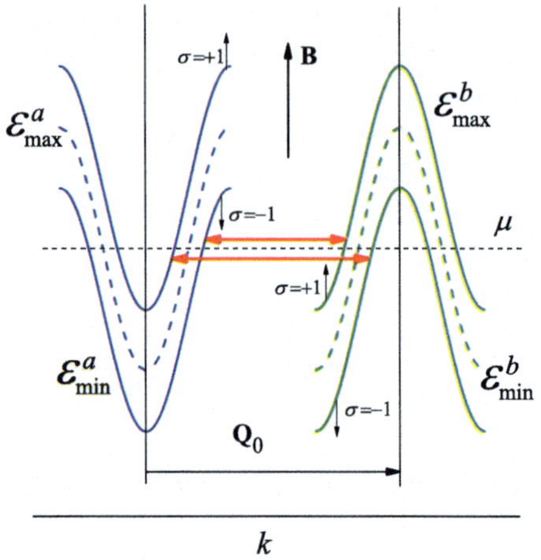

of high magnetic fields (which is also considered in Ref. [39]), there exists a set of oscillation phenomena, which highly complicates the phase diagram even of this simple model.

In the case of low magnetic field, we should add only the Zeeman terms in the Hamiltonian. As a result, we have instead of Eq. (8.2) the single-electron spectrum in the form

$$\begin{aligned} \epsilon_\sigma^a(\mathbf{k}) &= \hbar v_F(k-k_F) - \mu + g_a\sigma\hbar\omega_a, \\ \epsilon_\sigma^b(\mathbf{k}+\mathbf{Q}_0) &= -\hbar v_F(k-k_F) - \mu + g_b\sigma\hbar\omega_b. \end{aligned} \tag{8.12}$$

The applied magnetic field lifts the degeneracy with respect to the spin projection, and now we have two electron and two hole bands, see Fig. 8.3. In such case, we have to introduce a two-component order parameter corresponding to the nesting vectors shown by the arrows in Fig. 8.3. Neglecting for simplicity the incommensurate SDW ordering, these components can be written as [39]

$$\Delta_\uparrow = \frac{V}{\mathcal{V}}\sum_{\mathbf{k}}\langle a^\dagger_{\mathbf{k}\uparrow} b_{\mathbf{k}\downarrow}\rangle, \quad \Delta_\downarrow = \frac{V}{\mathcal{V}}\sum_{\mathbf{k}}\langle a^\dagger_{\mathbf{k}\downarrow} b_{\mathbf{k}\uparrow}\rangle. \tag{8.13}$$

The case when the electron and hole bands are perfectly symmetric is the simplest. In such a situation, however, the effect of Zeeman field on the electron spectrum of the model is absent, since it only shifts the ground state energy by a constant value. Thus, some electron–hole asymmetry is necessary to obtain a non-trivial result. Following Ref. [39], we consider as an example the case when $m_a = m_b = m$, hence, $\omega_a = \omega_b = \omega_H$, but $g_a \neq g_b$. Following the procedure described in the previous subsection, now we can find eigenenergies and calculate the grand potential Ω. Minimization of Ω gives the expression for components of the order parameter [39]

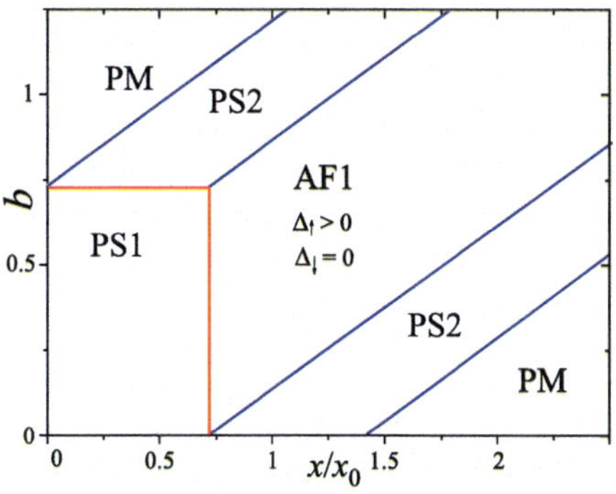

Fig. 8.4 Phase diagram of the model at zero temperature in the $(b, x/x_0)$ plane [39]. The value x_0 in Ref. [39] corresponds to $4x_0$ in the present notation. In the diagram, we see the homogeneous SDW phase AF1 (the definition is shown in the figure), two regions with the uniform paramagnetic phase, and the phase-separated states PS1 (mixture of two SDW phases, see the text) and PS2 (mixture of AF1 and the paramagnetic phases)

$$\ln\frac{\Delta_0}{\Delta_\sigma} = \int_0^\infty d\xi\, \frac{f(\sqrt{\Delta_\sigma^2+\xi^2}+\mu_\sigma)+f(\sqrt{\Delta_\sigma^2+\xi^2}-\mu_\sigma)}{\sqrt{\Delta_\sigma^2+\xi^2}}, \tag{8.14}$$

where $\mu_\sigma = \mu - \sigma(g_a - g_b)\hbar\omega_H/2$. Neglecting the existence of the nonmagnetic part of the Fermi surface, one can write the expression for doping analogous to Eq. (8.11)

$$\frac{x}{x_0} = \sum_\sigma \int_0^\infty \frac{d\xi}{4\Delta_0}\left[f(\sqrt{\Delta_\sigma^2+\xi^2}-\mu_\sigma) - f(\sqrt{\Delta_\sigma^2+\xi^2}+\mu_\sigma)\right]. \tag{8.15}$$

In Ref. [39], the phase diagram of the model in the plane (magnetic field, doping) was constructed in the limit $T = 0$. The obtained results are illustrated in Fig. 8.4, where the dimensionless magnetic field is defined as $b = |g_a - g_b|\hbar\omega_H/2\Delta_0$. The homogeneous paramagnetic state exists in the system at high doping ($x/4x_0 > 2b + \sqrt{2}$) or at high magnetic field ($b > 1/\sqrt{2} + x/8x_0$) (the value x_0 in Ref. [39] corresponds to $4x_0$ in the present notation). In the range of the low doping and low magnetic field, the ground state of the system is the phase-separated one, which we refer to as PS1. The phase PS1 is a mixture of the two antiferromagnetic phases. The first phase corresponds to SDW with zero doping, $x = 0$, and $\Delta_\uparrow = \Delta_\downarrow = \Delta_0$. The second phase with nonzero doping, $\Delta_\uparrow = \Delta_0$, and $\Delta_\downarrow = 0$ will be further referred to as AF1. The phase PS1 is the ground state of the model if $|b| < 1/\sqrt{2}$ and $|x/4x_0| < 1/\sqrt{2}$. The uniform SDW phase AF1 is stable within the intermediate magnetic fields range, $x/8x_0 - 2\sqrt{2} < b < x/8x_0 + 2\sqrt{2}$. The inhomogeneous state PS2 is stable in the regions between homogeneous paramagnetic and AF1 phases. The PS2 state consists of the mixture of the SDW phase AF1 and PM paramagnetic phase. It is interesting to note that the phase AF1 is some non-common type of the half-metal, since in this phase only electrons with a single spin projection exist at the Fermi surface [40].

Thus, we see that the magnetic field lifts the spin-projection degeneracy of the SDW order parameter in the system with imperfect nesting. As a result the number of possible homogeneous states increases, which, in turn, drastically affects the picture of electronic phase separation.

8.4 Phase Separation in the Organic Metals Under Pressure

In the above consideration, we demonstrated that the system with nesting is very sensitive to the shift of the Fermi level from the position of the perfect nesting due to the doping and/or to the change of the electron spectrum due to application of the magnetic field. This conclusion can be formulated in a more general form: any impact that affects the value of the chemical potential or electron spectrum in a system with nesting can give rise to the crucial changes in its thermodynamic state. Here we discuss the effect of the pressure. This case is of a special interest due to the experiments with an archetypical Bechgaard salt $(TMTSF)_2PF_6$ under pressure [14–17]. This system is commonly considered as a system with the nesting of the Fermi surface. In these experiments, the coexistence of the SDW and paramagnetic metallic (superconducting at low temperatures) phases was clearly observed. Moreover, in Ref. [14] the SDW and paramagnetic metallic phases manifest itself in the form of phase separation with macroscopic domains of the high-pressure paramagnetic metallic phase embedded into the insulating SDW host, which are aligned along certain crystallographic axes. A natural explanation of this effect is the same as in the case of doping or magnetic field: the application of the pressure shifts the electron system from the position of a perfect nesting.

A qualitative explanation of the experiment [14] can be obtained by an incorporation of the pressure effect in the Rice model [23]. In the experiments [14], an evolution with pressure of the geometry of the SDW-paramagnetic phase was also observed. To trace this evolution theoretically the electrostatic energy of the inhomogeneous state and the energy of interfaces between SDW and paramagnetic phases should be taken into account as it was described in Sect. 5.5.

In general, the pressure P may affect all parameters of the system, e.g., electron coupling V, electron and hole effective masses, band positions $\epsilon^{\alpha}_{\text{min}}$, $\epsilon^{\alpha}_{\text{max}}$, etc. However, if the pressure is not very high, it is natural to expect that the main effect on the electron properties of the material with nesting occurs due to the shift of the Fermi energy closer or farther from the ideal nesting position. Following Ref. [23], we assume that the change in the Fermi energy arises due to the shift of the energy bands with respect to each other. For simplicity, in the cited work only the shift of the nonmagnetic (reservoir) band is taken into consideration while the shift of the magnetic bands is ignored. Below we discuss only the case of the commensurate SDW, that is, $\mathbf{Q} = 0$. We restrict our consideration by this simple case since it occurs that such a minimal model is sufficient to reproduce all main features of the experimental observations reported in Ref. [14].

If the position of the bottom of the reservoir band is shifted by $\delta\epsilon$ due to pressure, then, the corresponding variation of the the chemical potential is $\mu(P) - \mu(0)$. It is natural to assume that the total number of electrons per unit cell remains unchanged. The variation in the number of electrons per unit cell in the nonmagnetic band is

$$\delta n_r = 2n_r[\mu(P) - \mu(0) - \delta\epsilon]\,, \tag{8.16}$$

while the variation in the number of magnetic electrons is

$$\delta n_m = 4n_m \int_0^\infty d\xi \{f[\eta - \mu(P)] - f[\eta + \mu(P)] - f[\eta - \mu(0)] - f[\eta + \mu(0)]\}. \tag{8.17}$$

We assume that the system is perfectly nested when $P = P_n$, that is, $\mu(P_n) = 0$ and $x = 0$. In the range of not too high pressures, the value $\delta\epsilon$ depends linearly on the pressure

$$\delta\epsilon(P) = \frac{\partial\epsilon^c_{\text{min}}}{\partial P}\,(P - P_n)\;. \tag{8.18}$$

We substitute Eqs. (8.16)–(8.18) into Eq. (8.11) with $Q = 0$ and make the replacement $\mu + \delta\mu \to \mu$. Then, taking into account the condition $x(P) = const = 0$, one derives [23]

$$r\bar{P} = \frac{r\mu}{\Delta_0} + \frac{1}{\Delta_0}\int_0^\infty d\xi\,[f(\eta - \mu) - f(\eta + \mu)]\,, \tag{8.19}$$

where the dimensionless variable $\bar{P}$ is

$$\bar{P} = \frac{\partial\epsilon^c_{\text{min}}}{\partial P}\,\frac{P - P_n}{\Delta_0}\;. \tag{8.20}$$

The sign of the derivative $\partial\epsilon^c_{\text{min}}/\partial P$ depends on the particular properties of the system. Below we assume that this value is positive. In the case of zero doping, $x = 0$, Eqs. (8.10) at $Q = 0$ and (8.19) form a closed system of equations determining the functions $\Delta(\bar{P}, T)$ and $\mu(\bar{P}, T)$. If $x \neq 0$, one has to substitute $r\bar{P} \to r\bar{P} + x/x_0$ in the left-hand side of Eq. (8.19), taking into account that under doping, pressure $P = P_n$ corresponds to a perfect nesting.

Comparing Eq. (8.19) with Eq. (8.11), we see that the pressure acts as some effective doping. The only difference is that this 'effective doping' is proportional to the density of electron states in the reservoir bands, r, and when the reservoir is absent, the pressure changes nothing. This is an artifact of the approximation used, where possible changes in the shapes of the Fermi surfaces of the magnetic bands were neglected. Under used here assumptions and $r = 0$, the only possible way to destroy the ideal nesting is to change the difference in the Fermi momenta$k_{Fa} - k_{Fb}$ of the magnetic bands, which is prohibited due to the charge conservation law. When the reservoir is present, thc imbalancc in the magnetic electrons, that appears due to the unequal k_{Fa} and k_{Fb}, can be compensated by the reservoir electrons.

The equation for the gap, Eq. (8.10), at $Q = 0$ and (8.19) can be solved numerically to obtain the functions $\Delta(\bar{P}, T)$ and $\mu(\bar{P}, T)$. Then, assuming that the ground state of the system is homogeneous, one can find the grand potential Ω, the free energy $F = \Omega + \mu n$, and construct the phase diagram of the model in the P–T plane. Such calculations were performed in Ref. [23] and the phase diagram of the model in the (P, T) plane was obtained. The calculated dependence of the gap Δ on the pressure occurs similarly to the dependence of the gap versus doping: the gap decreases with $P - P_n$ and goes to zero at some value of the applied pressure. However, the true ground state of the model is inhomogeneous similar to the cases considered in Sects. 8.2 and 8.3.

In the case of the finite doping x, Eq. (8.19) becomes [23]

$$\frac{x_{\text{eff}}(P)}{x_0} = \frac{r\mu}{\Delta_0} + \frac{1}{\Delta_0}\int\limits_0^\infty d\xi\,[f(\eta-\mu) - f(\eta+\mu)]\,, \tag{8.21}$$

where

$$x_{\text{eff}}(P) = x + r x_0 \bar{P}\,. \tag{8.22}$$

First, we have to calculate the chemical potential μ as a function of x_{eff}. The obtained dependence $\mu(x_{\text{eff}})$ is similar to $\mu(x)$ shown in Fig. 8.1. Thus, the value $\mu(x_{\text{eff}})$ behaves nonmonotonically indicating an instability of the homogeneous state toward the phase separation. The separated phases are the SDW phase with effective doping $x_{\text{eff}}^{(1)}$ and the paramagnetic phase with effective doping $x_{\text{eff}}^{(2)}$. As above, the values $x_{\text{eff}}^{(1,2)}$ can be obtained using the Maxwell construction, see Fig. 8.1. These values depend on the model parameters and temperature, but do not depend on pressure, while "bare" doping in the phase-separated phases, $x_{1,2}$, depend on P. From Eq. (8.22) it follows that $x_{1,2} = x_{\text{eff}}^{(1,2)} - r x_0 \bar{P}$. If the volume fraction of the paramagnetic phase is p, then, from the charge conservation law, we have

$$(1-p)x_1 + p x_2 = x\,. \tag{8.23}$$

In the doping range, $x_1 < x < x_2$, the ground state of the model is a mixture of SDW phase with the electron density $x_1 = x_{\text{eff}}^{(1)} - r x_0 \bar{P}$ and paramagnetic phase with the electron density $x_2 = x_{\text{eff}}^{(2)} - r x_0 \bar{P}$. This doping range shifts toward the smaller densities when the pressure increases. Suppose that $x < x_{\text{eff}}^{(1)}$, thus, at $P = P_n$ (when $\bar{P} = 0$), the ground state is a homogeneous SDW phase. Increasing $\bar{P}$ above the value $\bar{P}_1 = (x_{\text{eff}}^{(1)} - x)/(r x_0)$ makes the system inhomogeneous. The fraction of the paramagnetic phase, p, gradually increases with $\bar{P}$ and at $\bar{P} = \bar{P}_2 = (x_{\text{eff}}^{(2)} - x)/(r x_0)$, the system becomes a homogeneous paramagnetic. As an illustration, the

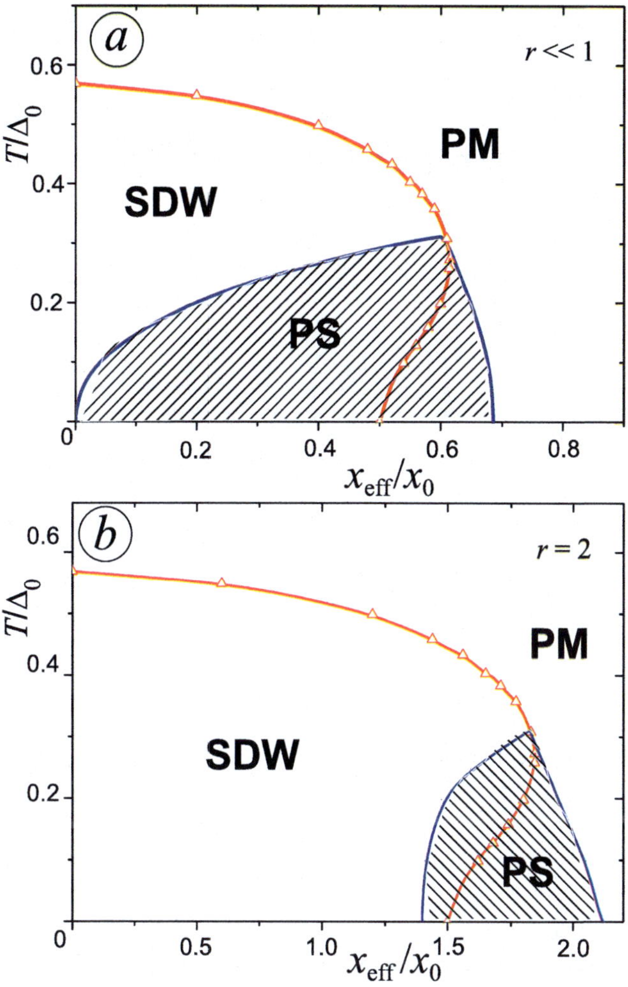

Fig. 8.5 Phase diagram of the model in the $(x_{\text{eff}}(P), T)$ plane in the cases $r \ll 1$ (**a**) and $r = 2$, (**b**). Solid (red) line with up triangles shows the boundary of the uniform stable SDW phase. The shaded area corresponds to the region with tendency to the phase separation (PS) [23]

phase diagram of the model in the $(x_{\text{eff}}(P), T)$ plane is shown in Fig. 8.5 for $r \ll 1$ (a) and for $r = 2$ (b). The region of the possible inhomogeneous phase occupies a significant part of the phase diagram.

As it was discussed in Sect. 5.5, the change in the phases volume fractions is accomplished by the evolution of the inhomogeneous phase geometry, see Fig. 5.9. This evolution occurs due to change of the ratio of the long-range Coulomb interaction E_c arising due to the violation of the local charge neutrality and the energy E_s of the interface between SDW and paramagnetic phases [41, 42].

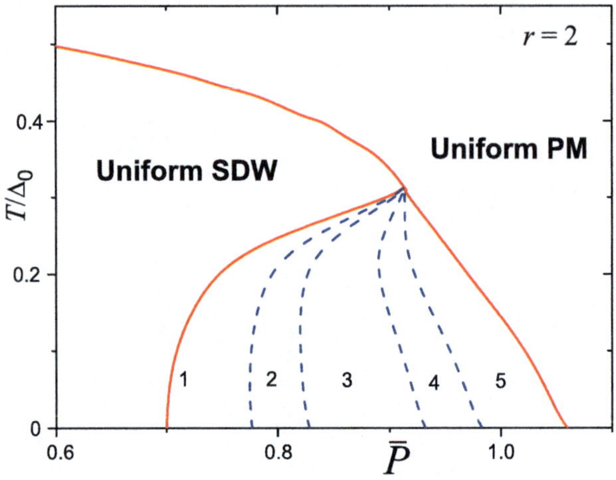

Fig. 8.6 (P, T) phase diagram in the case $x = 0$ and $r = 2$. Solid (red) lines show the boundary of the uniform stable phases. Blue dashed lines show the boundaries of the inhomogeneous phases with different shapes of the inhomogeneities: paramagnetic droplets in the SDW host exist in region 1; paramagnetic pillars in the SDW host exist in region 2; alternating paramagnetic and SDW slabs exist in region 3; SDW pillars in the paramagnetic host exist in region 4; and SDW droplets in the paramagnetic host exist in region 5 [23]

The evolution of the inhomogeneous state geometry with pressure was analyzed in Ref. [23] The content of the paramagnetic phase p can be expressed as (for $x = 0$)

$$p = \frac{r x_0 \bar{P} - x_{\text{eff}}^{(1)}}{x_{\text{eff}}^{(2)} - x_{\text{eff}}^{(1)}}, \tag{8.24}$$

where the effective dopings $x_{\text{eff}}^{(1,2)}$ are calculated according to the Maxwell rule. Then, following the procedure described in Sect. 5.5 the phase diagram of the system can be calculated. The results of such calculations are illustrated in Fig. 8.6. The volume fraction of the paramagnetic phase, p, increases monotonically with pressure. As a result, the sequence of the geometries of inclusions changing with the growth of pressure is the following: uniform SDW phase at low pressure is changed by the phase separated state with paramagnetic islands in the SDW host (area 1 in Fig. 8.6), then, by paramagnetic rods or pillars (region 2), paramagnetic slabs (region 3) and, finally, there arise inclusions of the SDW phase in the paramagnetic host (regions 4 and 5). The latter inhomogeneous states could be seen as a uniform paramagnetic metallic state in the transport measurements due to metal-insulator percolation transition [23].

In the above consideration, the pressure changes only the values of k_{Fa} and k_{Fb}, but does not change the spherical geometry of the Fermi surface pockets. Naturally, it is a simplification. Below we discuss the phase separation observed in the Bechgaard salt [14] following Ref. [23]. The Bechgaard salts are highly anisotropic, and one may expect that its Fermi surface is also anisotropic, and its nested parts are probably not spherical. The pressure can affect not only the chemical potential, but also the geometry of the nested pockets. Thus, we need to include into consideration some additional geometrical factor, which governs the nesting. In this case, the de-nesting under pressure can occur even in the absence of the nonmagnetic reservoir. At the same time, if the electron and hole pockets are nested only partially, one can suggest that the SDW order parameter arises only in the nested parts of the Fermi surfaces, while other parts remain ungapped and play a role of the reservoir. Thus, one can expect that the model considered in this section could be relevant, at least, qualitatively, even in the case of nonspherical Fermi surfaces. (A more detailed analysis of the effects of the Fermi surface shape is given in the next section, where the phase separation in iron pnictide superconductors is discussed.)

The de-nesting of any nature gives rise to the decrease of the SDW order parameter and to the phase separation. With the increase of the pressure, the uniform SDW order is destroyed, and the phase separation arises at some $P = P_0$. The volume concentration of the metallic phase increases with pressure and the described above evolution of the inhomogeneous phase geometry is observed, see Fig. 8.6. The metallic phase may be superconducting at low temperature. In Fig. 8.7a, the experimental results on the phase separation in the Bechgaard salt under pressure taken from Ref. [14] are shown. In Fig. 8.7b, for comparison, a part of the phase diagram shown in Fig. 8.6 redrawn in the proper coordinates is presented. The inhomogeneous phases at high pressure where the sample is a metal from the experimental point of view are omitted. Qualitatively, the pictures illustrating the experimental and theoretical results are quite similar.

8.5 Phase Separation in Iron Pnictide Superconductors

Superconducting iron-based pnictides form another system with the imperfect nesting, which we consider in this section. The phase diagram of iron pnictides is very rich and contains the regions of superconductivity and spin-density-wave order, both commensurate and incommensurate. Moreover, these materials often exhibit spin and charge inhomogeneity with the characteristic features of the systems with electronic phase separation [22, 43–48]. Superconducting iron chalcogenides also exhibit a phase separation [49, 50]. However, the physics related to the Fermi surface nesting may not be directly applicable to iron chalcogenides since some chalcogenides have no hole pockets.

Unlike superconducting cuprates, which are believed to be in the strong electron–electron interaction regime, iron pnictides may be described by a weak-interaction model (see, e.g., Sec.IIIA of Ref. [51]).

Currently, the Fermi surface of iron pnictides is typically described using two related representations briefly outlined below. In Fig. 8.8a, we plot the Fermi surface within the so-called unfolded Brillouin zone [51–53], which corresponds to the square lattice of iron atoms, with one Fe atom per unit cell and lattice constant a. In this representation, two quasi-two-dimensional nearly circular hole pockets

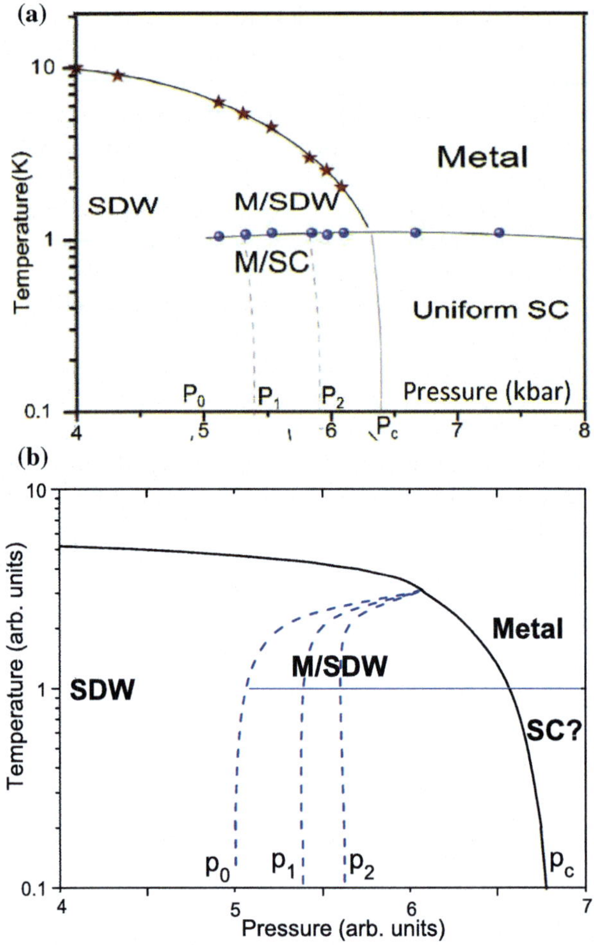

Fig. 8.7 **a** Experimental observation of the phase separation in the Bechgaard salt [14] (Fig. 1 from Ref. [14]). The uniform SDW phase is observed if pressure $P < P_0$. The inhomogeneous mixture of the SDW and paramagnetic metallic phases exists if $P_0 < P < P_c$. The PS state includes metallic droplets in the SDW host if $P_0 < P < P_1$, the metallic pillars in the SDW matrix if $P_1 < P < P_2$, and the metallic slabs in the SDW host if $P_2 < P < P_c$. When $P > P_c$ the sample is in the metallic state. The metal is superconducting if $T < T_c$, the dependence $T_c(P)$ is shown by (blue) solid line with circles. (**b**) A part of the phase diagram shown in Fig. 8.6 redrawn in different coordinates, the notation corresponds to panel (**a**) [23]

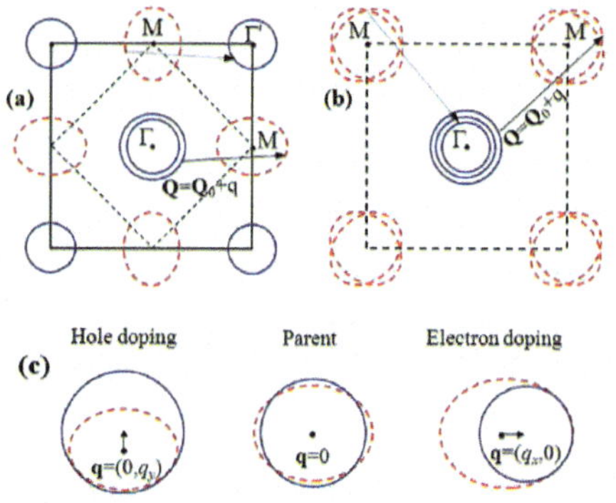

Fig. 8.8 Schematic illustration of the Fermi surface of iron pnictides in the unfolded **a** and folded **b** Brillouin zone (BZ). We have three hole pockets located near the center of the folded BZ [shown by (blue) solid curves]. In the unfolded BZ, one of the hole pockets moves to the corner of the BZ. For the unfolded BZ (**a**), the electron pockets are elliptic [(red) dashed curves] and located near the $(0, \pi/a)$ and $(\pi/a, 0)$ points, while in the folded BZ, they are represented by overlapping ellipses located at the corners. Arrows show the possible nesting vectors between hole and electron pockets, which give rise to the SDW order. **c** Illustration of the possible nesting at different doping levels. (Fig. 1 from Ref. [22])

are centered at the $\Gamma(0, 0)$ point, one more circular hole pocket is located near the $\Gamma'(\pi/a, \pi/a)$ point, and two elliptically-shaped electron pockets are centered at the $M(0, \pi/a)$ and $M(\pi/a, 0)$ points [see Fig. 8.8a]. At the same time, the actual unit cell of pnictides contains two Fe atoms since in the crystal lattice the pnictogen atoms are located at nonequivalent positions. The (folded) Brillouin zone corresponding to this unit cell is obtained by folding the Brillouin zone shown in Fig. 8.8a by dashed lines and consequent rotation by $\pi/4$. The Fermi surface in the folded Brillouin zone is shown in Fig. 8.8b. In this figure, all three hole pockets are located near the $\Gamma(0, 0)$ point, whereas the electron pockets represented by overlapping ellipses are located near the $M(\pm\pi/\bar{a}, \pi/\bar{a})$ points, with $\bar{a} = a/\sqrt{2}$. Formulating this model, we will make several simplifications. First, we neglect the effects associated with nonequivalent positions of the Fe atoms. Consequently, the use of unfolded Brillouin zone is sufficient. Second, we will neglect the 3D structure of the material and study only the 2D model.

Following Ref. [22], the Hamiltonian of the model can be written as $\hat{H} = \hat{H}_0 + \hat{H}_{\text{int}}$, where the kinetic energy term has the form

$$\hat{H}_0 = \sum_{\mathbf{k}\lambda\sigma} \epsilon^h_{\lambda\mathbf{k}} a^\dagger_{\mathbf{k}\lambda\sigma} a_{\mathbf{k}\lambda\sigma} + \sum_{\mathbf{k}s\sigma} \epsilon^e_{s\mathbf{k}} b^\dagger_{\mathbf{k}s\sigma} b_{\mathbf{k}s\sigma}. \tag{8.25}$$

Here $a^\dagger_{\mathbf{k}\lambda\sigma}$, $a_{\mathbf{k}\lambda\sigma}$ ($b^\dagger_{\mathbf{k}s\sigma}$, $b_{\mathbf{k}s\sigma}$) are the creation and annihilation operators for charge carriers in the holelike (electronlike) bands $\lambda = 1, 2, 3$ ($s = 1, 2$) with spectra $\epsilon^h_{\lambda\mathbf{k}}$ ($\epsilon^e_{s\mathbf{k}}$). Assuming that the bands have quadratic dispersion laws near the Fermi level, for circular holelike bands we have

$$\epsilon^h_{1\mathbf{k}} = -\frac{\hbar v^h_F(k^2 - k^2_F)}{2k_F} - \mu, \quad \epsilon^h_{2\mathbf{k}} = -\frac{\hbar v^h_{2F}(k^2 - k^2_F)}{2k_F} - \Delta\epsilon_2 - \mu,$$
$$\epsilon^h_{3\mathbf{k}+\bar{\mathbf{Q}}} = -\frac{\hbar v^h_{3F}(k^2 - k^2_F)}{2k_F} - \Delta\epsilon_3 - \mu. \tag{8.26}$$

Here v_F, v_{2F}, and v_{3F} are the Fermi velocities for the hole bands, $\bar{\mathbf{Q}} = (\pi/a, \pi/a)$, and the energy shifts $\Delta\epsilon_{2,3}$ determine the difference in radii of the hole pockets. The electronlike components of the Fermi surface are elliptic and can be presented as follows [22]

$$\epsilon^e_{1\mathbf{k}+\mathbf{Q}_0} = \epsilon^e_{1\mathbf{k}+\mathbf{Q}'_0} = \frac{\hbar v^e_F(k^2 - k^2_F)}{2k_F} + \frac{\alpha\hbar v^e_F}{k_F}(k^2_x - k^2_y) - \mu, \tag{8.27}$$

where the centers of the elliptic bands are $\mathbf{Q}_0 = (\pi/a, 0)$ and $\mathbf{Q}'_0 = (0, \pi/a)$, and v^e_F is the Fermi velocity for the electron bands averaged over the Fermi surface. The parameter α defines the ellipticity of the electron pockets. The major axes of the ellipses are directed toward the Γ point when $\alpha > 0$ and perpendicular to this direction when $\alpha < 0$.

Due to the multisheet structure of the Fermi surface, the interaction Hamiltonian $\hat{H}_{\text{int}}$, in general, includes a number of contributions. However, since we are interested in the SDW order and the phase separation, one can ignore the electron–electron and hole–hole interactions as in the previous analysis of the Rice model. It is known [43] that the value of local magnetic moment in pnictides oscillates along one of the crystal axes. To reproduce this stripy magnetic structure, it is sufficient to couple one electron and one hole band [54]. For the definiteness sake, one can assume that these are the hole band ϵ^h_1 and the electron band ϵ^e_1. Thus, we can write

$$\hat{H}_{\text{int}} = \frac{V}{\mathcal{V}} \sum_{\mathbf{k}\mathbf{k}'\mathbf{q}\sigma\sigma'} a^\dagger_{\mathbf{k}+\mathbf{q}1\sigma} a_{\mathbf{k}1\sigma} b^\dagger_{\mathbf{k}'-\mathbf{q}1\sigma'} b_{\mathbf{k}'1\sigma'}. \tag{8.28}$$

The described model for pnictides is a generalization of the Rice model. As in the latter case, two bands participate in the magnetic transition, and there is a "reservoir" (nonmagnetic bands). However, the perfect nesting is impossible for pnictides since the forms of the hole and electron pockets are different for any doping level.

As we pointed out above, the Coulomb interaction in iron pnictides is weak and one can assume that $V/\epsilon_F \ll 1$, where the Fermi energy $\epsilon_F = \hbar(v^e_F + v^h_F)k_F/2$. Then, Hamiltonian (8.28) can be considered in the mean-field approximation [22] and, as in the previous section, we will search the SDW order parameter in the form (8.4). In general, different parts of each pocket of the Fermi surface have different orbital composition. Such feature can be accounted by introducing a momentum-dependent coupling V and would lead to a pronounced wave-vector dependence of the SDW order parameter. However, unless the variations of V are extremely strong, they do not affect the phase separation qualitatively.

The magnetization corresponding to the chosen SDW order lies in the xy plane. For the commensurate SDW case, the magnetization direction remains constant when one moves along the direction normal to $\mathbf{Q}_0$. When one moves parallel to $\mathbf{Q}_0$, the magnetization reverses its direction from one iron atom to the next iron atom. For incommensurate SDW, this "stripy" pattern slowly rotates in the xy plane [22].

Performing mean-field transformations in the interaction Hamiltonian (8.28), we obtain the energy spectrum of the model similar to Eq. (8.6) and grand potential in the form (8.7), where energies $\epsilon^{a,b}$ should be replaced by $\epsilon^{e,h}$. As usual, the minimization of the grand potential gives the equations for the SDW order parameter Δ and the nesting vector $\mathbf{Q} = \mathbf{Q}_0 + \mathbf{q}$. The expression for Δ was derived in Ref. [22] at zero temperature

$$\ln \frac{\Delta_0}{\Delta} = \int \frac{\epsilon_F d^2\mathbf{k}}{2\pi k_F^2} \frac{\Theta[E_1(\mathbf{k})] + \Theta[-E_2(\mathbf{k})]}{\sqrt{\Delta^2 + \frac{1}{4}\left(\epsilon^e_{1\mathbf{k}+\mathbf{Q}_0} - \epsilon^h_{1\mathbf{k}-\mathbf{q}}\right)^2}}, \tag{8.29}$$

where $\Theta(x)$ is the step function, $\epsilon^{e,h}_{1\mathbf{k}}$ are defined by Eqs. (8.26) and (8.27), and the usual limits Δ/ϵ_F, $|\mathbf{q}|/k_F \ll 1$ are assumed.

In the weak-coupling limit under discussion, the SDW order can exist only if the deviation from the perfect nesting is small, that is, $\alpha \ll 1$. Taking this into account and using Eqs. (8.26) and (8.27), we can rewrite Eq. (8.29) in the form [22]

$$\ln \frac{1}{\delta} = \int_0^{2\pi} \frac{d\varphi}{2\pi} Re\left\{\cosh^{-1}\left[\frac{\nu_0(\mathbf{p},\varphi) - \nu}{\delta}\right]\right\}, \tag{8.30}$$

where

$$\nu_0(\mathbf{p},\varphi) = p_x \cos\varphi + p_y \sin\varphi - \frac{\bar{\alpha}}{2}\cos 2\varphi, \quad \bar{\alpha} = \frac{\alpha\kappa\epsilon_F}{\Delta_0}, \quad \kappa = \frac{2\sqrt{v_F^e v_F^h}}{v_F^e + v_F^h},$$
$$\delta = \frac{\Delta}{\Delta_0}, \quad \nu = \frac{\mu}{\kappa\Delta_0}, \quad \mathbf{p} = \frac{\kappa\hbar v_F}{2\Delta_0}\mathbf{q}. \tag{8.31}$$

Similarly, we obtain the equation for the nesting vector $\mathbf{Q} = \mathbf{Q}_0 + 2\Delta_0\mathbf{p}/v_F\kappa$:

$$\begin{pmatrix} p_x \\ p_y \end{pmatrix} = \int_0^{2\pi} \frac{d\varphi}{\pi} \begin{pmatrix} \sin\varphi \\ \cos\varphi \end{pmatrix} \operatorname{sgn}[\nu_0(\mathbf{p},\varphi) - \nu] Re\sqrt{[\nu_0(\mathbf{p},\varphi) - \nu]^2 - \delta^2}. \tag{8.32}$$

Equations (8.30) and (8.32) determine the SDW band gap and the nesting vector as functions of μ. Experiments are performed at fixed doping, and we have to relate the electron density and μ. Following the previously described approach [see Eq. (8.11) and the text above it], we find in the approximation under discussion [22]

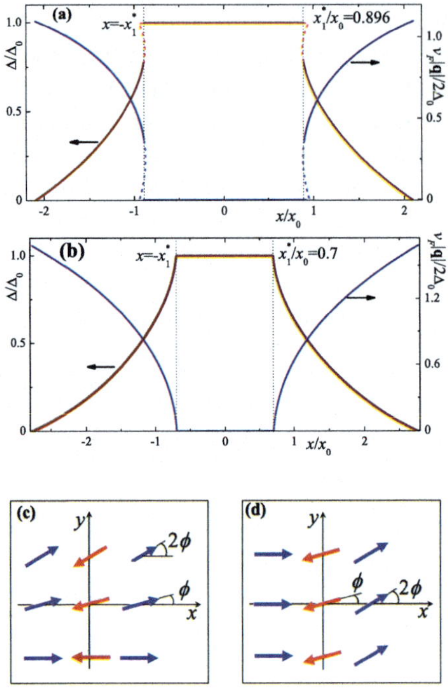

Fig. 8.9 Dependence of Δ and $|\mathbf{q}|$ on doping x, $n = \kappa = 1$. **a** $\bar{\alpha} = 0.1$. The transition from commensurate to incommensurate SDW state is of the first order: both Δ and $|\mathbf{q}|$ rapidly change at $x = x_1^*$. Dashed lines show the functions $\Delta(x)$ and $|\mathbf{q}(x)|$ corresponding to the metastable state. **b** $\bar{\alpha} = 0.6$. The transition from commensurate to incommensurate SDW state is of the second order. Panels (**c**) and (**d**) show the schematics of the incommensurate SDW spin structure. If $\alpha > 0$, then, **c** corresponds to the hole doping ($x < 0$) and **d** to the electron doping ($x > 0$). If $\alpha < 0$, then panel **c** corresponds to the electron doping, while panel **d** corresponds to the hole doping [22]

$$\frac{x}{x_0} = n\nu - \int_0^{2\pi} \frac{d\varphi}{2\pi} \operatorname{sgn}[\nu_0(\mathbf{p}, \varphi) - \nu] Re\sqrt{[\nu_0(\mathbf{p}, \varphi) - \nu]^2 - \delta^2}, \tag{8.33}$$

where x_0 is the characteristic doping level, n is the density of electrons in nonmagnetic bands

$$x_0 = \frac{2v_0 k_F^2 \Delta_0}{\pi \kappa \epsilon_F}, \quad n = \frac{(v_{2F}^h)^{-1} + (v_{3F}^h)^{-1} + (v_F^e)^{-1}}{(v_F^h)^{-1} + (v_{2F}^e)^{-1}}, \tag{8.34}$$

and v_0 is the unit cell volume. Note, that $x \sim \Delta_0/\epsilon_F \ll 1$ and considered below phases, existing when $x \sim x_0$, correspond to the low-doping regimes.

The computed functions $\Delta(x)$ and $q(x)$ are shown in Fig. 8.9a and b for $n = \kappa = 1$ and two different values of elipticity α. The numerical analysis reveals that the SDW order arises if $|\bar{\alpha}| < 2$ or $\alpha < \alpha_c = 2\Delta_0/\kappa\epsilon_F \ll 1$. At low doping, $|x| < x_1^*$, all the extra charge goes to the nonmagnetic bands. As a result, the order parameter is

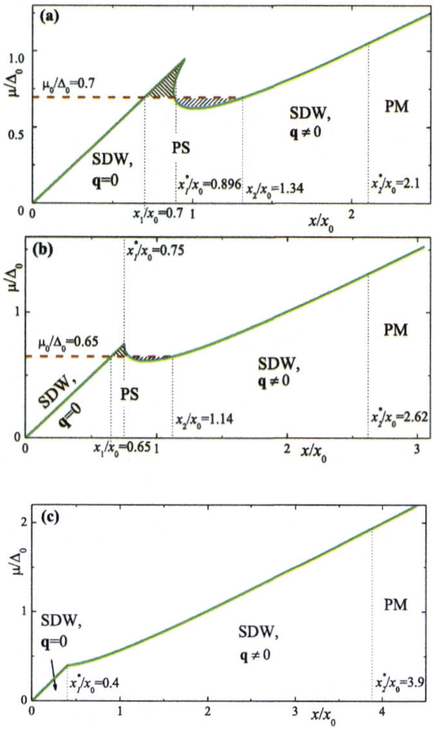

Fig. 8.10 Chemical potential $\mu(x)$ calculated at $\bar{\alpha} = 0.1$ (**a**), $\bar{\alpha} = 0.5$ (**b**), and $\bar{\alpha} = 1.2$ (**c**). Here, $n = \kappa = 1$. **a, b** The homogeneous state is unstable toward the phase separation if $x_1 < x < x_2$. The dashed (red) line corresponds to the μ_0 found using the Maxwell construction. The shaded areas above and below μ_0 are equal to each other. **c** $\mu(x)$ monotonically increases with x, no phase separation occurs. The homogeneous commensurate ($q = 0$) and incommensurate ($q \neq 0$) SDW, paramagnetic (PM), and inhomogeneous commensurate–incommensurate SDW (PS) states are separated by vertical dotted lines [22]

independent of x, $q(x) = 0$ (a commensurate SDW order), and the chemical potential increases linearly with x. When $|x| > x_1^*$, electrons (holes) appear in the band E_2 (E_1). The SDW order becomes incommensurate, $q \neq 0$. If $n < n_1 \approx 0.38$, the commensurate–incommensurate transition is of the second order for any $\bar{\alpha}$, while for larger n it becomes the first order one if $\bar{\alpha} < \bar{\alpha}_1(n)$. The spin configurations for the incommensurate SDW are schematically shown in Fig. 8.9c and d (for details, see Ref. [22]). For positive (negative) α, Fig. 8.9c corresponds to $x > 0$ ($x < 0$), and Fig. 8.9d corresponds to $x < 0$ ($x > 0$).

The computed in Ref. [22] plots of the chemical potential $\mu(x)$ are shown in Fig. 8.10 for three different values of elipticity $\bar{\alpha}$ and $n = \kappa = 1$. The $\mu(x)$ dependence for the hole doping can be determined using the particle–hole symmetry. For low $\bar{\alpha}$, the function $\mu(x)$ is nonmonotonic and multivalued near x_1^*, Fig. 8.10a. At higher $\bar{\alpha}$, the multivaluedness disappears; however, the nonmonotonicity remains, Fig. 8.10b. The latter vanishes at higher values of $\bar{\alpha}$, Fig. 8.10c. Thus, if the elipticity is not too large, there are finite ranges of doping where the chemical potential decreases with doping and the homogeneous state is unstable with respect to the separation into two phases. However, the phase transitions between homogeneous commensurate and incommensurate SDW phases, which we described above, can be masked, at least partially, by the phase separation. The range of doping x, where

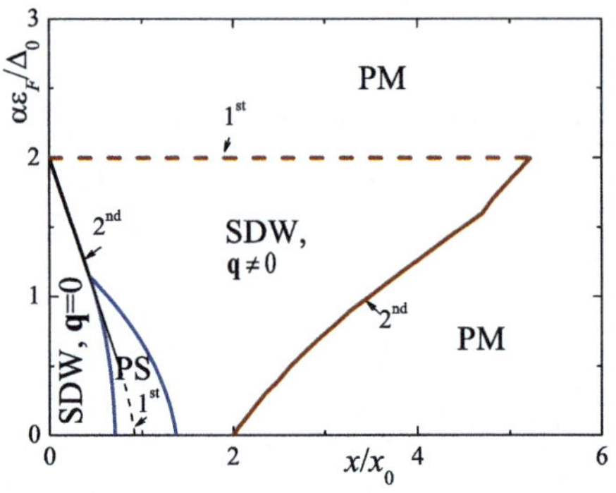

Fig. 8.11 Phase diagram of the model for pnictides in the (x, α) plane, for $n = \kappa = 1$ and $\alpha, x > 0$. It is symmetric with respect to the replacement $x \to -x$ and/or $\alpha \to -\alpha$. The boundary between the incommensurate SDW and paramagnetic (PM) states shown by the dashed (red) line corresponds to the first order phase transition and that shown by the solid (red) curve corresponds to the second order phase transition. Solid (blue) lines indicate the boundaries of the phase-separated (PS) state. The solid (black) curve (second order transition) and dashed (black) curve (first order transition) show the boundaries between commensurate and incommensurate homogeneous SDW phases. Note that the phase separation partially masks the transition line between the homogeneous SDW states [22]

the phase separation exists, is the largest when $\alpha = 0$, that is, when the model under study is identical to the two-dimensional Rice model. The range of the phase separation $x_1 < x < x_2$ shrinks if the elipticity increases, and disappears at the critical value $\bar{\alpha}_c = 1.15$.

In the phase-separated state, there exist two phases, phase 1 with a lower charge carrier density is the commensurate one and phase 2 with a higher charge carrier density corresponds to the incommensurate SDW state. The phase diagram of the system can be constructed following the approach described in the previous subsections (see, e.g., Sect. 8.2). The obtained results are summarized in the phase diagram in the (x, α) plane shown in Fig. 8.11 [22]. This phase diagram is calculated for $n = \kappa = 1$. It remains qualitatively the same if $n \neq 0$. If the nonmagnetic bands are absent, the homogeneous commensurate SDW phase exists only when doping is zero. Consequently, the electronic density x_1 vanishes in phase 1 of the phase-separated state for any α.

The phase separation in iron-based superconductors was observed in several experiments [44–47]. For example, the inhomogeneous state with a commensurate antiferromagnetic and nonmagnetic domains with characteristic sizes about 65 nm was observed in the hole-doped $Ba_{1-x}K_xFe_2As_2$ compound [44]. The described above model predicts that the second phase is an incommensurate SDW rather than a nonmagnetic one. However, the proposed mechanism of phase separation can, in general, be consistent with the observations. The thermodynamic potentials of the incommensurate SDW and the metastable paramagnetic phases are very close to each other in the doping range $x > x_2$. The incommensurate SDW phase can be destroyed by an additional cause not taken into account in the model, e.g., by a disorder. In this case, the phase separation might occur between the commensurate SDW and the paramagnetic phases.

Here, we consider the SDW order parameter in the form (8.4). As it was mentioned in the previous section, in the literature a different order parameter is also discussed (see, e.g., Ref. [35]). We must remember that the relative stability of different order parameters is likely a non-universal quantity, which depends on a variety of microscopic parameters (e.g., details of the band structure, interaction, disorder). Therefore, the type of suitable order parameter cannot be assuredly deduced without an input from experiments.

The calculations above show that there are only two possible equilibrium directions of the vector $\mathbf{q}$ characterizing the incommensurate SDW phase: it can be either parallel or perpendicular to the nesting vector $\mathbf{Q}_0$, depending on the type of doping and the sign of the ellipticity parameter α. It is clear that both the magnitude and direction of $\mathbf{q}$ are sensitive to the shape of the Fermi surface, which is simplified in the present calculations. In actual pnictides, the shape of the hole pockets deviates from perfect circles and the spectrum of the bands depends on the transverse momentum k_z. However, the observation of the incommensurate SDW phase with $\mathbf{q}$ perpendicular to $\mathbf{Q}_0$ in the electron-doped $Ba(Fe_{1-x}Co_x)_2As_2$ was reported in Refs. [55–57] and double-peaked spin fluctuations corresponding to $\mathbf{q}$ parallel to $\mathbf{Q}_0$ have been observed in the hole-doped $Ba_{1-x}K_xFe_2As_2$ compound in a wide doping range [58, 59]. These observations are in agreement with the prediction of the model.

Thus, a simple model approximating the Fermi surface of pnictides implies the existence of electronic phase separation even in the weak-coupling regime. This is an important finding for the interpretation of the experimental data on phase inhomogeneity of iron pnictides: it proves that a purely electronic model with moderate interaction is sufficient to explain the observed inhomogeneities.

8.6 Competition of Inhomogeneous States in Anisotropic Hubbard Model

Here we consider an anisotropic 3D Hubbard model, which has a nested Fermi surface in the regime of the weak coupling [18]. The goal of this chapter is not only to demonstrate one more example of the important model with the nesting, but also to address a question on a competition of different inhomogeneous phases, namely the phase separation and domains, against each other. It will be shown that theoretically estimated energies of competing inhomogeneous states may be very close to each other. Consequently, which of these phases becomes the true ground state in a real system requires taking into account factors external to the considered minimal models, such as disorder, phonons, etc.

We consider antiferromagnetic state of anisotropic Hubbard model on a 3D cubic lattice in the weak coupling regime. The Hamiltonian of the model is

$$\hat{H} = \hat{H}_0 + \hat{H}_{\text{int}}, \tag{8.35}$$
$$\hat{H}_0 = \sum_{\langle ij\rangle,\sigma} t_{ij}\left(a^{\dagger}_{i\sigma}a_{j\sigma} + H.c.\right) - \mu\sum_{i,\sigma} n_{i\sigma},$$
$$\hat{H}_{\text{int}} = U\sum_i \left(n_{i\uparrow} - \frac{1}{2}\right)\left(n_{i\downarrow} - \frac{1}{2}\right),$$

where the hopping amplitudes t_{ij} are different for different orientations of the neighboring ij-bond: when the bond is parallel to the α-axis ($\alpha = x, y, z$), the amplitude is $t_{ij} = t_\alpha$. The term $\hat{H}_{\text{int}}$ describes on-site Coulomb repulsion of electrons with opposite spin projections, $U > 0$. We will consider the system near the half-filling, that is, with one electron per site. The terms $1/2$ are added in order to chemical potential μ would equal to zero at half-filling.

The ground state of the Hubbard model at half-filling is antiferromagnetic [60]. It is reasonable to suppose that small doping does not destroy the antiferromagnetic ordering. In the antiferromagnetic state, the averaged number of spin-up electrons on a site i is not equal to the averaged number of spin-down electrons and we can introduce the position-dependent order parameter

$$\Delta_i = \frac{U}{2}\left(\langle n_{i\uparrow}\rangle - \langle n_{i\downarrow}\rangle\right). \tag{8.36}$$

We also define a local doping as $x_i = \langle n_{i\uparrow}\rangle + \langle n_{i\downarrow}\rangle - 1$.

We study the model Eq. (8.35) in the weak coupling regime, $U < W$, where $W = 4\Sigma_\alpha t_\alpha$ is the bandwidth In this case we can apply the , mean-field approach which is based on the following replacement $n_{i\uparrow}n_{i\downarrow} \to \langle n_{i\uparrow}\rangle n_{i\downarrow} + n_{i\uparrow}\langle n_{i\downarrow}\rangle - \langle n_{i\uparrow}n_{i\downarrow}\rangle$ in the Hamiltonian (8.35). As a result we obtain

$$\hat{H}^{\text{MF}}_{\text{int}} = \sum_i \left[\Delta_i\left(n_{i\downarrow} - n_{i\uparrow}\right) - U\left(\frac{x_i(x_i+2)}{4} - \frac{\Delta_i^2}{U^2}\right)\right] \tag{8.37}$$

and μ in the term $\hat{H}_0$ is replaced by local chemical potential $\mu'_i = \mu - Ux_i/2$. Values Δ_i and x_i should be found selfconsistently.

When electrons or holes are added, the homogeneous state of the model Eq. (8.35) may become unstable. We can consider different possible inhomogeneous states. The simplest one is the phase separation into commensurate antiferromagnetic and paramagnetic phases. The next step is to take into account the incommensurate antiferromagnetic state. The phase separation into incommensurate and commensurate antiferromagnetic phases is more energetically favorable than the commensurate antiferromagnetic/paramagnet separation [18].

At half-filling $\mu'_i = 0$ and, if we neglect the interaction, the kinetic energy of the system is

$$\varepsilon(\mathbf{k}) = -2\left[t_x\cos(k_x) + t_y\cos(k_y) + t_z\cos(k_z)\right], \tag{8.38}$$

where $\mathbf{k} = (k_x, k_y, k_z)$ is the wave vector. The Fermi surface $\varepsilon(\mathbf{k})$ nests perfectly with the nesting vector $\mathbf{Q}_0 = (\pi, \pi, \pi)$ (we assume here that the lattice constant $a = 1$). The doping and electron–electron interaction destroys the ideal nesting. First, we consider the homogeneous ($x_i = x$) commensurate antiferromagnetic state, when $\Delta_i = (-1)^{i_x+i_y+i_z}\Delta$ and $\mathbf{i} = (i_x, i_y, i_z)$ is the position of the lattice site. We perform the Fourier transformation in the Hamiltonian and introduce the four-component vector $\psi_{\mathbf{k}}^\dagger = (c_{\mathbf{k}\uparrow}^\dagger, c_{\mathbf{k}+\mathbf{Q}_0\uparrow}^\dagger, c_{\mathbf{k}\downarrow}^\dagger, c_{\mathbf{k}+\mathbf{Q}_0\downarrow}^\dagger)$. The Hamiltonian of the system $\hat{H} = \hat{H}_0 + \hat{H}_{\text{int}}^{\text{MF}}$ now can be rewritten as

$$\hat{H} = \sum_{\mathbf{k}} \psi_{\mathbf{k}}^\dagger \hat{H}_{\mathbf{k}} \psi_{\mathbf{k}} - U\mathcal{N}\left[\frac{x(x+2)}{4} - \frac{\Delta^2}{U^2}\right], \tag{8.39}$$

where $\mathcal{N}$ is the number of the lattice sites and

$$\hat{H}_{\mathbf{k}} = \begin{pmatrix} \varepsilon(\mathbf{k}) - \mu' & \Delta & 0 & 0 \\ \Delta & \varepsilon(\mathbf{k}) - \mu' & 0 & 0 \\ 0 & 0 & \varepsilon(\mathbf{k}) - \mu' & -\Delta \\ 0 & 0 & -\Delta & \varepsilon(\mathbf{k}) - \mu' \end{pmatrix}. \tag{8.40}$$

The eigenenergies of this Hamiltonian and corresponding grand potential per site at zero temperature are:

$$E_{1,2}(\mathbf{k}) = -\mu' \pm \sqrt{\varepsilon(\mathbf{k})^2 + \Delta^2}, \quad \varepsilon(\mathbf{k}) = -\varepsilon(\mathbf{k} + \mathbf{Q}_0), \tag{8.41}$$

$$\Omega = \sum_s \int \frac{d^3\mathbf{k}}{(2\pi)^3} E_s(\mathbf{k})\Theta[-E_s(\mathbf{k})] - \left[\frac{x(x+2)}{4} - \frac{\Delta^2}{U^2}\right]. \tag{8.42}$$

At half-filling, the minimization of the grand potential Eq. (8.42), with respect to Δ gives the following equation

$$\frac{2}{U} - \int_{-W/2}^{W/2} d\varepsilon \frac{N(\varepsilon)}{\sqrt{\varepsilon^2 + \Delta^2}} = 0, \tag{8.43}$$

where we introduce the electron density of states

$$N(\varepsilon) = \int \frac{d^3\mathbf{k}}{(2\pi)^3} \delta\left[\varepsilon - \varepsilon(\mathbf{k})\right].$$

In the weak coupling regime $\Delta << W$, we can put $N(\varepsilon) = N(0)$ and the integral in Eq. (8.43) can be taken analytically. In so doing, we obtain a usual BCS-like result $\Delta(x = 0) = \Delta_0 = W \exp(-1/N(0)U)$. In the case of non-zero doping, we obtain in the same approximation from Eqs. (8.41) and (8.42) the equation for Δ in the form

$$2\ln\frac{\Delta_0}{\Delta} = \int_{-W/2}^{W/2} d\varepsilon \frac{\Theta\left(\mu' - \sqrt{\varepsilon^2+\Delta^2}\right)}{\sqrt{\varepsilon^2+\Delta^2}}. \tag{8.44}$$

After integration we get

$$|\mu'| + \sqrt{\mu'^2 - \Delta^2} = \Delta_0. \tag{8.45}$$

The doping x is given by the evident relation:

$$x = \sum_s \int \frac{d^3\mathbf{k}}{(2\pi)^3} \Theta\left(-E_s(\mathbf{k})\right) - 1.$$

In the weak coupling regime, we derive $x = 2N(0)\mathrm{sign}(\mu')\sqrt{\mu' - \Delta^2}$. Neglecting insignificant correction $Ux/2$, we obtain from the latter relation and Eq. (8.45)

$$\Delta = \Delta_0\sqrt{1 - \frac{|x|}{N(0)\Delta_0}}, \qquad |\mu| = \Delta_0\left(1 - \frac{|x|}{2N(0)\Delta_0}\right). \tag{8.46}$$

We see that $\partial\mu/\partial x = -2N(0) < 0$ and in the weak coupling regime the considered homogeneous antiferromagnetic state is unstable. From the same relations in the paramagnetic phase $\Delta = 0$, we have $\mu = 1/2N(0) > 0$.

In the case of the inhomogeneous state consisting of the commensurate antiferromagnetic and paramagnetic phases, the Maxwell rule gives for the chemical potential $\mu_{\mathrm{cAFM}} = \Delta_0/\sqrt{2} \approx 0.7\Delta_0$. The main goal of this section is to determine which of the inhomogeneous states is energetically favorable. This can be found out by comparison of the free energies. In the considered case, the chemical potential is independent of the doping, then, we have

$$F_1 = F(0) + 0.707\Delta_0 x, \tag{8.47}$$

where $F(0)$ is the free energy of undoped antiferromagnetic insulator.

Following the line of the previous parts of these chapter, now we consider the incommensurate homogeneous antiferromagnetic state with the nesting vector $\mathbf{Q} = \mathbf{Q}_0 + \mathbf{q}$, where $|\mathbf{q}|$ is small. In this case, the order parameter has a form $\Delta_i = (-1)^{i_x+i_y+i_z}\Delta e^{i\mathbf{qr}}$. Taking into account that in the incommensurate state $\varepsilon(\mathbf{k}+\mathbf{Q}) = -\varepsilon(\mathbf{k}+\mathbf{q})$, we derive the eigenenergies of the mean-field Hamiltonian (8.39) in the form

$$E_{1,2}(\mathbf{k}) = -\mu' + \frac{\varepsilon(\mathbf{k}) - \varepsilon(\mathbf{k}+\mathbf{q})}{2} \pm \sqrt{\left[\frac{\varepsilon(\mathbf{k}) - \varepsilon(\mathbf{k}+\mathbf{q})}{2}\right]^2 + \Delta^2}. \tag{8.48}$$

According to the standard procedure, we substitute the eigenenergies in the grand potential (8.42) and obtain equations for Δ and $|\mathbf{q}|$ from the minimization conditions $\partial\Omega/\partial\Delta = 0$ and $\partial\Omega/\partial|\mathbf{q}| = 0$. The thermodynamical relation $\partial\Omega/\partial\mu = 0$ allows us to find the dependencies of Δ and $|\mathbf{q}|$ on doping. Then, we can calculate the free energy of the phase separated state consisting of the incommensurate and commensurate antiferromagnetic phases using the Maxwell rule. Such a problem can be solved numerically, see [18]. Under the same conditions as Eq. (8.47), we find the chemical potential of the phase separated state $\mu_{\text{iAFM}} \approx 0.698\Delta_0$ and the free energy

$$F_2 = F(0) + 0.698\Delta_0 x. \tag{8.49}$$

Comparing Eqs (8.47) and (8.49) we see that the phase separation into the commensurate and incommensurate antiferromagnetic phases is more favorable than the separation into the commensurate antiferromagnetic state and the paramagnetic state. However, this difference is very small, $F_1 - F_2 \approx 0.009\Delta_0 x$.

Another type of inhomogeneous phase competing to become the true ground state is the phase with domain walls. In this case the problem can be analyzed only numerically. To perform such a study in general form in this book is unreal. Then, following Ref. [18], we confine ourselves to the special case of low doping, when we can neglect the interaction between domain walls. The next important characteristic of the domain wall is its orientation relative to the crystal axis. In general, it could be arbitrary [61]. However, we expect that, in agreement with Ref. [62], the domain walls whose normal vectors are parallel to one of the crystal axis are the most stable in the case of anisotropic system.

We consider two types of domain walls, bond-centered and site-centered. Their structures can be schematically represented with the help of the following one-dimensional pictures

↑↓↑ ↓↓ ↑↓↑	bond-centered domain wall,
↑↓↑ ↓ ∘ ↑↓↑	site-centered domain wall.

The arrows represent the direction of the on-site spin, the symbol '∘' corresponds to a site with zero magnetization. Despite the differences in structures, the numerical simulations show that the energies of the both types of the domain walls are very close to each other.

For definiteness, we assume that the domain wall is perpendicular to the x axis. For such an orientation, the translation invariance in y and z directions is preserved, while it is broken in x-direction. It is convenient to switch to the mixed representation: in the x-direction, we continue using real space coordinate i_x, while in the transverse directions the 2D quasi-momentum $\mathbf{p} = (p_y, p_z)$ is introduced. The density of electrons, the order parameter and the partial Fourier transform of $c^\dagger_{i\sigma}$ are

$$\langle n_{i\uparrow}\rangle + \langle n_{i\downarrow}\rangle = n_{i_x}, \qquad \Delta_i = \Delta_{i_x}(-1)^{i_y+i_z},$$
$$c^\dagger_{i_x\mathbf{p}\sigma} = \frac{1}{\sqrt{N_y N_z}} \sum_{i_y,i_z} c^\dagger_{i\sigma} e^{i(p_y i_y + p_z i_z)}, \tag{8.50}$$

where $\mathcal{N}_\alpha \gg 1$ is the number of unit cells along axis $\alpha = x, y, z$.The mean-field Hamiltonian in the mixed representation reads

$$\hat{H} = \hat{H}_{\text{kin}} + \hat{H}_{\text{int}} \tag{8.51}$$
$$\hat{H}_{\text{kin}} = \sum_{i_x,\mathbf{p},\sigma} \left[t_x \left(c^\dagger_{i_x\mathbf{p}\sigma} c_{i_x+1\mathbf{p}\sigma} + H.c. \right) + \left(\varepsilon^\perp_{\mathbf{p}} - \mu'_{i_x} \right) c^\dagger_{i_x\mathbf{p}\sigma} c_{i_x\mathbf{p}\sigma} \right],$$
$$\hat{H}_{\text{int}} = -\sum_{i_x,\mathbf{p}} \Delta_{i_x} \left(c^\dagger_{i_x\mathbf{p}\uparrow} c_{i_x\mathbf{p}+\mathbf{P}_0\uparrow} - c^\dagger_{i_x\mathbf{p}\downarrow} c_{i_x\mathbf{p}+\mathbf{P}_0\downarrow} \right) - U\frac{\mathcal{N}}{\mathcal{N}_x} \sum_{i_x} \left(\frac{n^2_{i_x} - 1}{4} - \frac{\Delta^2_{i_x}}{U^2} \right),$$

where $\mu' = \mu - U(n_{i_x} - 1)/2$, $\varepsilon^\perp_{\mathbf{p}} = -2t_y \cos(p_y) - 2t_z \cos(p_z)$, and $\mathbf{P}_0 = (\pi, \pi)$.

Following Ref. [18], we introduce $2\mathcal{N}_x$-component vectors

$$\psi^\dagger_{\mathbf{p}\sigma} = (c^\dagger_{1\mathbf{p}\sigma}, \ldots, c^\dagger_{\mathcal{N}_x\mathbf{p}\sigma}, \ldots, c^\dagger_{1\mathbf{p}+\mathbf{P}_0\sigma}, \ldots, c^\dagger_{\mathcal{N}_x\mathbf{p}+\mathbf{P}_0\sigma})$$

and rewrite the Hamiltonian (8.51) in the matrix form $\hat{H} = \sum_{\mathbf{p}\sigma} \psi^\dagger_{\mathbf{p}\sigma} \hat{H}_{\mathbf{p}\sigma} \psi_{\mathbf{p}\sigma}$, where

$$\hat{H}_{\mathbf{p}\uparrow} = \begin{pmatrix} \hat{H}_{0\mathbf{p}} & \hat{\Delta} \\ \hat{\Delta} & \hat{H}_{0\mathbf{p}} \end{pmatrix}, \qquad \hat{H}_{\mathbf{p}\downarrow} = \begin{pmatrix} \hat{H}_{0\mathbf{p}+\mathbf{P}_0} & -\hat{\Delta} \\ -\hat{\Delta} & \hat{H}_{0\mathbf{p}+\mathbf{P}_0} \end{pmatrix},$$

$\hat{\Delta} = \text{diag}(\Delta_1, \ldots, \Delta_{\mathcal{N}_x})$, and

$$\hat{H}_{0\mathbf{p}} = \begin{pmatrix} \varepsilon^\perp - \mu'_1 & t & 0 & \ldots & t \\ t & \varepsilon^\perp - \mu'_2 & t & \ldots & 0 \\ 0 & t & \varepsilon^\perp - \mu'_3 & \ldots & 0 \\ \vdots & \vdots & \vdots & \ddots & \vdots \\ t & 0 & 0 & \ldots & \varepsilon^\perp - \mu'_{\mathcal{N}_x} \end{pmatrix}.$$

We impose periodic boundary conditions in the x direction constructing the matrix $\hat{H}_{\mathbf{p}\sigma}$. Then, it is convenient to introduce the grand potential $\Omega_{\mathcal{N}_x}$ of the system with $\mathcal{N}_x$ unit cells along x axis per unit area in the (y, z) plane

$$\Omega_{\mathcal{N}_x} = \sum_{\sigma,S} \int \frac{d^2\mathbf{P}}{(2\pi)^2} E^S_{\mathbf{p}\sigma} \Theta(-E^S_{\mathbf{p}\sigma}) - U \sum_{i_x} \left(\frac{n^2_{i_x} - 1}{4} - \frac{\Delta^2_{i_x}}{U^2} \right).$$

Here $S = 1, ..., \mathcal{N}_x$ and $E^S_{\mathbf{p}\sigma}$ are eigenenergies of the Hamiltonian $\hat{H}_{\mathbf{p}\sigma}$. A numerical minimization of $\Omega_{\mathcal{N}_x}$ is used to calculate the spatial dependence of the order parameter and the number of electrons per site.

Once $\Omega_{\mathcal{N}_x}$ is known, the energy of a single domain wall E_{dw} can be calculated. Consider the cases even and odd number of $\mathcal{N}_x$. A system with even $\mathcal{N}_x$ after periodic continuation is antiferromagnetically ordered and $\Omega^{\text{even}}_{\mathcal{N}_x} = \Omega_0 \mathcal{N}_x$, where Ω_0 is the grand potential per site of the system with homogeneous antiferromagnetic ordering.

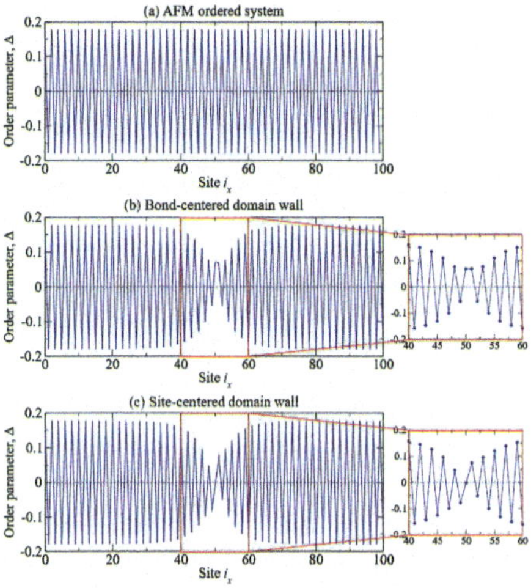

Fig. 8.12 Spatial variation of the order parameter along x-axis for even (panel (**a**), $\mathcal{N}_x = 100$) and odd [panels (**b**) and (**c**), $\mathcal{N}_x = 101$] values of $\mathcal{N}_x$. Panel (**b**) shows the bond-centered domain wall. In the center of such a structure, we see two sites with identical values of the order parameters. The site-centered domain wall is in panel (**c**). This domain wall is centered on a site with vanishing order parameter. The simulations are performed for U/W = 0.17, $(t_x, t_y, t_z) = (1.0, 1.0, 0.7)$ [18]

A system with odd $\mathcal{N}_x$ unavoidably contains a domain wall after the periodic continuation and $\Omega^{\text{odd}}_{\mathcal{N}_x} = \Omega_0 \mathcal{N}_x + E_{\text{dw}}$. Such a behavior is illustrated in Fig. 8.12, where the spatial dependence of the order parameter are shown for even and odd $\mathcal{N}_x$. The above consideration allows us to extract E_{dw} analyzing the dependence of $\Omega_{\mathcal{N}_x}$ on $\mathcal{N}_x$ [18].

In Fig. 8.13 we present the spatial dependences of the absolute value of the order parameter and the electron density calculated for different values of $\mu' = \mu - U(n - 1)/2$, where n is the averaged number of electrons per site. We see that Δ_{i_x} and n_{i_x} are even function of μ'. A small deviation of the chemical potential from zero sharply changes the order parameter and the electron density inside the domain wall. The accumulation of the injected charge carriers in the domain wall is illustrated in Fig. 8.13b. At half-filling there is one electron per site. When the chemical potential changes, the carriers, electron or holes, pile up in the domain wall.

Figure 8.14 presents the domain wall energy and the total charge accumulated inside the domain walls per unit area in the plane (y, z) versus the chemical potential. The energy E_{dw} is the even function of the chemical potential, while the accumulated charge is odd function, see Figs. 8.12b and 8.13b. These features are consequences of the charge conjugation symmetry of the considered model.

At zero doping (zero chemical potential), the domain wall energy is positive and the state with the domain walls is energetically unfavorable. As the doping (and the chemical potential) grows, charges dope the domain wall, improving its stability, see Fig. 8.13. At some threshold value of $\mu = \mu_{\text{dw}}$ the domain wall energy vanishes,

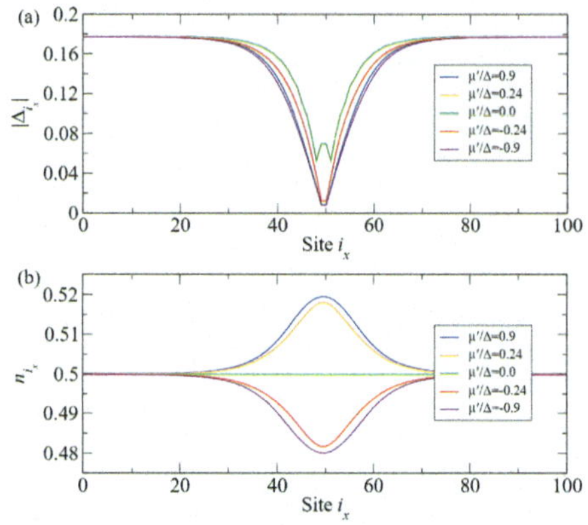

Fig. 8.13 Spatial dependence of **a** the absolute values of the order parameter, and **b** the electron density calculated for the bond-centered domain wall at different values of the chemical potential. The simulations are performed for $U/W = 0.17$, $(t_x, t_y, t_z) = (1.0, 1.0, 0.7)$ [18]

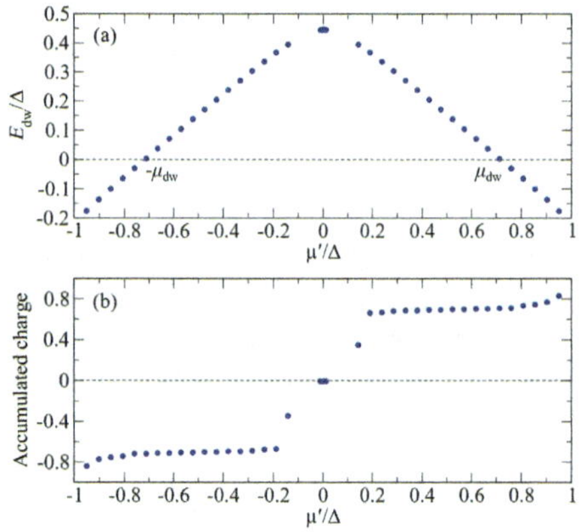

Fig. 8.14 Chemical potential dependence of **a** the bond-centered domain wall energy and **b** the number of charge carriers in the domain wall. In panel (**a**) the critical chemical potential μ_{dw} is marked. It corresponds to the value of μ at which the domain wall energy vanishes. Computation was performed at $N_x = 101$, $U/W = 0.17$, and $(t_x, t_y, t_z) = (1.0, 1.0, 0.7)$ [18]

$E_{dw} = 0$. When $|\mu| > \mu_{dw}$ the domain wall energy becomes negative, and domain walls carrying finite charge density enter the bulk of the system. In this case, similar to Eqs. (8.47) and (8.49) we have

$$F_{dw} \approx F(0) + \mu_{dw} x. \tag{8.52}$$

This equation remains applicable at low doping as long as the distance between the domain walls is large and the interaction between them may be neglected.

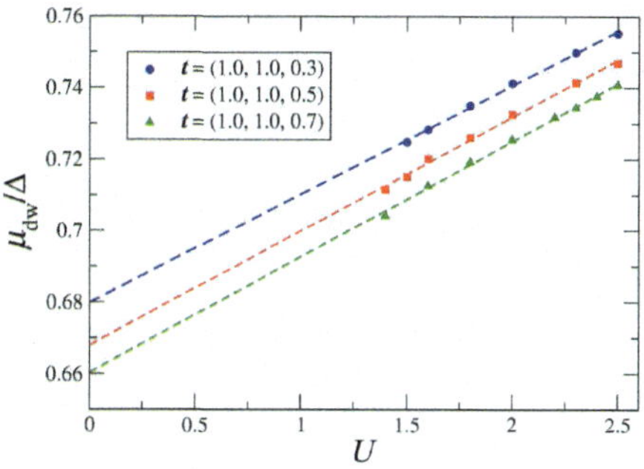

Fig. 8.15 Critical chemical potential versus interaction constant U for the systems with different anisotropy (see legend). Linear fits are shown as dashed lines. The value of U is formally normalized on $t_x = 1$. (Fig. 7 from Ref. [18])

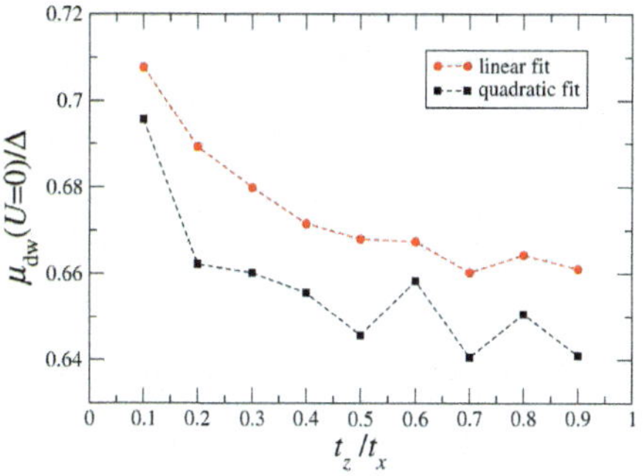

Fig. 8.16 Critical value of the chemical potential μ_{dw} versus the hopping anisotropy parameter t_z/t_x in the limit $U/t_x \to 0$. The data obtained by linear fit are shown by red circles and by quadratic function fit are shown by black squares [18]

The critical value of the chemical potential calculated numerically is shown in Fig. 8.15. The calculations were performed for the system with the anisotropy of easy-plane type ($t_x = t_y > t_z$) for three different t_z. We see that μ_{dw} increases when the anisotropy increases. We have to perform a consistent comparison of μ_{dw} with μ_{cAFM} and μ_{iAFM} in the weak-coupling regime, $U/t \ll 1$. The desired value of μ_{dw} was found in Ref. [18]) by linear extrapolation $\mu_{dw}(U)$ to $U = 0$, see dashed lines in Fig. 8.15. The results are shown in Fig. 8.16. Alternatively, the same data can be approximated by a quadratic function, which are also plotted in Fig. 8.16. The results of different fits are close.

In the considered model at low doping, the critical values of the chemical potential for different phases are ordered according to $\mu_{dw} < \mu_{iAFM} < \mu_{cAFM}$. Thus, the state with the domain walls has lower energy than the phase-separated states. However, the energy difference separating the most stable and metastable phases is very small, $\mu_{iAFM} - \mu_{dw} \lesssim 0.06\Delta_0$ for all anisotropy parameters [18], accordingly,

$$F_2 - F_{dw} \lesssim 0.06\Delta_0 x. \tag{8.53}$$

In conclusion, we see that the state with the domain walls is the most stable at low doping. However, the energies of metastable inhomogeneous states are found to be very close to the ground state energy, which introduce a significant uncertainty into theoretical modeling of the inhomogeneous states of many-body systems. Reliable description appears to be impossible without taking into account different factors characteristic of a particular system (e.g., disorder, electron-lattice interaction, etc.) and/or experimental input.

References

1. D. Khomskii, *Basic Aspects of the Quantum Theory of Solids* (Cambridge University Press, 2010)
2. G. Grüner, Rev. Mod. Phys. **60**, 1129 (1988)
3. P. Monceau, Adv. Phys. **60**, 325 (2012)
4. A.W. Overhauser, Phys. Rev. **128**, 1437 (1962)
5. G. Grüner, Rev. Mod. Phys. **66**, 1 (1994)
6. J. Ruvalds, C.T. Rieck, S. Tewari, J. Thoma, A. Virosztek, Phys. Rev. B **51**, 3797 (1995)
7. A. Gabovich, A.I. Voitenko, J.F. Annett, M. Ausloos, Supercond. Sci. Technol **14**(4), R1 (2001)
8. K. Terashima, Y. Sekiba, J. Bowen, K. Nakayama, T. Kawahara, T. Sato, P. Richard, Y.M. Xu, L. Li, G. Cao et al., PNAS **106**(18), 7330 (2009)
9. Y.D. Chuang, A.D. Gromko, D.S. Dessau, T. Kimura, Y. Tokura, Science **292**(5521), 1509 (2001)
10. A. Shibatani, J. Phys. Soc. Jpn. **26**(2), 299 (1969)
11. A. Shibatani, K. Motizuki, T. Nagamiya, Phys. Rev. **177**, 984 (1969)
12. A. Shibatani, J. Phys. Soc. Jpn. **29**(1), 93 (1970)
13. T.M. Rice, Phys. Rev. B **2**, 3619 (1970)
14. A. Narayanan, A. Kiswandhi, D. Graf, J. Brooks, P. Chaikin, Phys. Rev. Lett. **112**(14), 146402 (2014)
15. I.J. Lee, S.E. Brown, W. Yu, M.J. Naughton, P.M. Chaikin, Phys. Rev. Lett. **94**, 197001 (2005)
16. T. Vuletić, P. Auban-Senzier, S.T.D.J.C. Pasquier, M. Húritier, K. Bechgaard, Eur. Phys. J. B **25**, 319 (2002)
17. N. Kang, B. Salameh, P. Auban-Senzier, D. Jérome, C.R. Pasquier, S. Brazovskii, Phys. Rev. B **81**, 100509 (2010)
18. S.V. Kokanova, P.A. Maksimov, A.V. Rozhkov, A.O. Sboychakov, Phys. Rev. B **104**, 075110 (2021)
19. N.I. Kulikov, V.V. Tugushev, Phys. Usp. **27**(12), 954 (1984). [Usp. Fiz. Nauk **144**, 643 (1984)]
20. E. Fawcett, Rev. Mod. Phys. **60**, 209 (1988)
21. A.L. Rakhmanov, A.V. Rozhkov, A.O. Sboychakov, F. Nori, Phys. Rev. B **87**, 075128 (2013)
22. A.O. Sboychakov, A.V. Rozhkov, K.I. Kugel, A.L. Rakhmanov, F. Nori, Phys. Rev. B **88**, 195142 (2013)
23. A.L. Rakhmanov, K.I. Kugel, A.O. Sboychakov, J. Supercond. Novel Magn. **33**, 2405 (2020)
24. A.O. Sboychakov, A.L. Rakhmanov, A.V. Rozhkov, F. Nori, Phys. Rev. B **87**, 121401 (2013)
25. A.O. Sboychakov, A.V. Rozhkov, A.L. Rakhmanov, F. Nori, Phys. Rev. B **88**, 045409 (2013)
26. G. Campi, A. Bianconi, N. Poccia, G. Bianconi, L. Barba, G. Arrighetti, D. Innocenti, J. Karpinski, N.D. Zhigadlo, S.M. Kazakov, M. Burghammer, M. von Zimmermann, M. Sprung, A. Ricci, Nature **525**(7569), 359 (2015)
27. R.Y. Chen, B.F. Hu, T. Dong, N.L. Wang, Phys. Rev. B **89**(7), 075114 (2014)
28. A.L. Rakhmanov, K.I. Kugel, M.Yu. Kagan, A.V. Rozhkov, A.O. Sboychakov, JETP Lett. **105**(12), 806 (2017)
29. A.A. Gorbatsevich, Yu. V.. Kopaev, I.V. Tokatly, Zh. Eksp. Teor. Fiz. **101**, 971 (1992). [Sov. Phys. JETP 74, 521 (1992)]
30. P.A. Igoshev, M.A. Timirgazin, V.F. Gilmutdinov, A.K. Arzhnikov, V.Y. Irkhin, J. Phys.: Condens. Matter **27**(44), 446002 (2015)
31. P.A. Igoshev, M.A. Timirgazin, V.F. Gilmutdinov, A.K. Arzhnikov, V.Yu. Irkhin, J. Magn. Magn. Mater. **383**, 2 (2015)

32. A. Pankratova, P. Igoshev, V.Yu. Irkhin, J. Phys.: Condens. Matter **33**(37), 375802 (2021)
33. P. Fulde, R.A. Ferrell, Phys. Rev. **135**(3A), A550 (1964)
34. A.I. Larkin, Y.N. Ovchinnikov, Sov. Phys. JETP **20**, 762 (1965)
35. L.P. Gor'kov, G.B. Teitel'baum, Phys. Rev. B **82**(2), 020510 (2010)
36. M. Le Bellac, F. Mortessagne, G.G. Batrouni, *Equilibrium and Non-Equilibrium Statistical Thermodynamics* (Cambridge University Press, 2004)
37. D.E. Sheehy, L. Radzihovsky, Ann. Phys. **322**(8), 1790 (2007)
38. S. Takada, Progr. Theor. Phys. **43**(1), 27 (1970)
39. A.O. Sboychakov, A.L. Rakhmanov, K.I. Kugel, A.V. Rozhkov, F. Nori, Phys. Rev. B **95**, 014203 (2017)
40. A.V. Rozhkov, A.L. Rakhmanov, A.O. Sboychakov, K.I. Kugel, F. Nori, Phys. Rev. Lett. **119**, 107601 (2017)
41. J. Lorenzana, C. Castellani, C. Di Castro, Phys. Rev. B **64**(23), 235127 (2001)
42. K.I. Kugel, A.L. Rakhmanov, A.O. Sboychakov, F. Kusmartsev, N. Poccia, A. Bianconi, Supercond. Sci. Technol. **22**(1), 014007 (2008)
43. P. Dai, J. Hu, E. Dagotto, Nat. Phys. **8**(10), 709 (2012)
44. J.T. Park, D.S. Inosov, C. Niedermayer, G.L. Sun, D. Haug, N.B. Christensen, R. Dinnebier, A.V. Boris, A.J. Drew, L. Schulz, T. Shapoval, U. Wolff, V. Neu, X. Yang, C.T. Lin, B. Keimer, V. Hinkov, Phys. Rev. Lett. **102**(11), 117006 (2009)
45. D.S. Inosov, A. Leineweber, X. Yang, J.T. Park, N.B. Christensen, R. Dinnebier, G.L. Sun, C. Niedermayer, D. Haug, P.W. Stephens, J. Stahn, O. Khvostikova, C.T. Lin, O.K. Andersen, B. Keimer, V. Hinkov, Phys. Rev. B **79**(22), 224503 (2009)
46. G. Lang, H.J. Grafe, D. Paar, F. Hammerath, K. Manthey, G. Behr, J. Werner, B. Büchner, Phys. Rev. Lett. **104**(9), 097001 (2010)
47. B. Shen, B. Zeng, G.F. Chen, J.B. He, D.M. Wang, H. Yang, H.H. Wen, EPL (Europhysics Letters) **96**(3), 37010 (2011)
48. Q. Luo, G. Martins, D.X. Yao, M. Daghofer, R. Yu, A. Moreo, E. Dagotto, Phys. Rev. B **82**(10), 104508 (2010)
49. A. Ricci, N. Poccia, G. Campi, B. Joseph, G. Arrighetti, L. Barba, M. Reynolds, M. Burghammer, H. Takeya, Y. Mizuguchi et al., Phys. Rev. B **84**(6), 060511 (2011)
50. A. Ricci, N. Poccia, B. Joseph, G. Arrighetti, L. Barba, J. Plaisier, G. Campi, Y. Mizuguchi, H. Takeya, Y. Takano et al., Supercond. Sci. Technol. **24**(8), 082002 (2011)
51. G.R. Stewart, Rev. Mod. Phys. **83**(4), 1589 (2011)
52. J. Hu, N. Hao, Phys. Rev. X **2**(2), 021009 (2012)
53. P. Richard, T. Sato, K. Nakayama, T. Takahashi, H. Ding, Rep. Progr. Phys. **74**(12), 124512 (2011)
54. I. Eremin, A.V. Chubukov, Phys. Rev. B **81**(2), 024511 (2010)
55. P. Bonville, F. Rullier-Albenque., D. Colson, A. Forget, EPL (Euro Phys. Lett.) **89**(6), 67008 (2010)
56. Y. Laplace, J. Bobroff, F. Rullier-Albenque, D. Colson, A. Forget, Phys. Rev. B **80**(14), 140501 (2009)
57. D.K. Pratt, M.G. Kim, A. Kreyssig, Y.B. Lee, G.S. Tucker, A. Thaler, W. Tian, J.L. Zarestky, S.L. Bud'ko, P.C. Canfield, B.N. Harmon, A.I. Goldman, McQueeney. Phys. Rev. Lett. **106**(25), 257001 (2011)
58. H. Luo, R. Zhang, M. Laver, Z. Yamani, M. Wang, X. Lu, M. Wang, Y. Chen, S. Li, S. Chang et al., Phys. Rev. Lett. **108**(24), 247002 (2012)
59. C.H. Lee, K. Kihou, H. Kawano-Furukawa, T. Saito, A. Iyo, H. Eisaki, H. Fukazawa, Y. Kohori, K. Suzuki, H. Usui et al., Phys. Rev. Lett. **106**(6), 067003 (2011)
60. D. Baeriswyl, D.K. Campbell, J.M. Carmelo, F. Guinea, E. Louis, *The Hubbard Model: Its Physics and Mathematical Physics*, vol. 343 (Springer Science & Business Media, 2013)
61. M. Kato, K. Machida, H. Nakanishi, M. Fujita, J. Phys. Soc. Jpn. **59**(3), 1047 (1990)
62. B.X. Zheng, C.M. Chung, P. Corboz, G. Ehlers, M.P. Qin, R.M. Noack, H. Shi, S.R. White, S. Zhang, G.K.L. Chan, Science **358**(6367), 1155 (2017)

Chapter 9
Phase Separation in Graphene-Based Systems

Abstract We consider the phase separation of hydrogen atoms adsorbed onto graphene. We analyze also the instability of the Fermi liquid of the AA stacked bilayer graphene and the phase separation in twisted bilayer graphene.

The formation of non-uniform electron states is a rather common feature for many complicated systems. In this section, we show that the reviewed above ideas of the phase separation can be implemented in the physics of graphene. We consider three examples. The first one deals with a graphene sheet with attached hydrogen atoms, the second one is the so-called AA bilayer graphene, and the last one is the twisted bilayer graphene. It turns out that AA bilayer graphene has the Fermi surface with evident nesting, whereas the electron system of hydrogenized graphene can be described in terms of the Falicov–Kimball model on the honeycomb lattice. Thus, the phase separation scenarios described in previous chapters can be applied to these systems. In hydrogenized twisted bilayer graphene, one graphene layer is rotated with respect to another one by some twist angle θ forming a superstructure in the sample. At the so-called first magic angle $\theta_c \cong 1^\circ$, the low-energy spectrum is described by the flat bands. As a result, the system is very susceptible to interaction which leads to the formation of an order parameter and in phase separation under doping.

9.1 Phase Separation of Hydrogen Atoms Adsorbed on Graphene

In this section, we consider the tendency to phase separation in partially hydrogenated graphene following Ref. [1]. Fully hydrogenated graphene is commonly referred to as graphane [2]. Graphane is an insulator. Its gap is about 5–6 eV. In a sheet of graphane, we can create a graphene patch of desired shape using a local dehydrogenation. In such a system, low-energy charge carriers from graphene cannot penetrate the graphane host. Thus, the boundary between graphene and graphane can be considered as an effective edge of the graphene structure. Therefore, the stability of the

M. Yu. Kagan et al., *Electronic Phase Separation in Magnetic and Superconducting Materials*, Springer Series in Solid-State Sciences 201,
https://doi.org/10.1007/978-3-031-55467-4_9

graphene–graphane interface is an important issue. The numerical study indicates that such an interface is stable and that the adsorbed hydrogen atoms tend to join in clusters [3–5]. These features can be described in terms of the phase separation into hydrogen-rich and hydrogen-free regions. A specific type of coupling between electrons and adatoms is assumed in Ref. [6] to demonstrate the phase separation. This assumptions are supported by some experimental and numerical arguments. In Ref. [1], the phase separation is discussed within the framework of a modified Falicov–Kimball model. The advantage of the latter approach is its generality since the phase separation is a well-known property of a ground state of the Falicov–Kimball model (Chap. 5) and this property is robust against variation of the model details.

9.1.1 Hamiltonian of Graphane and Partially Hydrogenated Graphene

Graphene is the carbon monolayer having a hexagonal crystal lattice. Such a lattice can be considered as consisting of two triangular sublattices, $\mathcal{A}$ and $\mathcal{B}$. Thus, elementary unit cell of graphene contains 2 carbon atoms, A and B. Graphane is fully hydrogenated graphene. Its crystal structure is shown in Fig. 9.1. Hydrogen atoms on one side of the graphene sheet are located above carbon atoms in the sublattice A, while hydrogen atoms located on the opposite side of the graphene are placed below the carbon atoms in the sublattice B. The s-electrons of hydrogen atoms are hybridized with p_z-electrons of carbon with hybridization energy [7] about $t_{sp} \cong 5.8\,\text{eV}$. We also should take into account that on-site energies of s-electrons of hydrogen and p_z-electrons of carbon are different with energy difference [7] $\epsilon_H \cong 0.4\,\text{eV}$. As a result, we obtain the following tight-binding Hamiltonian of graphane:

$$H_{\text{A}} = -t\sum_{\langle\mathbf{nm}\rangle\sigma}\left(p^{\dagger}_{\mathbf{n}A\sigma}p_{\mathbf{m}B\sigma} + H.c.\right) - \sum_{\mathbf{n}a\sigma}\left[t_{sp}\left(p^{\dagger}_{\mathbf{n}a\sigma}s_{\mathbf{n}a\sigma} + H.c.\right) + \epsilon_H s^{\dagger}_{\mathbf{n}a\sigma}s_{\mathbf{n}a\sigma}\right]. \tag{9.1}$$

In this equation, $p^{\dagger}_{\mathbf{n}a\sigma}$ and $p_{\mathbf{n}a\sigma}$ are creation and annihilation operators of p_z-electrons of carbon with spin projection σ located in the unit cell $\mathbf{n}$ in sublattice $a = A,\ B$. Similarly, $s^{\dagger}_{\mathbf{n}a\sigma}$ and $s_{\mathbf{n}a\sigma}$ are electronic operators of s-electrons of hydrogen. The first term in Eq. (9.1) describes the nearest-neighbor electron hopping amplitude of p_z electrons with amplitude $t \cong 2.8\,\text{eV}$, the second term describes the hybridization between s-electrons of hydrogen and p_z-electrons of carbon, while the last term is the on-site energy of s-electrons (corresponding on-site energy of p_z electrons is taken to be zero).

Let us introduce four-component spinors

$$\Psi_{\mathbf{n}\sigma} = \begin{pmatrix} P_{\mathbf{n}\sigma} \\ S_{\mathbf{n}\sigma} \end{pmatrix}, \tag{9.2}$$

Fig. 9.1 Graphane. Black balls, forming a hexagonal lattice, denote carbon atoms, while gray balls denote hydrogen atoms

where $P_{\mathbf{n}\sigma}$ and $S_{\mathbf{n}\sigma}$ are two-component spinors of the form

$$P_{\mathbf{n}\sigma} = \begin{pmatrix} P_{\mathbf{n}A\sigma} \\ P_{\mathbf{n}B\sigma} \end{pmatrix}, \quad S_{\mathbf{n}\sigma} = \begin{pmatrix} S_{\mathbf{n}A\sigma} \\ S_{\mathbf{n}B\sigma} \end{pmatrix}. \tag{9.3}$$

Using such a form for electronic operators and going to momentum representation, one can rewrite Hamiltonian (9.1) in the block-matrix form

$$H_{\mathrm{A}} = \sum_{\mathbf{k}\sigma} \Psi^{\dagger}_{\mathbf{k}\sigma} \hat{H}^{\mathrm{A}}_{\mathbf{k}} \Psi_{\mathbf{k}\sigma}, \quad \hat{H}^{\mathrm{A}}_{\mathbf{k}} = \begin{pmatrix} \hat{H}_{\mathbf{k}} & -t_{sp}\hat{\sigma}_0 \\ -t_{sp}\hat{\sigma}_0 & -\epsilon_H \hat{\sigma}_0 \end{pmatrix}, \tag{9.4}$$

where $\hat{\sigma}_0$ is the unit 2×2 matrix. The matrix $\hat{H}_{\mathbf{k}}$ in Eq. (9.4) is [11]

$$\hat{H}_{\mathbf{k}} = -t \begin{pmatrix} 0 & f_{\mathbf{k}} \\ f^*_{\mathbf{k}} & 0 \end{pmatrix}, \tag{9.5}$$

where

$$f_{\mathbf{k}} = e^{-ia_0 k_x} \left[1 + 2 \exp\left(\frac{3ia_0 k_x}{2} \right) \cos \frac{\sqrt{3}a_0 k_y}{2} \right] \tag{9.6}$$

and $a_0 = 1.42$Å is the nearest-neighbor carbon–carbon distance. For the matrix $\hat{H}^{\mathrm{A}}_{\mathbf{k}}$, analytical expressions for eigenvalues and eigenvectors can be obtained. Thus, in the limit of $\epsilon_H = 0$, we have

$$E_{\mathbf{k}}^{(s)} = \frac{1}{2}\left(\pm t\zeta_{\mathbf{k}} \pm \sqrt{4t_{sp}^2 + t^2\zeta_{\mathbf{k}}^2}\right), \quad (s = 1, 2, 3, 4)\,. \tag{9.7}$$

where $\zeta_{\mathbf{k}} = |f_{\mathbf{k}}|$.

Despite the simplicity of the proposed model, Hamiltonian (9.1) adequately describes all the basic electronic properties of graphane. At half-filling, the model (9.1) corresponds to a dielectric with a direct gap at Γ point. As it follows from the Eq. (9.7), the value of the gap can be estimated as $\Delta = \sqrt{9t^2 + 4t_{sp}^2} - 3t$ ($\zeta_{\mathbf{k}}|_{\mathbf{k}=0} = 3$). With the characteristic values of the model parameters mentioned above, $t = 2.7\,\text{eV}$, $t_{sp} = 5.8\,\text{eV}$, the gap is $\Delta \approx 6\,\text{eV}$, which corresponds to the estimates found in the literature [8, 9]. Note, however, that there is currently no unambiguous consensus on the size of the graphane gap, and the value of Δ may vary in different papers.

Hamiltonian (9.1) is valid for graphane when the number of hydrogen and carbon atoms is the same. In partially hydrogenated graphene, some of the hydrogen atoms are missing, and, consequently, the corresponding s-electrons are also missing in these sites. This circumstance can be taken into account by introducing an infinitely strong on-site repulsion between the "hydrogen hole" and the s electron:

$$H_{\text{EA}} = H_{\text{A}} + U \sum_{\mathbf{n}\sigma} S_{\mathbf{n}\sigma}^{\dagger} \hat{N}_{\mathbf{n}}^{\text{hh}} S_{\mathbf{n}\sigma}\,, \tag{9.8}$$

$$\hat{N}_{\mathbf{n}}^{\text{hh}} = \begin{pmatrix} n_{\mathbf{n}A}^{\text{hh}} & 0 \\ 0 & n_{\mathbf{n}B}^{\text{hh}} \end{pmatrix}, \tag{9.9}$$

where $n_{\mathbf{n}a}^{\text{hh}}$ $(a = A, B)$ is the operator of the number of "hydrogen holes" in the cell $\mathbf{n}$ in the sublattice a, and $U \to +\infty$. In fact, the operators $n_{\mathbf{n}a}^{\text{hh}}$ can be considered as random variables taking the values 0 or 1 and the average value $n^{\text{hh}} \equiv \langle n_{\mathbf{n}a}^{\text{hh}} \rangle = 1 - n^{\text{H}}$, where n^{H} is the number of hydrogen atoms in the sample per carbon atom. Thus, we see that the H_{EA} Hamiltonian is an analogue of the Falikov–Kimbal Hamiltonian, in which mobile p- and s-electrons interact with "hydrogen holes" whose concentration is fixed. So, the limiting case $n^{\text{hh}} = 1$ corresponds to graphene, while the concentration $n^{\text{hh}} = 0$ corresponds to graphane. Below, we will analyze the intermediate case $0 < n^{\text{hh}} < 1$.

9.1.2 Simple Consideration Showing Phase Separation in Partially Hydrogenated Graphene

In this subsection, we would like to give simple arguments indicating a tendency to phase separation in the system under consideration. Since t_{sp} exceeds t, let us consider the limit of $t_{sp} \gg t$. To do this, we introduce new electronic operators a and b that diagonalize the one-site part of the Hamiltonian (9.8):

$$H_{\mathrm{EA}} - H_{\mathrm{graphene}} = \sum_i a_i^\dagger a_i \left[t_{sp}(1 - n_i^{\mathrm{hh}}) + U n_i^{\mathrm{hh}}\right] - t_{sp} \sum_i b_i^\dagger b_i (1 - n_i^{\mathrm{hh}}), \tag{9.10}$$

where H_{graphene} corresponds to the first term in Eq. (9.1), and we also put for simplicity $\epsilon_H = 0$. We will consider the Hamiltonian H_{graphene} describing electronic hoppings by perturbation theory on t/t_{sp}. The old operators are expressed through the new ones according to the formulas:

$$p_i = \frac{b_i - a_i}{\sqrt{2}}(1 - n_i^{\mathrm{hh}}) + n_i^{\mathrm{hh}} b_i\,, \quad s_i = \frac{b_i + a_i}{\sqrt{2}}(1 - n_i^{\mathrm{hh}}) + n_i^{\mathrm{hh}} a_i\,. \tag{9.11}$$

In this formula, and in the formula above, we, for simplicity, omitted the spin and sublattice indices for all electronic operators, while the index i marks carbon atoms regardless of the sublattice. As follows from the Eq. (9.10), the energy of fermions a is much greater than the energy of fermions b for any n_i^{hh}. Thus, in the lowest order by t/t_{sp}, the states with fermions a are empty, and they can be neglected. In this approximation, the complete Hamiltonian of the partially hydrogenated graphene will have the form:

$$H_{\mathrm{EA}} \approx -t_{sp} \sum_i b_i^\dagger b_i (1 - n_i^{\mathrm{hh}}) - \frac{t}{2} \sum_{\langle ij \rangle} b_i^\dagger b_j \left[1 + \gamma(n_i^{\mathrm{hh}} + n_j^{\mathrm{hh}}) + \gamma^2 n_i^{\mathrm{hh}} n_j^{\mathrm{hh}}\right], \tag{9.12}$$

where $\gamma = \sqrt{2} - 1 \approx 0.41$, and $\langle \ldots \rangle$ means summation over nearest neighbors. From this expression, it can be seen that in order to separate two hydrogen holes sitting on neighboring sites, it is necessary to spend the energy of the order of $t\gamma^2 \langle b_i^\dagger b_j \rangle$. This corresponds to the *attraction* between hydrogen holes, and, consequently, between hydrogen atoms. A similar statement (based on somewhat different considerations) was made in the work [10]. This attraction leads to the phase separation.

9.1.3 Hubbard I Approximation and Exact Diagonalization on Finite Cluster

In Ref. [1], Hamiltonian (9.8) is studied in the Hubbard I approximation. The latter is based on the decoupling in the infinite set of equations on the electron Green's functions. The equations for one-particle Green's functions read

$$\begin{aligned}
&(\omega + \mu - \hat{H}_{\mathbf{k}})\hat{G}_{pp} + t_{sp}\hat{G}_{sp} = 1\,,\\
&(\omega + \mu + \epsilon_H)\hat{G}_{sp} + t_{sp}\hat{G}_{pp} - U\hat{F}_{sp} = 0\,,\\
&(\omega + \mu - \hat{H}_{\mathbf{k}})\hat{G}_{ps} + t_{sp}\hat{G}_{ss} = 0\,,\\
&(\omega + \mu + \epsilon_H)\hat{G}_{ss} + t_{sp}\hat{G}_{ps} - U\hat{F}_{ss} = 0\,,
\end{aligned} \tag{9.13}$$

where $\hat{G}_{pp}$, $\hat{G}_{sp}$, $\hat{G}_{ps}$, $\hat{G}_{ss}$, $\hat{F}_{sp}$, and $\hat{F}_{ss}$ are Fourier transformed 2×2 matrix propagators

$$\begin{aligned}
\hat{G}_{pp}(\mathbf{n}-\mathbf{m},\, t-t') &= -i\left\langle \hat{T}\, P_{\mathbf{n}\sigma}(t) P^{\dagger}_{\mathbf{m}\sigma}(t')\right\rangle ,\\
\hat{G}_{sp}(\mathbf{n}-\mathbf{m},\, t-t') &= -i\left\langle \hat{T}\, S_{\mathbf{n}\sigma}(t) P^{\dagger}_{\mathbf{m}\sigma}(t')\right\rangle ,\\
\hat{G}_{ps}(\mathbf{n}-\mathbf{m},\, t-t') &= -i\left\langle \hat{T}\, P_{\mathbf{n}\sigma}(t) S^{\dagger}_{\mathbf{m}\sigma}(t')\right\rangle ,\\
\hat{G}_{sp}(\mathbf{n}-\mathbf{m},\, t-t') &= -i\left\langle \hat{T}\, S_{\mathbf{n}\sigma}(t) S^{\dagger}_{\mathbf{m}\sigma}(t')\right\rangle ,\\
\hat{F}_{sp}(\mathbf{n}-\mathbf{m},\, t-t') &= -i\left\langle \hat{T}\, \hat{N}^{\mathrm{hh}}_{\mathbf{n}}(t) S_{\mathbf{n}\sigma}(t) P^{\dagger}_{\mathbf{m}\sigma}(t')\right\rangle ,\\
\hat{F}_{ss}(\mathbf{n}-\mathbf{m},\, t-t') &= -i\left\langle \hat{T}\, \hat{N}^{\mathrm{hh}}_{\mathbf{n}}(t) S_{\mathbf{n}\sigma}(t) S^{\dagger}_{\mathbf{m}\sigma}(t')\right\rangle .
\end{aligned} \tag{9.14}$$

The equations for two-particle Green's functions, $\hat{F}_{sp}$ and $\hat{F}_{ss}$, in turn, have the form:

$$\begin{aligned}
(\omega + \mu + \epsilon_H - U)\hat{F}_{sp} + t_{sp}\hat{F}_{pp} &= 0 ,\\
(\omega + \mu + \epsilon_H - U)\hat{F}_{ss} + t_{sp}\hat{F}_{ps} &= n^{hh} ,
\end{aligned} \tag{9.15}$$

where $n^{\mathrm{hh}} = 1 - n^{\mathrm{H}}$ is the average number of oxygen holes, while

$$\begin{aligned}
\hat{F}_{pp}(\mathbf{n}-\mathbf{m},\, t-t') &= -i\left\langle \hat{T}\, \hat{N}^{\mathrm{hh}}_{\mathbf{n}}(t) P_{\mathbf{n}\sigma}(t) P^{\dagger}_{\mathbf{m}\sigma}(t')\right\rangle ,\\
\hat{F}_{ps}(\mathbf{n}-\mathbf{m},\, t-t') &= -i\left\langle \hat{T}\, \hat{N}^{\mathrm{hh}}_{\mathbf{n}}(t) P_{\mathbf{n}\sigma}(t) S^{\dagger}_{\mathbf{m}\sigma}(t')\right\rangle .
\end{aligned} \tag{9.16}$$

In the Hubbard I approximation, the Green's functions $\hat{F}_{pp}$ and $\hat{F}_{ps}$ are replaced by

$$\hat{F}_{pp} \to \langle \hat{N}^{\mathrm{hh}}\rangle \hat{G}_{pp} = n^{\mathrm{hh}}\hat{G}_{pp} , \quad \hat{F}_{ps} \to \langle \hat{N}^{\mathrm{hh}}\rangle \hat{G}_{ps} = n^{\mathrm{hh}}\hat{G}_{ps} . \tag{9.17}$$

As a result, one obtains a closed system of equations for $\hat{G}_{pp}$, $\hat{G}_{sp}$, $\hat{G}_{ps}$, $\hat{G}_{ss}$, $\hat{F}_{sp}$, and $\hat{F}_{ss}$. The solution is

$$\begin{aligned}
\hat{G}_{pp} &= \frac{\omega + \mu + \epsilon_H}{(\omega + \mu + \epsilon_H)(\omega + \mu - \hat{H}_{\mathbf{k}}) - n^{\mathrm{H}} t_{sp}^2} ,\\
\hat{G}_{ps} = \hat{G}_{sp} &= -\frac{n^{\mathrm{H}} t_{sp}}{(\omega + \mu + \epsilon_H)(\omega + \mu - \hat{H}_{\mathbf{k}}) - n^{\mathrm{H}} t_{sp}^2} ,\\
\hat{G}_{ss} &= \frac{n^{\mathrm{H}}(\omega + \mu - \hat{H}_{\mathbf{k}})}{(\omega + \mu + \epsilon_H)(\omega + \mu - \hat{H}_{\mathbf{k}}) - n^{\mathrm{H}} t_{sp}^2} ,
\end{aligned}$$

$$\hat{F}_{sp} = -\frac{n^{\text{hh}} t_{sp} \hat{G}_{pp}}{\omega + \mu + \epsilon_H - U},$$
$$\hat{F}_{ss} = -\frac{n^{\text{hh}} \left(t_{sp} \hat{G}_{ps} - 1 \right)}{\omega + \mu + \epsilon_H - U}. \tag{9.18}$$

Knowing these Green's functions one can calculate the system's energy per one carbon atom according to

$$E = \frac{\langle H^{EA} \rangle}{N_{\text{sites}}} = -\frac{i}{2} \int_{-\infty}^{\infty} \frac{d\omega}{2\pi} \int \frac{d^2\mathbf{k}}{v_{BZ}} e^{i\omega 0^+} \left\{ \text{Tr} \left[\hat{G}(\omega, \mathbf{k}) \hat{H}^{\text{A}}_{\mathbf{k}} \right] + U \, \text{Tr} \left[\hat{F}_{ss}(\omega, \mathbf{k}) \right] \right\}, \tag{9.19}$$

where $v_{BZ} = 8\pi^2/(3\sqrt{3}a_0^2)$ is the graphene's Brillouin zone area and $\hat{G}$ is the 4×4 propagator of the form

$$\hat{G} = \begin{pmatrix} \hat{G}_{pp} & \hat{G}_{ps} \\ \hat{G}_{sp} & \hat{G}_{ss} \end{pmatrix}. \tag{9.20}$$

The chemical potential in Eq. (9.18) is related to the average number of oxygens (per one carbon atom) according to

$$1 + n^{\text{H}} = -2 \frac{1}{4} i \int_{-\infty}^{\infty} \frac{d\omega}{2\pi} \int \frac{d^2\mathbf{k}}{v_{BZ}} e^{i\omega 0^+} \, \text{Tr} \, \hat{G}(\omega, \mathbf{k}). \tag{9.21}$$

Equations (9.19) and (9.21) form a closed system of equations to find the system's energy as function of the oxygen concentration. The result of calculations (for $U \to +\infty$) is shown in Fig. 9.2. We see that the energy $E(n^{\text{H}})$ has a negative curvature in the range $0 < n^{\text{H}} < 1$ indicating the instability of the homogeneous state toward the phase separation.

Besides Hubbard I approximation, Ref. [1] also studies the Hamiltonian (9.8) numerically by exact diagonalization of the Hamiltonian on the finite cluster. Namely, a sample having the shape of a rhombus and containing from 3×3 to 10×10 elementary graphene cells (from 18 to 200 carbon atoms) is considered. With a fixed concentration of n^{H} and a fixed cluster's size, one considers up to 1500 of various configurations in which hydrogen atoms are randomly distributed in the sample. For each such configuration, a numerical diagonalization of the Hamiltonian (9.8) with periodic boundary conditions is performed and the energy of the system is calculated. After that, the averaging over configurations is performed, and the mean value of the energy and its dispersion is evaluated. The result of calculation of $E(n^{\text{H}})$ by exact diagonalization of 10×10 cluster is shown in Fig. 9.2. We see the same behavior of $E(n^{\text{H}})$ as for the Hubbard I approximation; the energy has negative curvature in the whole range of n^{H}. This feature indicates the possible instability of the homogeneous state and the phase separation. The energy of the phase-separated state can be calculated using the Maxwell construction. In the considered case, this

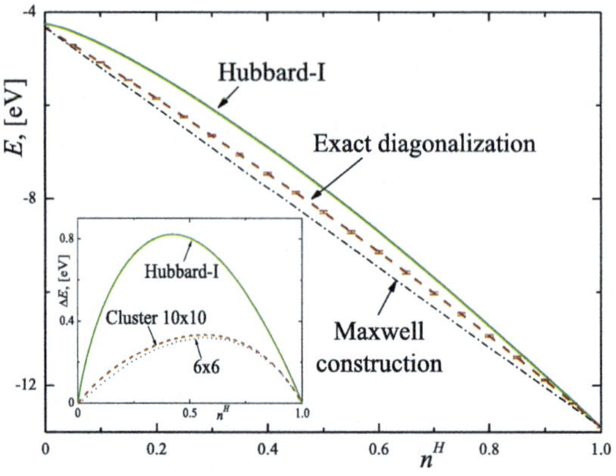

Fig. 9.2 Electron energy E as function of the concentration of hydrogen adatoms n^H; green solid curve—calculation in the Hubbard I approximation and red dashed curve—exact diagonalization of the cluster consisting of 10×10 unit cells. The negative curvature of $E(n^H)$ indicates the possible instability of the homogeneous system toward the phase separation into phases with $n^H = 0$ and $n^H = 1$. The blue dot-and-dash line shows the Maxwell construction. Inset shows the energy difference between the homogeneous and phase-separated states; green solid curve—calculation in the Hubbard I approximation, red dashed curve—exact diagonalizations of 10×10 cluster, and blue dotted curve—exact diagonalizations of 6×6 cluster. The model parameters are: $t_{sp} = 5.8$ eV, $t = 2.8$ eV, and $\varepsilon_H = 0.4$ eV. For exact diagonalizations, the value $U = 400$ eV is used [1]

construction is simply a single straight line which connects the energy of the graphene ($n^{\mathrm{H}} = 0$) and the energy of the graphane ($n^{\mathrm{H}} = 1$). Therefore, the separated states are graphene and graphane.

The phase separation occurs due to the redistribution of neutral hydrogen atoms. Thus, there is no violation of the charge neutrality and the geometrical structure of the inhomogeneous state depends only on the value and sign of the interface tension σ_0 between graphene and graphane. In the case of $\sigma_0 < 0$, it is favorable for the existence of small clusters in the inhomogeneous state to maximize the boundary length. If $\sigma_0 > 0$, it is favorable to minimize the length of the graphene–graphane boundary.

Electrons in graphane are mainly localized at the C–H valence bonds since $t \ll t_{sp}$ and their contribution to the interface tension σ_0 is small. The electrons in graphene move from one carbon atom to its nearest neighbors but they cannot penetrate to graphane since electrons have to overcome the energy gap, which is about t_{sp}. Therefore, each carbon–carbon bond connecting an atom in graphene with an atom in graphane does not contribute to the kinetic energy of the charge carriers in graphene. Effectively, this is equivalent to an increase in the electron kinetic energy in graphene. The number of broken bonds increases with the interface length. Thus, we have an estimate $\sigma_0 \sim \varepsilon_b/a_0$, where ε_b is the kinetic energy per single carbon–carbon bond

$$\varepsilon_b = \frac{2t}{3} \int d^2\mathbf{k} \frac{V_c |f_{\mathbf{k}}|}{(2\pi)^2} \sim t \,. \tag{9.22}$$

In this equation, the integration is carried out over the first Brillouin zone, $V_c = 3\sqrt{3}a_0^2/2$, the factor 2 arises due to two spin projections, and the factor of $1/3$ appears here since there are three bonds in a graphene unit cell. Thus, $\sigma_0 \sim t/a_0 > 0$. An accurate numerical computation gives $\sigma_0 \approx 0.2t/a_0 \approx 0.6\,\text{eV}$ [1]. In the case under study, we can neglect the effects of temperature on the phase separation since the characteristic energies of the problem are much higher than $k_B T$ for any realistic T.

Thus, the theoretical description of the partially hydrogenated graphene can be reduced to a Falicov–Kimball-like model. The thermodynamically stable state of this system is inhomogeneous. All hydrogen atoms are clustering together. As a result in the ground state, we have two phases: pure graphane and pure (hydrogen-free) graphene. The graphane-graphene interface tension is large and positive. This means that the graphane-graphene boundary is stable and flat (provided the number of hydrogen adatoms in the sample is not small). Note that relaxation time of the partially hydrogenated graphene to the ground state could be large.

9.2 AA Stacked Bilayer Graphene

In this section, we consider another graphene-based system—the so-called AA stacked bilayer graphene. In fact, there are several modifications of bilayer graphene, which are different in the arrangement of the atoms of one layer relative to the atoms of the other. The most studied type of bilayer graphene is the so-called AB or Bernal stacking. In this configuration, atoms from one sublattice, say B, of the upper layer are located above the atoms of the sublattice A of the lower layer, while the atoms of the sublattice A of the upper layer are located above the center of the hexagon formed by the atoms of the lower layer. This configuration is typical for adjacent graphite layers. There is also another configuration, in which the atoms from the upper layer are located above the equivalent atoms of the lower layer. This configuration is called AA stacked bilayer graphene (or simply AA graphene). The crystal structure of AA graphene is shown in Fig. 9.3. In both configurations, the unit cell of the bilayer contains four carbon atoms, $A1$, $A2$, $B1$, and $B2$.

In addition to these two types of bilayer graphene, there is another configuration (or rather a set of configurations)—the so-called twisted bilayer graphene. In this configuration, the crystal axes of different layers are not parallel to each other, but form a certain twist angle θ. For a commensurate set of angles, twisted bilayer graphene has a superstructure. Its superlattice can contain an extremely large number of carbon atoms. It turns out that when θ is small enough, the supercell can be represented as consisting of regions with AA, AB, and BA stackings, while the electrons at the Fermi level are localized in regions with AA stacking. Detailed information about all types of the bilayer graphene can be found in the review paper [11].

Since elementary unit cell of the AA stacked bilayer graphene contains four carbon atoms, its band structure at low energies consists four bands corresponding to p_z electrons of carbon. It turns out, that the spectrum at low energies is linear one. However, unlike single-layer graphene, AA stacked bilayer graphene has a Fermi

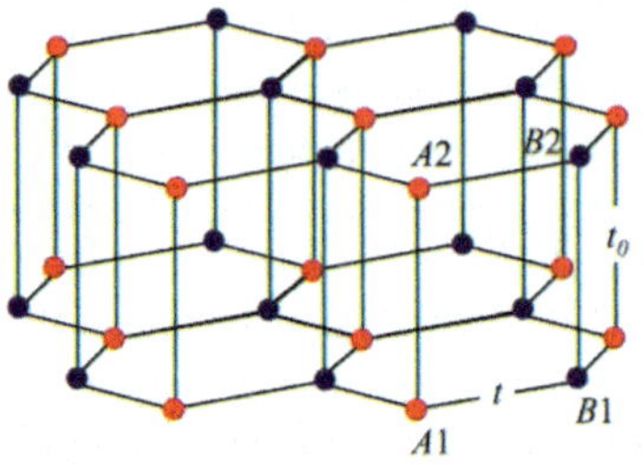

Fig. 9.3 Crystal lattice of the AA stacked bilayer graphene consists of two sublattices A (red circles) and B (blue circles). The unit cell of the crystal structure includes four atoms $A1$ and $B1$ in the top (1) layer and $A2$ and $B2$ in the bottom (2) layer. The values t and t_0 are the in-plane and inter-plane nearest-neighbor hopping integrals, respectively

surface, which is formed by one hole and one electron band. In the undoped case, the Fermi surface is characterized by the ideal nesting with the nesting vector $\mathbf{Q}_0 = 0$. This makes the Fermi liquid unstable with respect to spontaneous symmetry breaking and the appearance of some order parameter in the system. Below we analyze which symmetry should be broken, and what types of order parameters are possible. Next, we show that the most stable order parameter is G-type antiferromagnetism (the spins of any two neighboring atoms are antiparallel), which is stabilized by on-site Coulomb repulsion of electrons. We analyze the behavior of the antiferromagnetic state with a change of temperature and doping, and show the tendency to phase separation, when we dope the system. In presenting the material of this section, we follow Refs. [12–14].

9.2.1 Instability of the Fermi Liquid of the AA Stacked Bilayer Graphene. Symmetries of Possible Order Parameters

The simplest tight-binding Hamiltonian of the AA stacked bilayer graphene has the form

$$H_0 = -t \sum_{\langle \mathbf{nm} \rangle i\sigma} \left(p^{\dagger}_{\mathbf{n}iA\sigma} p_{\mathbf{m}iB\sigma} + H.c. \right) - t_0 \sum_{\mathbf{n}a\sigma} \left(p^{\dagger}_{\mathbf{n}1a\sigma} p_{\mathbf{n}2a\sigma} + H.c. \right), \qquad (9.23)$$

where $p^{\dagger}_{\mathbf{n}ia\sigma}$ and $p_{\mathbf{n}ia\sigma}$ are the creation and annihilation operators of the electron with spin σ located in the unit cell $\mathbf{n}$ of the layer $i = 1, 2$ in the sublattice $a = A, B$. The first term corresponds to the intralayer hopping amplitude, and the second term corresponds to the hopping between layers. Only the nearest-neighbor hoppings are considered. The parameters t and t_0 are the amplitudes of the electron hopping between neighboring atoms inside the layer and between layers, respectively. The distance between adjacent graphene layers is estimated as [15] $c \cong 3.32\,\text{Å}$, which is more than twice the distance between neighboring carbon atoms in a graphene layer, $a_0 \cong 1.42\,\text{Å}$. Thus, it is natural to expect that the parameter t will exceed t_0. The first-principle calculations give the following estimates [16]: $t \approx 2.57\,\text{eV}$, $t_0 \approx 0.36\,\text{eV}$.

Let us introduce the 4-component spinors: $\psi^\dagger_{\mathbf{k}\sigma} = (\psi^\dagger_{\mathbf{k}A\sigma}, \psi^\dagger_{\mathbf{k}B\sigma})$, where two-component spinors $\psi^\dagger_{\mathbf{k}a\sigma}$ have the form: $\psi^\dagger_{\mathbf{k}A\sigma} = (p^\dagger_{\mathbf{k}1A\sigma}, p^\dagger_{\mathbf{k}2A\sigma})$, $\psi^\dagger_{\mathbf{k}B\sigma} = e^{-i\phi_{\mathbf{k}}}(p^\dagger_{\mathbf{k}1B\sigma}, p^\dagger_{\mathbf{k}2B\sigma})$. In the latter expression, $\phi_{\mathbf{k}} = \arg(f_{\mathbf{k}})$, where the function $f_{\mathbf{k}}$ is given by Eq. (9.6). We added the phase factor $e^{-i\phi_{\mathbf{k}}}$ to the definition of $\psi^\dagger_{\mathbf{k}B\sigma}$ for convenience as it will be seen below. Spinors $\psi^\dagger_{\mathbf{k}A\sigma}$ and $\psi^\dagger_{\mathbf{k}B\sigma}$ have different sublattice indices.

In the basis $\psi_{\mathbf{k}\sigma}$, Hamiltonian (9.23) takes the form $H_0 = \sum_{\mathbf{k}\sigma} \psi^\dagger_{\mathbf{k}\sigma} \hat{H}_{0\mathbf{k}} \psi_{\mathbf{k}\sigma}$, where matrix $\hat{H}_{0\mathbf{k}}$ is

$$\hat{H}_{0\mathbf{k}} = -\begin{pmatrix} 0 & t_0 & t\zeta_{\mathbf{k}} & 0 \\ t_0 & 0 & 0 & t\zeta_{\mathbf{k}} \\ t\zeta_{\mathbf{k}} & 0 & 0 & t_0 \\ 0 & t\zeta_{\mathbf{k}} & t_0 & 0 \end{pmatrix}. \tag{9.24}$$

Let us introduce the Pauli matrices $\hat{\tau}_\alpha$ acting on the layer index, as well as the second set of Pauli matrices $\hat{\sigma}_\alpha$ acting on the sublattice index. Using these Pauli matrices, the matrix $\hat{H}_{0\mathbf{k}}$ can be represented as

$$\hat{H}_{0\mathbf{k}} = -t_0\hat{\tau}_x - t\hat{\sigma}_x\zeta_{\mathbf{k}}\,. \tag{9.25}$$

Note that the phase factor introduced in the definition of the bispinor $\psi_{\mathbf{k}\sigma}$ makes the matrix $\hat{H}_{0\mathbf{k}}$ real. This simplifies the subsequent analysis. Hamiltonian (9.24) is invariant with respect to permutation operations, both for sublattices and layers. Mathematically, this means that $[\hat{\sigma}_x, \hat{H}_{0\mathbf{k}}] = [\hat{\tau}_x, \hat{H}_{0\mathbf{k}}] = 0$. Thus, the eigenvectors and eigenvalues of the matrix (9.24) can be classified by the quantum numbers $\sigma = \pm 1$ and $\tau = \pm 1$, which characterize the parity of the eigenvector when the operators σ_x and τ_x act on it. Using these symmetry properties, it is easy to find a unitary transformation diagonalizing $\hat{H}_{0\mathbf{k}}$:

$$\hat{U} = \frac{1}{2}\left(\hat{\tau}_x + \hat{\tau}_z\right)\left(\hat{\sigma}_x + \hat{\sigma}_z\right) = \hat{U}^{-1}. \tag{9.26}$$

Electron spectrum $\varepsilon^{(s)}_{\mathbf{k}}$ consists of four bands, each of them is characterized by its own pair (σ, τ):

$$\varepsilon^{(1)}_{\mathbf{k}} = -t_0 - t\zeta_{\mathbf{k}}\,, \quad \sigma = 1, \quad \tau = 1\,, \tag{9.27}$$

$$\varepsilon^{(2)}_{\mathbf{k}} = +t_0 - t\zeta_{\mathbf{k}}\,, \quad \sigma = 1, \quad \tau = -1\,, \tag{9.28}$$

$$\varepsilon^{(3)}_{\mathbf{k}} = -t_0 + t\zeta_{\mathbf{k}}\,, \quad \sigma = -1, \quad \tau = 1\,, \tag{9.29}$$

$$\varepsilon^{(4)}_{\mathbf{k}} = +t_0 + t\zeta_{\mathbf{k}}\,, \quad \sigma = -1, \quad \tau = -1\,. \tag{9.30}$$

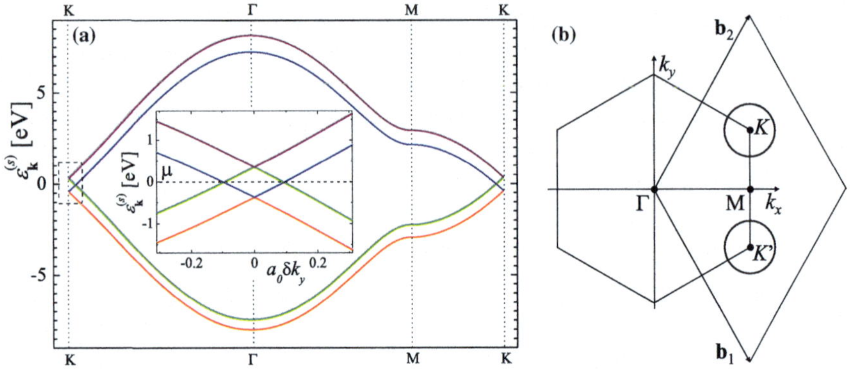

Fig. 9.4 **a** Band structure of the AA stacked bilayer graphene. Spectrum is calculated along the contour connecting the points **K**–$\boldsymbol{\Gamma}$–**M**–**K**. Inset shows the spectrum near the Dirac point **K**; $\mathbf{k} = \mathbf{K} + \delta k_y \mathbf{e}_y$. Bands $s = 2$ and $s = 3$ cross the Fermi level $\mu = 0$, forming doubly degenerate Fermi surface sheet. **b** Fermi surface of the (undoped) AA stacked bilayer graphene consisting of two closed curves, approximately having a circular shape with the center in Dirac points **K** and **K**$'$ [12]

The band structure of the AA stacked bilayer graphene is shown in Fig. 9.4. The bands $s = 2$ (hole) and $s = 3$ (electronic) intersect the Fermi level near the Dirac points **K** and **K**$'$ (see Fig. 9.4b). The most interesting property of this band structure is that at half-filling (which corresponds to undoped sample), the Fermi surfaces formed by the bands $s = 2$ and $s = 3$ coincide with each other. In other words, in the AA bilayer graphene, there is an ideal nesting of the Fermi surface with the nesting vector $\mathbf{Q}_0 = 0$. The Fermi level at half filling is equal to $\mu = 0$. The Fermi surface is given by the equation $\zeta_{\mathbf{k}} = t_0/t$ and consists of two closed curves located around the Dirac points **K** and **K**$'$. With good accuracy, the Fermi surface sheets have the shape of a circle with radius $k_r = 2t_0/(3ta_0)$ (see Fig. 9.4b).

In the simplest model of the stacked bilayer AA graphene considered here, we take into account only electron hoppings between the nearest neighbors. An important circumstance that allows us to safely use this approximation is that the property of the ideal nesting is preserved when we take into account more distant hoppings. This was proved numerically by considering hoppings inside the first 7 coordination spheres [12]. Moreover, the ideal nesting is preserved even in the case of nonequivalent layers. At the same time, various kinds of interaction (electron–electron, electron–phonon) can destabilize such a degenerate spectrum, leading to the appearance of an order parameter in the system. Below, based on symmetry considerations, we will consider the general properties of possible order parameters.

9.2.2 Mean-Field Hamiltonian

The presence of a hole and an electron bands with coinciding Fermi surfaces makes the system unstable with respect to spontaneous symmetry breaking. Below we demonstrate that the symmetries of the Hamiltonian (commutativity with the operators $\hat{\sigma}_x$ and $\hat{\tau}_x$) restrict the choice of possible order parameters.

In the mean-field approximation, the two-particle interaction operator $H_{\text{int}} \propto \psi^\dagger\psi^\dagger\psi\psi$ is replaced by the single-particle operator $H_{\text{int}}^{\text{MF}} \propto \langle\psi^\dagger\psi\rangle\psi^\dagger\psi$, where the mean value $\langle\psi^\dagger\psi\rangle$ represents various types of non-superconducting order parameters. These mean values are found from the self-consistency equations. Being at least metastable, the order parameter should open a gap on the Fermi level. The most general form of $H_{\text{int}}^{\text{MF}}$ (diagonal on spin index), which can open a gap, has the form:

$$H_{\text{int}}^{\text{MF}} = \sum_{\mathbf{k}\sigma} \psi_{\mathbf{k}\sigma}^\dagger \hat{\Delta}_{\mathbf{k}\sigma}\psi_{\mathbf{k}\sigma}\,,$$

$$\hat{\Delta}_{\mathbf{k}\sigma} = \sum_{\alpha}\left(\Delta_{AB\mathbf{k}\sigma}^{\alpha}\hat{\sigma}_\alpha + \Delta_{12\mathbf{k}\sigma}^{\alpha}\hat{\tau}_\alpha\right) + \sum_{\alpha\beta}\Delta_{\mathbf{k}\sigma}^{\alpha\beta}\hat{\tau}_\alpha\hat{\sigma}_\beta\,, \tag{9.31}$$

where $\Delta_{AB\mathbf{k}\sigma}^{\alpha}$, $\Delta_{12\mathbf{k}\sigma}^{\alpha}$, and $\Delta_{\mathbf{k}\sigma}^{\alpha\beta}$ are real functions of the momentum $\mathbf{k}$. In order to open a gap, the Hamiltonian $H_{\text{int}}^{\text{MF}}$ must couple the bands crossing the Fermi level, $\varepsilon_{\mathbf{k}}^{(2)}$ and $\varepsilon_{\mathbf{k}}^{(3)}$. Since these bands are characterized by different values of σ and τ (see formulas (9.28) and (9.29)), only the terms $\Delta_{\mathbf{k}\sigma}^{\alpha\beta}$ with $\alpha, \beta \neq x$ can hybridize these bands. The remaining terms commute either with $\hat{\sigma}_x$ or with $\hat{\tau}_x$.

In order to calculate renormalized spectra near the Fermi level, we need to diagonalize the matrix $\hat{H}_{\mathbf{k}\sigma} = \hat{H}_{0\mathbf{k}} + \hat{\Delta}_{\mathbf{k}\sigma}$. Performing the unitary transformation $\hat{H}_{\mathbf{k}\sigma}$ using the matrix $\hat{U}$ defined by the Eq. (9.26), one finds

$$\hat{U}^{-1}\hat{H}_{\mathbf{k}\sigma}\hat{U} = \begin{pmatrix} \ddots & \cdots & \cdots & \cdots \\ \vdots & \varepsilon_{\mathbf{k}}^{(2)} + \delta\varepsilon_{\mathbf{k}\sigma}^{(2)} & \Delta_{\mathbf{k}\sigma} & \vdots \\ \vdots & \Delta_{\mathbf{k}\sigma}^{*} & \varepsilon_{\mathbf{k}}^{(3)} + \delta\varepsilon_{\mathbf{k}\sigma}^{(3)} & \vdots \\ \cdots & \cdots & \cdots & \ddots \end{pmatrix}, \tag{9.32}$$

where $\delta\varepsilon_{\mathbf{k}\sigma}^{(2)} = \Delta_{AB\mathbf{k}\sigma}^{x} - \Delta_{12\mathbf{k}\sigma}^{x} - \Delta_{\mathbf{k}\sigma}^{xx}$ and $\Delta_{\mathbf{k}\sigma} = \Delta_{\mathbf{k}\sigma}^{zz} + \Delta_{\mathbf{k}\sigma}^{yy} + i(\Delta_{\mathbf{k}\sigma}^{zy} + \Delta_{\mathbf{k}\sigma}^{yz})$. Other elements of this matrix are not essential for this consideration. Solving the secular equation for the 2×2 matrix in the Eq. (9.32), we find a renormalized spectrum of the bands $s = 2, 3$:

$$E_{\mathbf{k}\sigma}^{(2,3)} = \frac{1}{2}\left(\varepsilon_{\mathbf{k}}^{(2)} + \delta\varepsilon_{\mathbf{k}\sigma}^{(2)} + \varepsilon_{\mathbf{k}}^{(3)} + \delta\varepsilon_{\mathbf{k}\sigma}^{(3)}\right) \mp \frac{1}{2}\sqrt{\left(\varepsilon_{\mathbf{k}}^{(2)} + \delta\varepsilon_{\mathbf{k}\sigma}^{(2)} - \varepsilon_{\mathbf{k}}^{(3)} - \delta\varepsilon_{\mathbf{k}\sigma}^{(3)}\right)^2 + 4\left|\Delta_{\mathbf{k}\sigma}\right|^2}\,. \tag{9.33}$$

The band gap is $\Delta_0 = 2\min_{\mathbf{k}} |\Delta_{\mathbf{k}\sigma}|$. We see that the contribution to the gap is given only by the order parameters $\Delta^{zz}_{\mathbf{k}\sigma}$, $\Delta^{yy}_{\mathbf{k}\sigma}$, $\Delta^{zy}_{\mathbf{k}\sigma}$, and $\Delta^{yz}_{\mathbf{k}\sigma}$, which violate both symmetry with respect to permutations of sublattices and symmetry with respect to permutations of layers. It is easy to show that the remaining elements of the 4×4 matrix (9.32), give only second-order corrections to the result obtained. Neglecting all the terms in the Eq. (9.31), except those that open a gap at the Fermi level, we can write the matrix $\hat{\Delta}_{\mathbf{k}\sigma}$ in the form

$$\hat{\Delta}_{\mathbf{k}\sigma} = \begin{pmatrix} \Delta^{zz}_{\mathbf{k}\sigma} & -i\Delta^{yz}_{\mathbf{k}\sigma} & -i\Delta^{zy}_{\mathbf{k}\sigma} & -\Delta^{yy}_{\mathbf{k}\sigma} \\ i\Delta^{yz}_{\mathbf{k}\sigma} & -\Delta^{zz}_{\mathbf{k}\sigma} & \Delta^{yy}_{\mathbf{k}\sigma} & i\Delta^{zy}_{\mathbf{k}\sigma} \\ i\Delta^{zy}_{\mathbf{k}\sigma} & \Delta^{yy}_{\mathbf{k}\sigma} & -\Delta^{zz}_{\mathbf{k}\sigma} & i\Delta^{yz}_{\mathbf{k}\sigma} \\ -\Delta^{yy}_{\mathbf{k}\sigma} & -i\Delta^{zy}_{\mathbf{k}\sigma} & -i\Delta^{yz}_{\mathbf{k}\sigma} & \Delta^{zz}_{\mathbf{k}\sigma} \end{pmatrix}. \tag{9.34}$$

In order to determine the ground state of the system, it is necessary to calculate the total energy and minimize it with respect to $\Delta^{\alpha\beta}_{\mathbf{k}\sigma}$. To do this, it is necessary to choose a specific type of the interaction Hamiltonian. Below we will show that the order parameter $\Delta^{zz}_{\mathbf{k}\sigma}$ can correspond to the G-type antiferromagnetism (the spins of all nearest neighbors are antiparallel), which is controlled by the on-site (Hubbard) Coulomb repulsion. The order parameter $\Delta^{yy}_{\mathbf{k}\sigma}$ may be related to the instability to a homogeneous shift of one graphene layer relative to another. The order parameters $\Delta^{zy}_{\mathbf{k}\sigma}$ and $\Delta^{yz}_{\mathbf{k}\sigma}$ can correspond to excitons with flowing current inside the layer and between layers, respectively. In this case, they are controlled by the Coulomb repulsion between the nearest neighbors. In general, all these order parameters can either coexist or compete with each other.

9.2.3 *Antiferromagnetism*

It is known that the Coulomb interaction of electrons in graphene is quite strong. Estimations of the magnitude of the on-site Coulomb repulsion U, made in Ref. [17] gives a value of the order of 10 eV. However, graphene remains a semimetal because in the graphene the density of states at the Fermi level is zero. In contrast, AA stacked bilayer graphene has a Fermi surface and a finite density of states. Thus, it can be expected that the electron–electron interaction in AA graphene plays a greater role and affects the properties of its ground state. In this section, we will take into account the interaction using the Hubbard Hamiltonian

$$H_{\text{int}} = U\sum_{\mathbf{n}ia}\left(n_{\mathbf{n}ia\uparrow} - \frac{1}{2}\right)\left(n_{\mathbf{n}ia\downarrow} - \frac{1}{2}\right), \tag{9.35}$$

It is known that the ground state of the Hubbard-type models at half-filling can be antiferromagnetic. Generally speaking, several types of antiferromagnetic state (having different spin configurations inside the unit cell) are possible in the AA stacked bilayer graphene. However, only G-type antiferromagnetism opens a gap at the Fermi level at arbitrarily weak U. The other two types of antiferromagnetism do not simultaneously break the symmetries with respect to permutations of layers and sublattices, and therefore, according to the analysis made above, they cannot open a gap at the Fermi level. Consequently, such phases are either unstable at low U, or metastable at strong interaction.

If we choose the z-axis being a spin quantization axis, the antiferromagnetic order parameter will take the form

$$\Delta_{ia} = \frac{U}{2}\left(\langle n_{\mathbf{n}ia\uparrow}\rangle - \langle n_{\mathbf{n}ia\downarrow}\rangle\right), \tag{9.36}$$

where, for G-type antiferromagnetism, we have

$$\Delta_{1A} = -\Delta_{1B} = -\Delta_{2A} = \Delta_{2B} \equiv \Delta\,. \tag{9.37}$$

In the mean-field approximation, in Eq. (9.35), we should write $n_{\mathbf{n}ia\sigma} = n_{ia\sigma} + \delta n_{\mathbf{n}ia\sigma}$, where $n_{ia\sigma} = \langle n_{\mathbf{n}ia\sigma}\rangle$, and $\delta n_{\mathbf{n}ia\sigma} = n_{\mathbf{n}ia\sigma} - n_{ia\sigma}$. The mean-field Hamiltonian is obtained by neglecting the terms which are quadratic in $\delta n_{\mathbf{n}ia\sigma}$. As a result, we obtain

$$H_{\text{int}}^{\text{MF}} = \sum_{\mathbf{n}ia}\left[-\Delta_{ia}\left(n_{\mathbf{n}ia\uparrow} - n_{\mathbf{n}ia\downarrow}\right) + \frac{\Delta_{ia}^2}{U}\right], \tag{9.38}$$

Such a Hamiltonian, with the exception of the c-number term proportional to Δ^2/U, corresponds to the Eq. (9.31), where $\Delta_{\mathbf{k}\uparrow}^{zz} = -\Delta$, $\Delta_{\mathbf{k}\downarrow}^{zz} = +\Delta$, and the other terms are zero.

The electronic spectrum of the mean-field Hamiltonian $H_0 + H_{\text{int}}^{\text{MF}}$ can be found analytically. The spectra for electrons with spin up and spin down coincide. As a result, we have 4 bands doubly degenerate on the electron spin:

$$E_{\mathbf{k}}^{(1)} = -\sqrt{\Delta^2 + (t_0 + t\zeta_{\mathbf{k}})^2}\,, \tag{9.39}$$

$$E_{\mathbf{k}}^{(2)} = -\sqrt{\Delta^2 + (t_0 - t\zeta_{\mathbf{k}})^2}\,, \tag{9.40}$$

$$E_{\mathbf{k}}^{(3)} = +\sqrt{\Delta^2 + (t_0 - t\zeta_{\mathbf{k}})^2}\,, \tag{9.41}$$

$$E_{\mathbf{k}}^{(4)} = +\sqrt{\Delta^2 + (t_0 + t\zeta_{\mathbf{k}})^2}\,. \tag{9.42}$$

Thermodynamic potential Ω per one unit cell is equal to

$$\Omega = \frac{\Delta^2}{U} - 2T\sum_{s=1}^{4}\int\frac{d^2\mathbf{k}}{v_{\text{BZ}}}\ln\left[1 + e^{(\mu - E_{\mathbf{k}}^{(s)})/T}\right]. \tag{9.43}$$

Let us consider first the undoped case corresponding to $\mu = 0$. Minimization of Ω with respect to Δ at $T = 0$ leads to the equation

$$\int_0^{3t} dE \left[\frac{\rho_0(E)}{\sqrt{\Delta_0^2 + (E + t_0)^2}} + \frac{\rho_0(E)}{\sqrt{\Delta_0^2 + (E - t_0)^2}} \right] = \frac{4t}{U}, \tag{9.44}$$

where $\Delta_0 \equiv \Delta(\mu = 0, T = 0)$ is the AFM order parameter at half-filling and zero temperature, and

$$\rho_0(E) = \int \frac{d^2\mathbf{k}}{v_{\mathrm{BZ}}} \left[\delta(E - t\zeta_{\mathbf{k}}) + \delta(E + t\zeta_{\mathbf{k}})\right] \tag{9.45}$$

is the density of states of a single graphene layer per one carbon atom. The integral of the second term on the left hand side of Eq. (9.44) diverges logarithmically when $\Delta_0 \to 0$. In the limit of small Δ_0, we obtain

$$\Delta_0 = 2\sqrt{t_0(3t - t_0)} \exp\left\{-\frac{4 - U\eta(t_0)}{2U\rho_0(t_0)}\right\}, \tag{9.46}$$

where

$$\eta(t_0) = \int_0^{3t} dE \left[\frac{\rho_0(E)}{E + t_0} + \frac{\rho_0(E) - \rho_0(t_0)}{|E - t_0|} \right]. \tag{9.47}$$

Figure 9.5 shows the dependence of Δ_0 on U. Curves corresponding to both the numerical solution of the Eq. (9.44) and the asymptotics (9.46) are shown. It can be seen that formula (9.46) works well when $U \lesssim 5.5\,\mathrm{eV}$.

The value of Δ_0 significantly depends on the parameter t_0, which determines the density of states at the Fermi level. From Eq. (9.46), it can be seen that $\Delta_0 \to 0$ at $t_0 \to 0$, if the value of U does not exceed the threshold value of U_c, at which the exponent in Eq. (9.46) becomes positive. The same value can be obtained from the Eq. (9.44) in the limit of $\Delta_0 = t_0 = 0$. Thus, we have $U_c = 4/\eta(0) \cong 2.231t$, or $U_c \cong 5.73\,\mathrm{eV}$ when $t = 2.57\,\mathrm{eV}$. For $U > U_c$, the Eq. (9.44) has a solution even for completely decoupled layers when $t_0 = 0$. In other words, the ground state of, the Hubbard model for a single layer graphene is antiferromagnetic at $U > U_c$. This is a well-known result. Since it follows from the experiment that graphene remains a semi-metal even at low temperatures, we should consider only the values $U < U_c$. Note that the value of the critical U was obtained here in the mean-field approximation. The estimate $U_c/t \cong 2.231$ coincides with the result of Ref. [18]. At the same time, numerical Monte Carlo simulations give a higher value [19, 20]: $U_c/t \approx 4.5$ (or $U_c \cong 11.57\,\mathrm{eV}$ at $t = 2.57\,\mathrm{eV}$.). First-principle calculations made in Ref. [17], give for the Hubbard U in graphene the value approximately equal to 9.3 eV, which is close, but below the critical value obtained by the Monte Carlo. Therefore, when constructing the phase diagram of the AA stacked bilayer graphene, we should take the value of U close, but somewhat smaller than $U_c = 2.231t$ obtained by the mean-field theory used here.

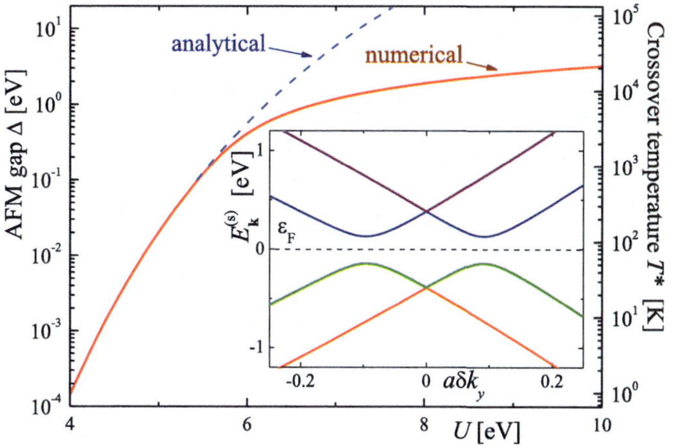

Fig. 9.5 Antiferromagnetic gap Δ, calculated at $T = 0$ (left y-axis) and crossover temperature T^* (right y-axis) depending on U. Solid curve corresponds to the numerical solution of the Eq. (9.44), while dashed curve corresponds to analytical formula (9.46). Crossover temperature T^* is proportional to Δ. Inset shows the electrons spectrum near the **K** point, calculated at $U = 5.5\,\text{eV}$ ($\Delta \approx 0.12\,\text{eV}$). The picture is taken from Ref. [12]

9.2.4 Crossover Temperature

Let us estimate the magnitude of the characteristic temperature at which the antiferromagnetic order in the system is completely destroyed. It should be noted that strictly speaking, for 2D systems (which is graphene), the AFM order is destroyed at an arbitrarily low temperature. The mean-field solution in this case characterizes the short-range order, which is preserved at finite temperatures. In this section, we first obtain the mean-field equation for $\Delta(T)$ and the transition temperature T_{MF}, and then consider the fluctuation effects beyond the mean-field approximation.

In the mean-field approximation, the equation for the gap at finite temperatures, as well as at $T = 0$, is obtained by minimization of the potential Ω, the Eq. (9.43), with respect to Δ. As a result, instead of Eq. (9.44), we will have

$$\int_0^{3t} dE\,\rho_0(E)\left[F\left(\sqrt{\Delta(T)^2+(E-t_0)^2}\right)+F\left(\sqrt{\Delta(T)^2+(E+t_0)^2}\right)\right]=\frac{4t}{U}, \tag{9.48}$$

where

$$F(\varepsilon)=\frac{n_F(-\varepsilon-\mu)-n_F(\varepsilon-\mu)}{\varepsilon}, \quad n_F(\varepsilon)=\frac{1}{e^{\varepsilon/T}+1}. \tag{9.49}$$

The equation for the Néel temperature can be obtained in the limit $\Delta \to 0$ and $T \to T_{MF}$. In the case of small Δ, when the asymptotic (9.46) is valid, we can obtain an analytical expression connecting T_{MF} and $\Delta_0 \equiv \Delta(\mu = 0, T = 0)$. To do this, we rewrite the Eq. (9.48) in the following form, valid for small Δ:

$$\ln\frac{\Delta_0}{\Delta} = \frac{1}{4T}\int\limits_{\Delta}^{\infty} d\varepsilon\, \operatorname{acosh}\left(\frac{\varepsilon}{\Delta}\right)\left[\cosh^{-2}\left(\frac{\varepsilon-\mu}{2T}\right) + \cosh^{-2}\left(\frac{\varepsilon+\mu}{2T}\right)\right]. \tag{9.50}$$

At half-filling and in the limit $\Delta \to 0$, we obtain

$$T_{MF} = \varkappa\Delta_0\,, \quad \varkappa = \frac{1}{4}\exp\left(\int\limits_0^{\infty} dx\frac{\ln x}{\cosh^2 x}\right) = \frac{e^{\gamma}}{\pi} \cong 0.567\,, \tag{9.51}$$

where $\gamma = 0.577\ldots$ is the Euler constant. This result is similar to that obtained in the BCS theory for a superconducting gap and the transition temperature to the normal state. The numerical analysis shows that the relation (9.51) connecting T_{MF} and Δ_0 works well even in the case of large Δ_0, when the asymptotic (9.46) and the formula (9.50) are not applicable.

In 2D systems, temperature fluctuations destroy the long-range antiferromagnetic order. Thus, the results obtained above in the mean-field approximation determine the short-range order. At the same time, they are valid only if the correlation length in the mean-field approximation $\xi(T) = v_F/\Delta(T)$ ($v_F = 3ta_0/2$ is the Fermi velocity in our model) is less than the spin-wave correlation length $\xi_{\text{sw}}(T)$. Otherwise, the short-range order is destroyed, and we cannot determine antiferromagnetism even locally.

In the case when $\xi_{\text{sw}} > \xi$, spin fluctuations can be described in the framework of the so-called nonlinear σ-model with a Lagrangian [21, 22]

$$\mathcal{L}_{\text{sw}} = \frac{\rho}{2}\left[(\partial_t \mathbf{D})^2 - c_{\text{sw}}^2(\partial_{\mathbf{r}}\mathbf{D})^2\right], \tag{9.52}$$

where $\mathbf{D}$ is a unit vector determining the direction of the antiferromagnetism, and the parameters ρ and c_{sw}, determining the stiffness and velocity of spin waves, respectively, can be calculated using the Eqs. (7.89) and (7.90) of Ref. [23]. This gives

$$c_{\text{sw}} = \frac{v_{\text{F}}}{\sqrt{2}}, \quad \rho = \begin{cases} t_0/(8\pi v_{\text{F}}^2), & t_0 \gg \Delta\,, \\ \Delta/(16\pi v_{\text{F}}^2), & t_0 \ll \Delta\,. \end{cases} \tag{9.53}$$

To determine the spin-wave correlation length, we calculate the correlation function $K(\mathbf{r}) = \langle \mathbf{D}(\mathbf{r})\mathbf{D}(0)\rangle$, where averaging is performed using Lagrangian (9.52) at finite temperatures. At large distances it can be calculated analytically [21, 22], which gives

$$K(\mathbf{r}) \approx 1 - \frac{T}{\pi\rho v_{\mathrm{F}}^2} \ln\left(e^{\gamma} \frac{\sqrt{2}T\,r}{\sqrt{3}t\,a}\right), \tag{9.54}$$

where $a = \sqrt{3}a_0$ is the length of the unit cell vectors of the graphene. The spin-wave correlation length ξ_{sw} can be determined from the condition $K(\xi_{\mathrm{sw}}) = 0$. As a result, one obtains

$$\xi_{\mathrm{sw}}(T) = \frac{at\sqrt{3}}{T\sqrt{2}} \exp\left(\frac{2\pi\rho c_{\mathrm{sw}}^2}{T} - \gamma\right). \tag{9.55}$$

Solving numerically the equation $\xi_{\mathrm{sw}}(T) = \xi(T) \equiv v_F/\Delta(T)$, one can find the crossover temperature T^* between the short-range antiferromagnetic order and paramagnetism. The short-range AFM order exists at $T < T^*$ and is destroyed if $T > T^*$. Numerical calculations show that $T^*/T_{MF} \approx 0.8$–0.9 for any values of Δ. Thus, the mean-field transition temperature gives an adequate estimate for the crossover temperature. Taking into account Eq. (9.51), it is possible to estimate T^* knowing the AFM gap at zero temperature. The curve in Fig. 9.5 allows to determine the crossover temperature as a function of the Hubbard U. There, it is assumed that $T^* = 0.85\varkappa\Delta_0 \approx 0.48\Delta_0$ for all values of U.

9.2.5 Doped AA Graphene. Incommensurate Antiferromagnetism and Phase Separation. Phase Diagram in the x–T Plane

Doping of graphene-based materials, as well as other 2D materials, is possible both by chemical way, by absorption of atoms or molecules onto the surface of the substance [24], and in an electrostatic way, by applying gate voltage to the substrate [25]. Doping of the AA stacked bilayer graphene introduces extra electrons or holes into the system, and destroys the ideal nesting of the Fermi surface. With electronic doping, the Fermi radius of the surfaces around the Dirac points created by the electron band will be greater than the Fermi radii of the surfaces created by the hole band. As we showed in Chap. 8 in the framework of the Rice model [26], this can lead to the appearance of an incommensurate antiferromagnetic phase, which is characterized by a nesting vector $\mathbf{q}$ incommensurate with the vectors of the reciprocal lattice.

In real space, the incommensurate phase is characterized by a gradual rotation of the vector of antiferromagnetism in some plane when moving along the direction of the vector $\mathbf{q}$. The Hubbard model of the AA stacked bilayer graphene is invariant with respect to full spin rotations. Therefore, for a commensurate AFM phase, we can choose the quantization axis arbitrarily. For the same reason, the plane of spin rotation of the incommensurate AFM phase is also can be chosen arbitrarily. In the previous section, it was assumed that the spins at all sites are parallel or antiparallel to the z axis. As a result, the mean-field Hamiltonian was diagonal on the spin index, which simplified calculations. In the case of incommensurate AFM phase, the mean-field Hamiltonian cannot be diagonal in the spin index for any choice of the spin rotation plane. However, it will have the simplest form if we choose this plane as the xy plane. In this case, the incommensurate AFM order parameter will have the form

$$\Delta_{\mathbf{n}ia} \equiv U\left\langle p^{\dagger}_{\mathbf{n}ia\uparrow} p_{\mathbf{n}ia\downarrow}\right\rangle = e^{i\mathbf{q}\mathbf{r}^a_{\mathbf{n}}}\Delta_{ia}\,, \tag{9.56}$$

where the radius vector $\mathbf{r}^a_{\mathbf{n}}$ ($a = A, B$) marks the position of the atom in the unit cell $\mathbf{n}$ of the sublattice a of the layer 1, and the parameters Δ_{ia} are given by the formula (9.37). The mean value of the magnetization of an atom located in the unit cell $\mathbf{n}$ in the layer i in the sublattice a, $\mathbf{S}_{\mathbf{n}ia}$, can be found from the formula $\langle d^{\dagger}_{\mathbf{n}ia\uparrow} d_{\mathbf{n}ia\downarrow}\rangle = S^x_{\mathbf{n}ia} + iS^y_{\mathbf{n}ia}$. As a result, we obtain

$$\mathbf{S}_{\mathbf{n}ia} = \frac{\Delta_{ia}}{U}\left(\cos(\mathbf{q}\mathbf{r}^a_{\mathbf{n}}),\ \sin(\mathbf{q}\mathbf{r}^a_{\mathbf{n}})\right). \tag{9.57}$$

Interaction Hamiltonian (9.35) in the mean-field approximation takes the form

$$\begin{aligned} H^{\text{MF}}_{\text{int}} &= \mathcal{N}\left[\frac{4\Delta^2}{U} - U(n^2-1)\right] + \frac{Ux}{2}\sum_{\mathbf{n}ia\sigma} n_{\mathbf{n}ia\sigma} - \\ &\sum_{\mathbf{k}ia}\Delta_{ia}\left(p^{\dagger}_{\mathbf{k}+\frac{\mathbf{q}}{2}ia\uparrow} p_{\mathbf{k}-\frac{\mathbf{q}}{2}ia\downarrow} + p^{\dagger}_{\mathbf{k}-\frac{\mathbf{q}}{2}ia\downarrow} p_{\mathbf{k}+\frac{\mathbf{q}}{2}ia\uparrow}\right), \end{aligned} \tag{9.58}$$

where $\mathcal{N}$ is the number of elementary unit cells in the sample, $n = \sum_{\sigma}\langle n_{\mathbf{n}ia\sigma}\rangle$ is the average number of electrons per site, and $x = n - 1$ is the doping level. The second term in Eq. (9.58) renormalizes the chemical potential. Thus, we define the renormalized chemical potential as $\mu' = \mu - Ux/2$. Further, it is convenient to redefine the 4-component spinor $\psi_{\mathbf{k}\sigma}$ introduced in the Sect. 9.2.1, removing the phase factor $e^{-i\phi_{\mathbf{k}}}$. Now it will have the form $\psi^{\dagger}_{\mathbf{k}\sigma} = \left(p^{\dagger}_{\mathbf{k}1A\sigma},\ p^{\dagger}_{\mathbf{k}2A\sigma},\ p^{\dagger}_{\mathbf{k}1B\sigma},\ p^{\dagger}_{\mathbf{k}2B\sigma}\right)$. Next, we define the 8-component spinor of the form

$$\Psi^{\dagger}_{\mathbf{k}\mathbf{q}} = \left(\psi^{\dagger}_{\mathbf{k}+\mathbf{q}/2\uparrow},\ \psi^{\dagger}_{\mathbf{k}-\mathbf{q}/2\downarrow}\right). \tag{9.59}$$

With this spinor, we can rewrite the total mean-field Hamiltonian as follows:

$$H = \mathcal{N}\left[\frac{4\Delta^2}{U} - U(n^2-1)\right] + \sum_{\mathbf{k}}\Psi^{\dagger}_{\mathbf{k}\mathbf{q}}\begin{pmatrix}\hat{H}_{0\mathbf{k}+\frac{\mathbf{q}}{2}} & \hat{\Delta}\\ \hat{\Delta} & \hat{H}_{0\mathbf{k}-\frac{\mathbf{q}}{2}}\end{pmatrix}\Psi_{\mathbf{k}\mathbf{q}}\,, \tag{9.60}$$

where matrices $\hat{H}_{0\mathbf{k}}$ and $\hat{\Delta}$ have the form

$$\hat{H}_{0\mathbf{k}} = -\begin{pmatrix} 0 & t_0 & tf_{\mathbf{k}} & 0 \\ t_0 & 0 & 0 & tf_{\mathbf{k}} \\ tf^*_{\mathbf{k}} & 0 & 0 & t_0 \\ 0 & tf^*_{\mathbf{k}} & t_0 & 0 \end{pmatrix}, \quad \hat{\Delta} = -\begin{pmatrix} \Delta & 0 & 0 & 0 \\ 0 & -\Delta & 0 & 0 \\ 0 & 0 & -\Delta & 0 \\ 0 & 0 & 0 & \Delta \end{pmatrix}. \tag{9.61}$$

The matrix $\hat{H}_{0\mathbf{k}}$ in Eq. (9.61) is complex. It can be made real by introducing the phase factors $e^{-i\phi_{\mathbf{k}}}$ into the definition of $\psi_{\mathbf{k}\sigma}$, as described above. In this case, however, the matrix $\hat{\Delta}$ becomes complex. The spectrum of quasiparticles is obtained by diagonalizing the 8×8 matrix in Eq. (9.60). Due to the fact that this matrix is complex, simple analytical expressions for the spectra in the general case cannot be obtained. Moreover, unlike the commensurate AFM phase, the spectrum is no longer degenerate in the electron spin. The band structure consists of eight bands $E^{(s)}_{\mathbf{kq}}$ ($s = 1, 2, \ldots, 8$), which are determined numerically. In the limit $a|\mathbf{q}| \ll 1$ it is possible to obtain an analytical expression for the spectra:

$$\frac{E^{(s)}_{\mathbf{kq}}}{t} \approx \pm\frac{\zeta_{\mathbf{k}+\frac{\mathbf{q}}{2}} - \zeta_{\mathbf{k}-\frac{\mathbf{q}}{2}}}{2} \pm \sqrt{\frac{\Delta^2}{t^2} + \left[\frac{t_0}{t} \pm \frac{\zeta_{\mathbf{k}+\frac{\mathbf{q}}{2}} + \zeta_{\mathbf{k}-\frac{\mathbf{q}}{2}}}{2}\right]^2}. \tag{9.62}$$

In the case of $\mathbf{q} = 0$ corresponding to the commensurate AFM phase, the spectrum becomes doubly degenerate and coincides with the formulas (9.39)–(9.42). The expression for the potential Ω will have the form

$$\Omega = \frac{\Delta^2}{U} - \frac{U}{4}(n^2 - 1) - T\sum_{s=1}^{8}\int\frac{d\mathbf{k}}{v_{\mathrm{BZ}}}\ln\left[1 + e^{(\mu' - E^{(s)}_{\mathbf{kq}})/T}\right]. \tag{9.63}$$

Minimizing Ω by Δ and $\mathbf{q}$ together with the condition connecting the doping level x and the chemical potential μ gives a closed equation system for determining Δ and $\mathbf{q}$ as functions of x and T:

$$\frac{\partial\Omega}{\partial\Delta} = 0, \quad \frac{\partial\Omega}{\partial\mathbf{q}} = 0, \quad 1 + x = -\frac{\partial\Omega}{\partial\mu'}. \tag{9.64}$$

The functions $\Delta(x, T)$ and $\mathbf{q}(x, T)$ are calculated numerically for different values of U. Typical curves $\Delta(x)$ and $|\mathbf{q}(x)|$ are shown in Fig. 9.6. For comparison, the dependence $\Delta(x)$ calculated for a commensurate AFM phase is also presented. The incommensurate phase occurs for arbitrarily small doping in the case of $T = 0$. At a finite T, commensurate phase exists until the doping reaches a certain threshold value, while the transition between phases is of the second order. The curve

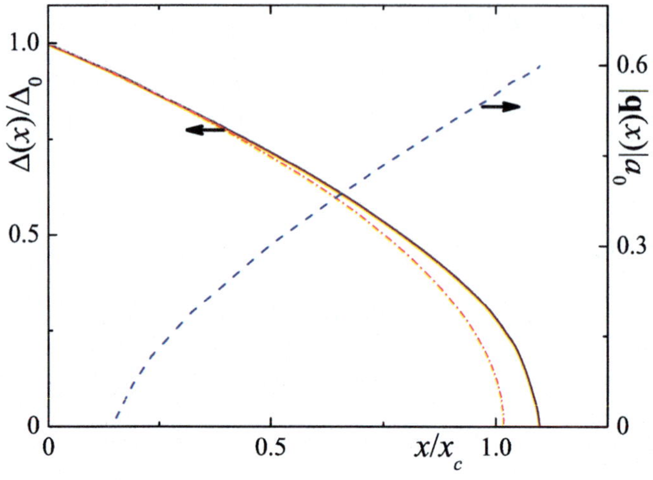

Fig. 9.6 Dependence of Δ (red solid curve) and $|\mathbf{q}|$ (blue dashed curve) on the doping level x, calculated at $T/\Delta_0 = 0.06$. The thin red dot-dashed curve corresponds to Δ calculated at $\mathbf{q} = 0$. Parameter $x_c = \Delta_0\rho_0(t_0)/2$. The picture is taken from Ref. [14]

$T^q(x)$ separates the commensurate and incommensurate AFM phases. It is calculated numerically. The mean-field transition temperature $T_{MF}(x)$ separating the commensurate or incommensurable AFM phase from the PM phase is also calculated.

The stability of the homogeneous state is investigated by analyzing the dependence of the free energy $F(x, T) = \Omega(\mu(x, T), T) + \mu(x, T)(1 + x)$ on doping level. At temperatures below a certain critical value, the function $F(x)$ has a negative second derivative, indicating the instability of the homogeneous state toward the phase separation. As well as for the Rice model, the separated phases are the commensurate and incommensurate AFM phases. The calculation results are summarized in the phase diagram in the x–T plane, shown in Fig. 9.7. The phase diagram is constructed at $U = 5.5\,\text{eV}$, which corresponds to $U/U_c = 0.96$. Qualitatively, it is very similar to the phase diagram of the Rice model in the absence of an electron reservoir. At lower U, the shape of the phase diagram does not change qualitatively. For large U ($U \gtrsim 1.15U_c$), the phase diagram is qualitatively different from that shown in Fig. 9.7; there appears a reentrance region—the transition from the PM to the AFM phase with an increase of temperature. We do not analyze in detail the case of $U > U_c$ because it does not correspond to the experiment. In addition, for large U, the question arises about the applicability of the mean-field approximation used here.

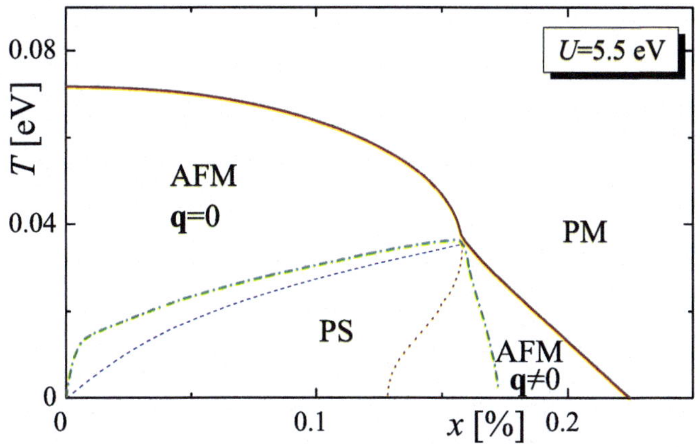

Fig. 9.7 Phase diagram of the AA stacked bilayer graphene in the x–T plane, calculated at $U = 5.5\,\text{eV}$ ($U/U_c = 0.96$). The red solid curve corresponds to $T_{MF}(x)$. The blue dashed curve corresponds to $T^q(x)$ and shows the boundary between the commensurate and incommensurate AFM phases without taking into account the phase separation. The green dot-dashed curve shows the region of the phase separation. Finally, the red dotted curve shows the transition temperature between the commensurate AFM and PM phases in the absence of the phase separation and incommensurate AFM phase. For hole ($x < 0$) doping, the results are similar. The picture is taken from Ref. [14]

9.3 Twisted Bilayer Graphene

Besides two simple stacking configurations of the bilayer graphene, AA and AB, there is another stacking configuration (more precisely a set of them) called twisted bilayer graphene (tBLG). Such a material attracts a great attention nowadays from both theorists and experimentalists. In twisted bilayer graphene, one layer is rotated with respect to another by the twist angle θ. Twisting produces a moiré pattern, which can be visualized by STM [27, 28]. Measuring the moiré period L one can extract the twist angle according to the formula

$$L = \frac{a}{2\sin(\theta/2)}, \tag{9.65}$$

where a is the length of the unit vector of the graphene. Moiré pattern exists for arbitrary twist angle, however, theorists consider mainly the so-called commensurate angles, when the system has a superstructure. It can be shown also that when the twist angle is small enough, the supercell can be considered as consisting of regions with almost AA, AB, and BA stackings.

The low-energy electronic properties of the twisted bilayer graphene substantially depend on the twist angle. When this angle is not too small, the spectrum at low energies consists of two doubly degenerate Dirac cones located at two non-equivalent Dirac points of the superlattice Brillouin zone. These Dirac cones intersect

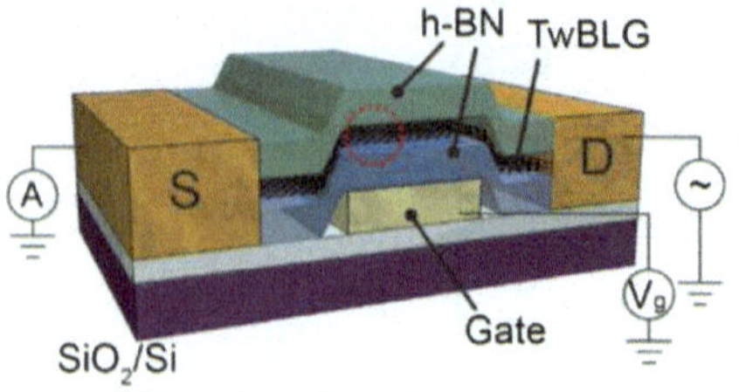

Fig. 9.8 Schematics of the tBLG devices used for measuring the conductivity as function of doping n. The picture is taken from Ref. [32]

at lower and higher energies producing low-energy van Hove singularities [27, 28]. Fermi velocity of these Dirac cones is smaller than that for single-layer graphene. It decreases with the decrease of the twist angle [29, 30]. This Fermi velocity reduction can be substantial but the sample remains semimetal. However, at the so-called first magic angle, the system acquires a Fermi surface even at the charge neutrality point [11]. The spectrum at magic angle consists of four almost flat bands separated from lower and higher dispersive bands by energy gaps. The width of the flat bands has minimum at magic angle. Both theoretical predictions [29, 30] and experimental observations [31–33] show that the value of the first magic angle $\theta_c \approx 1^\circ$.

The sufficient progress in the experimental studying of the magic angle twisted bilayer graphene was done in Refs. [32, 33], where the authors observed a Mott insulating states [32] and superconductivity in slightly doped samples. The device under consideration is schematically shown in Fig. 9.8. The tBLG sample with measured twist angle is placed between two boron nitride substrates. Two electrodes, source and drain ones are placed on opposite sides of the sample. The third, gate, electrode is located below the sample. Applying the gate voltage allows one to introduce an extra charge, electrons or holes, to the bilayer. This device allows measuring the conductivity σ of the bilayer as a function of doping n. The authors of Refs. [32, 33] measured a conductivity curves $\sigma(n)$ for several devices with slightly different twist angle of the tBLG samples around the value $\theta_c \cong 1.08^\circ$. Conductivity curves showed minima at $\nu \equiv n/n_s = 0, \pm 2 \pm 4$, where concentration n_s corresponds to 1 extra electron per one supercell (different samples in [32] showed also minima at $\nu = \pm 3$). The minima of conductivity at half-filling, $\nu = \pm 2$ (that is for two extra electrons per supercell), cannot be described within single-particle theory and many-body effects should be taken into account. Besides insulating states, the authors of Ref. [33] also observed two superconductivity "domes" slightly below and above half-filling, $\nu = -2$. The maximum superconducting transition temperature T_c observed in [33] is 1.7 K. It seems to be low, but there are few charge carriers in the miniband (the supercell is very large for $\theta = \theta_c$). Therefore, the ratio of T_c to the Fermi energy turns out to be higher than in cuprates, which indicates the relative strength of superconducting pairing. In similar studies of different groups, Ref. [34], the authors observed insulating states for all integer-valued doping levels, $\nu = 0, \pm 1, \pm 2, \pm 3$. Together with superconducting domes near $\nu = -2$, they observed also three new superconducting domes near $\nu = 0$ and $\nu = \pm 1$. Thus, the phase diagram of the magic angle twisted bilayer graphene can be very rich. Moreover, the screening of electron–electron interaction by introducing a metallic layer located close to twisted bilayer graphene provides a tool for controlling and tuning the form of the phase diagram [35]. Besides doping the electronic properties of the tBLG can be tuned by pressure. The applied pressure can be used for controlling the interparticle distance and thus the conditions for the formation of the flat bands. Indeed, more than 1 GPa of hydrostatic pressure-induced superconductivity in a twisted bilayer device with a larger twist angle of 1.27° that did not otherwise exhibit any correlated behavior [36].

Whenever you have flat bands, the interactions become very important. Interactions break the symmetry of the single-particle Hamiltonian giving rise to some order parameter in the system. The nature of the many-body insulating states in the magic angle twisted bilayer graphene was considered in several papers [37–45]. Different types of spin density wave states [39, 40, 42, 43, 45], ferromagnetic state [41], and other symmetry broken phases [46] have been proposed to be the ground state of the system. Potential mechanisms of the superconductivity (phonons [47, 48], electron correlations [39, 40, 49–51]), as well as various symmetries of the superconducting order parameters, have been considered. Since interactions are strong compared to the bandwidth, one can expect the phase separation via mechanisms similar to that considered in previous Chapters. The instability of the system toward the phase separation has been studied theoretically in Ref. [44]. The authors assumed the spin density wave to be the ground state of the system. They showed that there are four regions of the phase separation in the doping range $-4 < \nu < 4$. The phases in the inhomogeneous states are characterized by an even number ($\nu = 0, \pm 2, \pm 4$) of electrons per supercell. This allows us to explain the fact that the conductivity minima that could occur at doping concentrations corresponding to an odd number ($\nu = \pm 1, \pm 3$) of electrons per supercell are absent in some samples [32] (phase separation occurs) and are observed in others [34] (phase separation is suppressed by long-range Coulomb repulsion). In this section, we follow Ref. [44] considering the theory of the phase separation in magic-angle twisted bilayer graphene.

9.3.1 Geometry of the Twisted Bilayer Graphene

For the convenience of readers, we will briefly describe the main geometric features of the system under consideration (a more detailed description can be found in the review paper [11]). The graphene monolayer has a hexagonal crystal structure consisting of two triangular sublattices A and B. The coordinates of the atoms in layer 1 are $\mathbf{r}_{\mathbf{n}}^{1A} = \mathbf{r}_{\mathbf{n}}^{1} \equiv n\mathbf{a}_1 + m\mathbf{a}_2$ and $\mathbf{r}_{\mathbf{n}}^{1B} = \mathbf{r}_{\mathbf{n}}^{1} + \boldsymbol{\delta}$, where $\mathbf{n} = (n, m)$ (n and m are integers), $\mathbf{a}_{1,2} = a(\sqrt{3}, \mp 1)/2$ are lattice vectors ($a = 2.46$ Å), and $\boldsymbol{\delta} = a(1/\sqrt{3}, 0)$. The atoms in the 2 layer are located at the points $\mathbf{r}_{\mathbf{n}}^{2B} = \mathbf{r}_{\mathbf{n}}^{2} \equiv d\mathbf{e}_z + n\mathbf{a}'_1 + m\mathbf{a}'_2$ and $\mathbf{r}_{\mathbf{n}}^{2A} = \mathbf{r}_{\mathbf{n}}^{2} - \boldsymbol{\delta}'$, where the vectors $\mathbf{a}'_{1,2}$ and $\boldsymbol{\delta}'$ are obtained from the vectors $\mathbf{a}_{1,2}$ and $\boldsymbol{\delta}$ by rotation on angle θ and $\mathbf{e}_z$ is the unit vector in the z direction. The distance between the layers is $d = 3.35$ Å. The limiting case $\theta = 0$ corresponds to AB stacking (Bernal phase). A superstructure in twisted bilayer graphene occurs when the following condition is met

$$\cos\theta = \frac{3m_0^2 + 3m_0 r + r^2/2}{3m_0^2 + 3m_0 r + r^2}, \tag{9.66}$$

where m_0 and r are mutually prime positive integers. The number of elementary cells of each graphene layer in a supercell is $N_{sc} = (3m_0^2 + 3m_0 r + r^2)/g$, where $g = 1$ if $r \neq 3n$ or $g = 3$ otherwise. The reciprocal lattice vectors of layer 1 (layer 2) are denoted as $\mathbf{b}_{1,2}$ ($\mathbf{b}'_{1,2}$). For layer 1, we have $\mathbf{b}_{1,2} = 2\pi(1/\sqrt{3}, \mp 1)/a$, whereas $\mathbf{b}'_{1,2}$ are obtained from vectors $\mathbf{b}_{1,2}$ by rotation on the angle θ. Let $\mathcal{G}_{1,2}$ be the reciprocal vectors of the superlattice. It can be shown that the following equalities are valid: $\mathbf{b}'_1 = \mathbf{b}_1 + r(\mathcal{G}_1 + \mathcal{G}_2)$ and $\mathbf{b}'_2 = \mathbf{b}_2 - r\mathcal{G}_1$ if $r \neq 3n$ or $\mathbf{b}'_1 = \mathbf{b}_1 + r(\mathcal{G}_1 + 2\mathcal{G}_2)/3$ and $\mathbf{b}'_2 = \mathbf{b}_2 - r(2\mathcal{G}_1 + \mathcal{G}_2)/3$ otherwise. The Brillouin zone of the superlattice has a hexagonal shape. Two nonequivalent corners of the Brillouin zone (Dirac points of the superlattice), $\mathbf{K}_1$ and $\mathbf{K}_2$, can be expressed in terms of the vectors $\mathcal{G}_{1,2}$ by equalities $\mathbf{K}_1 = (\mathcal{G}_1 + 2\mathcal{G}_2)/3$ and $\mathbf{K}_2 = (2\mathcal{G}_1 + \mathcal{G}_2)/3$. We also define the $\mathbf{M}$ point as $\mathbf{M} = (\mathcal{G}_1 + \mathcal{G}_2)/2$.

9.3.2 Model Hamiltonian

The model Hamiltonian, studied in Ref. [44] is

$$\begin{aligned} H = &\sum_{\substack{\mathbf{nm}ij \\ ab\sigma}} t(\mathbf{r}_\mathbf{n}^{ia}; \mathbf{r}_\mathbf{m}^{jr}) p^\dagger_{\mathbf{n}ia\sigma} p_{\mathbf{m}jb\sigma} + U \sum_{\mathbf{n}ia} n_{\mathbf{n}ia\sigma} n_{\mathbf{n}ia\bar{\sigma}} + \\ &\frac{1}{2} \sum_{\substack{\mathbf{nm}ij \\ ab\sigma\sigma'}}{}' V(\mathbf{r}_\mathbf{n}^{ia} - \mathbf{r}_\mathbf{m}^{jb}) n_{\mathbf{n}ia\sigma} n_{\mathbf{m}jb\sigma'} \, . \end{aligned} \tag{9.67}$$

In this expression, $p^\dagger_{\mathbf{n}ia\sigma}$ and $p_{\mathbf{n}ib\sigma}$ are creation and annihilation operators of the electron with spin projection σ in the unit cell $\mathbf{n}$ of the layer $i\,(=1, 2)$ in the sublattice $a\,(=\mathcal{A}, \mathcal{B})$, and $\bar{\sigma} = -\sigma$. Operator $n_{\mathbf{n}ia\sigma} = p^\dagger_{\mathbf{n}ia\sigma} p_{\mathbf{n}ia\sigma}$. The first term in the Eq. (9.67) is the single-particle tight-binding Hamiltonian of the system, where $t(\mathbf{r}_\mathbf{n}^{ia}; \mathbf{r}_\mathbf{m}^{jb})$ is the amplitude of the electron hopping from the site in position $\mathbf{r}_\mathbf{m}^{jb}$ to the site in position $\mathbf{r}_\mathbf{n}^{ia}$. Within the layer, only nearest-neighbor hoppings with amplitude $t = -2.57\,\text{eV}$ are considered. The function $t(\mathbf{r}_\mathbf{n}^{1s}; \mathbf{r}_\mathbf{m}^{2r})$ describing hoppings between layers is modeled as described in Ref. [42] (note, however, that the choice of other parameterizations [43] does not change the results qualitatively). The maximum value of the interlayer hopping amplitude is chosen to be $t_0 = 0.4\,\text{eV}$.

The second term in Eq. (9.67) describes the one-site (Hubbard) interaction of electrons with opposite spins, while the third term describes the Coulomb interaction of electrons at different sites (the prime means that elements with $\mathbf{r}_\mathbf{n}^{is} = \mathbf{r}_\mathbf{m}^{jr}$ should be discarded).

9.3.3 Single-Electron Spectrum at Magic Angle

In the single-particle approximation, the energy spectrum of twisted bilayer graphene is determined by numerical diagonalization of the first term in the Eq. (9.67). For any superstructure (m_0, r), the electron spectrum of the bilayer consists of $4N_{sc}$ bands $E_{\mathbf{k}}^{(S)}$, where the quasimomentum $\mathbf{k}$ lies in the Brillouin zone of the superlattice, and $1 \leq S \leq 4N_{sc}$. For given $\mathbf{k}$, the energies $E_{\mathbf{k}}^{(S)}$ are arranged in ascending order. The four bands lying in the middle of the spectrum are of the greatest interest, since they are located near the Fermi level of the undoped sample.

As it was explained above, at magic angle the spectrum at low energies consists of four almost flat bands separated from lower and higher dispersive bands by energy gaps. The width W of the flat bands has a minimum at magic angle. The value of the magic angle θ_c depends on the type of the parametrization of interlayer hopping amplitudes. For the chosen parametrization, we have $\theta_c \cong 1.08°$ [superstructure $(m_0, r) = (30, 1)$]. This value θ_c corresponds to the experiment [32, 34]. The width of the flat bands is equal to $W = 1.9$ meV. The energy spectrum calculated for this superstructure is shown in Fig. 9.9.

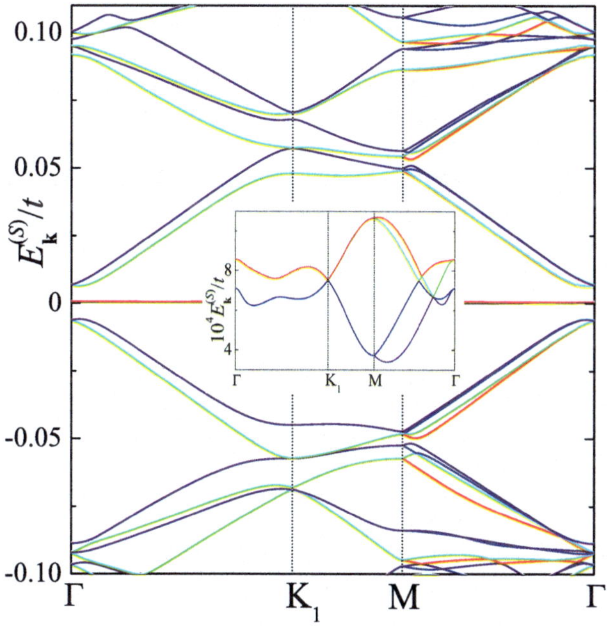

Fig. 9.9 The energy spectrum of the twisted bilayer graphene at the magic angle $\theta_c \cong 1.08°$ [superstructure $(m_0, r) = (30, 1)$]. Near zero energy, there are four flat bands separated by energy gaps from lower and higher dispersive bands. The insert shows the dispersions of flat bands on the enlarged scale. The picture is taken from Ref. [44]

9.3.4 The Structure of the Order Parameter

As already noted, a system having flat bands at the Fermi level is very susceptible to interactions. The interactions are described by the second and third terms in the Hamiltonian (9.67). Interactions lead to a spontaneous symmetry breaking and generate a certain order parameter. Here we assume that this order parameter is a spin density wave. This choice is not arbitrary. As has been shown in many papers (see, for example, Refs. [11, 52, 53]), at small twist angles electrons at the Fermi level occupy the regions of a supercell with AA stacking. At the same time, as we have shown in the previous Section the ground state of the AA stacked bilayer graphene should be antiferromagnetic. For this reason, we believe that the spin density wave is a good candidate for the ground state of the magic angle twisted bilayer graphene.

The order parameter under consideration is multicomponent one. First, it contains terms proportional to the local spins:

$$\Delta_{\mathbf{n}ia} = U\langle p^{\dagger}_{\mathbf{n}ia\sigma} p_{\mathbf{n}ia\bar{\sigma}}\rangle\,. \tag{9.68}$$

These components are controlled by the on-site Coulomb repulsion. The Hubbard U is taken to be $U = 2t$. This value is a bit less than the critical value for the transition of a single-layer graphene to an antiferromagnetic state in the mean-field approximation [20], $U_c = 2.23t$. That is, the chosen Hubbard interaction is quite strong, but not so strong to open a gap in a single-layer graphene [11].

Another order parameter describes the coupling of electrons at neighboring sites inside the layer. Each graphene atom in one sublattice has three nearest-neighbor belonging to another sublattice. As a result, there are three types of intra-layer nearest-neighbors order parameters: parameters $A^{(\ell)}_{\mathbf{n}i\sigma}$ ($\ell = 1, 2, 3$) correspond to three different links connecting the nearest sites within the layer. These order parameters are defined as follows:

$$A^{(\ell)}_{\mathbf{n}i\sigma} = V_{\text{nn}}\langle p^{\dagger}_{\mathbf{n}+\mathbf{n}_{\ell}iA\sigma} p_{\mathbf{n}iB\bar{\sigma}}\rangle\,, \tag{9.69}$$

where $\mathbf{n}_1 = (0, 0)$, $\mathbf{n}_2 = (1, 0)$, $\mathbf{n}_3 = (0, 1)$, and $V_{\text{nn}} = V(|\boldsymbol{\delta}|)$ is the Coulomb interaction of electrons located at neighboring sites inside the layer. According to Ref. [17], it is assumed that $V_{\text{nn}}/U = 0.59$.

Finally, we also take into account the order parameter describing the coupling of electrons in different layers. It is defined as:

$$B^{ab}_{\mathbf{m};\mathbf{n}\sigma} = V(\mathbf{r}^{1a}_{\mathbf{m}} - \mathbf{r}^{2b}_{\mathbf{n}})\langle p^{\dagger}_{\mathbf{m}1a\sigma} p_{\mathbf{n}2b\bar{\sigma}}\rangle\,. \tag{9.70}$$

Only such $B^{ab}_{\mathbf{m};\mathbf{n}\sigma}$ are considered for which the sites $\mathbf{r}^{1b}_{\mathbf{m}}$ and $\mathbf{r}^{2a}_{\mathbf{n}}$ are located close enough to each other. If the hopping amplitude between the sites $\mathbf{r}^{1b}_{\mathbf{m}}$ and $\mathbf{r}^{2a}_{\mathbf{n}}$ is negligible, then the order parameter $B^{ab}_{\mathbf{m};\mathbf{n}\sigma}$ is assumed to be zero. Assuming that the screening is negligible at short distances, the function $V(\mathbf{r})$ in the Eq. (9.70) is modeled as $V(\mathbf{r}) \propto 1/|\mathbf{r}|$ with $V(d) = V_{\text{nn}}|\boldsymbol{\delta}|/d = 0.25U$.

It is assumed that all order parameters have the same periodicity as the superlattice has, so no extra periodicity arises. The system of the mean-field equations for $\Delta_{\mathbf{n}ia}$, $A^{(\ell)}_{\mathbf{n}i\sigma}$ and $B^{ab}_{\mathbf{m};\mathbf{n}\sigma}$ is solved numerically for various doping levels in the range from -4 to $+4$ extra electrons per supercell. Details of the iteration scheme can be found in Ref. [43].

9.3.5 Results. Spectral Properties and Conductivity

The calculations show that all three types of order parameters reach their maximum values for the undoped sample (one electron per site). The order parameters $\Delta_{\mathbf{n}ia}$, $A^{(\ell)}_{\mathbf{n}i\sigma}$ and $B^{ab}_{\mathbf{m};\mathbf{n}\sigma}$ decrease monotonically with doping regardless of the type of doping (hole or electronic) and go to zero for doping $\nu = n/n_s = \pm 4$. The order parameters are inhomogeneous in space. The parameters $\Delta_{\mathbf{n}ia}$, $A^{(\ell)}_{\mathbf{n}i\sigma}$, and $B^{ab}_{\mathbf{m};\mathbf{n}\sigma}$ reach their largest values in the AA region of the superlattice cell. This is in good agreement with the fact that the wave functions of electrons near the Fermi level are localized in the AA region of the supercell. The maximum values of the order parameters at charge neutrality point are $\Delta_{\mathbf{n}ia} = 27\,\text{meV}$, $A^{(\ell)}_{\mathbf{n}i\sigma} = 3.5\,\text{meV}$, and $B^{ab}_{\mathbf{m};\mathbf{n}\sigma} = 0.2\,\text{meV}$.

When order parameters are found, one can calculate the perturbed energy spectrum. The spin density wave lifts the degeneracy of the bands on the electron spin. When spin degrees of freedom are taken into account, in the single-particle approximation, we have 8 almost degenerate flat bands near the Fermi level. The order parameter lifts (partially) this degeneracy, and it does it differently depending on the doping level. At charge neutrality point, $\nu \equiv n/n_s = 0$, eight bands are split into two quartets, the Fermi level lies between the quartets. At half-filling, $\nu = \pm 2$, each quartet is split into two doublets, the Fermi level lies between the two upper (lower) doublets for electronic (hole) doping. When we have one electron or hole per supercell, $\nu = \pm 1$, each quartet splits into a singlet and a triplet, and the Fermi level is between a singlet and a triplet. In the case of $\nu = \pm 3$, it turns out that each quartet splits into two singlets and a doublet, the Fermi level lies between two singlets, the upper in the case of electronic or the lower in the case of hole doping. It is not difficult to understand that these configurations of bands really should decrease the total energy of the system compared to the non-degenerate (no interaction) case. It should be noted that the above-mentioned splitting of bands is not actually observed in the entire Brillouin zone. Near the $\boldsymbol{\Gamma}$ point, the situation turns out to be more complicated (details can be found in Ref. [43]), however, this is not essential for present consideration.

Knowing the spectrum $E^{(S)}_{\mathbf{k}}$, it is not difficult to calculate the density of states at the Fermi level at different doping values. At finite temperature T the letter one is described by the formula:

$$\rho(\varepsilon_F) = \sum_S \int \frac{d^2\mathbf{k}}{v_{BZ}} \frac{1}{4T\cosh\left[(\varepsilon_F - E^{(S)}_{\mathbf{k}})/2T\right]^2}\,. \tag{9.71}$$

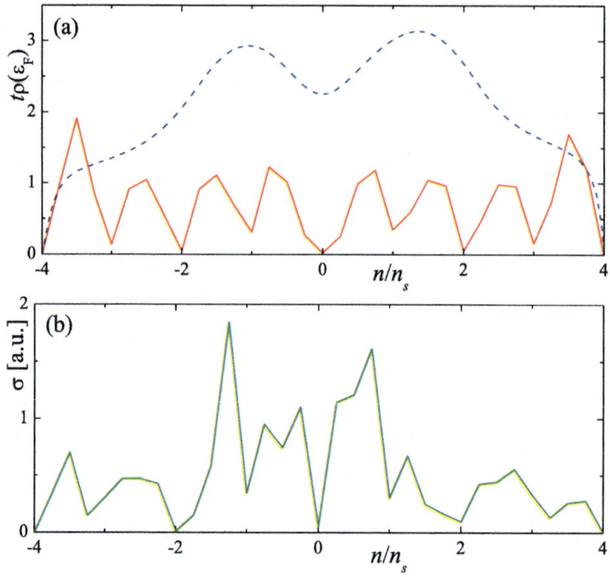

Fig. 9.10 **a** The density of states at the Fermi level as a function of doping. The solid (red) curve corresponds to the interacting system, whereas the dashed (blue) curve is constructed without taking into account the interaction. **b** The dependence of the conductivity of the system on doping. The pictures are taken from Refs. [42, 44]

The result is shown in Fig. 9.10a. For comparison, we plot also the density of states at the Fermi level, calculated for a non-interacting system. Without taking into account the interaction, the density of states has two maxima associated with the low-energy van Hove singularities, and three minima at $\nu = 0, \pm 4$. The interaction significantly modifies the dependence of ρ on n. Now the density of states has minima for all integer valued ν. This behavior is natural, it follows from the splitting of groups of bands at different doping levels as described above. The calculation of the conductivity in the τ approximation performed in Ref. [42], showed that the conductivity σ as a function of doping n has a similar behavior (we present the result of this calculation in Fig. 9.10b). Such a dependence $\sigma(n)$ is qualitatively consistent with the experimentally observed in Ref. [34]. At the same time, the experiments made in the Ref. [32] showed that the conductivity has minima only at $\nu = 0, \pm 2, \pm 4$ (note, however, that for some samples the authors of Ref. [32] also observed conductivity minima at $\nu = \pm 3$, see additional materials for [32]). Thus, we see that the sequence of minima of the conductivity of the system depends on the sample. This can be explained by the tendency to phase separation, which, as we will show, is characteristic of the interacting system.

9.3.6 Results. Phase Separation

To study the stability of the homogeneous state of the system, we analyze the dependence of the chemical potential on doping. The result of calculation is shown in Fig. 9.11. We see that near integer-valued doping levels, the chemical potential is a growing function of n. However, in intermediate doping, there are regions in which the chemical potential decreases, $\partial\mu(n)/\partial n < 0$, which indicates the instability of the homogeneous state with respect to the electronic phase separation, since the compressibility of the electronic system becomes negative. In order to understand into which phases the system is separated and in which areas the phase separation is realized, it is necessary to analyze the free energy of the system. If the dependence $\mu(n)$ is known, then the free energy of the homogeneous state (per one supercell) can be calculated by the formula:

$$f(n) = \frac{1}{n_s}\int\limits_{-4n_s}^{n} \mu(n')\,dn' - f_0\,, \tag{9.72}$$

where the constant f_0 is chosen such that the equality $f(0) = 0$ is fulfilled.

The dependence $f(n)$ is shown in Fig. 9.12. At each concentration n, it is necessary to compare the free energy of the homogeneous state $f(n)$ with the energy of the inhomogeneous state $f_{PS}(n)$. The latter can be obtained as follows. In the inhomogeneous state, the system with total electron concentration n is separated into the phase with electron concentration $n_1(<n)$ and a volume fraction p and the phase with electron concentration $n_2(>n)$ and a volume fraction $1 - p$. The

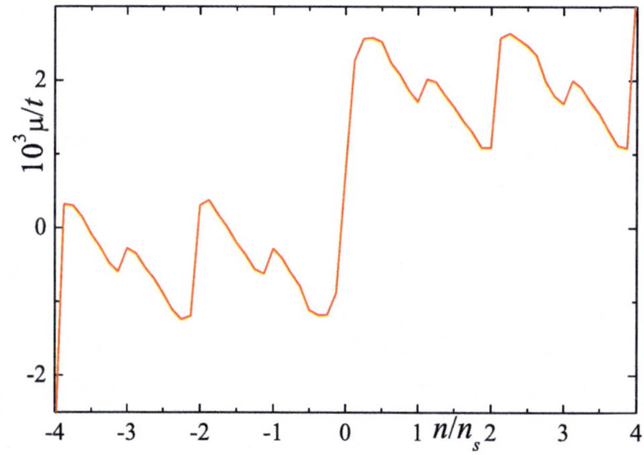

Fig. 9.11 Dependence of chemical potential on doping. Regions with $\partial\mu(n)/\partial n < 0$ signal instability of the homogeneous state. The calculation was performed at temperature $T/t = 1.5 \times 10^{-5}$. The picture is taken from Ref. [44]

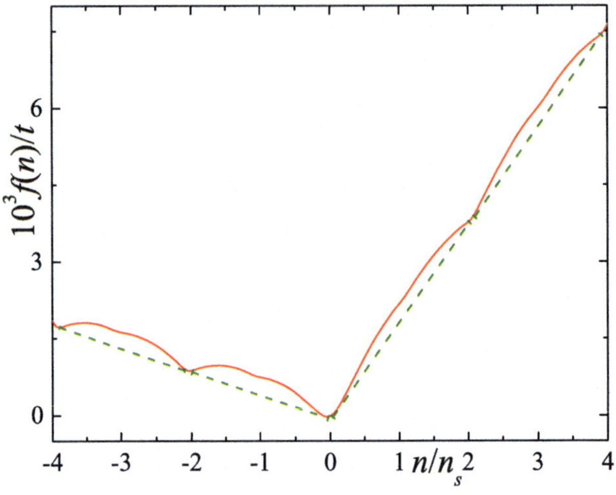

Fig. 9.12 The dependence of the free energy of the system (per supercell) on doping. The solid (red) curve corresponds to the free energy of the homogeneous state. The energies of inhomogeneous states obtained by Maxwell's construction are shown by dashed (green) straight lines. The calculation is performed at temperature of $T/t = 1.5 \times 10^{-5}$. The picture is taken from Ref. [44]

value of p is $p = (n - n_1)/(n_2 - n_1)$. The free energy of the inhomogeneous state depends linearly on n and is equal to $f_{PS}(n) = pf(n_1) + (1 - p)f(n_2)$. The equilibrium concentrations of electrons n_1 and n_2 are determined by the Maxwell construction. The analysis of the dependence $f(n)$ shows that there are four regions of concentration $n_1 < n < n_2$ in which the phase separation is realized. These the areas are ($n_1 \approx -4n_s$, $n_2 \approx -2n_s$), ($n_1 \approx -2n_s$, $n_2 \approx 0$), ($n_1 \approx 0$, $n_2 \approx 2n_s$), and ($n_1 \approx 2n_s$, $n_2 \approx 4n_s$).

Thus, we see that in the picture described above, homogeneous phases with concentrations near $\nu = \pm 1$ and $\nu = \pm 3$ are not realized. Note also that the phases in inhomogeneous states are (almost) dielectric, and such a simple approach does not allow to explain the behavior of the conductivity of the system with doping observed in the experiment. It should be kept in mind, however, that since phases in the inhomogeneous state have different concentrations of electrons, it is necessary to take into account the long-range Coulomb repulsion of electrons in the inhomogeneous state, which occurs due to violation of the local electroneutrality. This factor was not taken into account in the analysis described above. The long-range Coulomb repulsion narrows the region of the phase separation and may arrest it at any charge carrier concentrations [54, 55]. In this case, the system is always homogeneous, and the minima of the conductivity and density of states on the Fermi level exist for all integer ν, which is observed in the experiment in Ref. [34]. The magnitude of the long-range Coulomb repulsion can vary from sample to sample, since it depends on the substrate on which the sample is located, as well as on the quality of the sample. When the long-range Coulomb repulsion is not too large, the homogeneous state is more favorable near even values of ν, however, the phase separation still "hides" the regions near $\nu = \pm 1$ and $\nu = \pm 3$. In this case, we should see the minima of the conductivity only at $\nu = 0, \pm 2, \pm 4$, which corresponds to the experiment [32]. Finally, the long-range Coulomb repulsion can arrest the phase separation only for some of the four intervals $n_1 < n < n_2$. Then, in the dependence of the conductivity on the doping, additional minima may appear for some odd ν. Such minima were indeed observed for some samples in Ref. [32].

Another factor affecting the variability of the conductivity behavior for different samples is the twist angle. Indeed, the deviation of the twist angle θ from the magic value θ_c (in any direction) increases the width of the flat bands W. Increase in W decreases the value of the SDW order parameter. Therefore, the effects associated with SDW, such as the negative curvature of the free energy, become less pronounced. This, in turn, increases the role of the long-range Coulomb repulsion.

Thus, in the framework of the simple mean-field approach, the chemical potential of the magic angle twisted bilayer graphene exhibits a non-monotonic behavior, where the regions of positive $\partial\mu/\partial n$ alternate with the regions of doping where $\partial\mu/\partial n < 0$. Such a dependence correlates well with experiments [56, 57]. Thus, in the capacitance measurements performed in Ref. [57] the curve $\mu(n)$ flattens near the integer-valued doping levels. The measurements of $\mu(n)$ done in Ref. [56] showed that the derivative $\partial\mu/\partial n$ is negative in some doping regions in agreement with the theory described above. The curve $\partial\mu(n)/\partial n$ measured in Ref. [56] for tBLG sample with $\theta = 1.13°$ is shown in Fig. 9.13.

Note, that for the phase separation scenario described in this Section, the separated phases have different electron concentrations. As it was described in previous chapters, in this case, the long-range Coulomb repulsion tends to make the phase separation small-scale one. In the simplest situation, we have small droplets of one phase embedded into the matrix of another phase. The mechanism of the formation of small skyrmions—spin polarons in application to the twisted bilayer graphene has been considered in Ref. [58]. The properties of such a state as well as possible ways for experimental verifications of such spin polarons have been discussed there.

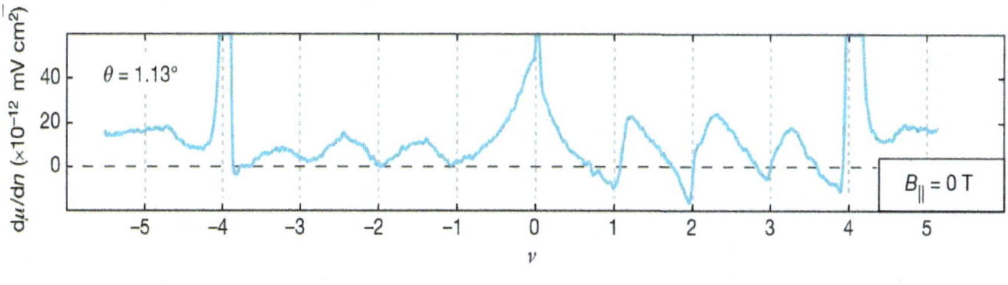

Fig. 9.13 The dependence $\partial\mu/\partial n$ versus ν measured for the sample with $\theta = 1.13°$ at $B_{\|} = 0$ and $T = 4$ K. The picture is taken from Ref. [56]

References

1. A.L. Rakhmanov, A. Rozhkov, A.O. Sboychakov, F. Nori, Phys. Rev. B **85**(3), 035408 (2012)
2. M.H.F. Sluiter, Y. Kawazoe, Phys. Rev. B **68**(8), 085410 (2003)
3. L.A. Openov, A.I. Podlivaev, JETP Lett. **90**(6), 459 (2009)
4. Z.M. Ao, A.D. Hernández-Nieves, F.M. Peeters, S. Li, Appl. Phys. Lett. **97**(23), 233109 (2010)
5. T. Roman, W.A. Dino, H. Nakanishi, H. Kasai, J. Phys.: Condens. Matter **21**(47), 474219 (2009)
6. A.V. Shytov, D.A. Abanin, L.S. Levitov, Phys. Rev. Lett. **103**(1), 016806 (2009)
7. M.J. Schmidt, D. Loss, Phys. Rev. B **81**, 165439 (2010)
8. J.O. Sofo, A.S. Chaudhari, G.D. Barber, Phys. Rev. B **75**, 153401 (2007)
9. S. Lebègue, M. Klintenberg, O. Eriksson, M.I. Katsnelson, Phys. Rev. B **79**, 245117 (2009)
10. A. Shytov, D.A. Abanin, L.S. Levitov, Phys. Rev. Lett. **103**, 016806 (2009)
11. A.V. Rozhkov, A.O. Sboychakov, A.L. Rakhmanov, F. Nori, Phys. Rep. **648**(1), 1 (2016)
12. A.L. Rakhmanov, A.V. Rozhkov, A.O. Sboychakov, F. Nori, Phys. Rev. Lett. **109**(20), 206801 (2012)
13. A.O. Sboychakov, A.L. Rakhmanov, A.V. Rozhkov, F. Nori, Phys. Rev. B **87**, 121401 (2013)
14. A.O. Sboychakov, A.V. Rozhkov, A.L. Rakhmanov, F. Nori, Phys. Rev. B **88**, 045409 (2013)
15. J. Borysiuk, J. Soltys, J. Piechota, J. Appl. Phys. **109**, 093523 (2011)
16. J.C. Charlier, J.P. Michenaud, X. Gonze, Phys. Rev. B **46**, 4531 (1992)
17. T.O. Wehling, E. Şaşığlu, C. Friedrich, A.I. Lichtenstein, M.I. Katsnelson, S. Blügel, Phys. Rev. Lett. **106**, 236805 (2011)
18. N.M.R. Peres, M.A.N. Araújo, D. Bozi, Phys. Rev. B **70**, 195122 (2004)
19. T.C. Meng, Z.Y. Lang, S. Wessel, F.F. Assaad, A. Muramatsu, Nature **464**(7290), 847 (2010)
20. S. Sorella, E. Tosatti, EPL (Euro Phys. Lett.) **19**, 699 (1992)
21. S. Chakravarty, B.I. Halperin, D.R. Nelson, Phys. Rev. B **39**, 2344 (1989)
22. E. Manousakis, Rev. Mod. Phys. **63**, 1 (1991)
23. A.M. Schakel, *Boulevard of Broken Symmetries: Effective Field Theories of Condensed Matter* (World Scientific, 2008)
24. S.Y. Zhou, D.A. Siegel, A.V. Fedorov, A. Lanzara, Phys. Rev. Lett. **101**, 086402 (2008)
25. J.B. Oostinga, H.B. Heersche, X. Liu, A.F. Morpurgo, L.M.K. Vandersypen, Nat. Mater. **7**, 151 (2008)
26. T.M. Rice, Phys. Rev. B **2**, 3619 (1970)
27. G. Li, A. Luican, J.M.B. Lopes dos Santos, A.H. Castro Neto, A. Reina, J. Kong, E.Y. Andrei, Nat. Phys. **6**(2), 109 (2010)
28. W. Yan, M. Liu, R.F. Dou, L. Meng, L. Feng, Z.D. Chu, Y. Zhang, Z. Liu, J.C. Nie, L. He, Phys. Rev. Lett. **109**, 126801 (2012)
29. R. Bistritzer, A.H. MacDonald, Proc. Natl. Acad. Sci. USA **108**(30), 12233 (2011)
30. J.M.B. Lopes dos Santos, N.M.R. Peres, A.H. Castro Neto, Phys. Rev. Lett. **99**, 256802 (2007)
31. L.J. Yin, J.B. Qiao, W.J. Zuo, W.T. Li, L. He, Phys. Rev. B **92**, 081406 (2015)
32. Y. Cao, V. Fatemi, A. Demir, S. Fang, S.L. Tomarken, J.Y. Luo, J.D. Sanchez-Yamagishi, K. Watanabe, T. Taniguchi, E. Kaxiras, R.C. Ashoori, P. Jarillo-Herrero, Nature **556**, 80 (2018)
33. Y. Cao, V. Fatemi, S. Fang, K. Watanabe, T. Taniguchi, E. Kaxiras, P. Jarillo-Herrero, Nature **556**, 43 (2018)
34. X. Lu, P. Stepanov, W. Yang, M. Xie, M.A. Aamir, I. Das, C. Urgell, K. Watanabe, T. Taniguchi, G. Zhang, A. Bachtold, A.H. MacDonald, D.K. Efetov, Nature **574**, 653 (2019)
35. P. Stepanov, X. Das, Ipsita ans Lu, A. Fahimniya, K. Watanabe, T. Taniguchi, F.H.L. Koppens, J. Lischner, L. Levitov, D.K. Efetov, Nature **583**, 375 (2020)
36. M. Yankowitz, S. Chen, H. Polshyn, Y. Zhang, K. Watanabe, T. Taniguchi, D. Graf, A.F. Young, C.R. Dean, Science **363**(6431), 1059 (2019)
37. B. Padhi, C. Setty, P.W. Phillips, Nano Lett. **18**(10), 6175 (2018)
38. M. Ochi, M. Koshino, K. Kuroki, Phys. Rev. B **98**, 081102 (2018)
39. C.C. Liu, L.D. Zhang, W.Q. Chen, F. Yang, Phys. Rev. Lett. **121**, 217001 (2018)

40. T. Huang, L. Zhang, T. Ma, Sci. Bull. **64**(5), 310 (2019)
41. K. Seo, V.N. Kotov, B. Uchoa, Phys. Rev. Lett. **122**, 246402 (2019)
42. A.O. Sboychakov, A.V. Rozhkov, A.L. Rakhmanov, F. Nori, Phys. Rev. B **100**, 045111 (2019)
43. A.O. Sboychakov, A.V. Rozhkov, A.L. Rakhmanov, F. Nori, Phys. Rev. B **102**, 155142 (2020)
44. A.O. Sboychakov, A.V. Rozhkov, K.I. Kugel, A.L. Rakhmanov, JETP Lett. **112**(10), 651 (2020)
45. A.O. Sboychakov, A.V. Rozhkov, A.L. Rakhmanov, JETP Lett. **116**, 729 (2022)
46. T. Cea, F. Guinea, Phys. Rev. B **102**, 045107 (2020)
47. F. Wu, A.H. MacDonald, I. Martin, Phys. Rev. Lett. **121**, 257001 (2018)
48. B. Lian, Z. Wang, B.A. Bernevig, Phys. Rev. Lett. **122**, 257002 (2019)
49. H. Guo, X. Zhu, S. Feng, R.T. Scalettar, Phys. Rev. B **97**, 235453 (2018)
50. J. González, T. Stauber, Phys. Rev. Lett. **122**, 026801 (2019)
51. B. Roy, V. Juričić, Phys. Rev. B **99**, 121407 (2019)
52. J.M.B. Lopes dos Santos, N.M.R. Peres, A.H. Castro Neto, Phys. Rev. B **86**, 155449 (2012)
53. P. San-Jose, J. González, F. Guinea, Phys. Rev. Lett. **108**, 216802 (2012)
54. J. Lorenzana, C. Castellani, C. Di Castro, Phys. Rev. B **64**(23), 235127 (2001)
55. A.O. Sboychakov, K.I. Kugel, A.L. Rakhmanov, Phys. Rev. B **76**, 195113 (2007)
56. U. Zondiner, A. Rozen, D. Rodan-Legrain, Y. Cao, R. Queiroz, T. Taniguchi, K. Watanabe, Y. Oreg, F. von Oppen, A. Stern, E. Berg, P. Jarillo-Herrero, S. Ilani, Nature **582**, 203 (2020)
57. S.L. Tomarken, Y. Cao, A. Demir, K. Watanabe, T. Taniguchi, P. Jarillo-Herrero, R.C. Ashoori, Phys. Rev. Lett. **123**, 046601 (2019)
58. E. Khalaf, A. Vishwanath, Nat. Commun. **13**(1), 6245 (2022)

Chapter 10
Electron Polaron Effect, Anomalous Resistivity, and the Origin of a Heavy Mass in the Two-Band Model with One Narrow Band

Abstract We discuss the electron polaron effect, the origin of a heavy mass, and a tendency toward phase separation in the two-band Hubbard model with one narrow band. We also discuss anomalous resistivity characteristics in the 2D and 3D two-band Hubbard models in the regime of a destroyed heavy band. The nature of the electron polaron effect is related to the formation of the anti-adiabatic cloud of soft electron–hole pairs of the light band, in which a heavy electron is dressed in the course of its tunneling from one unit cell to the neighboring one.

10.1 Introductory Remarks

Here, we present the results on the origin of a heavy mass, marginal Fermi-liquid behavior [1] and the anomalous resistivity characteristics in the 3D and 2D (or layered) strongly correlated electron systems generally described by the two-band Hubbard model with one narrow band (the details see in Refs. [2–6]). This model contains very rich and elegant physics closely related to the problems discussed above in Chap. 5. It describes adequately different systems such as uranium-based and Yb-based heavy-fermion (HF) compounds and possibly also some other novel superconductors and transition-metal systems with orbital degeneracy such as complex magnetic oxides (possibly including layered manganites) in an optimally doped case. Moreover, the model contains such highly nontrivial effects as electron polaron effect (EPE) already in the homogeneous case, as well as the tendency toward the phase separation in the case of a strong mismatch between the densities of heavy and light components [3, 7, 8].

In uranium-based HF compounds, we are usually dealing with the mixed valence limit [9–14] with a strong hybridization between heavy (f electrons or f-d electrons) and light (s-p electrons) components. The origin of a heavy mass, $m_h^* \gtrsim 200 m_e$, for f electrons in uranium-based HF is possibly very different from the physics of the cerium-based HF, where the Kondo effect is dominant [15–19]. According to the ideas of Refs. [20, 21] on the level of the two-particle hybridization, it is just sufficiently strong interband Hubbard interaction, which leads to an additional

M. Yu. Kagan et al., *Electronic Phase Separation in Magnetic and Superconducting Materials*, Springer Series in Solid-State Sciences 201,
https://doi.org/10.1007/978-3-031-55467-4_10

enhancement of the heavy electron mass due to the EPE. Physically, the EPE is related to a nonadiabatic part of the many-body wave function describing a heavy electron and a cloud of virtual electron–hole pairs of light charge carriers. These pairs are mixed with the wave function of heavy electrons but do not follow it, when a heavy electron tunnels from one unit cell to a neighboring one. It is shown in Refs. [20, 21] that in the unitary limit of the strong Hubbard interaction between heavy and light electrons, the effective heavy mass can reach the values $m_h^*/m_L \sim (m_h/m_L)^2$, and if we start from the ratio $m_h/m_L \sim 10$ of the bare masses of heavy and light electrons, for example, at the level of LDA approximation we could eventually achieve the effective mass as high as $m_h^* \sim 100 m_L$, which is typical of uranium-based HF compounds. A similar effect can also be described using the strong one-particle hybridization between heavy and light bands [20, 21].

A natural question arises, whether the two-band Hubbard model with one narrow band is a simple toy model to observe a non-Fermi-liquid behavior and well-known marginal Fermi-liquid behavior, in particular [1]. We recall that in the marginal Fermi-liquid (MFL) theory, the quasiparticles are strongly damped ($\mathrm{Im}\varepsilon \sim \mathrm{Re}\varepsilon \sim T$). According to [1], a strong damping $\gamma \sim T$ of quasiparticles (instead of the standard damping $\gamma \sim T^2/\varepsilon_F$ for a Landau-type Fermi liquid) can explain numerous experiments in high-T_c cuprates including the linear resistivity $R(T) \sim T$ for $T > T_c$ at optimum doping levels. The MFL picture was also proposed to describe the properties of UPt_3 doped by Pd including specific heat measurements [22]. The two-band Hubbard model with one narrow band is a natural generalization of the well-known Falicov–Kimball model [23] discussed in Chap. 5, but contains much richer physics.

10.2 Two-Band Model with One Narrow Band

The Hamiltonian of the two-band Hubbard model can be written as

$$\hat{H}' = -t_h \sum_{\langle i,j\rangle\sigma} a_{i\sigma}^\dagger a_{j\sigma} - t_L \sum_{\langle i,j\rangle\sigma} b_{i\sigma}^\dagger b_{j\sigma} - \varepsilon_0 \sum_{i\sigma} n_{i\sigma}^h - \mu \sum_{i\sigma} (n_{i\sigma}^L + n_{i\sigma}^h)$$
$$+ U_{hh} \sum_i n_{ih}^\uparrow n_{ih}^\downarrow + U_{LL} \sum_i n_{iL}^\uparrow n_{iL}^\downarrow + \frac{U_{hL}}{2} \sum_i n_{ih} n_{iL}, \tag{10.1}$$

where U_{LL} and U_{hh} are the intraband Hubbard interactions [9] for heavy and light electrons, respectively, U_{hL} is the interband Hubbard interaction between light and heavy electrons, t_h and t_L are the hopping integrals for heavy and light electrons, $n_{ih}^\sigma = a_{i\sigma}^\dagger a_{i\sigma}$ and $n_{iL}^\sigma = b_{i\sigma}^\dagger b_{i\sigma}$ are the densities of heavy and light electrons at site i with spin projection σ, and μ is the chemical potential. We note that $-\varepsilon_0$ is the center of gravity of the heavy band, and the difference Δ between the bottoms of the bands is given by

$$\Delta = -\varepsilon_0 + \frac{W_L - W_h}{2} = E^h_{min} - E^L_{min}\,. \tag{10.2}$$

Note that it is the most general form of the two-band model for the mixed valence situation. In manganites (see Chap. 7), we deal with the two-band degenerate Hubbard model, which can be reduced within some range of parameters region to the orbital t–J model and describes the Kugel–Khomskii physics of orbital ordering [24] together with the beautiful physics of orbital ferrons in the nanoscale phase-separated state [25].

Returning back to the generic two-band model, let us make the Fourier transformation of Hamiltonian (10.1). As a result, we find [3, 26]

$$\begin{aligned}\hat{H}' &= \sum_{\mathbf{p}\sigma} \varepsilon_h(\mathbf{p}) a^\dagger_{\mathbf{p}\sigma} a_{\mathbf{p}\sigma} + \sum_{\mathbf{p}\sigma} \varepsilon_L(\mathbf{p}) b^\dagger_{\mathbf{p}\sigma} b_{\mathbf{p}\sigma} + U_{hh} \sum_{\mathbf{p}\mathbf{p}'\mathbf{q}} a^\dagger_{\mathbf{p}\uparrow} a^\dagger_{\mathbf{p}'\downarrow} a_{\mathbf{p}-\mathbf{q}\downarrow} a_{\mathbf{p}'+\mathbf{q}\uparrow} \\ &+ U_{LL} \sum_{\mathbf{p}\mathbf{p}'\mathbf{q}} b^\dagger_{\mathbf{p}\uparrow} b^\dagger_{\mathbf{p}'\downarrow} b_{\mathbf{p}-\mathbf{q}\downarrow} b_{\mathbf{p}'+\mathbf{q}\uparrow} + \frac{U_{hL}}{2} \sum_{\mathbf{p}\mathbf{p}'\mathbf{q}\sigma\sigma'} a^\dagger_{\mathbf{p}\sigma} b^\dagger_{\mathbf{p}'\sigma'} b_{\mathbf{p}-\mathbf{q}\sigma'} a_{\mathbf{p}'+\mathbf{q}\sigma}\,, \end{aligned} \tag{10.3}$$

where in D dimensions for the hypercubic lattice, the quasiparticle energies for heavy and light bands are

$$\varepsilon_h(\mathbf{p}) = -2t_h \sum_{\alpha=1}^{D} \cos(p_\alpha a) - \varepsilon_0 - \mu$$

and

$$\varepsilon_L(\mathbf{p}) = -2t_L \sum_{\alpha=1}^{D} \cos(p_\alpha a) - \mu\,,$$

and $p_\alpha = \{p_x, p_y, \ldots\}$ are the Cartesian projections of the momentum. For low densities of heavy and light components $n_{tot}a^D = (n_h + n_L)a^D \ll 1$, the quasiparticle spectra are

$$\begin{aligned}\varepsilon_h(\mathbf{p}) &\approx -\frac{W}{2} + 2t_h(pa)^2 - \varepsilon_0 - \mu\,, \\ \varepsilon_L(\mathbf{p}) &\approx -\frac{W}{2} + 2t_L(pa)^2 - \mu\,. \end{aligned} \tag{10.4}$$

Note that throughout this chapter (in which we consider the low-density expansions mostly), in contrast to the previous chapters, we will work not with dimensionless, but with conventional charge carrier density n, so that the product na^D is a dimensionless parameter in the D-dimensional case. We hope that this will not lead to any misunderstanding. Note also, that we included the chemical potential μ in the definition of $\varepsilon_h(\mathbf{p})$ and $\varepsilon_L(\mathbf{p})$ in Eq. (10.4). In Eq. (10.4), $W_h = 4Dt_h$ and $W_L = 4Dt_L$ are the bandwidths of heavy and light electrons for the D-dimensional hypercubic lattice and a is an intersite distance, as usual (see Fig. 10.1).

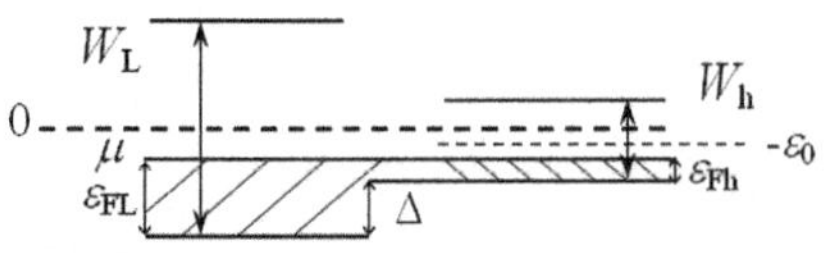

Fig. 10.1 Band structure in the two-band model with one narrow band. W_h and W_L are the bandwidths of heavy and light electrons, ε_{Fh} and ε_{FL} are the Fermi energies, and $\Delta = -\varepsilon_0 + (W_L - W_h)/2 = E^h_{min} - E^L_{min}$ is the energy difference between the bottoms of heavy and light bands, while $-\varepsilon_0$ is the center of gravity of the heavy band. The center of gravity of the light band is at zero. Finally, μ is the chemical potential [3]

For the future calculations, it is necessary to introduce the bare masses of heavy and light components

$$m_h = \frac{1}{2t_h a^2}, \quad m_L = \frac{1}{2t_L a^2}, \tag{10.5}$$

and the Fermi energies

$$\varepsilon_{Fh} = \frac{p_{Fh}^2}{2m_h} = \frac{W_h}{2} + \mu + \varepsilon_0, \quad \varepsilon_{FL} = \frac{p_{FL}^2}{2m_L} = \frac{W_L}{2} + \mu\,. \tag{10.6}$$

Using Eqs. (10.5) and (10.6), we finally get the quasiparticle spectra at low energies, $T \to 0$

$$\varepsilon_h(p) = \frac{p^2}{2m_h} - \varepsilon_{Fh}, \quad \varepsilon_L(p) = \frac{p^2}{2m_L} - \varepsilon_{FL}\,. \tag{10.7}$$

In deriving expressions (10.5)–(10.7), we implicitly assume that the difference in energy Δ between the bottoms of the bands in Fig. 10.1 is not too large, and hence the parabolic approximation for the spectra of both bands is still valid. We note that there is no one-particle hybridization in Hamiltonians (10.1) and (10.3), but there is the strong two-particle hybridization $\frac{U_{hL}}{2}\sum_i n_i^h n_i^L$.

We assume that $m_h \gg m_L$, and therefore

$$W_h / W_L = m_L / m_h \ll 1\,. \tag{10.8}$$

We will work in the most instructive strong-coupling limit assuming that

$$U_{hh} \sim U_{LL} \sim U_{hL} \gg W_L \gg W_h\,. \tag{10.9}$$

Note that U_{hL} can be large enough since actually the light particles are subjected to a strong scattering by heavy ones as if by a quasiresonance level. Finally, for the homogeneous state, we consider the simplest case when the electron densities for both bands are of the same order $n_h \sim n_L$ (a strong mismatch between the densities, which leads to the instability of the homogeneous case, will be considered separately).

Note that in the parabolic approximation, the density has the form $n = \frac{p_F^3}{3\pi^2}$ in 3D and $n = \frac{p_F^2}{2\pi}$ in 2D for both spin projections.

10.3 Kanamori T-Matrix Approximation

According to the Kanamori renormalization scheme [27], the strong Hubbard interaction in the case of low electron density (almost empty lattice) in all the observables can be replaced by the corresponding vacuum T-matrices (see Fig. 10.2). In the 3D case, the solution of the Bethe–Salpeter integral equations in vacuum (or Lippmann–Schwinger equations) [28, 29] yields the following expressions for the T-matrices, which correspond to "heavy-heavy", "heavy-light", and "light-light" Hubbard interactions, respectively (see [3, 26, 30–34])

$$T_{hh} = \frac{U_{hh}a^3}{1 - U_{hh}a^3 K_{hh}^{vac}(0,0)} \approx \frac{U_{hh}a^3}{1 + U_{hh}/8\pi t_h},$$
$$T_{hL} \approx \frac{U_{hL}a^3}{1 + U_{hL}/8\pi t_{hL}^*}, \quad T_{LL} \approx \frac{U_{LL}a^3}{1 + U_{LL}/8\pi t_L}. \tag{10.10}$$

In Eq. 10.10, we introduced the "heavy-heavy", "heavy-light", and "light-light" Cooper loops in vacuum (the product of two vacuum Green's functions in the Cooper channel for the total frequency and total momentum equal to zero) according to

$$K_{hh}^{vac}(0,0) = -\int \frac{d^3\mathbf{p}}{(2\pi)^3} \frac{m_h}{p^2}, \quad K_{hL}^{vac}(0,0) = -\int \frac{d^3\mathbf{p}}{(2\pi)^3} \frac{m_{hL}^*}{p^2},$$
$$K_{LL}^{vac}(0,0) = -\int \frac{d^3\mathbf{p}}{(2\pi)^3} \frac{m_L}{p^2}. \tag{10.11}$$

Note that on the lattice, the upper cutoff in the formally divergent linearly expressions for the vacuum Cooper loops is just an inverse intersite distance $\sim 1/a$.

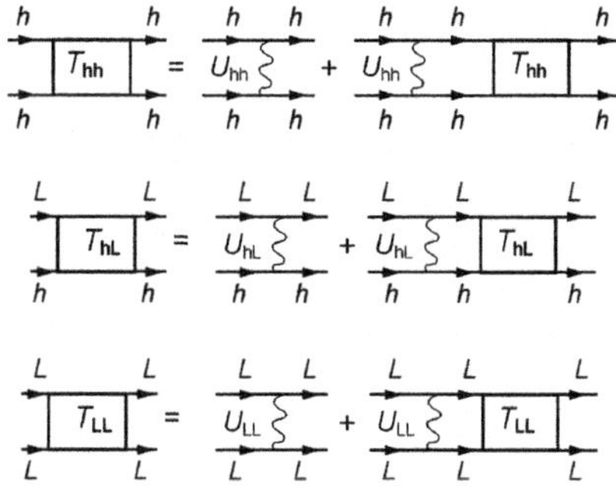

Fig. 10.2 Lippmann–Schwinger integral equations for the T-matrices T_{hh}, T_{LL}, and T_{hL} in the two-band model with heavy (h) and light (L) electrons. U_{hh} and U_{LL} are the intraband Hubbard interactions, and U_{hL} is the interband Hubbard interaction between heavy and light electrons [3]

Note also that for the reduced "heavy-light" mass in the corresponding Cooper loop (10.11) for $m_h \gg m_L$, we have

$$m^*_{hL} = \frac{1}{2t^*_{hL}a^2} = \frac{m_h m_L}{m_h + m_L} \approx m_L \,. \tag{10.12}$$

From Eq. (10.12), it immediately follows that the effective hopping integral is $t^*_{hL} \approx t_L$.

Let us emphasize that the product Ua^3 in Eq. (10.10) plays the role of the zeroth Fourier component of the Hubbard interaction in 3D. Finally using Eq. (10.12) and relations (10.9), we obtain the T-matrices in the strong-coupling limit

$$T_{hh} \approx 8\pi t_h a^3, \quad T_{hL} \approx T_{LL} \approx 8\pi t_L a^3 \,. \tag{10.13}$$

The s-wave scattering length for the Hubbard model [18] is defined as $a_0 = mT/4\pi = T/8\pi t a^2$, and hence in the strong-coupling case, we have

$$a_0^{hh} \approx a_0^{hL} \approx a_0^{LL} \approx a \,. \tag{10.14}$$

Correspondingly, the gas parameter of Galitskii $f_0 = 2a_0 p_F/\pi$ [35, 36] (which is convenient for the calculations) in the case of equal densities of heavy and light electrons $n_h = n_L$ is given by

$$f_0^{3D} \approx (f_0^L \approx 2ap_{FL}/\pi) \approx (f_0^h \approx 2ap_{Fh}/\pi) \approx 2ap_F/\pi \,. \tag{10.15}$$

In the 2D case, for strong Hubbard interaction and low densities, the vacuum T-matrices are written with the logarithmic accuracy as [3, 26, 31–34]

$$\begin{aligned} T_{hh} &\approx \frac{U_{hh}a^2}{1 + \frac{U_{hh}}{8\pi t_h}\int_{p_F^2}^{1/a^2}\frac{dp^2}{p^2}} \approx \frac{U_{hh}a^2}{1 + \frac{U_{hh}}{8\pi t_h}\ln\frac{1}{p_F^2 a^2}}\,, \\ T_{LL} &\approx \frac{U_{LL}a^2}{1 + \frac{U_{LL}}{8\pi t_L}\ln\frac{1}{p_F^2 a^2}}\,, \quad T_{hL} \approx \frac{U_{hL}a^2}{1 + \frac{U_{hL}}{8\pi t_L}\ln\frac{1}{p_F^2 a^2}}\,. \end{aligned} \tag{10.16}$$

In Eq. (10.16), the product Ua^2 plays the role of zeroth Fourier component of the Hubbard interaction in 2D. As a result, in the strong-coupling case, the 2D gas parameter of Bloom [37] for equal densities $n_h = n_L$ is

$$f_0^{2D} \approx f_0^h \approx f_0^L \approx \frac{1}{2\ln(1/p_F a)}\,. \tag{10.17}$$

10.4 Evaluation of the Self-energies of Heavy and Light Bands

In this section, we evaluate the imaginary parts $\mathrm{Im}\Sigma$ of the self-energies in the two-band Hubbard model considering the clean case (no impurities) and taking only the electron–electron scattering into account. These calculations will be important for the evaluation of the scattering times for heavy and light electrons and the subsequent calculation of the electrical resistivity $R(T)$. In the two-band model, we obtain (see Fig. 10.3)

$$\Sigma_h = \Sigma_{hh} + \Sigma_{hL} \text{ and } \Sigma_L = \Sigma_{LL} + \Sigma_{Lh} . \tag{10.18}$$

In the 3D case, the full T-matrix in substance, T_{hh}, which enters the first diagram for the self-energy Σ_{hh} of the heavy particles in Fig. 10.3, has the form

$$T_{hh}(\Omega, \mathbf{p}) = \frac{U_{hh}a^3}{1 - U_{hh}a^3 K_{hh}(\Omega, \mathbf{p})}, \tag{10.19}$$

where

$$K_{hh}(\Omega, \mathbf{p}) = \int \frac{d^3\mathbf{p}'}{(2\pi)^3} \frac{1 - f_F[\varepsilon_h(\mathbf{p}' + \mathbf{p})] - f_F[\varepsilon_h(-\mathbf{p}')]}{\Omega - \varepsilon_h(\mathbf{p}' + \mathbf{p}) - \varepsilon_h(-\mathbf{p}') + i0} . \tag{10.20}$$

In Eq. 10.20, $K_{hh}(\Omega, \mathbf{p})$ is a Cooper loop in the substance and $f_F[\varepsilon_h]$ is the Fermi–Dirac distribution function for heavy particles. The similar expressions describe the full T-matrices T_{hh}, T_{LL}, and T_{Lh}, and the Cooper loops K_{hh}, K_{LL}, and K_{Lh}.

If we expand the T-matrix for the heavy particles up to the second order (inclusively) in terms of the gas parameter, then according to Galitskii [35], we obtain

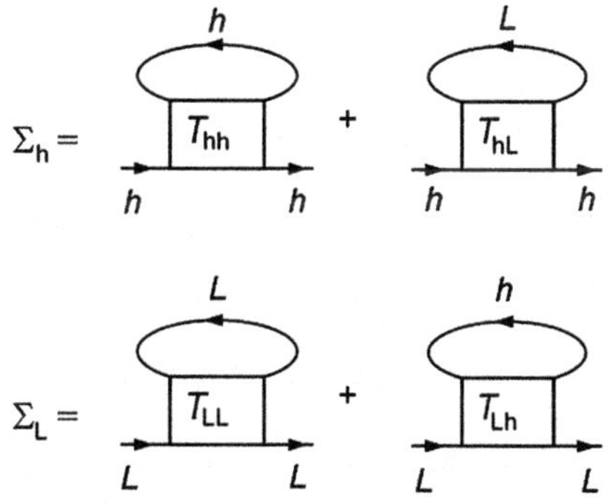

Fig. 10.3 Self-energies of heavy and light particles in the T-matrix approximation. T_{hh}, T_{hL}, T_{LL}, and T_{Lh} and are the full T-matrices of heavy and light particles in the substance [3]

$$T_{hh}(\Omega, \mathbf{p}) \approx \frac{4\pi a_0^{hh}}{m_h} + \left(\frac{4\pi a_0^{hh}}{m_h}\right)^2 [K_{hh}(\Omega, \mathbf{p}) - K_{hh}^{vac}(\Omega, \mathbf{p})]$$
$$+ O\left[\left(\frac{4\pi a_0^{hh}}{m_h}\right)^3 [K_{hh}(\Omega, \mathbf{p}) - K_{hh}^{vac}(\Omega, \mathbf{p})]^2\right]. \quad (10.21)$$

In expansion (10.21), we introduced the s-wave scattering length for the "heavy-heavy" interaction

$$\frac{4\pi a_0^{hh}}{m_h} \approx \frac{U_{hh}a^3}{1 - U_{hh}a^3 K_{hh}^{vac}(\Omega, \mathbf{p})}. \quad (10.22)$$

Note that in the right-hand side of Eq. (10.22), we have the Cooper loop in vacuum

$$K_{hh}^{vac}(\Omega, \mathbf{p}) \approx \int \frac{d^3\mathbf{p}'}{(2\pi)^3} \frac{1}{\Omega - \frac{(\mathbf{p}'+\mathbf{p})^2}{2m_h} - \frac{p'^2}{2m_h}}. \quad (10.23)$$

Let us emphasize that rigorously speaking the scattering length a_0^{hh} in Eq. (10.22) should be defined via the vacuum Cooper loop $K_{hh}^{vac}(0, 0)$ rather than by $K_{hh}^{vac}(\Omega, \mathbf{p})$. However, it is possible to show that $\frac{4\pi a_0^{hh}}{m_h}[K_{hh}^{vac}(\Omega, \mathbf{p}) - K_{hh}^{vac}(0, 0)] \propto f_0^h \ll 1$, and thus this difference is not important for our expansion up to the second order of the gas parameter. Analogously in the T-matrix expansion (10.21), we have $K_{hh}^{vac}(\Omega, \mathbf{p})$ instead of more conventional $K_{hh}^{vac}(0, 0)$, but if we restrict ourselves to the case of low densities and the energies $p^2/2m$ close to ε_F, then it is not important again. Note that K_{hh} in Eq. (10.21) is the full Cooper loop (cooperon) for heavy particles in the substance, which is defined by Eq. (10.20).

Now we are ready to perform the low-density expansion of the self-energies. The self-energy Σ_{hh} (which corresponds to the first term in the right-hand side of Eq. (10.18) for heavy particles) in the first two orders of the gas parameter symbolically can be represented as

$$\Sigma_{hh}(p) = \sum_k T_{hh}(k + p)G_h(k) \approx \frac{4\pi a_0^{hh}}{m_h} \sum_k G_h(k)$$
$$- \left(\frac{4\pi a_0^{hh}}{m_h}\right)^2 \sum_k (K_{hh} - K_{hh}^{vac})G_h(k) + O[(f_0^h)^3], \quad (10.24)$$

where $p = (\Omega, \mathbf{p})$ is the 4-momentum.

The first term in the self-energy expansion (10.24) corresponds just to the Hartree–Fock contribution [36] and at low temperatures reads $2\pi a_0^{hh} n_h/m_h$.

In the second term, we can make an analytic continuation $i\omega \to \omega + i0$ for the bosonic (two-particle) propagator K_{hh} and the fermionic propagator G_h. As a result (having in mind, that $\mathrm{Im}K_{hh}^{vac} = 0$, we obtain the imaginary part of the self-energy

$\Sigma_{hh}^{(2)}$ (expanded up to the second order of the gas parameter, inclusively) in the following form

$$\begin{aligned}\mathrm{Im}\Sigma_{hh}^{(2)}(\varepsilon,\mathbf{p}) &= \left(\frac{4\pi a_0^{hh}}{m_h}\right)^2 \sum_k \mathrm{Im}K_{hh}[\varepsilon_h(\mathbf{k})+\varepsilon;\mathbf{k}+\mathbf{p})]\{f_B[\varepsilon_h(\mathbf{k})+\varepsilon]+f_F[\varepsilon_h(\mathbf{k})]\} \\ &= -\left(\frac{4\pi a_0^{hh}}{m_h}\right)^2 \pi \int \frac{d^3\mathbf{k}}{(2\pi)^3}\int \frac{d^3\mathbf{p}'}{(2\pi)^3}(1-f_F[\varepsilon_h(\mathbf{p}+\mathbf{p}'+\mathbf{k})]-f_F[\varepsilon_h(-\mathbf{p}')] \\ &\times (f_B[\varepsilon_h(\mathbf{k})+\varepsilon]+f_F[\varepsilon_h(\mathbf{k})])\delta[\varepsilon+\varepsilon_h(\mathbf{k}) \\ &-\varepsilon_h(\mathbf{p}+\mathbf{p}'+\mathbf{k})-\varepsilon_h(-\mathbf{p}')]) \qquad (10.25)\end{aligned}$$

and similarly for the real part of $\Sigma_{hh}^{(2)}$, we have

$$\begin{aligned}\mathrm{Re}\Sigma_{hh}^{(2)}(\varepsilon,\mathbf{p}) &= \left(\frac{4\pi a_0^{hh}}{m_h}\right)^2 \sum_k \{\mathrm{Re}K_{hh}[\varepsilon_h(\mathbf{k})+\varepsilon;\mathbf{k})+\mathbf{p}]-\mathrm{Re}K_{hh}^{vac}[\varepsilon_h(\mathbf{k})+\varepsilon_h(\mathbf{p}),\mathbf{k}+\mathbf{p}]\} \\ &\times(f_B[\varepsilon_h(\mathbf{k})+\varepsilon]+f_F[\varepsilon_h(\mathbf{k})]), \qquad (10.26)\end{aligned}$$

where for the real part of the Cooper loop in vacuum, we get

$$\mathrm{Re}K_{hh}^{vac}[\varepsilon_h(\mathbf{k})+\varepsilon_h(\mathbf{p}),\mathbf{k}+\mathbf{p}] = \int \frac{d^3\mathbf{p}'}{(2\pi)^3} P \frac{2m_h}{\mathbf{k}^2+\mathbf{p}^2-(\mathbf{p}+\mathbf{p}'+\mathbf{k})^2-\mathbf{p}'^2}. \qquad (10.27)$$

Note that $\mathrm{Re}K_{hh}^{vac}$ in Eq. (10.27) is calculated in resonance for the total frequency $\Omega = \varepsilon_h(\mathbf{k})+\varepsilon_h(\mathbf{p})$ in the Cooper channel (or for $\varepsilon = \varepsilon_h(\mathbf{p})$) and P is the principal value. Note also that in Eqs. 10.25 and 10.26, $f_B(\varepsilon) = 1/[\exp(\varepsilon/T)-1]$ and $f_F(\varepsilon) = 1/[\exp(\varepsilon/T)+1]$ are the bosonic and fermionic distribution functions, and hence

$$f_B[\varepsilon_h(\mathbf{k})+\varepsilon]+f_F[\varepsilon_h(\mathbf{k})] = \frac{1}{2}\left[\coth\frac{\varepsilon_h(\mathbf{k})+\varepsilon}{2T} - \tanh\frac{\varepsilon_h(\mathbf{k})}{2T}\right]. \qquad (10.28)$$

The real part of the Cooper loop for heavy particles in the substance is given by

$$\mathrm{Re}K_{hh}[\varepsilon_h(\mathbf{k})+\varepsilon;\mathbf{k}+\mathbf{p}] = \int \frac{d^3\mathbf{p}'}{(2\pi)^3}\frac{1-f_F[\varepsilon_h(\mathbf{p}+\mathbf{p}'+\mathbf{k})]-f_F[\varepsilon_h(-\mathbf{p}')]}{\varepsilon+\varepsilon_h(\mathbf{k})-\varepsilon_h(\mathbf{p}+\mathbf{p}'+\mathbf{k})-\varepsilon_h(-\mathbf{p}')}. \qquad (10.29)$$

In the 2D case, the analytic continuation for $\Sigma_{hh}^{(2)}$ is similar.

We stress that for $\varepsilon/T \gg 1$, the bosonic distribution function $f_B(\varepsilon) \to 0$ and $f_F(\varepsilon) \to \theta(\varepsilon)$, where θ is the step function. Hence, at low temperatures, $\mathrm{Im}\Sigma_{hh}$ and $\mathrm{Re}\Sigma_{hh}$ acquire the standard form [30, 35, 36]. In fact, for low temperatures, $T \ll W_h \ll W_L$, the most convenient way is to evaluate $\Sigma_{hh}^{(2)}(\varepsilon)$ for $T \to 0$, which yields the standard Fermi-liquid result $\mathrm{Im}\Sigma_{hh}^{(2)}(\varepsilon) \propto \varepsilon^2$, and then to perform the temperature averaging with the corresponding fermionic distribution function $f_F(\varepsilon)$.

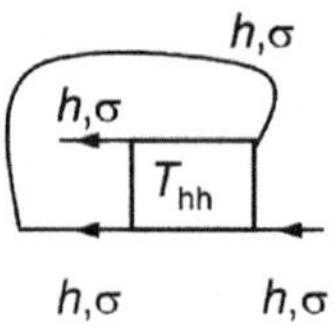

Fig. 10.4 Exchange-type diagram for the self-energy $\Sigma_{hh}^{(2)}$, which contains the matrix element $a_\sigma^\dagger a_\sigma^\dagger a_\sigma a_\sigma$ and therefore is absent in the Hubbard model [3]

Therefore, $\varepsilon \sim T$ for the scattering times of quasiparticles. The evaluation of Σ_{hL}, Σ_{Lh}, and Σ_{LL} at low temperatures in the first two orders in the gas parameter is similar to the evaluation of Σ_{hh} both in 3D and 2D cases.

However, for higher temperatures, we should keep in mind that $f_B(\varepsilon) \to T/\varepsilon$ for $T \gg \varepsilon$. The fermionic distribution function is smeared by temperature. Accordingly, we have $f_F(\varepsilon) \approx \frac{1}{2}(1 - \varepsilon/2T)$. These approximations will be important for the evaluation of $\mathrm{Im}\Sigma$ for higher temperatures $T > W_h$ [3, 26].

Let us stress also, that in contrast to the model of a slightly non-ideal Fermi gas, considered in [35, 37, 38], the Hubbard model does not contain an exchange-type diagram for Σ_{hh} (see Fig. 10.4) since the T-matrix in this diagram corresponds to the incoming and outgoing heavy particles with the same spin projection, $a_\sigma^\dagger a_\sigma^\dagger a_\sigma a_\sigma$, while the Hubbard model contains only the matrix elements $a_\uparrow^\dagger a_\downarrow^\dagger a_\downarrow a_\downarrow$.

10.5 Electron Polaron Effect (EPE)

The Green's function of heavy particles for $T \to 0$ reads

$$G_h(\omega, \mathbf{q}) = \frac{1}{\omega - \varepsilon_h(\mathbf{q}) - \Sigma_h(\omega, \mathbf{q})} = \frac{Z_h}{\omega - \varepsilon_h^*(\mathbf{q}) + i0}, \tag{10.30}$$

and analogously, the Green's function of light particles

$$G_L(\omega, \mathbf{q}) = \frac{Z_L}{\omega - \varepsilon_L^*(\mathbf{q}) + i0}, \tag{10.31}$$

where

$$\varepsilon_h^*(\mathbf{q}) = \frac{\left(q^2 - p_{Fh}^2\right)}{2m_h^*}, \tag{10.32}$$

and

$$\varepsilon_L^*(\mathbf{q}) = \frac{\left(q^2 - p_{FL}^2\right)}{2m_L^*} \tag{10.33}$$

are renormalized quasiparticle spectra, m_h^* and m_L^* are effective masses of heavy and light particles.

Correspondingly

$$Z_h^{-1} = \left\{ 1 - \frac{\partial Re\Sigma_h^{(2)}(\omega, \mathbf{q})}{\partial \omega}|_{\omega \to 0, q \to p_{Fh}} \right\}, \tag{10.34}$$

and

$$Z_L^{-1} = \left\{ 1 - \frac{\partial Re\Sigma_L^{(2)}(\omega, \mathbf{q})}{\partial \omega}|_{\omega \to 0, q \to p_{FL}} \right\} \tag{10.35}$$

are inverse Z-factors of heavy and light electrons. Substitution of the leading contribution from $Re\Sigma_{hL}^{(2)}(\omega, \mathbf{q})$ (described by the formula similar to (10.26)) to $Re\Sigma_h^{(2)}(\omega, \mathbf{q})$ in (10.33) yields:

$$\lim_{\omega \to 0, q \to p_F} \frac{\partial Re\Sigma_h^{(2)}(\omega, \mathbf{q})}{\partial \omega} \sim -(\frac{4\pi a_0^{hL}}{m_{hL}^*})^2 \tag{10.36}$$
$$\iint \frac{d^D\mathbf{p}}{(2\pi)^D} \frac{d^D\mathbf{p}'}{(2\pi)^D} \frac{\left[1 - f_F\left[\varepsilon_L(\mathbf{p}+\mathbf{p}')\right] - f_F\left[\varepsilon_h(-\mathbf{p}')\right]\right] f_F\left[\varepsilon_L(\mathbf{p}-\mathbf{q})\right]}{\left[\varepsilon_L(\mathbf{p}-\mathbf{q}) - \varepsilon_L(\mathbf{p}+\mathbf{p}') - \varepsilon_h(-\mathbf{p}')\right]^2},$$

where for low temperatures $\varepsilon/T \gg 1$, the bosonic distribution function $f_B(\varepsilon) \to 0$, and fermionic distribution function $f_F(\varepsilon)$ is a step function, the s-wave scattering length $a_0^{hL} \approx a$ is connected with the vacuum T-matrix T_{hL} and an effective mass $m_{hL}^* \approx m_L$for $m_h \gg m_L$.

Replacing in (10.35) the measure of integration $\frac{d^D\mathbf{p}}{(2\pi)^D} \frac{d^D\mathbf{p}'}{(2\pi)^D}$ by $N_L^2(0)$ $d\varepsilon_L(\mathbf{p})\, d\varepsilon_L(\mathbf{p}')$ (where $N_L(0)$ is a density of light particles) and taking into account that the quasiparticle energy $\varepsilon_L(\mathbf{p}-\mathbf{q}) < 0$ while $\varepsilon_L(\mathbf{p}+\mathbf{p}') > 0$, we can easily check that for $m_h \gg m_L$ (or equivalently for $\varepsilon_{FL} \gg \varepsilon_{Fh}$), this expression contains a large logarithm (see [3, 21]). Thus, for the inverse Z-factor of the heavy particles in the leading approximation

$$Z_h^{-1} = 1 + 2f_0^2 ln\frac{m_h}{m_L}, \tag{10.37}$$

where $f_0 = 2p_{FL}a/\pi$ is the gas parameter in 3D and equivalently $f_0 = \frac{1}{2\ln(1/p_{FL}a)}$ in 2D.

Note that the contribution to Z_h from $Re\Sigma_{hh}^{(2)}(\omega, \mathbf{q})$ does not contain a large logarithm. Correspondingly, for effective mass of a heavy particle in (10.32), we get

$$\frac{m_h}{m_h^*} = Z_h \left\{ 1 + \left. \frac{\partial Re \Sigma_{hL}^{(2)} \left(\varepsilon_h (\mathbf{q}), \mathbf{q} \right)}{\partial \varepsilon_h (\mathbf{q})} \right|_{\varepsilon_h(\mathbf{q}) \to 0} \right\}, \tag{10.38}$$

where in agreement with (10.36)

$$Z_h = \frac{1}{1 + 2 f_0^2 ln \frac{m_h}{m_L}} \approx 1 - 2 f_0^2 ln \frac{m_h}{m_L}. \tag{10.39}$$

The analogous calculation for Z_L with $Re\Sigma_{Lh}$ and$Re\Sigma_{LL}$ yields only

$$\frac{m_L}{m_L^*} \sim Z_L^{-1} \sim 1 + f_0^2. \tag{10.40}$$

If the effective parameter in the expression (10.38) for Z_h is large

$$2 f_0^2 ln \frac{m_h}{m_L} = b \; ln \frac{m_h}{m_L} > 1, \tag{10.41}$$

where $b = 2 f_0^2$, then we are in the situation of strong electron polaron effect.

In this range of parameters, to get a correct polaron exponent diagrammatically we should sum up at least so-called maximally crossed diagrams for $Re\Sigma_{hL}$. The polaron exponent evaluation could be fulfilled, however, in a different technique, which is based on the non-adiabatic part of the many-particle wave function [20, 21] which describes a heavy particle dressed in a cloud of electron-hole pairs of light particles. This yields

$$\frac{m_h^*}{m_h} \propto Z_h^{-1} = \left(\frac{m_h}{m_L} \right)^{b/(1-b)}, \tag{10.42}$$

For $b = 1/2$ or, equivalently, for $f_0 = 1/2$ (as for the coupling constant of the screened Coulomb interaction in the RPA scheme), we are in the so-called unitary limit. In this limit, according to Refs. [20, 21], the polaron exponent is

$$\frac{b}{1-b} = 1, \tag{10.43}$$

and hence

$$\frac{m_h^*}{m_h} = \frac{m_h}{m_L}, \tag{10.44}$$

or, equivalently

$$\frac{m_h^*}{m_L} = \left(\frac{m_h}{m_L} \right)^2. \tag{10.45}$$

Thus, starting from the ratio $m_h/m_L \sim 10$ of the bare masses (obtained, for instance, using the local density approximation (LDA)), in the unitary limit, we eventually obtain $m_h/m_L \sim 100$ (due to the many-body electron polaron effect), which is a typical ratio for uranium-based heavy-fermion systems.

Let us say a few words about the possible nature of the narrow band near the Fermi level. In the systems with magnetic impurities, the many-body effects give rise to the well-known Abrikosov–Suhl resonance [39, 40], which can lead to an instability of the Fermi-liquid state [41]. A thorough analysis of this situation is given in Ref. [42] by the slave-boson treatment of the periodic Anderson model.

If we assume that the narrow band near the Fermi level is itself the Abrikosov–Suhl resonance and the EPE acts on top of the effective model, arising, say from the periodic Anderson model, the effective width of the hybridization peak according to, e.g., book of Fulde [13] is given by the expression

$$\Gamma^* = \pi N_L(0)(V^*)^2 = \pi N_L(0)V^2(1 - n_f), \tag{10.46}$$

where $N_L(0)$ is the density of states in the broad conduction band (light band in our notation), V is the hybridization matrix element, and n_f is the number of f electrons per site.

If on top of this model, we will add the Falicov–Kimball type Coulomb interaction between heavy electrons (originally inside the hybridization peak) and light electrons, which is precisely our interband Hubbard interaction $U_{hL}n_h n_L$; then, we will get an additional narrowing of the resonance width to the values

$$\Gamma^{**} = \Gamma^*(\Gamma^*/W_L)^{b/(1-b)}, \tag{10.47}$$

where $W_L \sim \varepsilon_{FL}$ is the width of the light band and $b = 2N_L^2(0)U_{hL}^2$ is the exponent characterizing EPE.

Thus, EPE identically acts on the width of the heavy band W_h^* in the two-band Hubbard model and on the effective width of the resonance peak Γ^{**} (located near the Fermi level of the conduction band in mixed valence limit) in the extended periodic Anderson model with the Falicov–Kimball term (without any double counting of these two effects). This result has been obtained in the important paper [21] (see also the first paper on this topic [20]). It was shown that in the two-band Hubbard model with one narrow band, we can completely neglect the one-particle hybridization in favor of the two-particle one (the interband Hubbard interaction).

As we have already mentioned above, the EPE and dressing of a heavy particle by the cloud of virtual electron–hole pairs of light particles is due to the nonadiabatic part of the heavy particle many-body wave function at the tunneling of the heavy particle between neighboring unit cells in the presence of Fermi liquid formed by light particles. The energy range for the nonadiabaticity was thoroughly estimated for different cases in the framework of the density matrix approach in the first fundamental works on this problem [20, 21]. It was demonstrated that the energy range for the nonadiabaticity corresponds to that between $\hbar/\tau$ and ε_{FL} or W_L, where τ is the effective tunneling time for the heavy particle (see [20]). For a more complicated problem of many heavy particles, the lower limit of nonadiabaticity is obtained in

[21]. At low temperatures, it corresponds to the substitution of $\hbar/\tau$ by the heavy particle bandwidth W_h.

Here, we would like to emphasize that in the main approximation, EPE plays in a concert with the strong phonon polaron effect (PPE), which has its own specific features in strongly correlated electron system (see, e.g., [43]). Namely, two exponents for PPE and EPE should be multiplied by each other to result in the effective width of the heavy band W_h^*. This is also the case in manganites. More complicated nonlinear amplification effects due to the coupling between PPE and EPE arise when the renormalized width of the heavy band becomes smaller than the Debye temperature.

10.6 Other Mechanisms of the Heavy Mass Enhancement

Note that rigorously speaking (see Eq. 10.38), the momentum dependence of $\mathrm{Re}\Sigma_{hL}^{(2)}(\varepsilon_h(\mathbf{q}), \mathbf{q})$ is also very important. The calculations [28, 30] show that in the principal approximation in the small parameter $m_L/m_h \ll 1$, in the 3D case close to the Fermi surface (for the energies $\varepsilon_h(\mathbf{q}) = (q^2 - p_{Fh}^2)/2m_h \to 0$ and correspondingly $q \to p_{Fh}$), we find

$$\mathrm{Re}\Sigma_{hL}^{(2)}(\varepsilon_h(\mathbf{q}), \mathbf{q}) \approx \left(\frac{4\pi a_0^{hL}}{m_L}\right)^2 \int \frac{d^3\mathbf{p}}{(2\pi)^3} \Pi_{LL}(0, \mathbf{p}) f_F[\varepsilon_h(\mathbf{p} - \mathbf{q})], \tag{10.48}$$

where

$$\Pi_{LL}(0, \mathbf{p}) = \int \frac{d^3\mathbf{p}'}{(2\pi)^3} \frac{f_F[\varepsilon_L(\mathbf{p} + \mathbf{p}')] - f_F[\varepsilon_L(\mathbf{p}')])}{\varepsilon_L(\mathbf{p}') - \varepsilon_L(\mathbf{p} + \mathbf{p}')} \tag{10.49}$$

is the static polarization operator for light particles. Having in mind that $|\mathbf{p} - \mathbf{q}| < p_{Fh}$ and $q \approx p_{Fh}$ in Eq. (10.48), we can see that $\mathbf{p} \to 0$ and thus we can use the asymptotic form for $\Pi_{LL}(0, \mathbf{p})$ at small momenta $p \ll p_{FL}$ (if the densities of heavy and light bands are not very different and $p_{FL} \sim p_{Fh}$), namely,

$$\lim \Pi_{LL}(0, \mathbf{p})\big|_{\mathbf{p} \to 0} \approx N_L^{3D}(0) \left(1 - \frac{p^2}{12 p_{FL}^2}\right). \tag{10.50}$$

Note that in Eq. (10.50), $N_L^{3D}(0) = \frac{m_L p_{FL}}{2\pi^2}$ is the density of states for light electrons in 3D. The substitution of expression (10.50) for $\lim \Pi_{LL}(0, \mathbf{p})|_{\mathbf{p} \to 0}$ in Eq. (10.48) yields

$$\mathrm{Re}\Sigma_{hL}^{(2)}(\varepsilon_h(\mathbf{q}), \mathbf{q}) \approx \mathrm{Re}\Sigma_{hL}^{(2)}(0, p_{Fh}) - \frac{(q^2 - p_{Fh}^2)}{2m_h} \frac{f_0^2}{9} \frac{m_h n_h}{m_L n_L}, \tag{10.51}$$

where $f_0 = f_0^{3D} \approx 2a p_{FL}/\pi$, $n_h = p_{Fh}^3/3\pi^2$, and $n_L = p_{FL}^3/3\pi^2$ are the densities of heavy and light electrons. The first term in Eq. 10.51 describes $\mathrm{Re}\Sigma_{hL}^{(2)}(\varepsilon_h(\mathbf{q}), \mathbf{q})$

at the Fermi surface (for $\varepsilon_h(\mathbf{q}) = 0$ and $q = p_{Fh}$). In the 3D case, it reads

$$\mathrm{Re}\Sigma_{hL}^{(2)}(0, p_{Fh}) \approx \frac{4 f_0^2}{3} \frac{n_h}{n_L} \varepsilon_{FL} \left(1 - \frac{2 p_{Fh}^2}{15 p_{FL}^2}\right) > 0 \text{ for } p_{Fh} \sim p_{FL} . \tag{10.52}$$

This term is a renormalization of the effective chemical potential of heavy electrons accurate to the second order in the gas parameter due to the interaction between light and heavy particles.

We note that according to [30, 35], the renormalized spectrum of heavy electrons is given by

$$\varepsilon_h^* = \left(\frac{q^2}{2m_h} - \mu_h^{eff}\right) + \frac{2\pi}{m_L} n_L(\mu) a_0^{hL} + \mathrm{Re}\Sigma_{hL}^{(2)}(\varepsilon_h(\mathbf{q}), \mathbf{q}) = \frac{q^2 - p_{Fh}^2}{2m_h^*} , \tag{10.53}$$

where the s-wave scattering length $a_0^{hL} \approx a$, the effective chemical potential $\mu_h^{eff} = \mu_h + \frac{W_h}{2} + \varepsilon_0$ is measured from the bottom of the heavy band, and the Hartree–Fock term $(2\pi/m_L)n_L(\mu)a_0^{hL}$ is the contribution to the self-energy $\mathrm{Re}\Sigma_{hL}^{(1)}$ in the first order in the gas parameter. From Eq. (10.53), we obtain summing the terms proportional to $\varepsilon_h(\mathbf{q}) = (q^2 - p_{Fh}^2)/2m_h$

$$\frac{q^2 - p_{Fh}^2}{2m_h^*} = \varepsilon_h(\mathbf{q}) \left[1 - \frac{f_0^2}{9} \frac{m_h n_h}{m_L n_L}\right] . \tag{10.54}$$

Correspondingly, the contribution of the momentum dependence of the self-energy to the effective mass of heavy electrons reads

$$\frac{m_h}{m_h^*} \propto \left(1 + \frac{\partial \mathrm{Re}\Sigma_{hL}^{(2)}(\varepsilon_h(\mathbf{q}), \mathbf{q})}{\partial \varepsilon_h(\mathbf{q})}\Big|_{\varepsilon_h \to 0}\right) = 1 - \frac{f_0^2}{9} \frac{m_h n_h}{m_L n_L} . \tag{10.55}$$

As a result, we obtain much more dramatic mass enhancement than that in the electron polaron effect, which gives only $m_h/m_h^* \approx 1 - 2 f_0^2 \ln{(m_h/m_L)}$ due to the Z-factor of a heavy particle. Note that the full expression for the effective heavy mass in the second order of the perturbation theory yields [6]

$$\frac{m_h^*}{m_h} = 1 + b \ln \frac{m_h}{m_L} + \frac{b}{18} \frac{m_h n_h}{m_L n_L}, \quad b = 2 f_0^2 . \tag{10.56}$$

If as in LDA, we have $m_h/m_L \sim 10$, then for the large density mismatch between heavy and light particles, $n_h \geq 5 n_L$, the second term in Eq. (10.56), which is linear in the bare mass ratio m_h/m_L, becomes dominant over the contribution of electron polaron effect (which is proportional only to $\ln{(m_h/m_L)}$). Let us stress again that the contribution to the effective heavy mass from $\mathrm{Re}\Sigma_{hh}^{(2)}(\varepsilon_h(\mathbf{q}), \mathbf{q})$ (which is associated with the "heavy-heavy" interaction) is small in comparison to the contribution from $\mathrm{Re}\Sigma_{hL}^{(2)}$ (associated with the "heavy-light" interaction) due to the smallness of the ratio $m_L/m_h \ll 1$. We can now collect the terms, which do not depend on $\varepsilon_h(\mathbf{q})$ in Eq. (10.53). As a result, we find the effective chemical potential of heavy electrons

$$\mu_h^{eff} = \frac{p_{Fh}^2}{2m_h} + \frac{2\pi}{m_L} n_L(\mu) a_0^{hL} + \mathrm{Re}\Sigma_{hL}^{(2)}(0, p_{FL}) . \tag{10.57}$$

We note that the contributions to μ_h^{eff} from the Hartree–Fock term $(2\pi/m_h)n_h(\mu)a_0^{hh}$ of heavy electrons as well as from "heavy-heavy" self-energy $\mathrm{Re}\Sigma_{hh}^{(2)}(0, p_{Fh})$ are small in comparison to the "heavy-light" contribution again due to the smallness of the ratio $m_L/m_h \ll 1$.

Let us now discuss for comparison the 2D case. In 2D, the static polarization operator for light electrons reads

$$\Pi_{LL}(0, \mathbf{p}) = \frac{m_L}{2\pi} \left[1 - \mathrm{Re}\sqrt{1 - \frac{4p_{FL}^2}{p^2}} \right] , \tag{10.58}$$

and hence for $p < 2p_{FL}$, we have $\Pi_{LL}(0, \mathbf{p}) = m_L/2\pi$, in contrast to the 3D case, the polarization operator does not contain any dependence on p^2. Thus, the electron polaron effect in 2D is a dominant effect in the heavy mass enhancement.

We note that for the light electrons, the momentum dependence of $\mathrm{Re}\Sigma_{Lh}^{(2)}$ and $\mathrm{Re}\Sigma_{LL}^{(2)}$ yields only $m_L^*/m_L \sim 1 + f_0^2$, and hence the light mass is not strongly enhanced in both 3D and 2D cases.

10.7 Tendency Toward Phase Separation

It is very interesting to emphasize that in the 3D case, the same parameter $b\frac{m_h n_h}{m_L n_L} \geq 1$ (which governs the dominance of the last term in Eq. (10.45) for the effective heavy mass) is responsible for the tendency toward the phase separation in the two-band Hubbard model. It leads to the negative partial compressibility for larger densities of heavy particles $n_h \sim n_C \leq 1$ and a large mismatch $n_h \gg n_L$ in the electron densities of the bands

$$\kappa_{hh}^{-1} \propto c_h^2 \propto (n_h/m_h)(\partial\mu_h^{eff}/\partial n_h) < 0 , \tag{10.59}$$

where μ_h^{eff} is given by Eq. 10.57 and c_h is the sound velocity for heavy particles. This result is in qualitative agreement with the predictions of the mean-field type variational analysis [7]. Note that in the 2D case, the contributions coming from the momentum dependence of the self-energy to m_h^* and μ_h^{eff} are absent and thus there is no tendency toward the phase separation in this case.

10.8 Temperature Dependence of the Electrical Resistivity in 3D

10.8.1 Imaginary Parts of the Self-energies in the Homogeneous Case for Low Temperatures $T < W_h$

In the homogeneous case, after the averaging of $\mathrm{Im}\Sigma^{(2)}(\varepsilon(\mathbf{q}), \mathbf{q})$ in Eqs. (10.18) and (10.25) with the fermionic distribution function $f_F(\varepsilon/T)$, we obtain the following expressions for $\mathrm{Im}\Sigma^{(2)}(T)$ of heavy and light electrons at low temperatures and for equal densities $n_h = n_L$ of the heavy and light bands

$$\mathrm{Im}\Sigma_{LL}^{(2)}(T) = f_0^2 \frac{T^2}{\varepsilon_{FL}}, \quad \mathrm{Im}\Sigma_{Lh}^{(2)}(T) = f_0^2 \frac{T^2}{\varepsilon_{Fh}} \frac{m_h}{m_L} \tag{10.60}$$

and accordingly

$$\mathrm{Im}\Sigma_{hh}^{(2)}(T) = f_0^2 \frac{T^2}{\varepsilon_{Fh}}, \quad \mathrm{Im}\Sigma_{hL}^{(2)}(T) = f_0^2 \frac{T^2}{\varepsilon_{Fh}}. \tag{10.61}$$

It follows that all the $\mathrm{Im}\Sigma^{(2)}(T)$ functions behave in the standard Fermi-liquid fashion $\mathrm{Im}\Sigma^{(2)}(T) \propto T^2$. Moreover, $\mathrm{Im}\Sigma_{Lh}^{(2)}(T) \gg \mathrm{Im}\Sigma_{LL}^{(2)}(T)$, and hence

$$\mathrm{Im}\Sigma_{L}^{(2)}(T) = \mathrm{Im}\Sigma_{Lh}^{(2)}(T) + \mathrm{Im}\Sigma_{LL}^{(2)}(T) \approx \mathrm{Im}\Sigma_{Lh}^{(2)}(T). \tag{10.62}$$

We can now estimate the Drude conductivity for the light band

$$\sigma_L = \frac{n_L e^2 \tau_L}{m_L}, \tag{10.63}$$

where a naive estimate for the inverse scattering time of light electrons $1/\tau_L$ gives

$$\gamma_L = 1/\tau_L = \mathrm{Im}\Sigma_{L}^{(2)}(T) \approx \mathrm{Im}\Sigma_{Lh}^{(2)}(T). \tag{10.64}$$

We conclude that $1/\tau_L \approx 1/\tau_{Lh}$. Correspondingly, we obtain for the conductivity of the light band

$$\sigma_L \approx \sigma_{Lh} = \frac{n_L e^2 \tau_{Lh}}{m_L} = \frac{n_L e^2}{f_0^2 T^2} \frac{\varepsilon_{Fh}}{m_h} \sim \frac{n_L e^2}{f_0^2 p_{Fh}^2} \left(\frac{\varepsilon_{Fh}}{T}\right)^2. \tag{10.65}$$

Introducing the minimal Mott–Regel conductivities [44, 45]

$$\sigma_{\min} = \left(\frac{e^2}{\hbar}\right) p_F \quad \text{in 3D and} \quad \sigma_{\min} = \left(\frac{e^2}{\hbar}\right) \quad \text{in 2D} \tag{10.66}$$

and working with the units where $\hbar = 1$, for equal densities $n_h = n_L$, we find

$$\sigma_L \approx \sigma_{Lh} \sim \frac{\sigma_{\min}}{f_0^2}\left(\frac{\varepsilon_{Fh}}{T}\right)^2 . \tag{10.67}$$

10.8.2 Role of Umklapp Processes

We note that rigorously speaking, the nondiagonal conductivity σ_{Lh} (which is governed by the scattering of light electrons by the heavy ones) is finite only due to the Umklapp processes

$$\mathbf{p}_{1L} + \mathbf{p}_{2h} = \mathbf{p}_{3L} + \mathbf{p}_{4h} + \mathbf{K}, \tag{10.68}$$

where $K \sim \pi/a$ is the wave vector of the reciprocal lattice. For $p_{FL} \sim p_{Fh}$, this means that the electron densities in the heavy and light bands cannot be very small (otherwise the conductivity σ_{Lh} would be exponentially large). Hence, within the accuracy of our estimates

$$\sigma_{Lh} \approx \frac{\sigma_{\min}}{f_0^2}\left(\frac{W_h}{T}\right)^2 . \tag{10.69}$$

The situation with the conductivity of the heavy band is a slightly more tricky since $\mathrm{Im}\Sigma_{hh}^{(2)}(T) \sim \mathrm{Im}\Sigma_{hL}^{(2)}(T)$, and hence $\sigma_{hh} \sim \sigma_{hL}$. However, for a crude estimate, we can again consider only the nondiagonal part of the conductivity σ_{hL} and take Umklapp processes 10.68 into account. Then, we have

$$\sigma_{hL} \sim \sigma_{Lh} \sim \frac{\sigma_{\min}}{f_0^2}\left(\frac{W_h}{T}\right)^2 . \tag{10.70}$$

We note that the estimates (10.70) for σ_{Lh} and σ_{hL} can be justified using the exact solution of the coupled kinetic equations for heavy and light particles with an account taken of the Umklapp processes.

The total resistivity at low temperatures $T < W_h$ is given by

$$R = \frac{1}{\sigma_L + \sigma_h} \sim \frac{f_0^2}{\sigma_{\min}}\left(\frac{T}{W_h}\right)^2 . \tag{10.71}$$

It behaves in a Fermi-liquid fashion $R(T) \propto T^2$. For $m_h^* \gg m_h$, we can replace W_h by W_h^* in Eqs. (10.70) and (10.71).

10.9 Chemical Potentials of Heavy and Light Bands at Higher Temperatures $W_h^* < T < W_L$

If $T > W_h^*$ (or more precisely for $f_0^2 T > W_h^*$, as we will see shortly) the coherent motion in the heavy band is totally destroyed. The effective chemical potential of the heavy band becomes large and negative [3, 4, 6]

$$\mu_h^{eff} = \mu + \frac{W_h}{2} + \varepsilon_0 \sim -T \ln\left(\frac{1}{n_h a^D}\right) . \tag{10.72}$$

In other words, we have the Boltzmann behavior of μ_h^{eff}. The fermionic distribution function for heavy particles has the form

$$f_F(\varepsilon_h/T) \approx \exp\left(\frac{\mu_h^{eff}}{T}\right) = const . \tag{10.73}$$

At the same time, since $m_h \gg m_L$, the effective chemical potential for light electrons

$$\mu_L^{eff} = \mu + \frac{W_L}{2} \approx \varepsilon_{FL} , \tag{10.74}$$

and hence they are still described by the step function

$$f_F(\varepsilon_L/T) \approx \theta\left(\frac{p^2}{2m_L} - \varepsilon_{FL}\right) . \tag{10.75}$$

10.10 Evaluation of Imaginary Parts of the Self-energies at Higher Temperatures $W_h^* < T < W_L$

For light electrons, the self-energy $\mathrm{Im}\Sigma_{LL}^{(2)}(T) = f_0^2 T^2/W_L$ does not change. However, we can write

$$\mathrm{Im}\Sigma_{Lh}^{(2)}(T) = f_0^2 W_h \frac{m_h}{m_L} \gg \mathrm{Im}\Sigma_{LL}^{(2)}(T) \quad \text{for} \quad W_h^* < T < W_L . \tag{10.76}$$

This behavior of $\mathrm{Im}\Sigma_{Lh}^{(2)}(T)$ in Eq. (10.76) corresponds to the almost elastic scattering of light electrons by heavy ones as if by immobile (static) impurities in the principal approximation in the small parameter $W_h/W_L \ll 1$ (we note the useful relation $W_h m_h = W_h^* m_h^*$ for the regime of strong electron polaron effect here).

Let us also emphasize that for heavy electrons (and "heavy-heavy" self-energy) at high temperatures $\varepsilon_h/T \ll 1$, we should take into account the bosonic contribution $f_B(\varepsilon_h/T) \approx T/\varepsilon_h$ and the fermionic contribution $f_F(\varepsilon_h/T) \approx \frac{1}{2}(1 - \varepsilon_h/2T)$ to $\mathrm{Im}\Sigma^{(2)}_{hh}$. This yields

$$\mathrm{Im}\Sigma^{(2)}_{hh}(T) = f_0^2 W_h \,. \tag{10.77}$$

Equation 10.77 describes the scattering of heavy electrons by each other in the situation when they uniformly occupy the heavy band and can transfer to each other only an energy $\sim W_h$ (see [80]). In the regime of strong electron polaron effect, we can replace W_h by W_h^* in Eq. (10.77). At the same time, for scattering of heavy particles by the light ones, we have:

$$\mathrm{Im}\Sigma^{(2)}_{hL}(T) = f_0^2 T \gg \mathrm{Im}\Sigma^{(2)}_{hh}(T) \,. \tag{10.78}$$

Equation 10.78 describes the marginal Fermi-liquid behavior [1] for the diffusive motion of heavy electrons in the surrounding of light electrons. Note that to derive Eq. (10.78) in the principal order in the small parameter $m_L/m_h \ll 1$, we should rewrite expression (10.25) for the self-energy $\mathrm{Im}\Sigma^{(2)}_{hL}(\varepsilon_h(\mathbf{p}), \mathbf{p})$ in the form

$$\mathrm{Im}\Sigma^{(2)}_{hL}(\varepsilon_h(\mathbf{p}), \mathbf{p}) = -\pi \left(\frac{4\pi a_0^{hL}}{m_L}\right)^2 N_L^2(0) \int\int d\varepsilon_L(\mathbf{p}')d\varepsilon_L(\mathbf{k})\delta[\varepsilon_L(\mathbf{p}') - \varepsilon_L(\mathbf{k})] f_F[\varepsilon_L(\mathbf{k})](1 - f_F[\varepsilon_L(\mathbf{p}')]) \,. \tag{10.79}$$

After the δ-functional integration, we obtain

$$\mathrm{Im}\Sigma^{(2)}_{hL}(\varepsilon_h(\mathbf{p}), \mathbf{p}) \sim \pi f_0^2 \int d\varepsilon_L(\mathbf{k}) f_F[\varepsilon_L(\mathbf{k})](1 - f_F[\varepsilon_L(\mathbf{k})]) \sim \pi f_0^2 \int\limits_{-\infty}^{+\infty} d\varepsilon \frac{\exp(\varepsilon/T)}{[1+\exp(\varepsilon/T)]^2} \sim f_0^2 T \,. \tag{10.80}$$

10.11 Resistivity for $T > W_h^*$ in the 3D Case

In agreement with Eqs. (10.76), (10.77), and (10.76) of the previous section, we get the following estimates for the inverse scattering times of heavy and light electrons at temperatures $T > W_h^*$

$$\frac{1}{\tau_L} \approx \frac{1}{\tau_{Lh}} = f_0^2 W_h \frac{m_h}{m_L} \,. \tag{10.81}$$

Having in mind that $W_h m_h/m_L = W_h^* m_h^*/m_L \sim W_L$, we obtain $1/\tau_L \sim f_0^2 W_L$. At the same time,

$$\frac{1}{\tau_h} \approx \frac{1}{\tau_{hL}} = f_0^2 T \,, \tag{10.82}$$

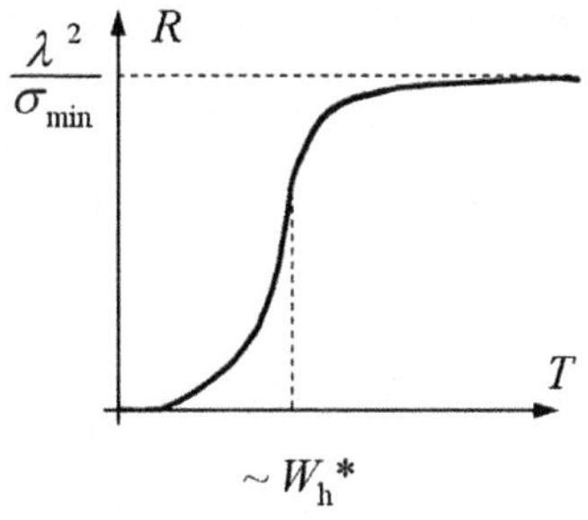

Fig. 10.5 Resistivity characteristics $R(T)$ in the two-band Hubbard model with one narrow band in 3D [3, 6]

and hence the heavy component is marginal, while light one is not.

The conductivity of the light band is given by

$$\sigma_L = \frac{n_L e^2 \tau_L}{m_L} \approx \frac{n_L e^2 \tau_{Lh}}{m_L} = \frac{\sigma_{\min}}{f_0^2}. \tag{10.83}$$

For the heavy band, the Drude formula at higher temperatures $T > W_h^*$ should be modified using the Einstein relation $\partial n_h/\partial T \sim n_h/T$. This yields

$$\sigma_h = \frac{\sigma_{\min}}{f_0^2}\left(\frac{W_h^*}{T}\right)^2. \tag{10.84}$$

Correspondingly, the resistivity

$$R = \frac{1}{\sigma_L + \sigma_h} = \frac{f_0^2}{\sigma_{\min}} \frac{1}{\left[1 + \left(\frac{W_h^*}{T}\right)^2\right]} \tag{10.85}$$

tends to the saturation in the 3D case (see Fig. 10.5). In the intermediate temperature range (on the intermediate asymptotics) between the Fermi-liquid behavior at low temperatures and saturation at high temperatures the system mimics the marginal behavior [3–6].

This behavior of the resistivity is typical of some uranium-based heavy-fermion compounds such as UNi_2AL_3. Note that rigorously speaking, in the regime of the destroyed heavy band, which takes place at $f_0^2 T > W_h^*$, the diagrammatic technique can be used only for the light particles and not for heavy ones. However, the exact solution of the density matrix equations obtained in [20, 21] shows that the inverse scattering times $1/\tau_{hL}$ and $1/\tau_{Lh}$ are qualitatively the same as for $f_0^2 T > W_h^*$ in our estimates. That is why, the conductivities σ_h and σ_L and hence $R(T)$ behave smoothly for $T \geq W_h^*$.

10.12 Weak Localization Corrections and Altshuler–Aronov Effect in 2D

In the 3D case at higher temperatures, the heavy component is marginal, while the light one is not. The natural question arises: whether it is possible to find the marginal behavior for the light component as well. The tendency toward the marginalization of the light component manifests itself in the 2D case. We know that logarithmic corrections to the classical Drude formula for conductivity arise in 2D due to the weak-localization effects [46]. However, according to our ideology, heavy electrons play the role of impurities for the scattering of light particles by them. That is why, the correct expression for the conductivity of the light band σ_L in the absence of the spin–orbit coupling is given by

$$\sigma_L^{loc} = \frac{\sigma_{\min}}{f_0^2}\left(1 - f_0^2 \ln \frac{\tau_\varphi}{\tau}\right). \tag{10.86}$$

According to the weak-localization theory in 2D, τ in Eq. (10.86) is the elastic time and τ_φ is the inelastic (decoherence) time, while $f_0 = f_0^{2D} = \frac{1}{2\ln(1/ap_F)}$ is the 2D gas parameter (we assume again that $n_h \sim n_L$). In our problem, we have

$$\tau = \tau_{ei} = \tau_{Lh}, \quad \tau_\varphi = \tau_{ee} = \tau_{LL}, \quad \tau_{LL} \gg \tau_{Lh}, \tag{10.87}$$

where τ_{ei} and τ_{ee} are the times describing the scattering of electrons by impurities and other electrons, respectively. Hence, between two scattering events of a light electron by another light one, it scatters many times by heavy electrons (see Fig. 10.6).

As a result, the motion of light electrons becomes much slower (also of the diffusion type) and two characteristic lengths arise in the theory, namely, the elastic length

$$l = v_{FL}\tau_{Lh} \tag{10.88}$$

and the diffusion length

$$L_\varphi = \sqrt{D_L \tau_\varphi}, \tag{10.89}$$

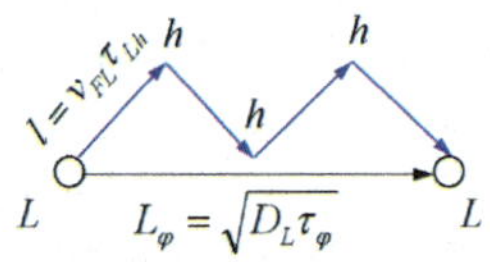

Fig. 10.6 Multiple scattering of light electrons by heavy ones in the interval between the scattering of a light electron by another light particles. L_φ is the characteristic diffusion length, l is the elastic length, D_L and v_{FL} are the diffusion coefficient and the Fermi velocity for light electrons, respectively, and τ_{Lh} and τ_φ are respectively the elastic time for scattering of light electrons by heavy ones and the inelastic (decoherence) time [3]

where D_L is the diffusion coefficient for light electrons and v_{FL} is the Fermi velocity for light electrons. That is why, in a more rigorous theory, according to [46], we should replace the inverse scattering time

$$\frac{1}{\tau_{LL}(\varepsilon)} \sim \frac{f_0^2}{m_L} \int_0^{\varepsilon} d\omega \int_0^{\omega} d\varepsilon' \int_0^{\infty} \frac{q dq}{(v_{FL}q)^2} \sim f_0^2 \frac{\varepsilon^2}{W_L} \ln \frac{W_L}{\varepsilon} \tag{10.90}$$

by the modified scattering time

$$\frac{1}{\tilde{\tau}_{LL}(\varepsilon)} \sim \frac{f_0^2}{m_L} \int_0^{\varepsilon} d\omega \int_0^{\omega} d\varepsilon' \int_0^{\infty} \frac{q dq}{(i\varepsilon' + D_L q^2)^2}, \tag{10.91}$$

where $f_0 = f_0^{2D}$ is the 2D gas parameter. In fact, in Eq. (10.91), we replace $v_{FL}q$ by the "cooperon" pole $i\varepsilon' + D_L q^2$ in the Altshuler–Aronov terminology. Hence, the characteristic wave vectors in the evaluation of $\tilde{\tau}_{LL}$ are $q \sim \sqrt{\varepsilon / D_L}$, where ε is an energy variable. Correspondingly, we have

$$\frac{1}{\tilde{\tau}_{LL}(\varepsilon)} \sim f_0^2 \frac{\varepsilon}{m_L D_L}. \tag{10.92}$$

For the diffusion coefficient in Eq. (10.92), we can use the estimate

$$D_L = v_{FL}^2 \tau_{Lh}, \tag{10.93}$$

where according to Eq. (10.81), we have $1/\tau_{Lh} \sim f_0^2 W_h m_h / m_L \sim f_0^2 W_L$. As a result, we obtain

$$\frac{1}{\tilde{\tau}_{LL}(\varepsilon)} \sim f_0^4 \varepsilon. \tag{10.94}$$

Therefore, the inverse scattering time $\tilde{\tau}_{LL}$ also becomes marginal for $\varepsilon \sim T$. For logarithmic corrections to the conductivity, we have

$$\frac{\tau_\varphi}{\tau} = \frac{\tilde{\tau}_{LL}}{\tau_{Lh}} = \frac{W_L}{f_0^2 T} \gg 1. \tag{10.95}$$

Hence, the conductivity of the light band with an account taken of the weak-localization corrections has the form

$$\sigma_L^{loc} = \frac{\sigma_{\min}}{f_0^2} \left(1 - f_0^2 \ln \frac{W_L}{f_0^2 T}\right). \tag{10.96}$$

For $f_0^2 T \sim W_h$, the large logarithm in Eq. (10.96) is $\ln (W_L / f_0^2 T) \sim \ln (W_L / W_h)$ and thus

$$Z_h \approx \frac{\sigma_L^{loc}}{\sigma_L} \approx 1 - f_0^2 \ln \frac{W_L}{W_h}. \tag{10.97}$$

Therefore, in 2D case at the higher temperatures $f_0^2 T \sim W_h$, a contribution to the Z-factor of heavy electrons due to the electron polaron effect and weak-localization corrections to the conductivity of light electrons are governed by the same parameter $f_0^2 \ln (m_h/m_L)$.

10.13 Justification of the Expression for Localization Corrections in 2D with an Account Taken of the Recoil Energy

In principle, heavy particles, which play the role of impurities in our problem, are mobile and have some recoil energy. That is why, the formula $\sigma_L^{loc}/\sigma_L = 1 - f_0^2 \ln (W_L/f_0^2 T)$ should be justified, at least, with regard to the temperature exponent under logarithm, T or T^α. For the justification, we need to estimate the loss of the energy by one light election before it collides with another light electron. The number of collisions with heavy particles between the scattering of a light particle by another light one is L_φ/l. The maximum loss of energy in one collision is W_h^*. Thus, the total loss of energy is $W_h^* L_\varphi/l = W_h^* \sqrt{W_L/T}$. The energy of the light particle itself is T. This means that for $W_h^* \sqrt{W_L/T} < T$ or, equivalently, for

$$T > W_h^* \left(\frac{W_L}{f_0^2 W_h^*} \right)^{1/3} \tag{10.98}$$

the loss of energy is small and heavy electrons can be regarded as immobile impurities. Hence, the exponent α under logarithm is equal to unity.

10.14 Resistivity in the Homogeneous 2D Case

The behavior of the resistivity in the 2D case can be qualitatively described by the expression

$$R \approx \frac{f_0^2}{\sigma_{\min}} \frac{1}{[1 + (W_h^*/T)^2 - f_0^2 \ln (W_L/f_0^2 T)]}. \tag{10.99}$$

It has a maximum at $T_{\max} \sim W_h^*/f_0$ and a localization tail at higher temperatures (see Fig. 10.7). Such resistivity characteristics resemble the curve for $R(T)$ in optimally doped layered CMR systems. However, as we have already mentioned in Chap. 2, the alternative explanations of this curve based on electron tunneling from one ferromagnetic droplet to a neighboring one via the paramagnetic barrier in the regime of

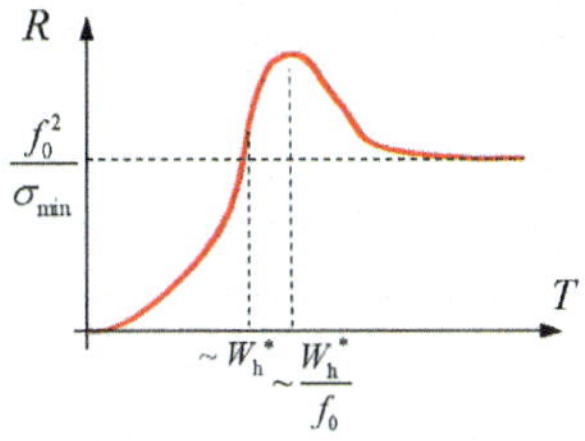

Fig. 10.7 Resistivity characteristics $R(T)$ for the 2D two-band Hubbard model with one narrow band. The resistivity exhibits a peak and a localization tail at higher temperatures $T > W_h^*$ [3]

the Coulomb blockade are also possible. As a problem for the future investigations, it would be very interesting to find the magnetoresistance in the 2D or layered case in the two-band Hubbard model with one narrow band for a strong quantizing magnetic field oriented perpendicular to the layer [3, 45].

Note, however, that in the case of layered manganites, the approach related to FM polarons, which is discussed in Chap. 2, seems to be more consistent with the scanning-tunneling experiments [47] and with the confinement of electronic motion to two dimensions, Nevertheless, we cannot exclude that for different types of layered materials, the explanation based on the localization tail in the homogeneous state of the two-band model is valid in the regime of a dynamically destroyed (e.g., by temperature) heavy band.

10.15 Superconductivity in the Two-Band Hubbard Model with One Narrow Band

At the end of this chapter, let us mention briefly that the leading superconductivity(SC) mechanism in the two-band Hubbard model at low electron densities corresponding to triplet p-wave pairing and is governed, especially in 2D, by the pairing of heavy electrons via polarization of light ones (see Fig. 10.8 and [3, 4, 6, 26, 31–34, 38, 48, 49]) in the framework of the enhanced Kohn–Luttinger mechanism [50].

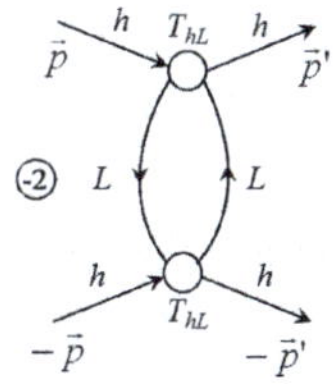

Fig. 10.8 Leading contribution to the effective interaction U_{eff} for the p-wave pairing of heavy particles via the polarization of light particles. The open circles stand for the vacuum T-matrix T_{hL} [3]

The corresponding p-wave critical temperature T_{C1} nonmonotonically depends on the relative doping of the bands n_h/n_L and has a broad and pronounced maximum at $n_h/n_L = 4$ in 2D, where it can reach the experimentally feasible values realistic for layered ruthenates Sr_2RuO_4 [51, 52], uranium-based heavy-fermion compounds like $U_{1-x}Th_xBe_{13}$ [53], as well as for layered dichalcogenides CuS_2, $CuSe_2$ and semimetallic superlattices InAs-GaSb, PbTe-SnTe with the spatially separated bands belonging to different layers [54]. For typical values of $\varepsilon^*_{Fh} \sim 30$–50 K, the critical temperatures T_{C1} can achieve 1–5 K, which is quite nice. Moreover, as it was shown in [30], the two superconducting gaps for heavy and light electrons are opened simultaneously below this temperature.

Finally, let us stress again that in 2D, where only the electron polaron effect contributes to the mass enhancement of heavy electrons, the restrictions on the stability of the homogeneous phase are much milder than in 3D, where a tendency toward the phase separation exists at a large density mismatch.

References

1. C.M. Varma, P.B. Littlewood, S. Schmitt-Rink, E. Abrahams, A.E. Ruckenstein, Phys. Rev. Lett. **63**, 1996 (1989)
2. M.Yu. Kagan, A.G. Aronov, Czechoslov. J. Phys. **46**(5), 2661 (1996)
3. M.Yu. Kagan, V.V. Val'kov, J. Exp. Theor. Phys. **113**(1), 156 (2011)
4. M.Yu. Kagan, V.V. Valkov, Low Temp. Phys. **37**(1), 69 (2011)
5. M.Yu. Kagan, V.V. Val'kov, in *A Lifetime in Magnetism and Superconductivity: A Tribute to Professor David Shoenberg*, ed. by G. Lonzarich, S.S. Saxena, V.A. Sirenko, V.V. Eremenko (Cambridge Scientific Publishers, 2012)
6. MYu. Kagan, V.V. Val'kov, J. Supercond. Nov. Magn. **25**(5), 1379 (2012)
7. A.O. Sboychakov, K.I. Kugel, A.L. Rakhmanov, Phys. Rev. B **76**, 195113 (2007)
8. A.V. Rozhkov, A.O. Sboychakov, A.L. Rakhmanov, F. Nori, Phys. Rep. **648**(1), 1 (2016)
9. D.M. Newns, N. Read, Adv. Phys. **36**(6), 799 (1987)
10. P. Coleman, Phys. Rev. B **35**, 5072 (1987)
11. A.M. Tsvelik, *Quantum Field Theory in Condensed Matter Physics*, 2nd edn. (Cambridge University, 2007)
12. C.M. Varma, Phys. B **378–380**(Supplement C), 17 (2006)
13. P. Fulde, *Electron Correlations in Molecules and Solids*, Springer Series in Solid-State Sciences, vol. 100, 3rd edn. (Springer-Verlag, Berlin, Heidelberg, 2002)
14. H. Keiter, N. Grewe, in *Valence Fluctuations in Solids*. ed. by L.M. Falicov, W. Hanke, M.B. Maple (North Holland, Amsterdam, 1981), p.451
15. C.M. Varma, Y. Yafet, Phys. Rev. B **13**, 2950 (1976)
16. A.C. Hewson, *The Kondo Problem to Heavy Fermions*. Cambridge Studies in Magnetism (Cambridge University Press, 1993)
17. K.G. Wilson, Rev. Mod. Phys. **47**, 773 (1975)
18. P.W. Anderson, G. Yuval, D.R. Hamann, Phys. Rev. B **1**, 4464 (1970)
19. P. Nozières, J. Low Temp. Phys. **17**(1), 31 (1974)
20. M.Yu. Kagan, N.V. Prokof'ev, J. Exp. Theor. Phys. **63**(6), 1276 (1986)
21. M.Yu. Kagan, N.V. Prokof'ev, J. Exp. Theor. Phys. **66**(1), 211 (1987)
22. J.S. Kim, B. Andraka, G.R. Stewart, Phys. Rev. B **45**, 12081 (1992)
23. L.M. Falicov, J.C. Kimball, Phys. Rev. Lett. **22**(19), 997 (1969)
24. K.I. Kugel, D.I. Khomskii, Sov. Phys. Usp. **25**, 231 (1982)

25. K.I. Kugel, A.L. Rakhmanov, A.O. Sboychakov, D.I. Khomskii, Phys. Rev. B **78**(15), 155113 (2008)
26. M.Yu. Kagan, *Modern Trends in Superconductivity and Superfluidity*, Lecture Notes in Physics, vol. 874. (Springer, Netherlands, 2014)
27. J. Kanamori, Prog. Theor. Phys. **30**(3), 275 (1963)
28. E.E. Salpeter, H.A. Bethe, Phys. Rev. **84**, 1232 (1951)
29. E.M. Lifshitz, L.P. Pitaevskii, *Physical Kinetics, Course of Theoretical Physics*, vol. 10 (Butterworth-Heinemann, Oxford, 1981)
30. A.A. Abrikosov, L.P. Gorkov, I.E. Dzyaloshinskii, *Methods of Quantum Field Theory in Statistical Physics* (Dover Books in Physics (Dover, New York, 1975)
31. M.A. Baranov, A.V. Chubukov, M.Yu. Kagan, Int. J. Mod. Phys. B **06**(14), 2471 (1992)
32. D.V. Efremov, M.S. Mar'enko, M.A. Baranov, M.Yu. Kagan, J. Exp. Theor. Phys. **90**(5), 861 (2000)
33. M.Yu. Kagan, Phys. Lett. A **152**(5), 303 (1991)
34. M.A. Baranov, M.Yu. Kagan, Sov. Phys. JETP **75**(1), 165 (1992)
35. V.M. Galitskii, Sov. Phys. JETP **7**(1), 104 (1958)
36. E.M. Lifshitz, L.P. Pitaevskii, *Statistical Physics, part II* (Pergamon, Oxford, 1980)
37. P. Bloom, Phys. Rev. B **12**, 125 (1975)
38. M.Yu. Kagan, A.V. Chubukov, JETP Lett. **47**(10), 614 (1988)
39. A.A. Abrikosov, Phys. **2**, 5 (1965)
40. H. Suhl, Phys. Rev. **138**, A515 (1965)
41. P. Coleman, E. Miranda, A. Tsvelik, Phys. B **199–200**, 197 (1994)
42. A.J. Millis, P.A. Lee, Phys. Rev. B **35**, 3394 (1987)
43. M. Grilli, C. Castellani, Phys. Rev. B **50**(23), 16880–16898 (1994)
44. N.F. Mott, E.A. Davis, *Electronic Processes in Non-crystalline Materials*. Monographs on Physics (Oxford University Press, 1971)
45. Y. Imry, *Introduction to Mesoscopic Physics*, 2nd edn. (Oxford University Press, 2008)
46. B.L. Altshuler, A.G. Aronov, in Modern Problems, in *Condensed Matter Systems*, vol. 10, ed. by A.L. Efros, M. Pollak (North Holland, Amsterdam, 1985), pp.1–153
47. H.M. Rønnow, C. Renner, G. Aeppli, T. Kimura, Y. Tokura, Nat. **440**(7087), 1025 (2006)
48. M.Yu. Kagan, V.A. Mitskan, M.M. Korovushkin, Phys. Usp. **58**(8), 733 (2015)
49. M.Yu. Kagan, JETP Lett. **103**(11), 728 (2016)
50. W. Kohn, J.M. Luttinger, Phys. Rev. Lett. **15**, 524 (1965)
51. T.M. Rice, M. Sigrist, J. Phys. Condens. Matter **7**(47), L643 (1995)
52. Y. Maeno, T.M. Rice, M. Sigrist, Phys. Today **54**(1), 42 (2001)
53. F. Kromer, R. Helfrich, M. Lang, F. Steglich, C. Langhammer, A. Bach, T. Michels, J.S. Kim, G.R. Stewart, Phys. Rev. Lett. **81**, 4476 (1998)
54. K. Murase, S. Ishida, S. Takaoka, T. Okumura, H. Fujiyasu, A. Ishida, M. Aoki, Surf. Sci. **170**(1), 486 (1986)

Chapter 11
Manifestation of the EPE in the Microcontact with Deep and Shallow Traps

Abstract We consider the manifestations of the electron polaron effect (EPE) in the microcontact with deep and shallow traps specified by the model with the quadrupole structure of four quantum dots (QD). In this model, two central QDs play the role of a deep two-level trap, while two other QDs at the periphery play the role of the shallow traps with one level in them. In the framework of the nonequilibrium Keldysh Green's function technique, we calculate the electric current in this system both in the isotropic and in the anisotropic case, which is close to the situation in the EPE.

11.1 Introduction. QQD Structure

In this chapter, we study the manifestation of the EPE in the reduced geometry of a microcontact with deep and shallow traps specified by the quadrupole quantum dot structure. We base our presentation on the results reported in Refs. [1–6].

Note that the quantum dots (QDs) and in particular dipole and quadrupole quantum dot (QQD) structures are very popular low-dimensional systems nowadays not only in the context of the fundamental quantum physics but also being very promising for possible future applications in nanoelectronics and quantum calculations.

Note also that the variety of different often coexisting processes ranging from Fano resonances [7], Kondo and Aharonov–Bohm effects [8] to Hubbard type of electron correlations play an important role and are deeply investigated in the physics of QDs [9–11].

Let us emphasize that already in single-electron regime (even neglecting the many-body Coulomb correlations) these structures can be possibly utilized as spin qubits [12, 13], as well as rectifiers, spin filters and valves [14]. Note that the QD structures containing three and more QDs have been actively studied only in recent years due to their more difficult experimental realization [15, 16]. Moreover, experimentally they were studied in the geometry of the linear (arrays-type) structures. In contrast to the structures with three QDs the quadrupole (QQD) structures were studied in the different square-like geometry [17].

It is important to emphasize that in all the geometries the Coulomb correlations inside the QD (intra-dot Coulomb repulsion) as well as Coulomb repulsion between

M. Yu. Kagan et al., *Electronic Phase Separation in Magnetic and Superconducting Materials*, Springer Series in Solid-State Sciences 201,
https://doi.org/10.1007/978-3-031-55467-4_11

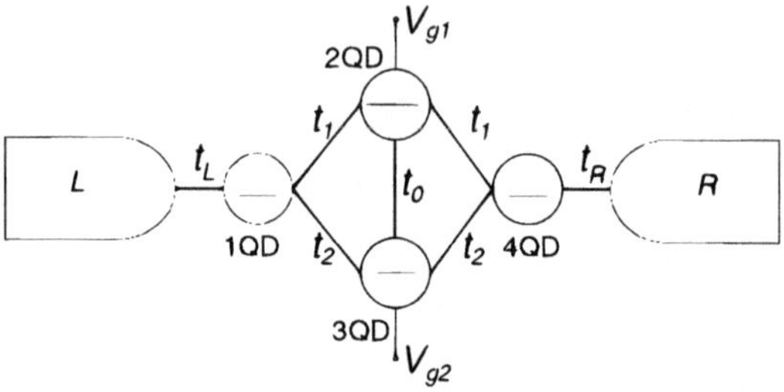

Fig. 11.1 QQD structure between metallic leads. The electron hopping t_1 takes place between the first and second quantum dots (1QD and 2QD) and between second and fourth quantum dots (2QD and 4QD) on the upper shoulder of the figure. The electron hopping t_2 corresponds to the lower shoulder and takes place between 1QD and 3QD and between 3QD and 4QD. Finally, the "vertical" hopping t_0 takes place between the upper and lower shoulders or equivalently between 3QD and 2QD. In the anisotropic case, we have $t_1 > t_2$. At the same time, the hoppings generally obey the inequality $t_0 \leq t_1$ [1]

the neighboring QDs (inter-dot Coulomb repulsion) play the crucial role in the determination of the electron spectrum and transport properties of the QD structures [18, 19].

Note that the most typical situation for quadrupole (QQD) structure corresponds to the geometry (to the topology of the microcontact) when left and right metallic contacts are coupled with two QDs while the remaining two quantum dots are situated in the central part of the microcontact (see Fig. 11.1).

The investigation of the Fano and Aharonov–Bohm effects in these geometry as well as other transport properties were fulfilled mainly in the strong-coupling Kondo regime or in the absence of any correlations at all [20, 21], utilizing the well-known Landauer formalism [22] for the transfer matrix.

The more general situation, when not only spin (Kondo channel [23]) but also charge (mixed valence) channel with the electron correlations of both intra-dot Hubbard type (specified by on-site Hubbard U) and inter-dot Coulomb repulsions (specified by nearest-neighbor Coulomb interaction V) are fully taken into account, is much less investigated.

The first theoretical analysis along these lines was performed in Ref. [1]. In this paper both intra-dot U and inter-dot V Coulomb repulsions as well as the anisotropy effects related to anisotropic hopping integrals t between the neighboring dots and on-site carrier energies ϵ_0 inside each dot were fully taken into account, while studying the transport properties and electron spectra of the QQD system in the framework of the famous Keldysh Green's function technique [24]. Note that the origin of the anisotropy can be connected for example with the gate fields always present in the experiments with the QD structures.

Thus, the analysis performed in Ref. [1] serves as a bridge between the physics of correlated electron transport in metals and in low-dimensional physics of the nanostructures (tunneling contacts and traps) and moreover allows us to study the manifestations of such highly nontrivial effects as the EPE in the reduced geometry of the microcontact with multiple QDs.

11.2 Interconnection of QQD Structure with EPE in the Two-Band Hubbard and Anderson Models

Let us emphasize here that the introduction of the anisotropy in the hopping integrals provides an important analogy between the QQD structure described in the introductory Sect. 11.1 and the physics of the electron polaron effect (EPE) considered in Chap. 10 in the framework of the two-band Hubbard model [25–29].

Remind that as we already discussed in the beginning of Chap. 10, the EPE is always present in the multiband strongly correlated electron systems with substantially different electron bandwidths (or effective masses) of heavy and light electrons.

We emphasized above that this situation can be realized in uranium-based heavy-fermion systems and other systems in the mixed valence regime [30–34]. Note that in the mixed valence case the most simple model to describe the EPE is the two-band Hubbard model with one narrow (or heavy) band and one broad (light) band intersecting with each other in the presence of sufficiently strong interband Hubbard interaction between them.

The situation in the microcontact with the QQD structure is more close to the Anderson model (AM) [35] with one-particle hybridization between conduction $s - p$ band and localized $f - d$ levels (in our case its role plays the "vertical" electron hopping t_0 between the upper and lower shoulders of Fig. 11.1) and two-particle Hubbard interaction (V in Fig. 11.1 and U_{hL} in Chap. 10) between electrons belonging to lower and upper shoulders in Fig. 11.1.

Remind that in the two-band Hubbard model, the EPE leads to the additional narrowing of the heavy bandwidth due to the dressing of the heavy particle in the virtual of soft electron-hole pairs of light particles. In the close similarity, the EPE in AM leads to an additional narrowing of the hybridization matrix element between the conduction electrons and electrons of the localized levels (see Refs. [1, 33, 36]).

Note that in the QQD scheme presented in Fig. 11.1, these hybridization matrix elements correspond to the electron hopping t_2 from the first quantum dot (1QD) to the third one (3QD) and to the reverse hopping from the third quantum dot (3QD) to the fourth one (4QD) along the lower shoulder of Fig. 11.1.

As a result, in all our estimates in the regime of strong EPE, which corresponds to large values of one-particle and two-particle hybridizations t_0 and V, we should effectively replace t_2 by $t_2' \ll t_2$. Hence, in our numerical calculations, we further assume that the anisotropy in our system is related to both the specific design of the QQD structure in our case and the many-body generalized EPE.

11.3 Coulomb Correlations and Conductance in the Microcontact

In this section, we will study manifestations of the Coulomb correlations in the tunneling conductance of the microcontact. Note that as we already discussed in Chap. 10, the EPE leads to anomalous resistivity characteristics both in 3D and especially in 2D or layered infinite systems. The manifestations of EPE in the restricted geometry of the tunnel junctions are also quite interesting.

The first pioneering results here belong to Matveev and Larkin [37]. Important results elucidating the role of strong Hubbard (intra-dot) and Coulomb (inter-dot) correlations in the charge sector for the dipole QD structure (for the central "vertical" part 2QD-3QD in Fig. 11.1) were obtained also by Arseev, Maslova et al. [38–40]. Finally, a formation of additional Fano–Feshbach resonances [7, 41] in the tunneling conductance and in effective one-particle density of states was demonstrated in Refs. [1–3].

Let us stress here one more important analogy between our system of quantum dots in the microcontact and the dielectric or semiconducting systems with impurities and defects. Note that in the actual experimental situation, when we investigate electric currents and hopping transport in dielectrics and semiconductors, we often deal with a small number of deep two-level traps in the surrounding of a large number of shallow one-level traps.

The basic results here belong to Bishop's group [42], who considered two-level traps randomly distributed between one-level traps. These results were obtained in connection with the physics of radiation-induced conductivity. They are important both for space applications and for more fundamental physics of Holstein type configurational polaron effect in molecularly doped polymers.

In our case, in terms of deep and shallow traps, we consider the QQD structure with one two-level trap in the central part of Fig. 11.1 (two levels correspond to quantum dots 2QD and 3QD) surrounded by two shallow one-level traps 1QD and 4QD in the restricted geometry of the junction. Of course, in Refs. [1–3], we solve a simpler problem then in a localization physics and neglect the peculiarities of the hopping conductivity on the Miller–Abrahams lattice [43].

11.4 Calculation Technique and the Results

In this section, we obtain the results for spectral and transport characteristics of the QQD structure in the nonequilibrium Keldysh [24] Green's function technique using the tight-binding approximation. We derive analytically the retarded Green's functions of the QQD structure taking into account both strong Hubbard and Coulomb correlations [1].

Based on this analysis, we investigate different mechanisms of the suppression and rectification of the electric current in the QQD structure connected first of all with Aharonov–Bohm [8] and Fano [7] effects, and the negative differential conductance (NDC) as well [3]. We study also the possibility of the perfect spin-filtering effect in the QQD structure induced by Coulomb interactions [2]. Note that according to the results of Ref. [3], the NDC effect manifests itself in the anisotropic QQD structure and can be described in terms of the bound states in the continuum (BIC) and the interaction between the Fano resonances, which are formed by the Coulomb interaction V between the electrons of central dots. We demonstrate that the corresponding peak-to-valley ratio in the $I - V$ characteristics related to NDC can be strongly enhanced, when we change the parameters of the QQD cell.

Note also that if we apply an external magnetic field to the QQD structure than we can get a perfect spin filter effect or, in other words, a full spin blockade (SB) in it. Namely the technique of the equation of motion presented in Ref. [2] for the retarded Green's functions can lead to the appearance of the spin-polarized windows (SPW) in the spin-dependent conductance when we take both intra-dot (Hubbard) and inter-dot (Coulomb) correlations into account. Effectively these correlations create the zones of high-spin polarization (P) in the current.

The width and magnitude of the spin-polarization windows in our geometry can be controlled by the gate fields and the ratio between different transfer integrals in the QQD cell. Moreover, it can be shown that perfect spin-filtering effect can be obtained both in anisotropic configuration (in the regime of strong EPE) and isotropic configuration, where EPE is absent.

11.5 Model Hamiltonian

The Hamiltonian of the QQD cell illustrated in Fig. 11.1 includes four terms and is given by

$$\hat{H} = \hat{H}_L + \hat{H}_R + \hat{H}_D + \hat{H}_T \, . \tag{11.1}$$

The first two terms correspond to the left and right metallic leads

$$\hat{H}_{L(R)} = \sum_{k,\sigma} (\xi_{k\sigma} \mp \frac{eV}{2}) c^{\dagger}_{L(R)k\sigma} c_{L(R)k\sigma} \, , \tag{11.2}$$

where $c^{\dagger}_{L(R)k\sigma}$ and $c_{L(R)k\sigma}$ are the creation (annihilation) operators in the left (right) lead with momentum k, spin projection σ, and the spin-dependent energy

$$\xi_{k\sigma} = \varepsilon_{k\sigma} - \mu \, . \tag{11.3}$$

In Eq. 11.3, μ is the chemical potential of the system.

The third and the fourth terms describe, respectively, the QQD structure itself and the interaction between the leads and the QQD structure. Namely, the third term reads

$$\hat{H}_D = \sum_{\sigma}\sum_{j=1}^{4} \xi_{j\sigma} a_{j\sigma}^{\dagger} a_{j\sigma} + U \sum_{j=1}^{4} n_{j\uparrow} n_{j\downarrow} + V \sum_{\sigma\sigma'} n_{2\sigma} n_{3\sigma'}$$
$$+ \sum_{\sigma} [t_1 (a_{1\sigma}^{\dagger} + a_{4\sigma}^{\dagger}) a_{2\sigma} + t_2 (a_{1\sigma}^{\dagger} + a_{4\sigma}^{\dagger}) a_{3\sigma} + t_0 a_{2\sigma}^{\dagger} a_{3\sigma} + h.c.], \quad (11.4)$$

where $a_{j\sigma}^{\dagger}$ and $a_{j\sigma}$ are the creation and annihilation operators for electrons with spin projection σ and energy $\xi_{k\sigma} = \varepsilon_{k\sigma} - \mu$ at the QD with number j , $(j = 1, 4)$, t_1 and t_2 are the are hopping integrals in the upper (1QD-2QD-4QD) or (1QD-3QD-4QD) lower shoulder of the QQD structure shown in Fig. 11.1, t_0 is the hopping integral between the shoulders, U is the intra-dot Hubbard interaction, and V is the inter-dot Coulomb interaction between electrons of the second (2QD) and third (3QD) dots in the central part of the cell.

Finally, the fourth term in Hamiltonian (11.1), which describes the interaction between the leads and the cell is given by

$$\hat{H}_T = t_L \sum_{k\sigma} c_{Lk\sigma}^{\dagger} a_{1\sigma} + t_R \sum_{k\sigma} c_{Rk\sigma}^{\dagger} a_{4\sigma} + h.c., \quad (11.5)$$

where $t_{L(R)}$ are the hopping integrals between the left (right) lead and the first (or the fourth) QD. Since the bias voltage can be explicitly taken into account in our case, it is convenient to make a unitary transformation (see Ref. [44])

$$\hat{U} = \exp[\frac{ieVt}{2} t_L \sum_{k\sigma} (n_{Rk\sigma} - n_{Lk\sigma})], \quad (11.6)$$

and then to insert this transformation to the tunneling Hamiltonian (11.5)

$$\hat{H}_T = T_L(t) \sum_{k\sigma} c_{Lk\sigma}^{\dagger} a_{1\sigma} + T_R(t) \sum_{k\sigma} c_{Rk\sigma}^{\dagger} a_{4\sigma} + h.c., \quad (11.7)$$

where

$$T_{L(R)}(t) = t_{L(R)} e^{\mp \frac{ieV}{2} t} \quad (11.8)$$

is an effective coupling parameter of the QQD cell with the left (or right) leads. Note that the linear (in eV) time dependence in the exponent of $T_{L(R)}(t)$ can be understood also from the nonequilibrium condition on the electrochemical potentials of the left and right leads in the presence of the voltage, which yields

$$\mu_L - \mu_R = eV. \quad (11.9)$$

11.6 Electric Current and Nonequilibrium Green's Functions

In this section, we analyze the transport properties of the system. We use the nonequilibrium Keldysh [24] Green's function technique in the tight-binding approximation [45, 46] (see also the more recent extended review by Arseev [47] for the update). We start with the definition of the current.

An operator of a steady-state electric current in the left lead is determined by the change of the corresponding charge per unit time

$$\hat{I}_L = e\dot{N}_L \,, \tag{11.10}$$

where

$$N_L = \sum_{k\sigma} c^{\dagger}_{Lk\sigma} c_{Lk\sigma} \tag{11.11}$$

is the operator of the number of charged particles in the left lead. Utilizing the equation of motion for the Heisenberg operator $\hat{I}_L$, we get ($\hbar = 1$)

$$I(t) = \langle \hat{I}_L(t,t)\rangle = e\langle \dot{N}_L\rangle = ie\sum_{k\sigma}[T_L^{+}(t)G^{+-}_{Lk1\sigma}(t,t) + T_L(t)G^{+-}_{1\,Lk\sigma}(t,t)]\,, \tag{11.12}$$

where we introduced the nonequilibrium mixed Green's functions

$$G^{+-}_{L(R)kj\sigma}(t,t') = \langle\langle c_{L(R)j\sigma}(t)|a^{\dagger}_{j,\sigma}(t')\rangle\rangle^{+-}\,. \tag{11.13}$$

Let us stress that the Heisenberg operators $c_{L(R)j\sigma}(t)$ and $a^{\dagger}_{j,\sigma}(t')$, which enter in the definitions of these Green's functions are ordered at the Keldysh contour C [24]. Moreover, superscripts (+-) in the so-called mixed lesser Green's functions $G^{+-}_{L(R)kj\sigma}(t,t)$ correspond to the lower and upper branches of the Keldysh contour, C_+ and C_-, respectively (see [47] and Fig. 11.2).

According to the general definition of the ordering of the Heisenberg operators $\hat{\Psi}$ and $\hat{\Psi}^{\dagger}$ on the Keldysh contour C (which is usually referred to in the literature as the T_C ordering) the lesser Green's function G^{+-} in general case can be represented as[47]

$$G^{+-}(rt, r't') = -i\langle\hat{\Psi}(rt)\hat{\Psi}^{\dagger}(r't')\rangle\,. \tag{11.14}$$

It is important to stress that the time t^{+} on the contour C is always "after" the time t^{-}, and hence the operator $\hat{\Psi}$ is always located to the left of the operator $\hat{\Psi}^{\dagger}$ in Green's function G^{+-} irrespectively to the relation between the times t and t' on the real axis.

Note that in the Keldysh nonequilibrium diagram technique, there exists a "doubling of variables", and four different Green's functions of the real-time variables t

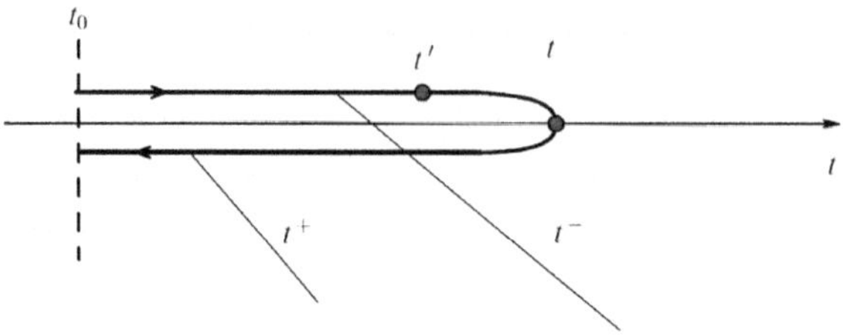

$$\mathrm{Sp}\left[\rho_0 \tilde{T}\exp\left[+\mathrm{i}\int_{t_0}^{t'} \hat{V}(t_2')\,\mathrm{d}t_2'\right] T\left\{\exp\left[-\mathrm{i}\int_{t_0}^{t} \hat{V}(t_1')\,\mathrm{d}t_1'\right] \hat{A}(t)\hat{B}(t')\right\}\right]$$

Fig. 11.2 Keldysh contour C with the lower branch C_- and the upper branch C_+. The time point t^+ on the contour is always "after" the time point t^- [47]

and t', namely G^{+-}, G^{-+}, G^{--}, and G^{++}. They enter in the 2×2 matrix Green's function G^{ab}

$$G^{ab} = \begin{pmatrix} G^{--} & G^{-+} \\ G^{+-} & G^{++} \end{pmatrix}, \tag{11.15}$$

in which G^{--} and G^{++} are the diagonal elements and G^{+-} and G^{-+} are the nondiagonal ones.

There exists an important sum rule for the elements of this matrix, namely the sum of nondiagonal elements is equal to the sum of diagonal ones

$$G^K = G^{+-} + G^{-+} = G^{--} + G^{++}, \tag{11.16}$$

where we introduced the Keldysh nonequilibrium Green's function G^K as a sum of nondiagonal elements. Let us emphasize that in momentum space, the Keldysh function for fermions is proportional to the density factor

$$G^K \propto 1 - n_p, \tag{11.17}$$

where n_p is the occupation number for electrons with momentum p.

The elements of this matrix are related to the more standard advanced and retarded Green's functions as follows

$$G^a = G^{--} - G^{+-} = G^{-+} - G^{++} \tag{11.18}$$

for the advanced Green's function G^a, and

$$G^r = G^{--} - G^{-+} = G^{+-} - G^{++} \tag{11.19}$$

for the retarded Green's function G^r.

Finally, the density of the system $n(\mathbf{r}, t)$ in the inhomogeneous and unsteady state is given by

$$n(\mathbf{r}, t) = -iG^{-+}(\mathbf{r}t, (\mathbf{r}t)\,. \tag{11.20}$$

Note that throughout this section and further on, we for the sake of simplicity consider the isotropic case with respect to left and right leads, namely

$$t_L = t_R = t\,, \tag{11.21}$$

retaining at the same time the anisotropic possibilities to other hopping integrals t_0, t_1, and t_2.

If the tunneling Hamiltonian $\hat{H}_T$ in Eq. (11.5) is treated as an interaction operator, then our analysis of the series of the perturbation expansion for the nonequilibrium Green's functions $G^{+-}_{Lk1\sigma}$ and $G^{+-}_{1Lk\sigma}$ leads to the following expression for the current

$$I(t) = e\sum_{\sigma}\int_C d\tau_1[\Sigma^{+a}_{L\sigma}(t-\tau_1)G^{a-}_{11\sigma}(\tau_1 - t) - G^{+a}_{11\sigma}(t-\tau_1)\Sigma^{a-}_{L\sigma}(\tau_1 - t)]\,, \tag{11.22}$$

where similarly to the matrix form for one-particle Green's function G^{ab} in Eq. (11.5), we introduced the 2×2 matrix for the self-energies $\Sigma^{ab}_{L\sigma}(t-\tau')$ defined as

$$\Sigma^{ab}_{L\sigma}(t-\tau') = T^{+}_{L}(\tau)\sum_{k} g^{ab}_{Lk\sigma}(t-\tau')T_L(\tau')\,. \tag{11.23}$$

The self-energies in Eq. (11.23) are related to the effect of the left lead on the QQD structure in the middle of the cell. They contain the factors $g^{ab}_{Lk\sigma}(\tau-\tau')$, which correspond to the 2×2 matrix for the bare one-electron Green's function of the left lead. The upper indices $(a, b = +, -)$ in the self-energies are related again to the lower and upper branches of the Keldysh contour.

At this point, it is convenient to use the Dyson equation for the full one-particle Green's functions $G^{ab}_{11\sigma}$

$$\begin{aligned} G^{ab}_{11\sigma}(\tau-\tau') &= g^{ab}_{11\sigma}(\tau-\tau') + \int_C\int_C d\tau_1 d\tau_2[g^{ac}_{11\sigma}(\tau-\tau_1)\Sigma^{cd}_{L\sigma}(\tau_1-\tau_2)G^{db}_{11\sigma}(\tau_2-\tau')] \\ &+ \int_C\int_C d\tau_1 d\tau_2[g^{ac}_{41\sigma}(\tau-\tau_1)\Sigma^{cd}_{L\sigma}(\tau_1-\tau_2)G^{db}_{41\sigma}(\tau_2-\tau')]\,. \end{aligned} \tag{11.24}$$

Note that in this subsection, while deriving Eq. (11.24), we use the nonmagnetic approximation. In the framework of this approximation, we neglect the spin-flip processes. Hence, we have $\langle a^{\dagger}_{j\sigma}a_{j-\sigma}\rangle = 0$.

As a result, deforming the Keldysh contour to that in the real-time one (via the Wick transformation $t \to i\tau$), then integrating over the new contour, and performing the Fourier transform, we obtain the following expression for the current

$$I = i\Gamma \sum_{\sigma} \int_{-\infty}^{+\infty} d\omega [f_L (G^a_{11\sigma} - G^r_{11\sigma}) - G^{+-}_{11\sigma}], \tag{11.25}$$

where $G^a_{11\sigma}$ and $G^r_{11\sigma}$ are the advanced and retarded one-particle Green's functions defined by Eqs. (11.18) and (11.19), and $f_L = f(\omega + eV/2)$ is the Fermi–Dirac distribution function. Parameter Γ in Eq. (11.25) is given by the relation

$$\Gamma/2 = \Gamma_L = \Gamma_R = \pi t^2 g\,, \tag{11.26}$$

where $\Gamma_L = \Gamma_R$ are the parameters, which describe the broadening of the QQD levels due to the coupling to the leads.

Note that in general the density of states of the leads (which enters in the definition of the broadening Γ depends on frequency ω and spin projection σ

$$g_\sigma(\omega) = \sum_k \delta(\omega - \xi_{k\sigma})\,. \tag{11.27}$$

However, in this section, we consider the paramagnetic leads and the limit of a broad band. In this case, we can ignore these dependences and assume that $g = const$ in Eq. (11.26). As a result, the Fourier transforms of the self-energy functions of the αth lead, which enter the expressions of the current (11.22)–(11.25) are given by simple expressions

$$\Sigma^r_{\alpha\sigma} = -i\Gamma/2, \quad \Sigma^{+-}_{\alpha\sigma} = -i\Gamma f_\alpha\,. \tag{11.28}$$

11.7 Details of the Calculation for the Current

In this subsection, we present the details of the calculation (see [3]) for the DC current in the paramagnetic regime. To calculate the current, let us find Green's functions of the QQD which enter Eq. (11.25). To do that, it is convenient to use the technique of equations of motion.

Note that the general form of the expressions of the retarded Green's function G^r and Green's function G^{+-} differ from each other since they are governed by different equations of motion, namely,

$$z\langle\langle a_{i\sigma} | a^\dagger_{j\sigma'} \rangle\rangle^r = \langle \{ a_{i\sigma} a^\dagger_{j\sigma'} \} \rangle + \langle\langle \big[a_{i\sigma}, \widehat{H} \big] | a^\dagger_{j\sigma'} \rangle\rangle^r\,, \tag{11.29}$$

$$z\langle\langle a_{i\sigma} | a^\dagger_{j\sigma'} \rangle\rangle^{+-} = \langle\langle \big[a_{i\sigma}, \widehat{H} \big] | a^\dagger_{j\sigma'} \rangle\rangle^{+-}\,, \tag{11.30}$$

where $z = \omega + i\delta$.

Moreover, taking into account the diagrammatic expansion of the mixed Green's function (see Eq. (11.13)

$$G_{L(R)kj\sigma}(t-t') = \int g_{L(R)kj\sigma}(t-\tau)T_{L(R)}(\tau)G_{1(4)j\sigma}(\tau-t')d\tau \tag{11.31}$$

the corresponding equations of motion for $G^r_{L(R)kj\sigma)}$ and $G^{+-}_{L(R)kj\sigma)}$ can be written as

$$z\langle\langle c_{L(R)k\sigma}(t)|a^\dagger_{j\sigma'}(t')\rangle\rangle^r = g^r_{L(R)k\sigma}t_{L(R)}\langle\langle a_{1(4)\sigma}|a^\dagger_{j\sigma'}\rangle\rangle^r\,, \tag{11.32}$$

$$\begin{aligned} z\langle\langle c_{L(R)k\sigma}(t)|a^\dagger_{j\sigma'}(t')\rangle\rangle^{+-} = t_{L(R)}(g^r_{L(R)k\sigma}\langle\langle a_{1(4)\sigma}|a^\dagger_{j\sigma}\rangle\rangle^{+-} \\ +g^{+-}_{L(R)k\sigma}\langle\langle a_{1(4)\sigma}|a^\dagger_{j\sigma}\rangle\rangle^a)\,, \end{aligned} \tag{11.33}$$

where the bare Green's functions $g^r_{L(R)k\sigma}$ and $g^{+-}_{L(R)k\sigma}$ are given by

$$g^r_{L(R)k\sigma} = (z-\xi_{k\sigma})^{-1}\,, \tag{11.34}$$

$$g^{+-}_{\alpha k\sigma} = 2\pi i f_\alpha \delta(\omega-\xi_{k\sigma})\,. \tag{11.35}$$

Our next step is to derive a closed set of equations to use the decoupling procedure for the nonmagnetic case developed in [1–3]. This approximation is valid for temperatures exceeding the Kondo temperature [48]. In this procedure, we should decouple the equations for the third-order Green's functions, e.g.,

$$\langle\langle n_{3\sigma}n_{2-\sigma}a_{2\sigma}|a^\dagger_{j\sigma}\rangle\rangle^{r,+-}\,. \tag{11.36}$$

The solution of the final set of the equations for the retarded Green's functions yields

$$\begin{aligned} G^r_{\beta\beta} = \frac{C_\beta Z_{-\beta}}{Z},\quad G^r_{\beta-\beta} = \frac{C_\beta C_{-\beta}x_2}{Z},\quad G^r_{\alpha\alpha} = \frac{C_\alpha \Delta_{-\alpha}}{Z}, \\ G^r_{\alpha-\alpha} = \frac{C_\alpha C_{-\alpha}\Delta_1}{Z},\quad G^r_{\beta\alpha} = \frac{C_\alpha C_\beta T_{-\beta}P_{-\alpha}}{Z}, \end{aligned} \tag{11.37}$$

where we introduced the following notation: $\beta = 1, 4$; $\alpha = 1, 3$; $t(\alpha) = t_{1,2}$;

$$\begin{aligned} \Delta_\alpha = D_\alpha T_\beta T_{-\beta} - t^2(\alpha)C_\alpha S,\quad \Delta_1 = t_0 T_\beta T_{-\beta} + t(\alpha)t(-\alpha)S, \\ S = C_\beta T_{-\beta} + C_{-\beta}T_\beta,\quad P_\alpha = t(-\alpha)D_\alpha + t_0 t(\alpha)C_{-\alpha}\,, \\ Z = T_\beta T_{-\beta}x_1 - Sx_2,\quad Z_\beta = T_\beta x_1 - C_\beta x_2,\quad T_\beta = D_\beta + i\Gamma C_\beta/2, \\ x_1 = \Delta_\alpha\Delta_{-\alpha} - t_0^2 C_\alpha C_{-\alpha},\quad x_2 = t(\alpha)C_\alpha P_{-\alpha} + t(-\alpha)C_{-\alpha}P_\alpha\,. \end{aligned} \tag{11.38}$$

The factors $C_{\alpha,\beta}$ and $D_{\alpha,\beta}$ in Eq. (11.37) contain the explicit dependences on the occupation numbers, correlation functions, and the strength of Coulomb interactions in the QQD [1–3]. To be more specific, we can write $C_\alpha = C_{\alpha 1} + C_{\alpha 2}$, where

$$C_{\alpha 1} = b_{\alpha 4}(b_{\alpha 2}b_{\alpha 3} + Ub_{\alpha 3}\langle n_\alpha\rangle + 2Vb_{\alpha 3}\langle n_{-\alpha}\rangle),$$
$$C_{\alpha 2} = UV(b_{\alpha 2}(b_{\alpha 2} + b_{\alpha 3})(2\langle n_\alpha\rangle\langle n_{-\alpha}\rangle - \langle a_\alpha^\dagger a_{-\alpha}\rangle^2),$$
$$C_\beta = b_{\beta 2} + U\langle n_\alpha\rangle, \quad D_\alpha = b_{\alpha 1}b_{\alpha 2}b_{\alpha 4}, \quad D_\beta = b_{\beta 1}b_{\beta 2}, \tag{11.39}$$

and correspondingly

$$b_{\alpha 1} = z - \xi_\alpha, \quad b_{\alpha 2} = b_{\alpha 1} - U,$$
$$b_{\alpha 3} = b_{\alpha 1} - V(1 + \langle n_{-\alpha}\rangle), \quad b_{\alpha 4} = b_{\alpha 3} - U\,. \tag{11.40}$$

Note that in all expressions (11.37)–(11.40), we omit for simplicity spin indices since in the paramagnetic case, the polarization is absent and thus

$$\langle a_{i\alpha}^\dagger a_{i\alpha}\rangle = \langle a_{i-\alpha}^\dagger a_{i-\alpha}\rangle\,. \tag{11.41}$$

At the same time, the solution of the system of equations for the lesser Green's functions G_{ij}^{+-} reads

$$G_{\beta\beta(-\beta)}^{+-} = i\Gamma\frac{C_\beta(f_\beta Z_\beta G_{\beta\beta(-\beta)}^a + f_{-\beta}C_{-\beta}x_2 G_{-\beta\beta(-\beta)}^a)}{Z},$$
$$G_{\alpha\alpha(-\alpha)}^{+-} = i\Gamma\frac{P_{-\alpha}C_\alpha(f_\beta C_\beta T_{-\beta}G_{\beta\alpha(-\alpha)}^a + f_{-\beta}C_{-\beta}T_\beta G_{-\beta\alpha(-\alpha)}^a)}{Z},$$
$$G_{\beta\alpha}^{+-} = i\Gamma\frac{C_\beta(f_\beta Z_{-\beta}G_{\beta\alpha(-\beta)}^a + f_{-\beta}C_{-\beta}x_2 G_{-\beta\alpha(-\alpha)}^a)}{Z}, \tag{11.42}$$

where for the distribution functions and Green's functions, we have

$$f_\beta = f_{L(R)}, \quad G_{ij}^a = (G_{ij}^r)^*, \quad G_{ij}^{+-} = -(G_{ji}^{+-})^*\,. \tag{11.43}$$

From the definitions of the lesser Green's functions (11.31) and (11.14), we can obtain the expressions for the correlation functions and the averaged occupation numbers, utilizing the self-consistent solution of the following integral equations [3]

$$\langle n_i\rangle = 2\int_{-\infty}^{\infty} G_{ii}^{+-}\frac{d\omega}{2\pi}, \quad \langle a_i^\dagger a_j\rangle = 2\int_{-\infty}^{\infty} G_{ji}^{+-}\frac{d\omega}{2\pi}\,. \tag{11.44}$$

After substitution of Green's functions calculated in (11.37)–(11.44) to Eq. (11.25), we find the final expression for the steady current of the QQD structure

$$I = 2e\Gamma^2\int_{-\infty}^{\infty} G_{41}^r G_{41}^a (f_L - f_R)d\omega = 2e\Gamma^2\int_{-\infty}^{\infty}\frac{C_1^2 C_4^2 x_2^2}{|Z|^2}(f_L - f_R)d\omega\,. \tag{11.45}$$

Let us stress that the factor 2 in expressions (11.44) and (11.45) appears as a result of the spin summation in the paramagnetic case. In further discussions, we scale all energy values in units of the width Γ. Additionally, we will analyze the strong-coupling regime for which $\Gamma = t_1$. In the subsequent calculations, we also assume that the energies of the edge dots are the same

$$\xi_{1\sigma} = \xi_{4\sigma} = \varepsilon_D \,. \tag{11.46}$$

Correspondingly, the difference in energy of two central dots is governed by the parameter Δ according to the relation

$$\xi_{2(3)\sigma} = \varepsilon_D \pm \Delta \,. \tag{11.47}$$

It is interesting that expanding the distribution functions in Eq. (11.45) for the current in the Taylor series up to linear terms for small bias voltages $eV \leq \Gamma$

$$f_L(\omega - eV/2) - f_R(\omega + eV/2) \approx eV \frac{df}{d\omega} \,, \tag{11.48}$$

we can get the following expression for the differential conductance $G = dI/dV$

$$\begin{aligned} G = \frac{dI}{dV} &= \frac{d}{dV}\left\{2e\Gamma^2 \int_{-\infty}^{\infty} G^r_{41} G^a_{41} (f_L - f_R) d\omega\right\} \\ &\approx -G_0\Gamma^2 \int_{-\infty}^{\infty} G^r_{14} G^a_{41} \frac{df}{d\omega} d\omega \,, \end{aligned} \tag{11.49}$$

where $G_0 = e^2/\hbar$ is the conductance quantum. Then, we can derive the famous Landauer–Büttiker formula [22, 49]. At finite temperatures, it reads

$$G = -G_0 \int_{-\infty}^{\infty} \mathrm{Tr}\, \hat{T}(\varepsilon = \varepsilon_D, \omega) \frac{df}{d\omega} d\omega \,, \tag{11.50}$$

where we introduced the trace of the transmission matrix coefficient $\mathrm{Tr}\, \hat{T}$.

At low temperatures, we can use the relation

$$-\frac{df(\omega)}{d\omega} \approx \delta(\omega) \,, \tag{11.51}$$

and obtain a simple expression for the conductance

$$G = -G_0\, \mathrm{Tr}\, \hat{T}(\varepsilon = \varepsilon_D, \omega = 0) \,. \tag{11.52}$$

In Eq. (11.52), the transition coefficient $\mathrm{Tr}\, \hat{T}$ is proportional to the product of retarded and advanced Green's functions and yields

$$\mathrm{Tr}\, \hat{T}(\varepsilon = \varepsilon_D, \omega = 0) = \Gamma^2 G^r_{14} G^a_{41} \,. \tag{11.53}$$

11.8 Results of Numerical Calculations for Nonequilibrium Transport in the Isotropic QQD Structure

In this section, we describe the nonequilibrium transport through the QQD structure. In Fig. 11.3, we present the results of numerical calculations for the conductance G (Fig. 11.3a) and the occupation numbers n (Fig. 11.3b) of the QQD as functions of the gate field ε_D at different bias voltages eV. We consider the isotropic case, for which $t_1 = t_2$ (see [3]) and analyze the location and splitting of the resonances in the conductance $G(\varepsilon_D)$ in Fig. 11.3a.

First of all, in Fig. 11.3, we can see that the resonances are located on the left- and right-hand sides of the band gap, which corresponds to the half-filling ($n = 1$) in Fig. 11.3b and where the conductance $G \to 0$ in Fig. 11.3a. Moreover, for $eV \neq 0$, the resonances are split in contrast to the equilibrium case ($eV = 0$) (see dotted and dashed curves in Fig. 11.3a).

The resonance splitting can be explained in the following way: $eV \neq 0$, the transmission of the electrons is increased if the energy level, which is governed by parameter ε_D coincides with the electrochemical potentials of the left or right leads $\mu_{L,R} = \mu \mp eV/2$.

Note that the band gaps obtained in the regime of the linear response are conserved at the bias voltage $eV = 0.5$. At the same time, the further increase in the bias voltage until $eV = 1$ leads to the decrease in the bandwidths (see the solid curve in Fig. 11.3a). Moreover, in the strongly nonequilibrium regime for $eV > \Gamma = 1$, we can observe the effects, which are not described by linear response regime and the Landauer–Büttiker formula [22, 50]. As a result, for some gate voltages ε_D in the situation, when the width $\Gamma \sim \{U, V\}$, we can have here a nontypical situation, for which $G > 2G_0$ (the doubling of the conductance quantum is due to the summation over spin projections in the unpolarized case, so usually $G \leq 2G_0$).

In Fig. 11.3b, the steps in the occupation numbers are also split at $eV \neq 0$. It is especially evident for the population of two internal dots (QD2 and QD3). In this case, each step for $n_{2,3}$ corresponds to the resonance in the conductance [3].

Let us stress that the Fano resonances [7], which emerge in the conductance of the QQD system due to the Coulomb interaction V between the central dots QD2 and QD3 are also modified and have nonzero intensities for $eV \neq 0$ [1–3].

Note that the qualitative explanation for the arising Fano (or Fano–Feshbach [7, 41]) resonances in the QQD structure is the following [1]. Let us consider the central dots QD2 and QD3 in Fig. 11.1 as an artificial molecule or a dimer [51, 52]. The dimer is effectively a two-level system (or a two-level trap in the central part of the microcontact) surrounded by two one-level traps QD1 and QD4 on the microcontact periphery. It has bonding and antibonding levels, which in principle can be coupled differently to the periphery of the QQD system in Fig. 11.1 (to the QD1 and QD4 and the leads).

Then the bonding and antibonding levels of the molecule are broadened differently due to this coupling. Hence, we can treat a more broadened level as a continuum or

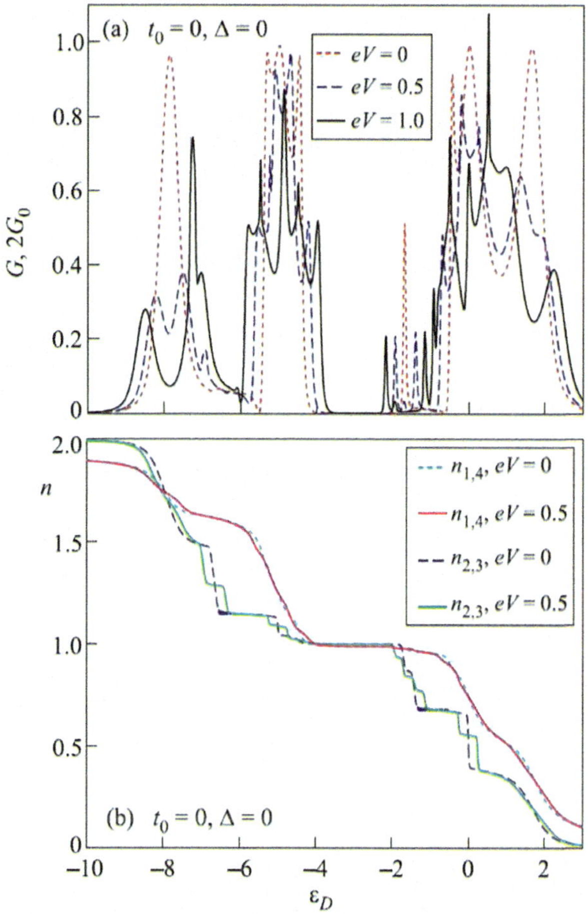

Fig. 11.3 Effect of the bias voltage eV on the dependence of conductance $G(\varepsilon_D)$ **a** and occupation number $n(\varepsilon_D)$ **b** on the gate voltage ε_D in the isotropic QQD. The calculations are performed for the following set of parameters: $U = 5$, $V = 1$, $t_1 = t_2 = 1$, $t_0 = \Delta = 0$, and $k_B T = 0.01$. All parameters are scaled in units of the width Γ. The bias voltage corresponds to $eV = 0$, 0.5, and 1 (dashed, dotted, and solid lines in panel a) and $eV = 0$ and 0.5 in panel b. The conductance G is scaled in the unpolarized case in units of the doubled conductance quantum $2G_0$ [3]

nonresonant channel in the original Fano–Feshbach picture, while the less broadened level plays the role of the discrete level or a resonant channel [1].

The phase of the wave function in the nonresonant channel changes only slightly as the energy passes a range of the order of the width of the discrete level Γ. At the same time, the phase in the resonant channel exhibits a shift of the order of π within the same range. Correspondingly, the Fano–Feshbach resonance peak appears in the QQD structure as a result of the interference processes.

In the equilibrium situation for $eV = 0$ and $\Delta = 0$ and in the isotropic case $t_1 = t_2$ the coupling with one of the molecular channels is absent (see [51]) and there is no Fano–Feshbach effect. However, the presence of the bias voltage $eV \neq 0$ and different electrochemical potentials of the leads $\mu_{L,R}$ changes the situation and the resonances can emerge even in the isotropic case.

11.9 Nonequilibrium Transport in the Anisotropic QQD Structure

In this section, we generalize the numerical results obtained in the previous section to the strongly anisotropic case, when $t_1 \gg t_2$.

Let us emphasize, that the energy structure of the anisotropic QQD, where $t_1 \gg \{t_2, t_0\}$, is closer to the physics of EPE then that in the isotropic case. Namely, in this case, we have broad band at the upper shoulder of Fig. 11.1 and narrow band at its lower shoulder hybridized at the one-particle level with the hopping integral t_0 and at the two-particle level by the interband Coulomb repulsion V between the central dots QD2 and QD3. Moreover, the anisotropy can be additionally increased in this case by the many-body effects governed by the interband Coulomb repulsion.

Let us start the analysis of the anisotropic QQD with the modification of the results for the dependence of the conductance on the gate voltage in the regime, when the bias voltage is switched on.

The results of the numerical calculations for the dependence of the conductance G on the gate voltage ε_D for different bias voltages eV and the current–voltage $(I - V)$ characteristics of the highly anisotropic QQD for different gate voltages are presented in Fig. 11.4a and b, respectively.

The characteristic features of Fig. 11.4 are related to the presence of the regions of the negative differential conductance (NDC) in Fig. 11.4a as well as the negative values of electric current $(I < 0)$ in Fig. 11.4b. Moreover, we observe the emergence of the additional narrow antiresonances in the negative ranges for the conductance $(G < 0)$ in the inset of Fig. 11.4a.

We can see that the antiresonance with the highest amplitude corresponds to the gate voltage $\varepsilon_D = -0.82$ in Fig. 11.4. We notice also that the $I - V$ curve in Fig. 11.4b exhibits four ranges, where the behavior of the conductance is substantially different. For the bias voltage $|eV| \leq 0.75$, the current practically does not increase reminding the effect of Coulomb blockade. At the same time, in the $0.75 \leq |eV| < 1$ range, we observe at first the significant growth of the current and then a sharp decrease at $|eV| \approx 1$. The current further grows within the third range $1 \leq |eV| \leq 1.5$. The peak-to-valley ratio in this range is about 1.4.

We can observe a similar scenario in Fig. 11.4b (the dashed curve) for the QQD with the occupation numbers exceeding the half-filling $(n > 1)$. Note that the peak-to-valley ratio can be additionally enhanced if we consider nonzero hopping integral $(t_0 \neq 0)$ between the central dots QD2 and QD3 and make the energies different in

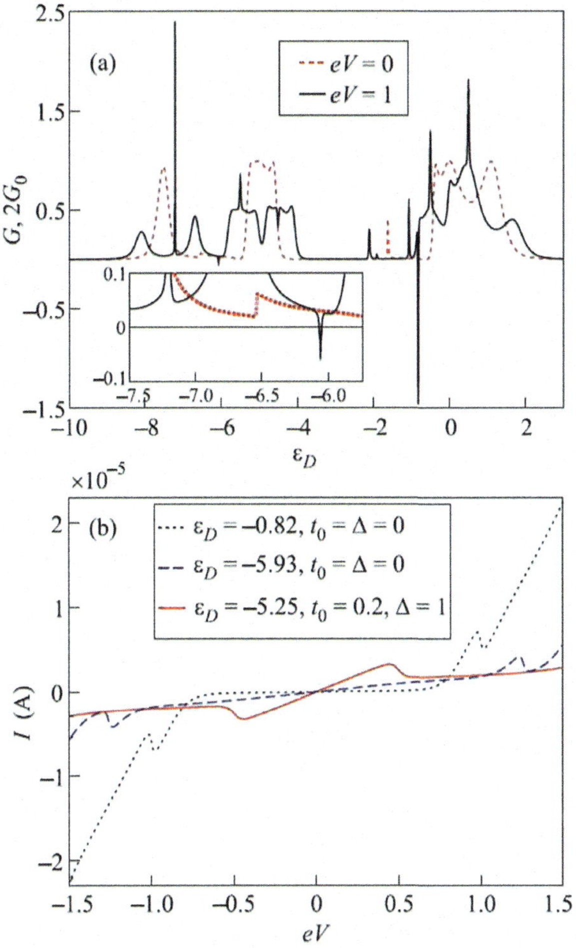

Fig. 11.4 Transport properties of the anisotropic QQD structure: **a** dependence of the conductance G on the gate voltage ε_D for different values of the bias voltage $eV = 0$ and 1. Inset: Fano antiresonance and its splitting for $eV \neq 0$. **b** current–voltage (I-V) characteristics for different values of the gate voltage ε_D, hopping integral t_0, and parameter Δ in the anisotropic case for $t_1 = 1$, $t_2 = 0.1$, $U = 5$, $eV = 0$ and 1, $k_B T = 0.01$. The conductance G is scaled in the unpolarized case in the units of the doubled conductance quantum $2G_0$ [3]

Eq. (11.47) for $\Delta \neq 0$ using the scheme with several gate electrodes. This case is described by the solid curve in Fig. 11.4b. It corresponds to the peak-to-valley ratio close to 1.9. In extremely anisotropic case when $t_2 \to 0$, the peak-to-valley ratio becomes as large as 2.6.

Finally in the strong-coupling case of large Coulomb repulsion, when not only U but also $V \gg \Gamma = 1$ (we do not plot this case in Fig. 11.4) the peak-to-valley ratio can be as high as 4.

11.10 Total Density of States (TDOS)

In this subsection, we discuss the total density of states (TDOS) in the anisotropic case taking into account Coulomb correlations.

Note that the effect of the negative differential conductance (NDC) observed in Fig. 11.4 presented in the previous section is related to the behavior of the total density of states (TDOS) of the QQD structure in the anisotropic regime.

In the Keldysh formalism, the TDOS is expressed in terms of the imaginary part of the retarded Green's functions of the QQD and can be written as

$$\mathrm{TDOS}(\omega) = \frac{1}{\pi}\sum_{i=1}^{4} \mathrm{Im}G_{ii}^{r}(\omega)\,. \tag{11.54}$$

Let us start first with the analysis of the isotropic situation $t_1 = t_2$. The corresponding density of states (DOS) is represented in Fig. 11.5a. If we neglect the Coulomb interaction, then the positions of the maxima of the $TDOS(\omega)$ are determined by the eigenvalues of the energies in the Hamiltonian H_{QQD} of the QQD cell given by Eq. (11.1). The dotted curve in Fig. 11.5a just corresponds to the eigenvalues of H_{QQD} for $U = V = 0$.

If we put the parameters $t_0 = \Delta = 0$, then there appear three levels, $\omega = \varepsilon_D$ and $\varepsilon_D \pm 2t_1$ (see Fig. 11.5a). In addition, due to the presence of degeneracy in the system [53], we also have a bound state in the continuum (BIC) at $\omega = 0$, which is described by the infinitely narrow peak with the width $i\delta \to 0$ in the retarded Green's functions $G_{ii}^{r}(\omega)$.

If we switch on now the intra-dot Hubbard repulsion U, each level in the individual dot will be additionally split: $\omega = \varepsilon_D$ and $\omega = \varepsilon_D + U$ (see the dashed curve in Fig. 11.5a). Moreover, an additional BIC generated by the Hubbard repulsion also emerges in the system [54, 55].

The inter-dot Coulomb repulsion V between central dots leads to the further splitting of the one-electron excitation energies. As a result, two new maxima and two BICs arise in the TDOS (solid curve in Fig. 11.5a).

Note that the peaks in the TDOS are the cause for emerging of the resonances in the conductance plotted in Fig. 11.3a in the linear response regime. In particular, the extra peaks for $V \neq 0$ in the TDOS generate the asymmetric Fano resonances. At the same time, the BICs do not manifest themselves in the transport characteristics of the QQD structure.

Let us stress that in the anisotropic case, the widths of two BICs generated by the inter-dot Coulomb repulsion V become finite. As a result, two narrow peaks of finite height arise in Fig. 11.4b and in the inset. Moreover, new Fano antiresonances appear in the conductance presented in Fig. 11.4. One of them is plotted by the dotted line in the inset of Fig. 11.4a and corresponds to the gate voltage $\varepsilon_D \approx -6.1$. A nonzero value of the conductance here can be explained by the temperature factor.

We already mentioned in the preceding subsections that at $V \neq 0$ the resonances in the conductance are split. At the same time, the antiresonance in this case is

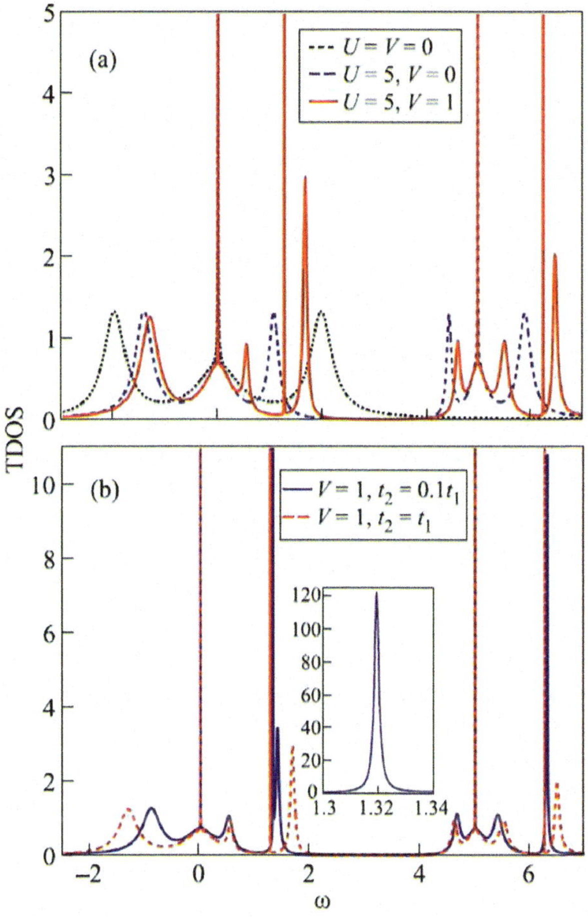

Fig. 11.5 **a** Effect of Coulomb correlation on the emergence of BICs in the total density of states (TDOS) of the isotropic QD structure. **b** Effect of the anisotropy of hopping integrals in the QQD on BICs. Inset: the width and the position of one of the peaks related to the BIC, which appears at $V \neq 0$. We use $\varepsilon_D = 0$, the other parameters coincide with those in Fig. 11.4 [3]

transformed into a narrow resonance with $G > 0$ and antiresonance with $G < 0$. The distance between them is of the order of eV (see the solid curve in the inset of Fig. 11.4a for the bottoms of the resonance and antiresonance).

If we increase the bias voltage, the antiresonance becomes shifted to the right-hand side. At the same time, the Fano asymmetric peak, which emerges at $V \neq 0$ in the isotropic case is shifted to the left.

We can thus conclude that the NDC effect is significantly enhanced when the antiresonance and Fano asymmetric peaks are close and interact with each other.

Concluding this section, note that Arseev et al. in [40] considered the NDC effect generated by Coulomb correlations, which arises in the parallel coupled double QD

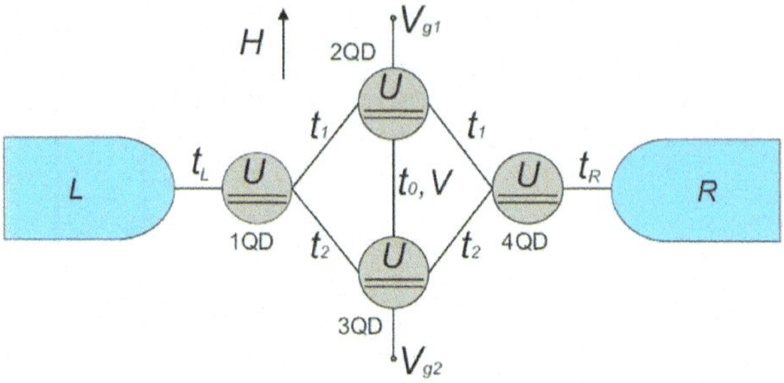

Fig. 11.6 QQD cell in the presence of applied external magnetic field H [2]

if the dots are connected asymmetrically with the leads. In this section, we on the contrary considered the NDC effect, which manifests itself in the symmetric coupling regime $t_L = t_R$. At the same time, the asymmetry of the hopping integrals $t_1 \gg t_2$, which leads to interesting peculiarities in the TDOS and specific redistribution in the occupation numbers of different dots is a property of the anisotropic QQD structure itself.

11.11 Perfect Spin Filtering Effect in the Polarized QQD Structure

In this section, we concentrate on spin-filtering properties of the QQD cell in the presence of the Coulomb interaction. The corresponding spin-polarization coefficient for the transport processes in the presence of applied magnetic field H reads (see Fig. 11.6)

$$P = \frac{G_\uparrow - G_\downarrow}{G_\uparrow + G_\downarrow}, \qquad (11.55)$$

where $G_\uparrow$ and $G_\downarrow$ are the differential conductances of "up" and "down" spins, respectively. We study again the symmetric situation $t_L = t_R$ and $\Gamma_L = \Gamma_R$ in the strong-coupling regime with respect to the interactions $\{U, V\}$, and we assume that $\Gamma = 1$.

We use again the same decoupling procedure as in the paramagnetic case in Sect. 11.7 and obtain the expression for the steady current I in the Keldysh formalism. The dependence of conductance on the gate voltage at finite temperatures is governed again by the Landauer–Büttiker formula. In the spin-polarized case, it has the form

$$G = G_\uparrow + G_\downarrow = G_0 \int_{-\infty}^{\infty} T_\sigma(\varepsilon_D, \omega) \frac{df(\omega)}{d\omega} d\omega\,, \qquad (11.56)$$

where we can express up (or down) conductance G_σ in terms of the transmission coefficient T_σ for the same spin projection. For T_σ, we have

$$T_\sigma = \Gamma^2 G^r_\sigma G^a_\sigma \,. \tag{11.57}$$

Correspondingly, the set of self-consistency equations for the averaged density $\langle n_{i\sigma}\rangle$ at site i with spin projection σ, and the correlation function $\langle a^\dagger_{i\sigma} a_{j\sigma}\rangle$ for different sites $i \neq j$ acquires the form

$$\begin{aligned}\langle n_{i\sigma}\rangle &= -\frac{1}{\pi}\int d\omega f(\omega)\mathrm{Im}G^r_{i\sigma,i\sigma}(\omega)\,,\\ \langle a^\dagger_{i\sigma} a_{j\sigma}\rangle &= -\frac{1}{\pi}\int d\omega f(\omega)\mathrm{Im}G^r_{i\sigma,j\sigma}(\omega)\,.\end{aligned} \tag{11.58}$$

11.11.1 Isotropic QQD Cell

Let us start again with the isotropic case when hopping integrals $t_1 = t_2$ and energy levels coincide in each dot $\xi_j = \varepsilon_D$. We can observe from Fig. 11.7a–c that if we apply the magnetic field to the cell, the spin-up (solid curve) and spin-down (dashed curve) conductances $G_\uparrow$ and $G_\downarrow$ are clearly separated as the functions of the gate field ε_D. Correspondingly at some value of the gate voltage and magnetic field, we can achieve the perfect spin-filtering effect, when the conductance for the down spins $G_\downarrow = 0$ and at the same time, the conductance for up-spins $G_\uparrow = G_0$ coincides with the conductance quantum (or, vice versa, when $G_\uparrow = 0$ and $G_\downarrow = G_0$). Hence, in this situation $P = \pm 1$ (see Fig. 11.7d), which is quite desirable for the spintronics applications.

Note that when we take into account only the intra-dot Hubbard repulsion U, we obtain two triple-peak structures (TPS) [1, 2], due to the electron–hole symmetry in this case (see Fig. 11.7a). Moreover, the number of the resonances in each TPS coincides with the number of QDs in the upper and lower shoulders of the QQD structure. In Fig. 11.7a, we can see as well the band gap with steep edges between two TPS for $U = 6$ and $V = 0$. As a result, stepwise features emerge in Fig. 11.7d (see the dotted curve) for the spin polarization P for the gate voltages within the range between $\varepsilon_D \approx -5$ and $\varepsilon_D \approx -1$.

We can conclude that at upper and lower plateaux of P, there exists the perfect spin polarization of charge carriers, which is governed by the spin-polarization windows (SPW) arising in the conductance G. If we take into account the inter-dot Coulomb interaction V between central dots QD2 and QD3 together with Hubbard repulsion U, we will have even more interesting picture. In this regime, the Fano–Feshbach resonance generates an additional wide conduction band with low values of the conductance for the gate voltages exceeding that corresponding to the half-filling ($\varepsilon_D \geq 3$ for $G_\uparrow$ [2]). We can see this band, e.g., in the spin-up conductance $G_\uparrow$, in the range between $\varepsilon_D = -8$ and $\varepsilon_D = -6$ for $U = 6$ and $V = t$ in Fig. 11.7b.

If due to the Zeeman shift [56] the position of the Fano antiresonance in the spin-down conductance $G_\downarrow$ coincides with the corresponding peaks of the spin-up

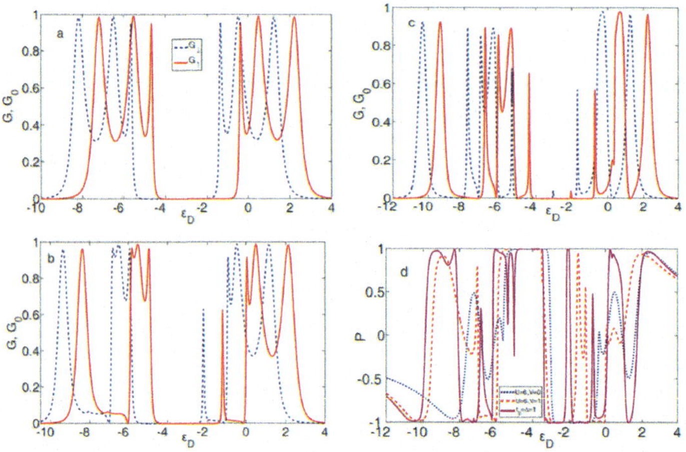

Fig. 11.7 Spin-up $G_\uparrow$ and spin-down $G_\downarrow$ conductance of the isotropic QQD cell for sets of parameters: **a** $U = 6$, $V = t_0 = \Delta = 0$, **b** $U = 6$, $V = 1$, $t_0 = \Delta = 0$, **c** $U = 6$, $V = t_0 = \Delta = 1$, (d)spin polarization P. Other parameters: temperature $k_B T = 0.01$ and applied magnetic field $H = 0.5$ for the (all parameters are scaled in the units of $\Gamma = 1$). In the case of polarization, $G_\uparrow$ and $G_\downarrow$ are scaled in the units of the conductance quantum G_0 [2]

conductance $G_\uparrow$, we find new stepwise features in P in the range between $\varepsilon_D = -7$ and $\varepsilon_D = -5$ (see the dashed curve in Fig. 11.7d).

Integrally, the inter-dot Coulomb interaction V produces three bands with high polarization of up spins ($P > 0.5$) and four bands with high polarization of down spins ($P < -0.5$). Thus, the Coulomb repulsion V violates the electron–hole symmetry [57] in the cell generating larger number of bands with the negative polarization ($P < 0$) and causing asymmetry in the widths and shapes of the peaks in $G_\uparrow$ and $G_\downarrow$ on Fig. 11.7b for gate voltages lower and higher than those corresponding to the half-filling.

Additional ways to manipulate the spin polarization P by creating new SPWs are related to the possibility to create additional asymmetry between the energy levels of the central dots switching on the parameter $\Delta \neq 0$. Then, according to Eq. (11.47), we have

$$\xi_{2\sigma} = \xi_{3\sigma} + \Delta\,. \tag{11.59}$$

Further on, we can include the one-particle hybridization to the game via the hopping integral t_0 between the central dots QD2 and QD3. In Fig. 11.7c, we demonstrate the total effect on spin-dependent conductance, which arises due to the presence of both these factors.

We can clearly see the formation of larger amount of spin-up and spin-down polarization windows in the spin-dependent conductance generated by new Fano–Feshbach resonances. Correspondingly, the spin polarization P rapidly varies

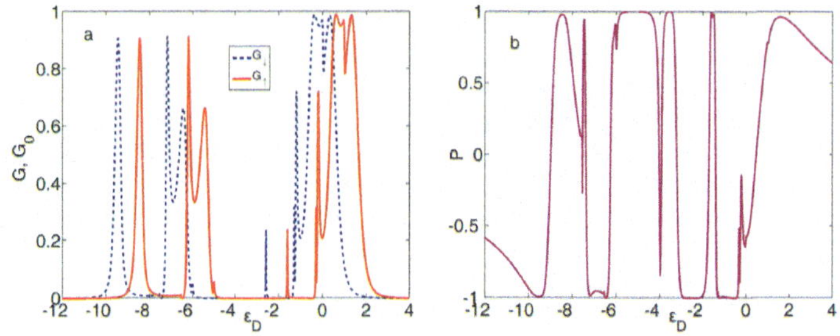

Fig. 11.8 **a** Spin-dependent conductances $G_{\uparrow}$ and $G_{\downarrow}$ for the set of parameters: $U = 6$, $V = 1$, $t_1 = 1$, $t_2 = 0.1$, $t_0 = 0.2$, $\Delta = 0.5$, $k_B T = 0.01$, and $H = 0.5$. **b** Spin polarization P for the strongly anisotropic QQD cell. $G_{\uparrow}$ and $G_{\downarrow}$ are scaled in the units of the conductance quantum G_0 [2]

("jumps") between two limiting values $P = \pm 1$ as the new perfectly spin-polarized conducting channels are switched on (see the solid curve in Fig. 11.7d).

11.11.2 Anisotropic QQD Cell

Concluding this section, let us consider again the anisotropic case $t_1 \gg \{t_2, t_0\}$, which is close to the physics of EPE. The decrease in t_2 and t_0 leads to the suppression of narrow peaks in the conductance, which are shown in the isotropic case for $t_1 = t_2$ in Fig. 11.7c. As a result, for the spin-dependent conductance of the anisotropic QQD cell, we obtain the picture presented in Fig. 11.8a. In this picture, we can observe six SPWs formed by the Zeeman splitting (three for spin-up, and three for spin-down conductances $G_{\uparrow}$ and $G_{\downarrow}$. Correspondingly, the spin polarization P in Fig. 11.8b exhibits a set of bands with high values of polarization ($P \to \pm 1$).

We can conclude that both isotropic and strongly anisotropic QQD structures can be used to produce a perfect spin-filtering effect, when we switch on both the intra-dot Hubbard repulsion U and the inter-dot Coulomb interaction V in the system.

References

1. M.Yu. Kagan, V.V. Val'kov, S.V. Aksenov, Phys. Rev. B **95**, 035411 (2017)
2. M.Yu. Kagan, V.V. Val'kov, S.V. Aksenov, J. Magn. Magn. Mater. **440**(Supplement C), 15 (2017)
3. M.Yu. Kagan, S.V. Aksenov, JETP Lett. **107**(8), 493 (2018)
4. V.V. Val'kov, M.Yu. Kagan, S.V. Aksenov, J. Phys. Condens. Matter **31**(22), 225301 (2019)
5. S.V. Aksenov, M.Yu. Kagan, JETP Lett. **111**(5), 286 (2020)
6. V.V. Val'kov, M.S. Shustin, S.V. Aksenov, A.O. Zlotnikov, A.D. Fedoseev, V.A. Mitskan, M.Yu. Kagan, Phys. Usp. **65**(1), 2 (2022)
7. U. Fano, Phys. Rev. **124**, 1866 (1961)
8. Y. Aharonov, D. Bohm, Phys. Rev. **115**, 485 (1959)

9. K. Kikoin, Y. Avishai, Phys. Rev. Lett. **86**, 2090 (2001)
10. A. Oguri, A.C. Hewson, J. Phys. Soc. Jpn. **74**(3), 988 (2005)
11. F. Delgado, Y.P. Shim, M. Korkusinski, L. Gaudreau, S.A. Studenikin, A.S. Sachrajda, P. Hawrylak, Phys. Rev. Lett. **101**, 226810 (2008)
12. R. Hanson, L.P. Kouwenhoven, J.R. Petta, S. Tarucha, L.M.K. Vandersypen, Rev. Mod. Phys. **79**, 1217 (2007)
13. I.P. Gimenez, C.Y. Hsieh, M. Korkusinski, P. Hawrylak, Phys. Rev. B **79**, 205311 (2009)
14. M.E. Torio, K. Hallberg, S. Flach, A.E. Miroshnichenko, M. Titov, Eur. Phys. J. B **37**(3), 399 (2004)
15. M.B. Haider, J.L. Pitters, G.A. DiLabio, L. Livadaru, J.Y. Mutus, R.A. Wolkow, Phys. Rev. Lett. **102**, 046805 (2009)
16. C.Y. Hsieh, Y.P. Shim, M. Korkusinski, P. Hawrylak, Rep. Prog. Phys. **75**(11), 114501 (2012)
17. T.A. Baart, N. Jovanovic, C. Reichl, W. Wegscheider, L.M.K. Vandersypen, Appl. Phys. Lett. **109**(4), 043101 (2016)
18. Y. Nisikawa, A. Oguri, Phys. Rev. B **73**, 125108 (2006)
19. I. Ozfidan, A.H. Trojnar, M. Korkusinski, P. Hawrylak, Solid State Commun. **172**, 15 (2013)
20. Z.T. Jiang, Q.Z. Han, Phys. Rev. B **78**, 035307 (2008)
21. A.M. Lobos, A.A. Aligia, Phys. Rev. Lett. **100**, 016803 (2008)
22. R. Landauer, IBM, J. Res. Dev. **32**(3), 306 (1988)
23. J. Kondo, Prog. Theor. Phys. **32**(1), 37 (1964)
24. L.V. Keldysh, Sov. Phys. JETP **20**, 1018 (1965)
25. M.Yu. Kagan, A.G. Aronov, Czech. J. Phys. **46**(5), 2661 (1996)
26. M.Yu. Kagan, V.V. Val'kov, J. Exp. Theor. Phys. **113**(1), 156 (2011)
27. M.Yu. Kagan, V.V. Valkov, Low Temp. Phys. **37**(1), 69 (2011)
28. M.Yu. Kagan, V.V. Val'kov, in *A Lifetime in Magnetism and Superconductivity: a Tribute to Professor David Shoenberg*, ed. by G. Lonzarich, S.S. Saxena, V.A. Sirenko, V.V. Eremenko (Cambridge Scientific Publishers, 2012)
29. M.Yu. Kagan, V.V. Val'kov, J. Supercond. Novel Magn. **25**(5), 1379 (2012)
30. P. Coleman, Phys. Rev. B **35**, 5072 (1987)
31. A.M. Tsvelik, *Quantum Field Theory in Condensed Matter Physics*, 2nd edn. (Cambridge University, 2007)
32. C.M. Varma, Physica B **378-380**(Supplement C), 17 (2006)
33. P. Fulde, *Electron Correlations in Molecules and Solids*, Springer Series in Solid-State Sciences, vol. 100, 3rd edn. (Springer-Verlag, Berlin, Heidelberg, 2002)
34. H. Keiter, N. Grewe, in *Valence Fluctuations in Solids*. ed. by L.M. Falicov, W. Hanke, M.B. Maple (North Holland, Amsterdam, 1981), p.451
35. P.W. Anderson, J. Phys. C Solid State Phys. **3**(12), 2436 (1970)
36. M.Yu. Kagan, K. Kugel, A. Rakhmanov, Phys. Rep. **916**, 1 (2021)
37. K.A. Matveev, A.I. Larkin, Phys. Rev. B **46**, 15337 (1992)
38. P.I. Arseev, N.S. Maslova, J. Exp. Theor. Phys. **75**(3), 575 (1992)
39. P.I. Arseyev, N.S. Maslova, V.N. Mantsevich, JETP Letters **94**(5), 390 (2011)
40. P.I. Arseev, N.S. Maslova, V.N. Mantsevich, J. Exp. Theor. Phys. **115**(1), 141 (2012)
41. H. Feshbach, Ann. Phys. **19**(2), 287 (1962)
42. Z.G. Yu, D.L. Smith, A. Saxena, R.L. Martin, A.R. Bishop, Phys. Rev. B **63**, 085202 (2001)
43. A. Miller, E. Abrahams, Phys. Rev. **120**, 745 (1960)
44. D. Rogovin, D.J. Scalapino, Ann. Phys. **86**(1), 1 (1974)
45. S. Datta, *Electronic Transport in Mesoscopic Systems*, Cambridge Studies in Semiconductor Physics and Microelectronic Engineering (Cambridge University Press, 1995)
46. S. Datta, *Quantum Transport: Atom to Transistor* (Cambridge University Press, 2005)
47. P.I. Arseev, Phys. Uspekhi **58**(12), 1159 (2015)
48. J.M. Foley, S.M. Young, J.D. Phillips, Phys. Rev. B **89**, 165111 (2014)
49. M.L.L.d. Guevara, F. Claro, P.A. Orellana, Phys. Rev. B **67**, 195335 (2003)
50. M. Büttiker, Y. Imry, R. Landauer, S. Pinhas, Phys. Rev. B **31**, 6207 (1985)
51. H. Lu, R. Lü, B.F. Zhu, Phys. Rev. B **71**, 235320 (2005)

52. Y.S. Liu, H. Chen, X.F. Yang, J. Phys. Condens. Matter **19**(24), 246201 (2007)
53. C.W. Hsu, B. Zhen, A.D. Stone, J.D. Joannopoulos, M. Soljačić, Nat. Rev. Mater. **1**(9), 16048 (2016)
54. W. Gong, Y. Han, G. Wei, J. Phys.: Condens. Matter **21**(17), 175801 (2009)
55. A.F. Sadreev, T.V. Babushkina, JETP Lett. **88**(5), 312 (2008)
56. A. Cottet, W. Belzig, Europhys. Lett. **66**(3), 405 (2004)
57. G. Chen, G. Klimeck, S. Datta, G. Chen, W.A. Goddard, Phys. Rev. B **50**, 8035 (1994)

Chapter 12
Space-Separated Fermi–Bose Mixtures in SC Bismuth Oxides (BaKBiO)

Abstract We discuss the superconducting and normal properties in the space-separated Fermi–Bose mixture of paired (bosonic) and normal (fermionic) BiO_6 octahedral clusters in bismuth oxides ($BaKBiO_3$). Note that superconductivity and the long-range order in this system is established by the tunneling of the local preformed pairs from one bosonic cluster to the neighboring one via the insulating fermionic barrier.

12.1 Introduction

In Chaps. 12 and 13, in the framework of the attractive-U Hubbard model with strong random potential [1–3] and Fermi–Bose mixture model [4–7], we analyze the electronic phase separation in superconducting bismuth oxides and in thin disordered films of the dirty metal close to superconductor–localization transition. In both models, we find superconducting (bosonic) droplets or clusters embedded in the fermionic matrix of normal (non-superconducting) metal or insulator. In this chapter, we consider the formation of space-separated Fermi–Bose mixture and superconductivity in the bismuthates, $BaKBiO_3$ or $BaKPbBiO_3$. We demonstrate that the formation of inhomogeneous states in SC bismuth oxides is intrinsically related to the peculiarities of their crystal and electronic structure specified by the presence of the two-well potentials in BiO_6 clusters and valence-skipping considerations in them, which lead to the possibility to form a local electron pair (bipolaron) on the cluster. It is possible to show also that the superconductivity in bismuth oxides is governed by the tunneling of the local electron pair from one SC (bosonic) cluster to the neighboring one via the normal (non-superconducting) fermionic barrier in the two-well structure of the effective ionic potential. In Chap. 13, we analyze disordered thin films of the granular SC on the basis of the 2D Hubbard model in the presence of random potential describing the interaction of electrons with impurities and defects in the dirty film. In the framework of the Bogoliubov–de Gennes approach (BdG), we observe in this model an emergence of inhomogeneous states of spatially separated Fermi–Bose mixture of Cooper pairs and unpaired electrons with the formation of bosonic droplets of different size in the matrix of the unpaired normal states for large

M. Yu. Kagan et al., *Electronic Phase Separation in Magnetic and Superconducting Materials*, Springer Series in Solid-State Sciences 201,
https://doi.org/10.1007/978-3-031-55467-4_12

values of Hubbard attraction and diagonal disorder. We demonstrate also that it is possible to increase the droplet size when we increase the electron density close to the percolation threshold.

In Chap. 14, we describe the tendency toward the phase separation and the possibility of the formation of the Fermi–Bose mixtures and quantum droplets in imbalanced ultracold resonance Fermi gas as well as in strongly dipolar Bose gas in the magnetic, dipolar, or optical traps of different geometry (of different shape) [8–13].

We also discuss briefly the possibility of the formation of the Fermi–Bose mixture (as well as Bose–Bose mixtures of bielectron and biproton Bose condensates [8] and the possibility of the formation of the metallic droplets close to the first order phase boundary in the phase diagram between the metallic liquid and crystalline hydrogen and its metallic alloys.

12.2 Formation of Space-Separated Fermi–Bose Mixture in Bismuthates

In this section, we provide an introduction to the interesting problem of the possible formation of spatially separated Fermi–Bose mixture and unconventional local pairs (bipolaron) type of superconductivity in $Ba(K)BiO_3$ oxide.

Note that the superconductivity in the related (to BaKBiO) compound BaPbBiO was discovered in 1975, that is 11 years earlier [14] than the discovery of the superconductivity in copper oxides. At the same time, the superconductivity in BaKBiO was discovered in 1985 [15] only one year before Müller and Bednorz [16, 17]. Interesting observation is that the critical temperature in $Ba_{1-x}K_xBiO_3$ for the optimum concentration of K, namely, at the concentration $x = 0.4$ is also of the same order of magnitude $T_c = 30$ K as the critical temperature $T_c = 36.5$ K in $La_{2-x}Ba_xCuO_4$ for the optimum hole doping (Ba concentration) $x = 0.15$.

Note that nevertheless the precise nature of superconductivity both in BaPbBiO and in BaKBiO is still not totally clear and is the subject of intensive debate in the literature (see, e.g., [4–6] and references therein) in spite of the fact that most of the researchers are convinced in the formation of local bielectron pairs in these compounds (see, e.g., [18–23]). They are fully aware also in the Varma concept [20] of valence skipping in Bi-based compounds (where a Bi-ion prefers the valences 3+ and 5+, thus skipping the valence 4+).

At the same time, the precise mechanism for the formation of local pairs and the interconnection between the mechanism of superconductivity and local crystalline and electronic structure in BaKBiO is still actively discussed. In this section, we advocate an ideology, according to which the local pair will be formed in BiO_6 cluster [4, 5], which serves as a building block in SC bismuth oxides similar to CuO_6 blocks in cuprates.

Moreover, the formation of the macroscopic wave function of the SC state is governed by the tunneling of local pairs from one (bosonic) BiO_6 cluster to the neighboring one via the normal (fermionic) barrier.

Thus, we can say that the spatially separated Fermi–Bose mixture is formed in these Bi-based compounds, promoting the superconductivity (as well as interesting phase diagram of the normal state) in them.

12.3 Peculiarities of the Local Crystal Structure in $BaKBiO_3$

Let us emphasize that the three-dimensional cubic structure of bismuthates differs from the quasi-two-dimensional structure of the layered high-T_c cuprates. As we already mentioned in Sect. (12.2), the building block in bismuthates is BiO_6 octahedral complex. while in the superconducting metallic CuO layers of high-T_c cuprates, the building block is the elementary plaquette CuO_4.

Note that the similar block CuO_6 in, e.g., $La_{2-x}Ba_xCuO_4$ arises only if we take into account the two additional oxygen ions (so-called apical ions) at LaO or BaO layers of charge reservoir neighboring to the conducting CuO layer.

Let us stress that the elementary plaquette CuO_4 in the metallic CuO layer plays a crucial role in the theory of superconductivity in high-T_c cuprates. Namely, the famous Zhang–Rice singlets [24] are formed on this plaquette. It is precisely the Zhang–Rice construction with copper spin in the surrounding of four oxygen spins, which governs the transition from the two-band scenario advocated by Emery et al. [25] to the effective one band theory of Anderson [26] described by the generalized t-J model and d-wave superconductivity in it [27, 28].

Returning back to bismuthates, note that the octahedral complexes BiO_6 are the most tightly bound items of the structure in BaKBiO compounds corresponding to the strong covalent nature of the Bi6s-O2p_σ chemical bond.

The crystalline data [29] suggest that the crystal structure of the parent compound $BaBiO_3$ corresponds to the alternative arrangement of the expanded and contracted BiO_6 octahedra.

The results of Menushenkov et al. [30–32] suggest that the larger soft octahedron corresponds to the BiO_6 complex with the totally filled Bi6s-O2p orbitals while the smaller rigid octahedron corresponds to the BiL^2O_6 complex, in which L^2 describes the free level in the antibonding Bi6s-O2p_σ orbital (see Fig. 12.1).

It is important to note that according to [31] and Fig. 12.1 the soft larger BiO_6 complex contains 20 electrons and corresponds to the formation of local electron pair on the upper antibonding Bi6s-O2p orbital while the rigid smaller BiL^2O_6 complex contains 18 electrons and the local hole pair is formed in the antibonding Bi6s-O2p_σ orbital there.

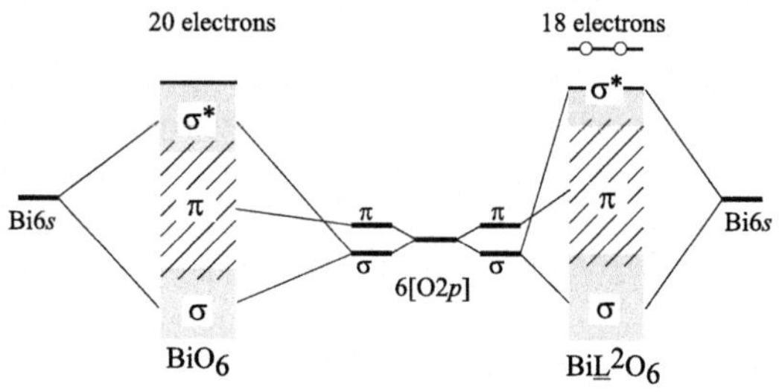

Fig. 12.1 The scheme of the formation of the electron structure for the different octahedra BiO_6 complexes [31].

The different electron fillings of the upper antibonding Bi6*s*-O2*p* orbital of the neighboring rigid and soft octahedra allow for the tunneling of the local electron pairs between them in the picture of the dynamic exchange $BiL^2O_6 \leftrightarrow BiO_6$.

The low temperature EXAFS results of [31] establish that the tunneling of local electron pairs is correlated with the vibrations (referred to as "the breathing distortions" in the barium lattice) of the corner-sharing oxygen ions between the neighboring octahedra.

It is interesting to note that if we consider two neighboring equivalent rigid BiL^2O_6 octahedra, then only standard harmonic potentials for the oscillations of the corner-sharing oxygen ions appear in the system (see Fig. 12.2). At the same time, if we have the tunneling of electron pairs between the nonequivalent rigid BiL^2O_6 and soft BiO_6 neighboring clusters, then the corner-sharing oxygen ions vibrate in the double-well potentials.

It is important to emphasize again that the peculiarities and distortions of the local lattice structure, which violate translational symmetry and demonstrate significant differences with the average one manifest themselves only in the local sensitive probes such as XPS [33], Raman scattering [34], and EXAFS [30, 31, 35, 36].

Note that the doping with K of the parent compound $BaBiO_3$ is equivalent to the hole doping and leads to the partial replacement of the smaller rigid BiL^2O_6 octahedra by the larger soft BiO_6 octahedra [31]. This fact leads to the decrease and to the eventual disappearance of the static breathing and tilting distortions.

As a result, the lattice contracts despite almost equal ionic radii of K^+ and Ba^{2+} and the average crystal structure becomes a simple cubic one at the optimum doping level $x = 0.37$ in $Ba_{1-x}K_xBiO_3$ in accordance with the sample averaged (integral) neutron diffraction data [37].

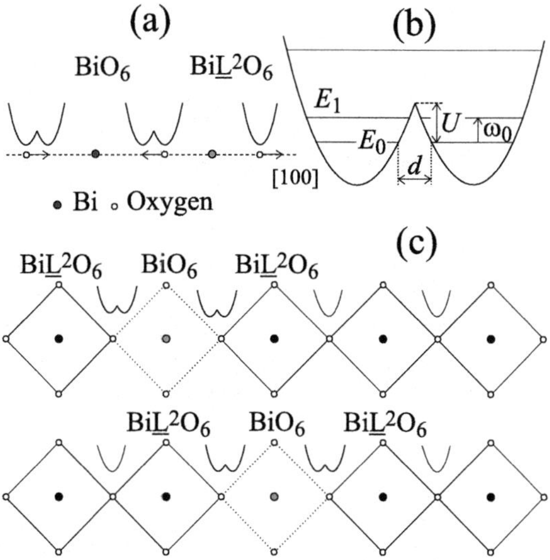

Fig. 12.2 A sketch of the dynamic exchange $BiL^2O6 \leftrightarrow BiO6$ in the BiO_2 plane. In panel a, we show the breathing mode of the vibrations along the [100]-type of direction for two neighboring octahedra with different electron structures. The soft BiO6 octahedra transforms to the rigid BiL^2O6 ones and vice versa due to the electron pair tunneling between them. The corner-sharing oxygen ions at the border between different (soft and rigid) octahedra vibrate in the double-well potential. The corner-sharing oxygen ions between two equivalent neighboring rigid BiL^2O6 octahedra oscillate in a simple parabolic potential. In panel b, we present important characteristics of the double-well potential for $Ba_{0.6}K_{0.4}BiO_3$ (see [32]): the tunneling frequency $\omega_0 \approx 200$K, the barrier height $U = 500$K, and the barrier width $d \approx 0.07$Å. In panel c, we show the tunneling of the electron pair centered on the BiO_6 complex through the BiL^2O6 Fermi cluster [4]

12.4 Local Electronic Structure in SC Bismuth Oxides

12.4.1 Local Electronic Structure of the Parent Compound $BaBiO_3$

The formation of the two types of rigid and soft octahedra in the parent compound $BaBiO_3$ with different lengths and strengths of the Bi–O bonds reflects the different electron structure and fillings of the BiO_6 and BiL^2O_6 complexes already at zero K doping.

Let us stress once more that the valence band of the parent compound $BaBiO_3$ is formed by the overlap of the Bi6s and O2p orbitals [14, 38]. Moreover, due to strong Bi6s–O2p_σ hybridization, both soft and rigid octahedra can be considered as quasimolecular complexes [39]. Note that each complex has ten electron levels, which represent four bonding–antibonding Bi6s–O2p_σ orbitals and six nonbonding O2p_σ orbitals.

Correspondingly, a unit cell (which is monoclinic in the parent compound) contains two octahedra and 38 valence electrons: ten from two Bi ions, four from two Ba ions, and 24 from six O ions. It seems from the first sight that all the lengths of Bi–O bonds should be equal to each other and local magnetic moments should be formed for the equal electron filling (in the configuration BiL^1O6-BiL^1O6) of the neighboring octahedra.

However, experiments [18, 40] reveal an opposite picture, namely the presence of two types of soft and rigid octahedra and the absence of any magnetic moments respectively. Trying to match these observations, Menushenkov et al. [31] suggested the following scheme of valence disproportionation in the parent compounds

$$2BiL^1O6 \rightarrow BiO6 + BiL^2O6\,. \tag{12.1}$$

In this scheme, the number of occupied sites is different in two neighboring octahedra. Moreover, the rigid BiL^2O_6 octahedron contains 18 electrons and has one free level (or a local hole pair L^2) in the upper antibonding orbital Bi6s–O2p_σ, while the soft BiO_6 octahedron contains 20 electrons and the upper antibonding orbital is filled in this complex (see Fig. 12.1).

The interplay between local crystal and electronic structure in the parent compound $BaBiO_3$ is presented in Fig. 12.3a. Note that free electronic carriers are absent for $x = 0$ and the conductivity is only due to the tunneling of local pairs in the system [18, 32]. Note also that experimentally the parent compound $BaBiO_3$ exhibits a semiconducting behavior with the activation energy gap $E_a = 0.24$ eV. It is natural

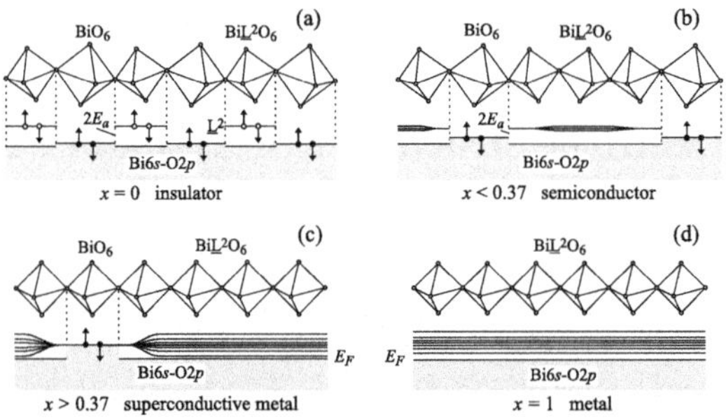

Fig. 12.3 Scheme of the insulator–metal phase transition in $Ba_{1-x}K_xBiO_3$ under K doping x due to the interplay between the local crystal and local electronic structure. In panel a, we show the monoclinic phase of the $BaBiO_3$ insulating parent compound in the absence of doping (for $x = 0$). In panel b, we present the orthorhombic phase of semiconducting $Ba_{1-x}K_xBiO_3$ for the doping levels $0 < x < 0.37$. We show also a sketch for the splitting of the free level L^2 at the spatial overlap of the rigid complexes of BiL^2O6. In panel c, we show the undistorted cubic phase of the SC metal for $x > 0.37$. Here, we sketch the formation of the Fermi-liquid state due to the overlap of the unoccupied fermionic band F with the occupied O2p_π valence band as the percolation threshold is reached. In panel d, we present the undistorted cubic phase of a non-superconducting metal at $x = 1$. Here, we depict the Fermi-liquid state and show the Fermi level E_F [4]

to assume that this gap is related to the two-particle transport with the activation energy $2E_a$ due to the delocalization (and percolation) of the local pairs.

At the same time, the transport gap E_a does not manifest itself in the photoconductivity, optical absorption, or photoacoustic measurements [41], where another, optical gap plays the dominant role. In agreement with Varma's suggestion [20], the optical gap is governed by the relation

$$E_g = 2E_a + E_b = 2E(BiL^1O6) - [E(BiO6) + E(BiL^2O6)], \tag{12.2}$$

where E_b is the binding energy of the pair related to the dissociation of local pairs. The optical gap E_g can be measured experimentally and corresponds to the peak in the optical conductivity at the photon energy $h\nu = 1.9$ eV [18].

Note that according to Eq. (12.2), the optical excitation transforms two different octahedra to equivalent ones

$$BiO6 + BiL^2O6 \stackrel{h\nu}{\rightarrow} 2(BiL^1O6) \tag{12.3}$$

and thus creates a local lattice distortion.

These local deformations (which are in fact dynamic ones) can be seen in the Raman spectra as an extremely large amplitude of the breathing type vibrations of oxygen octahedra when the laser frequency is in resonance with the optical gap $h\nu = E_g$ [22, 40, 42]. At the same time, when we leave the resonant situation, the amplitude of vibrations is strongly reduced [42].

Let us stress again that in the absence of laser radiation, there are no free fermionic charge carriers in the parent compound. The excited one-particle fermionic excitations can arise in the system only due to the destruction of local pairs, which have a large binding energy $E_b \gg k_B T$, and thus the fermionic contribution to the static (zero-frequency) charge transport is negligibly small for all temperatures.

We can say that in the parent compound, the bosonic and fermionic subsystems are separated both in space (belong to different octahedral complexes) and in energy (separated from each other by different energy gaps). That is why, the Fermi–Bose mixture cannot be formed in $BaBiO_3$ in the absence of doping. We illustrate this situation in Fig. 12.4a, where we show separately the bosonic and fermionic contributions to the one-particle density of states. In the absence of doping ($x = 0$) the filled fermionic band in Fig. 12.4a is separated from the empty fermionic band (namely, from the excited band F′) by large (optical) energy gap E_g and from the empty bosonic band B by a smaller transport gap $2E_a$.

Note that according to the ideas of [43] concerning the bosonic metal phase, the bosonic band effectively plays the role of a conduction band for normal (non-superconducting) transport of local pairs. Moreover, in [43], in the framework of simple attractive-U Hubbard model it was shown that the filled and empty bosonic bands typically have different (electron-like and hole-like) dispersions, if we project the two-particle physics on the one-particle representation used for the density of states.

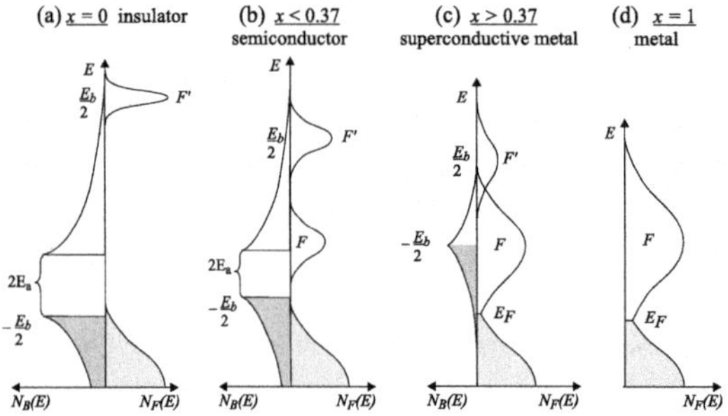

Fig. 12.4 Sketch of the one-particle density of states for $Ba_{1-x}K_xBiO_3$ at different doping levels x. The contributions of the bosonic $N_B(E)$ and fermionic $N_F(E)$ densities of states are shown separately since they are space separated. The filled (blue) and the unoccupied (colorless) bosonic bands correspond respectively to the contributions of the electron pairs and hole pairs. The bands are separated by the activation (transport) gap $2E_a$. The filled fermionic $O2p_\pi$ valence band is highlighted by yellow color. In panel a, we show the insulating phase of the parent compound $BaBiO_3$ in the absence of doping ($x = 0$).The empty fermionic band F′ corresponding to the destruction of pairs is separated from the occupied bosonic band by the optical gap E_g. In panel b, we show semiconducting behavior of $Ba_{1-x}K_xBiO_3$ for the doping levels $0 < x < 0.37$. The transport gap $2E_a$ decreases with the increase in the doping level x. An empty fermionic band F is formed due to the splitting of the L^2 levels caused by the spatial overlap of BiL^2O6 complexes. The fermionic band F grows with the doping due to the increase of free levels. In panel c, we demonstrate the formation of SC metal for $x > 0.37$. The Fermi-liquid metallic state is formed here as a result of the overlap between the empty fermionic band F and the filled $O2p_\pi$ fermionic valence band. The transport gap between bosonic band is absent as well ($E_a = 0$). The Fermi–Bose mixture is formed. In panel d, we depict the density of states in normal (non-superconducting) metal $KBiO_3$ (at $x = 1$). Here, the bosonic subsystem (and bosonic density of states) totally disappears [4]

12.4.2 Evolution of the Electronic Structure of $Ba_{1-x}K_xBiO_3$ Upon Doping ($x \neq 0$)

In this subsection, we study the evolution of the electronic structure of $Ba_{1-x}K_xBiO_3$ with doping level x and the formation of the Fermi–Bose mixture in the system, which is separated in space, but not in energy. Note that the K doping of $BaBiO_3$ is equivalent to a hole doping [4, 5]. Effectively it leads to a partial replacement of larger soft octahedral BiO6 complexes by smaller rigid complexes BiL^2O6 (see Fig. 12.5).

Thus, the number of BiL^2O6 complexes increases and becomes larger than the number of BiO6 ones. For the doping level x the number of BiL^2O6 octahedra is given by $n_0(1 + x)/2$, the number of BiO_6 complexes reads $n_0(1 - x)/2$. In these expressions, $n_0 = 1/a^3$ is the number of unit cells and a is the lattice parameter.

The increase in the number of the smaller complexes BiL^2O6 with rigid bonds causes the decrease and finally the disappearance of static breathing and tilting distortions. As a result, the lattice should contract despite almost equal ionic radii of K^+ and Ba^{2+}. Correspondingly, the average crystal structure becomes a simple cubic

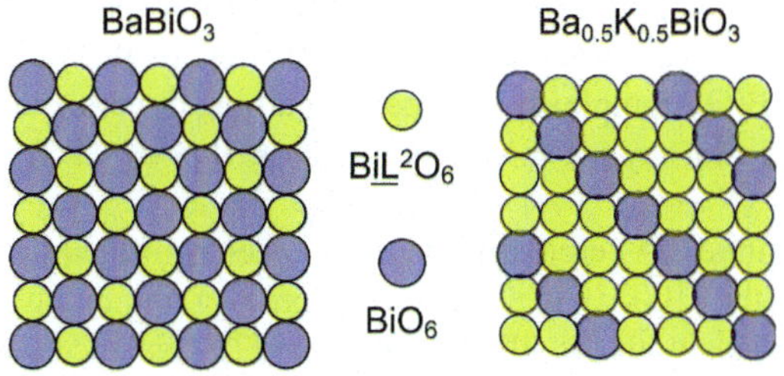

Fig. 12.5 Scheme of the local crystal structure in BiO_2 plane of the parent compound $BaBiO_3$, which is the CDW insulator with the checkerboard distribution of BiL^2O6 and BiO6 octahedra (left) and local structure of $Ba_{0.5}K_{0.5}BiO_3$ with the diagonal chains of BiO6 complexes embedded in the large percolating cluster of BiL^2O6 octahedra (right) [5]

one at the optimum doping level $x = 0.37$ in $Ba_{1-x}K_xBiO_3$ in accordance with the neutron diffraction data [37].

At the same time, the evolution of the local crystal structure within the $0 < x < 0.5$ range shown in the upper panel of Fig. 12.2 manifests the tendencies to form large diagonal chains of soft BiO_6 complexes embedded in the matrix of rigid BiL^2O6 octahedra.

These essential changes in the local crystal structure are accompanied by the important changes in the local electronic structure. Namely, a spatial overlap of the L^2 levels leads to their splitting into the empty fermionic band F due to the spatial overlap of rigid BiL^2O6 complexes (see Fig. 12.2b). For the doping levels $x < 0.37$, this band is rather narrow and is still separated from the filled $O2p_\pi$ valence band in Fig. 12.4b.

Note that the total number of empty states in the F band is governed by the relation

$$n_F = n_0(1 + x)\,. \tag{12.4}$$

At the same time, the number of local electron pairs decreases as

$$n_B = n_0(1 - x)/2\,. \tag{12.5}$$

Though the activation (transport) gap $2E_a$ decreases with the increase of doping level x, it is still nonzero for $x < 0.37$ due to the intersite Coulomb repulsion V. It is screened inside bosonic clusters but is still effective between neighboring bosonic clusters.

As a result, $Ba_{1-x}K_xBiO_3$ compounds at the doping levels $x < 0.37$ demonstrate a semiconducting-type conductivity, which changes from a simple exponential (activation-type) behavior to the Mott variable-range hoping [44].

Let us stress here that there are essential changes in the local crystal and local electronic structures of BKBO at the doping level $x = 0.37$ (see Figs. 12.3c and 12.4c).

First, the static breathing and rotational lattice distortions transform to dynamic ones. Moreover, for the corner-sharing oxygen ions at the border between rigid BiL^2O6 and soft BiO_6 complexes, the local breathing dynamic distortion manifests itself as a vibration of oxygen ions in a double-well potential (see Fig. 12.2, lower panel).

Second, an infinite 3D percolating Fermi cluster is formed by the rigid BiL^2O6 complexes along the [100]-type directions. Moreover, the empty fermionic band F in Fig. 12.4c overlaps with the filled fermionic $O2p_\pi$ valence band. As a result, a conduction (partially filled) fermionic band is formed. Hence, the insulator–metal phase transition takes place in the system and the Fermi-liquid metallic state is formed in it, when we overcome the percolation threshold for $x > 0.37$.

Third, the activation energy (the transport gap) $E_a = 0$ vanishes and hence, electron pairs (which are formed at BiO_6 complexes) can become delocalized and move freely, thus providing a bosonic contribution to the conductivity. Correspondingly, two types of charge carriers are present in the metallic phase: the conduction electrons of the partially filled fermionic band (formed by the percolation cluster of BiL^2O6 complexes) and delocalized bosonic band of electron pairs (from BiO6 octahedra).

It occurs that a normal state conductivity in BKBO is mainly due to the fermionic contribution. At the same time, the bosonic contribution was also observed in the experiments of Hellman and Hartford [45], where it manifests itself as a two-particle normal state tunneling.

Thus, we can say that at the doping levels $x \geq 0.37$, we have a new interesting state of matter namely the space-separated Fermi–Bose mixture, which can be explained in the unified manner both normal and SC properties of BKBO. Let us stress that since fermions and bosons belong to different complexes, the Fermi and Bose subsystems are spatially separated at any doping level. Nevertheless, we can write the simple sum rule for our system which represents the equation for the total density in the mixture

$$n_F + 2n_B = 2n_0 . \tag{12.6}$$

Correspondingly, we have the convenient ratio of fermionic and bosonic densities

$$\frac{2n_B}{n_F} = \frac{1-x}{1+x} . \tag{12.7}$$

Note that the destruction of pairs in bosonic clusters is prevented by a rather high value of the binding energy $E_b = E_g = 0.5$ eV (for $E_a = 0$) [46] in the SC composition for $x > 0.37$. The unpairing processes (which leads to the partial filling of the empty fermionic band F′ in Fig. 12.4c) can be important only in the optical experiments. At the same time, they do not play any role in the zero-frequency charge transport.

Finally, for $KBiO_3$ (at $x = 1$) all the BiO_6 octahedra are transformed to the BiL^2O_6 ones. The bosonic subsystem in Figs. 12.3d and 12.4d totally disappears together with the excited fermionic band F′. The bosonic density vanishes ($n_B = 0$), as well as the bosonic density of states, $N_B(E) = 0$. Hence, $KBiO_3$ corresponds to non-superconducting (normal) fermionic metal described by the standard Fermi-liquid theory.

Let us emphasize here that metallic $KBiO_3$ exists only hypothetically since the potassium solubility limit $x = 0.5$ is exceeded in this composition. However, its electron analog $PbBiO_3$ is experimentally available and exhibits also the non-superconducting metallic properties.

Note that experimental efforts to synthesize $KBiO_3$ at high pressure were pursued in [47]. Unfortunately, these efforts demonstrated only partial replacement of the K^+ ions by Bi^{3+} ones. As a result, the $K_{1-y}Bi_yBiO_3$ composition was obtained [47].

Let us stress that this type of partial replacement (instead of the full one) gives rise to some number of BiO_6 octahedra in the system (in addition to BiL^2O_6 ones) and to the formation of local electron pairs at them. Therefore, we can assume that the $K_{1-y}Bi_yBiO_3$ composition can be described as well in terms of the space-separated Fermi–Bose mixture and becomes superconducting again. This assumption is confirmed by the experimental results of Ref. [47], where the SC transition at the critical temperature $T_c = 10.2$ K was observed.

From our considerations, it also follows that the $BaPbO_3$ must be superconducting as well if we partially substitute the Ba^{2+} divalent ions by some trivalent ions (e.g., La^{3+}) since this substitution should produce local electron pairs as in the case of bismuth materials, such as $Ba_{1-x}K_xBiO_3$ and $K_{1-y}Bi_yBiO_3$. Indeed, the novel compound $Ba_{1-y}La_yPbO_3$ synthesized by Menushenkov et al. [48] has demonstrated the SC transition at $T_c = 11$ K.

12.5 Superconductivity in BaKBiO

In the previous sections, we advocated the idea of the space-separated Fermi–Bose mixture to explain the interplay of the local crystal and electron structures and the basic transport properties in the normal state of $Ba_{1-x}K_xBiO_3$ bismuthates, as well as in the other related compounds. In this section, we consider superconductivity in the metallic $BaKBiO_3$ within the $0.37 < x < 0.5$ doping range.

As we briefly mentioned already in Sect. 12.2 the existence of the double-well potentials allows us to consider superconductivity in $Ba_{1-x}K_xBiO_3$ as a formation of the macroscopic wave function and the long-range order in the system of the compact bosonic pairs.

The long-range order is established as a result of the tunneling of the compact electron pairs from one bosonic cluster to a nearest one through the potential barrier organized by the fermionic cluster (located between the two bosonic ones). The delocalization and percolation of the pairs from the left bank of the sample to the right bank in a sequence of the tunneling processes of this type helps us to form the coherence and the macroscopic wave function in the system.

Let us start our analysis of the superconductivity in $Ba_{1-x}K_xBiO_3$ with an estimate of the matrix element for the pair tunneling in the double-well potential presented in Fig. 12.2b. The matrix element is given by

$$t_B \propto \omega_0 \exp(-D), \tag{12.8}$$

where $\omega_0 = 200$ K is the tunneling frequency and

$$D = \frac{1}{\hbar}\int_{x_0}^{x_1} |p|dx \approx \frac{d}{\hbar}\sqrt{2MU}, \tag{12.9}$$

is the coefficient of the quasiclassical transparency of the barrier in the double-well potential. In Eq. (12.9), the parameters $U = 500$ K and $d \approx 0.07$ Åare the barrier height and width, and M is the oxygen ion mass. We remind here of the interconnection of the pair and oxygen tunneling processes illustrated in Fig. 12.2. Thus, a large oxygen mass enters in the estimate for the transparency coefficient D in Eq. (12.9). Note also that the tunneling frequency ω_0 in Eq. (12.8) already incorporates all the polaronic effects in bismuthates, and hence it is relatively small.

Effectively for the concentration range $0.37 < x < 0.5$, a local electron pair centered at soft BiO_6 complexes is transferred from one bosonic cluster to the nearest one via the fermionic cluster, which acts as a potential barrier and depending upon the doping level x, contains several rigid BiL^2O_6 octahedra.

It is natural to assume that the pairs overcome the fermionic cluster step by step. Each step corresponding to the transfer of a pair into a neighboring octahedron is accompanied by the tunneling of oxygen ions in the double-well potential.

The tunneling frequency ω_0 is thus the same for each step. We can treat each step as an independent event. Therefore, the probability to overcome the fermionic cluster as a whole can be estimated as a product of the probabilities of each step. Then, the resulting matrix element for the pair tunneling via the fermionic cluster, which in average contains $\langle N\rangle$ rigid BiL^2O_6 octahedra is given by

$$t_B^{eff} \propto \omega_0 \exp(-\langle N\rangle D), \tag{12.10}$$

where $\langle N\rangle$, in other words, is the average number of steps. Note that effectively $\langle N\rangle$ is proportional to the linear size of the fermionic cluster.

We can estimate $\langle N\rangle$ utilizing the ratio of concentrations of the fermionic BiL^2O_6 and bosonic BiO_6 octahedra given by the inverse proportion with respect to the one in Eq. (12.7). As a result, in 3D bismuthates, the linear size of the fermionic cluster reads

$$\langle N\rangle = \left(\frac{1+x}{1-x}\right)^{1/3}. \tag{12.11}$$

Relations (12.10) and (12.11) allow us to get a reasonable estimate for the superconducting critical temperature. Namely, we can identify the onset of superconductivity with the temperature of the Bose–Einstein condensation (BEC) in the bosonic subsystem. The BEC critical temperature is given by

$$T_c \propto \frac{n_B^{2/3}}{m_B^{eff}} \propto t_B^{eff} a^2 n_B^{2/3}, \tag{12.12}$$

where we introduced a heavy bosonic effective mass

$$m_B^{eff} \propto \frac{1}{t_B^{eff} a^2}, \tag{12.13}$$

which corresponds to the sequence of the pair tunneling processes.

We should keep in mind also that the bosonic density n_B in Eq. (12.12) for T_c is governed by the relation

$$a^3 n_B = \frac{n_B}{n_0} = (1-x)/2, \tag{12.14}$$

which is equivalent to Eq. (12.5).

Thus, the critical temperature can be written as

$$T_c \propto t_B^{eff} \left(\frac{1-x}{2}\right)^{2/3}, \tag{12.15}$$

where the effective matrix element is given by Eqs. (12.10) and (12.11).

Now, we can substitute in this expression the parameters of the double-well potential obtained in [31] by Menushenkov et al. and obtain $T_c \approx 50$ K for the doping level $x = 0.4$, which overestimates the experimentally measured $T_c \approx 30$ K in the $Ba_{1-x=0.6}K_{x=0.4}BiO_3$ compound. This difference can be qualitatively understood in the following manner: our simple estimates in Eqs. (12.10), (12.11), and (12.15) do not take into account the adjustment effects related to the formation of the phase coherence in the system, which effectively reduces T_c. These effects (resembling to some extent the physics of EPE considered in Chap. 10 and Anderson's orthogonality catastrophe [49, 50]) reduce the matrix element t_B^{eff}, which in a more rigorous treatment is proportional to the scalar product of many-particle wave functions of the system before and after each tunneling event.

The adjustment effects are governed by the parameter $\omega_0\tau$, where ω_0 is the tunneling frequency of the pair and τ is the relaxation time for the vibration modes of oxygen ions. In other words, we have to compare the tunneling time for the pair transfer from one octahedron to the neighboring one and the effective relaxation time of the phonon subsystem due to the soft phonon modes. Note that in the case of bismuthates, longitudinal phonons are the most effective ones.

Each time the soft phonon modes help to form a new elastic (deformational) configuration before the next tunneling event takes place. Thus, the many-particle wave function of the system (which includes both bosonic subsystem of electron pairs and elastic background due to soft phonon modes) can be changed considerably during the tunneling time, thus reducing the effective matrix element t_B^{eff} and T_c.

12.6 Summary of the Results

Summarizing this chapter, let us stress again that we predict a new type of the spatially separated Fermi–Bose mixture in $Ba_{1-x}K_xBiO_3$ at the doping level $x \geq 0.37$. The concentration $x = 0.37$ serves as a percolation threshold, and the infinite percolation cluster in the real space is formed here within the fermionic subsystem (see Fig. 12.5), leading to the insulator–metal phase transition.

In the energy space, the percolation phase transition manifests itself in the overlap of the empty fermionic band F with the occupied Bi6s-O2p valence band as shown in Fig. 12.4c as soon as the percolation threshold is achieved.

Thus, we conclude that in this region, we effectively have two bands in the system: fermionic and bosonic ones. The bosonic band is responsible for superconductivity at low temperatures, while the fermionic band governs the formation of the Fermi-liquid state in the system.

At higher temperatures, $T > T_c$, the fermionic contribution $\sigma_F(T)$ to the resulting normal state conductivity $\sigma(T) = \sigma_B(T) + \sigma_F(T)$ of the Fermi–Bose mixture strongly bridges the bosonic contribution $\sigma_B(T)$, which arises due to the two-particle transport.

It would also be very interesting to carry out more precise measurements of the electronic contribution to the specific heat $C_V(T)$ of $Ba_{1-x}K_xBiO_3$ at temperatures $T \sim T_c$ to reveal the bosonic and fermionic contributions to the thermodynamics of the normal state at $x \geq 0.37$.

Note that in a very simple approach, the specific heat of 3D almost ideal Bose gas behaves as $C_B(T) \propto (T/T_c)^{3/2}$ at $T < T_c$. At the same time, $C = const$ for $T \gg T_c$. Hence, we have a typical λ-point behavior of the specific heat for temperatures $T \sim T_c$. However, in the Fermi–Bose mixture, there exists an additional fermionic contribution to the specific heat, which is basically linear in temperature, $C_F \propto \gamma T$.

This contribution can in principle destroy the λ-point behavior of the specific heat for temperatures $T \sim T_c$. Note that the available measurements of the specific heat mostly yield the smooth behavior of the specific heat near T_c, since in all of them, the larger lattice contribution is indistinguishable from the smaller electronic one.

References

1. M.Yu. Kagan, E.A. Mazur, J. Exp. Theor. Phys. **132**, 596 (2021)
2. M.Yu. Kagan, E.A. Mazur, J. Phys. Conf. Ser. **1740**, 012045 (2021)
3. E.A. Mazur, R.S. Ikhsanov, M.Yu. Kagan, J. Phys. Conf. Ser. **2036**(1), 012019 (2021)
4. A.P. Menushenkov, K.V. Klementev, A.V. Kuznetsov, M.Yu. Kagan, J. Exp. Theor. Phys. **93**(3), 615 (2001)
5. A.P. Menushenkov, A.V. Kuznetsov, K.V. Klementiev, M.Yu. Kagan, J. Supercond. Novel Magn. **29**(3), 701 (2016)
6. P. Giraldo-Gallo, Y. Zhang, C. Parra, H. Manoharan, M. Beasley, T. Geballe, M. Kramer, I. Fisher, Nat. Commun. **6**(1), 8231 (2015)
7. M.Yu. Kagan, A. Bianconi, Condens. Matter **4**(2) (2019)

8. M.Yu. Kagan, JETP Lett. **103**(11), 728 (2016)
9. A.V. Turlapov, M.Yu. Kagan, J. Phys. Condens. Matter **29**(38), 383004 (2017)
10. M.Yu. Kagan, A.V. Turlapov, Phys. Usp. **62**(3), 215 (2019)
11. E. Burovski, R.S. Ikhsanov, A. Kuznetsov, M.Yu. Kagan, J. Phys. Conf. Ser. **1163**, 012046 (2019)
12. D.V. Efremov, L. Viverit, Phys. Rev. B **65**, 134519 (2002)
13. I. Ferrier-Barbut, H. Kadau, M. Schmitt, M. Wenzel, T. Pfau, Phys. Rev. Lett. **116**, 215301 (2016)
14. A. Sleight, J. Gillson, P. Bierstedt, Solid State Commun. **17**(1), 27 (1975)
15. R.J. Cava, J.J. Batlogg, B. Krajewski, R. Farrow, L.W. Rupp, A.E. White, K. Short, W.F. Peck, T. Kometani, Nature **332**(6167), 814 (1985)
16. J.G. Bednorz, K.A. Müller, Zs. Phys. B **64**(2), 189 (1996)
17. J.G. Bednorz, K.A. Müller, Rev. Mod. Phys. **60**(3), 585 (1988)
18. S. Uchida, K. Kitazawa, S. Tanaka, Phase Transit. **8**(2), 95 (1987)
19. T.M. Rice, L. Sneddon, Phys. Rev. Lett. **47**, 689 (1981)
20. C.M. Varma, Phys. Rev. Lett. **61**, 2713 (1988)
21. L. De Jongh, Phys. C **152**(2), 171 (1988)
22. S. Sugai, Solid State Commun. **72**(12), 1187 (1989)
23. J. Yu, X.Y. Chen, W.P. Su, Phys. Rev. B **41**, 344 (1990)
24. F.C. Zhang, T.M. Rice, Phys. Rev. B **37**, 3759 (1988)
25. P. Unger, P. Fulde, Phys. Rev. B **47**, 8947 (1993)
26. P.W. Anderson, in *Frontiers and Borderlines in Many-particle Physics: Varenna on Lake Como, Villa Monastero, 7–17 July 1987* (1988)
27. H.Q. Lin, Phys. Rev. B **44**, 4674 (1991)
28. M.Yu. Kagan, T.M. Rice, J. Phys. Condens. Matter **6**(20), 3771 (1994)
29. D.E. Cox, A.W. Sleight, Acta Cryst. B **35**(1), 1 (1979)
30. A.P. Menushenkov, K.V. Klement'ev, P.V. Konarev, A.A. Meshkov, JETP Lett. **107**(8), 493 (2018)
31. A.P. Menushenkov, K.V. Klementev, J. Phys. Condens. Matter **12**(16), 3767 (2000)
32. A.P. Menushenkov, Nucl. Instrum. Methods Phys. Res. Sect. A **405**(2), 365 (1998)
33. M. Qvarford, V.G. Nazin, A.A. Zakharov, M.N. Mikheeva, J.N. Andersen, M.K.J. Johansson, G. Chiaia, T. Rogelet, S. Söderholm, O. Tjernberg, H. Nylén, I. Lindau, R. Nyholm, U.O. Karlsson, S.N. Barilo, S.V. Shiryaev, Phys. Rev. B **54**, 6700 (1996)
34. N. Anshukova, A. Golovashkin, V. Gorelik, L. Ivanova, K. Mitsen, A. Rusakov, R. Khashimov, J. Mol. Struct. **219**, 147 (1990)
35. S. Salem-Sugui, Jr., E.E. Alp, S.M. Mini, M. Ramanathan, J.C. Campuzano, G. Jennings, M. Faiz, S. Pei, B. Dabrowski, Y. Zheng, D.R. Richards, D.G. Hinks, Phys. Rev. B **43**, 5511 (1991)
36. Y. Yacoby, S.M. Heald, E.A. Stern, Solid State Communications **101**(11), 801 (1997)
37. S. Pei, J.D. Jorgensen, B. Dabrowski, D.G. Hinks, D.R. Richards, A.W. Mitchell, J.M. Newsam, S.K. Sinha, D. Vaknin, A.J. Jacobson, Phys. Rev. B **41**, 4126 (1990)
38. L.F. Mattheiss, D.R. Hamann, Phys. Rev. B **28**, 4227 (1983)
39. S. Sugai, Jpn. J. Appl. Phys. **26**(S3–2), 1123 (1987)
40. Y.J. Uemura, B.J. Sternlieb, D.E. Cox, J.H. Brewer, R. Kadono, J.R. Kempton, R.F. Kiefl, S.R. Kreitzman, G.M. Luke, P. Mulhern, T. Riseman, D.L. Williams, W.J. Kossler, X.H. Yu, C.E. Stronach, M.A. Subramanian, J. Gopalakrishnan, Nat. **335**(6186), 151 (1988)
41. S. Tajima, S. Uchida, A. Masaki, H. Takagi, K. Kitazawa, S. Tanaka, A. Katsui, Phys. Rev. B **32**, 6302 (1985)
42. S. Tajima, M. Yoshida, N. Koshizuka, H. Sato, S. Uchida, Phys. Rev. B **46**, 1232 (1992)
43. M.Yu. Kagan, R. Frésard, M. Capezzali, H. Beck, Phys. Rev. B **57**, 5995 (1998)
44. E.S. Hellman, B. Miller, J.M. Rosamilia, E.H. Hartford, K.W. Baldwin, Phys. Rev. B **44**, 9719 (1991)
45. E.S. Hellman, E.H. Hartford, Phys. Rev. B **52**, 6822 (1995)
46. S.H. Blanton, R.T. Collins, K.H. Kelleher, L.D. Rotter, Z. Schlesinger, D.G. Hinks, Y. Zheng, Phys. Rev. B **47**, 996 (1993)

47. N. Khasanova, A. Yamamoto, S. Tajima, X.J. Wu, K. Tanabe, Phys. C **305**(3), 275 (1998)
48. A.P. Menushenkov, A.V. Tsvyashchenko, D.V. Eremenko, K.V. Klementev, A.V. Kuznetsov, V.N. Trofimov, L.N. Fomichev, Phys. Solid State **43**(4), 613 (2001)
49. P.W. Anderson, Phys. Rev. Lett. **18**, 1049 (1967)
50. P.W. Anderson, Phys. Rev. **164**, 352 (1967)

Chapter 13
Droplets of the Order Parameter and Superconductivity in a Low-Density Attracting Fermion System in the Presence of a Strong Random Potential

Abstract In the framework of the 2D attractive-U Hubbard model at low electron density and in the presence of strong diagonal disorder, we analyze the possibility of the formation of the paired (order parameter) droplets in the matrix of the unpaired (normal) states. We investigate also the nature of the percolative superconductor–insulator phase transition in this system when droplets are growing with the increase in electron density forming a large percolative cluster.

13.1 Introduction. Droplets of the Order Parameter and Formation of the Bosonic Metal

In this chapter, for an s-wave disordered superconductor, we consider the 2D low density ($n \ll 1$) attractive-U Hubbard model with strong onsite Hubbard attraction $|U| \gg W$ ($W = 8t$ being the bandwidth for the square lattice) and a strong random potential $V \gg W$ uniformly distributed in the range from $-V$ to $+V$.

The obtained results are important for the construction of the gross phase diagram and understanding the nature of the phase transition between superconducting and localized states in quasi-2D (thin) films of a dirty metal which is important for SC nanoelectronics and quantum information (construction of SC qubits—so-called SC flux qubits) [1].

Note that a "direct" transition from a superconducting state to an insulating state is possible in the 2D systems. In our model, at low electron densities, large Hubbard attraction, and large disorder, an interesting state of bosonic metal [2] (or bosonic insulator) can be realized in the framework of spatially separated Fermi–Bose mixture with paired bosonic states in the matrix of unpaired fermionic states.

Our results show that when we increase the density (at fixed values of Hubbard attraction and disorder) at first the transition from single electron pairs to larger droplets (containing larger number of pairs) takes place. Then, finally, a large percolating network of the paired chains is formed in our system.

M. Yu. Kagan et al., *Electronic Phase Separation in Magnetic and Superconducting Materials*, Springer Series in Solid-State Sciences 201,
https://doi.org/10.1007/978-3-031-55467-4_13

13.2 Model and Basic Equations

The 2D Hubbard model with the short-range (onsite) attraction between electrons and diagonal disorder has the following form on the square lattice

$$\tilde{H} = H - \mu N = -t \sum_{i,j,\sigma} c^{\dagger}_{i\sigma} c_{j\sigma} - |U| \sum_{i} n_{i\uparrow} n_{i\downarrow} + \sum_{i,\sigma} (V_i - \mu) n_{i\sigma} \,, \tag{13.1}$$

where $|U|$ is the absolute value of the Hubbard onsite attraction, t is the nearest-neighbor hopping, $c^{\dagger}_{i\sigma}$ and $c_{i\sigma}$ are the creation and annihilation operators at site i with spin projection σ, $n_{i\sigma}$ is the onsite (local) electron density at site i for one component of spin, and μ is the chemical potential. The random onsite potential V_i describes the diagonal (onsite) disorder and is uniformly distributed within the $[-V, V]$ range.

Following an approach of the Randeria group [3–5], we use the mean-field BdG equations for the electron-like and hole-like Bogoliubov functions $u_n(\mathbf{r}_i)$ and $v_n(\mathbf{r}_i)$ to describe the formation of the inhomogeneous Fermi–Bose mixture and the superconductor–insulator phase transition in this model. The BdG equations on the lattice have the form:

$$-t \sum_{\pm x, \pm y} u_n(\mathbf{r}_i \pm \hat{x} \pm \hat{y}) + (U(\mathbf{r}_i) + V_i - \mu) u_n(\mathbf{r}_i) + \Delta_i v_n(\mathbf{r}_i) = E_n u_n(\mathbf{r}_i) \,, \tag{13.2}$$

$$+t \sum_{\pm x, \pm y} v_n(\mathbf{r}_i \pm \hat{x} \pm \hat{y}) + (U(\mathbf{r}_i) + V_i - \mu) v_n(\mathbf{r}_i) + \Delta^{*}_i u_n(\mathbf{r}_i) = E_n v_n(\mathbf{r}_i) \,, \tag{13.3}$$

where E_n are the excitation energies of the system, $U_i = U(\mathbf{r}_i)$ is the mean-field potential at site $\mathbf{r}_i$, given by

$$U_i = U(\mathbf{r}_i) = (|U|/2) \sum_{n} [|u_n(\mathbf{r}_i)|^2 f_n + |v_n(\mathbf{r}_i)|^2 (1 - f_n)] \,, \tag{13.4}$$

where $f_n = 1/[\exp(E_n/T) + 1]$ is the Fermi–Dirac distribution function for the given value of the excitation energy E_n and $\Delta_i = \Delta(\mathbf{r}_i)$ is the superconducting order parameter at site i or the pairing potential in the terminology accepted for the BdG equations.

The normalization condition yields

$$\sum_{n} [|u_n(\mathbf{r}_i)|^2 + |v_n(\mathbf{r}_i)|^2] = 1 \,. \tag{13.5}$$

The chemical potential $\mu$is determined from the self-consistency equation on the average number of electrons per site

$$\frac{1}{N}\sum_i n_i = \sum_i \sum_n [|u_n(\mathbf{r}_i)|^2 f_n + |v_n(\mathbf{r}_i)|^2(1 - f_n)] = \sum_i n_{el}(\mathbf{r}_i) = n/2\,, \tag{13.6}$$

where we introduced the local electron density

$$n_{el}(\mathbf{r}_i) = \sum_n [|u_n(\mathbf{r}_i)|^2 f_n + |v_n(\mathbf{r}_i)|^2(1 - f_n)] \tag{13.7}$$

at site i averaged over different excitation energies E_n with the Fermi–Dirac distribution functions f_n. In fact, this averaged local density enters in the definition of the mean-field potential at the site $\mathbf{r}_i$. Namely

$$U_i = U(\mathbf{r}_i) = (|U|/2)\sum_n n_{el}(\mathbf{r}_i)\,, \tag{13.8}$$

Correspondingly, introducing the Hartree–Fock shift of the chemical potential

$$\mu_i = \mu - (|U|/2)n_{el}(\mathbf{r}_i) = \mu - (|U|/2)\sum_n [|u_n(\mathbf{r}_i)|^2 f_n + |v_n(\mathbf{r}_i)|^2(1 - f_n)]\,, \tag{13.9}$$

we can rewrite the BdG equations (13.2) and (13.3) in a form more convenient for the numerical simulations:

$$-t\sum_{\pm x,\pm y} u_n(\mathbf{r}_i \pm \hat{x} \pm \hat{y}) + (V_i - \mu_i)u_n(\mathbf{r}_i) + \Delta_i v_n(\mathbf{r}_i) = E_n u_n(\mathbf{r}_i)\,, \tag{13.10}$$

$$+t\sum_{\pm x,\pm y} v_n(\mathbf{r}_i \pm \hat{x} \pm \hat{y}) - (V_i - \mu)v_n(\mathbf{r}_i) + \Delta_i^* u_n(\mathbf{r}_i) = E_n v_n(\mathbf{r}_i)\,. \tag{13.11}$$

Note that in Eqs. (13.10) and (13.11), we have the renormalized chemical potential μ_i, which is site-dependent. At the same time, the pairing potential has the form

$$\Delta(\mathbf{r}_i) = |U|\sum_n u_n(\mathbf{r}_i)v_n^*(\mathbf{r}_i)(1 - f_n)\,. \tag{13.12}$$

13.3 Numerical Solutions of the BdG Equations

In this section, we discuss numerical solutions of the BdG equations (13.9)–(13.12) with the normalization condition (13.5). We solve the BdG equations on the square lattice of $N \times N$ sites with the periodic boundary conditions. We find the eigenvalues E_n and the eigenfunctions $u_n(\mathbf{r}_i)$ and $v_n(\mathbf{r}_i)$. After that, at low temperatures $T \to 0$, we calculate separately the contributions of the electron-like

$$n_{u,\sigma}(\mathbf{r}_i) = \sum_n |u_n(\mathbf{r}_i)|^2 f_n \tag{13.13}$$

and hole-like Bogoliubov excitations

$$n_{v,\sigma}(\mathbf{r}_i) = \sum_n |v_n(\mathbf{r}_i)|^2(1 - f_n) \tag{13.14}$$

in the averaged local density $n_{el}(\mathbf{r}_i)$, which is given in accordance with Eq. (13.7) by the sum of these two contributions. Namely, the local density at the given site i is equal to

$$n_{el}(\mathbf{r}_i) = n_{u,\sigma}(\mathbf{r}_i) + n_{v,\sigma}(\mathbf{r}_i)\,. \tag{13.15}$$

Then, we calculate the local pairing potential $\Delta(\mathbf{r}_i)$ at the site i with the eigenfunctions $u_n(\mathbf{r}_i)$ and $v_n(\mathbf{r}_i)$ using Eq. (13.12). As the next step of the iteration procedure, knowing the local pairing potential, we recalculate the electron-like and hole-like contributions in Eqs. (13.13) and (13.14) at sites $\mathbf{r}_i$. We repeat this type of iteration procedure until we achieve the full self-consistency between the averaged local density $n_{el}(\mathbf{r}_i)$ and the pairing potential $\Delta(\mathbf{r}_i)$.

13.3.1 *Method of Calculations*

In this subsection, we explain in more detail the method of our numerical calculations. Throughout this chapter, we work on the lattice with 24×24 sites with the periodic boundary conditions. We start our calculation with random values of the pairing potential Δ_i and of the chemical potential μ_i. Then, we numerically diagonalize Hamiltonian (13.1)and find the eigenvectors $u_n(\mathbf{r}_i)$, $v_n(\mathbf{r}_i)$, and eigenvalues E_n. After that, Δ_i and μ_i are recalculated using Eqs. (13.9) and (13.12).

We perform the iteration cycle until we reach the convergence of the values of Δ_i and μ_i at each site with the given accuracy. We analyze the model Hamiltonian for the parameter ranges $1 \leq |U|/t \leq 10$ and $0 \leq V/t \leq 12$ for the dimensionless Hubbard interaction $|U|/t$ and dimensionless diagonal disorder V/t. (t is the hopping integral). We start from low-density limit $n = 0.125 - 0.15$ (n is the average dimensionless electron density) at low temperatures, $T \to 0$.

In the numerical calculations, we introduce the convenient notation E_{gap} for the value of the energy gap and define this value as the smallest possible eigenvalue of the energy. The duration of the numerical procedure is several hours. We also use the Quantum Express software package while obtaining our results.

13.4 Different Tendencies Arising with an Increase in the Hubbard Attraction

In this section, we present the results and trends in the system in the limits of both weak and strong disorder arising with an increase in the Hubbard interaction $|U|$ from small to large values. We start our calculations with two very different values of $V/t = 2$ and $V/t = 10$ corresponding to the weak and very strong disorder, respectively. At the same time, we consider weak $|U|/t = 2$, moderate $|U|/t = 4$, and very strong $|U|/t = 10$ attraction between electrons. In the latter case, the Hubbard interaction is $|U| > W$, where $W = 8t$ is the bandwidth for the square lattice.

We present the results of the numerical calculation on Fig. 13.1, where we show the spatial distribution of the local electron density (left column) and electron gap (right column) in four different cases, namely, (a) for the weak Hubbard interaction and weak disorder: $|U|/t = 2$, $V/t = 2$, (b) for the moderate Hubbard interaction and weak disorder: $|U|/t = 4$, $V/t = 2$, (c) for the strong Hubbard interaction and weak disorder: $|U|/t = 10$, $V/t = 2$, and (d) for the strong Hubbard interaction and strong disorder: $|U|/t = 10$, $V/t = 10$.

We perform all the calculations for fixed electron density $n = 0.15$. We also illustrate the mixing of holes and electrons (at the level of the Bogoliubov coefficients u and v) in the middle column of Fig. 13.1. From the middle column in Fig. 13.1, we can see that electron–hole mixing correlates with the electron gap Δ.

Let us emphasize that effectively three first cases in Fig. 13.1, namely (a), (b), and (c), demonstrate the similar tendencies in the behavior of spatial distributions of electron density n and of the electron gap Δ with an increase of Hubbard attraction $|U|$ from the weak coupling limit $|U| = 2t$ to the strong coupling limit $|U| = 10t > W$ at the moderate disorder $V = 2t$.

At the same time, the last case in Fig. 13.1, i.e. (d), corresponds to the very interesting extreme case of both the very strong Hubbard attraction $|U| = 10t$ and the very strong disorder $V = 10t$. We can expect here the formation of compact pairs and their localization at very low electron density.

From Fig. 13.1, we can see that already at the moderate disorder $V = 2t$ (or more precisely at $-2t < V < 2t$ for the width and local values of disorder in the "dirty" band) and at a (moderately) low averaged electron density $n = 0.15$, the droplets of the order parameter are formed in our system. Moreover, the spatial distribution of the droplets is correlated with the spatial distribution of electron density in the left column and electron–hole mixing in the middle column.

We observe that electrons are predominantly localized in the regions with small values of local disorder. It is interesting, however, that our system is not in an insulating state since in this range of parameters, the spatial distribution of electrons is not split into separate isolated regions.

At the same time, at the strong Hubbard attraction $|U| = 10t > W$ and low electron densities $n \ll 1$, we can assume that the system has the properties of a bosonic metal. The concept for this new metallic state of the matter was introduced by Kagan et al. in [2, 7] for the Hubbard model with sufficiently strong Hubbard attraction in the 2D low-density electron system in the clean case (in the absence of impurities and disorder).

In this new metallic state, both in SC and normal domains, the charge carriers are local (compact) electron pairs, which can form more extended bosonic clusters (see e.g., the results of [4, 5] for bismuth oxides and the preceding Chap. 12). In our case, these clusters can contain one, two, or several compact (bielectron) pairs. More precisely, we have here the spatially separated Fermi–Bose mixture of compact Cooper

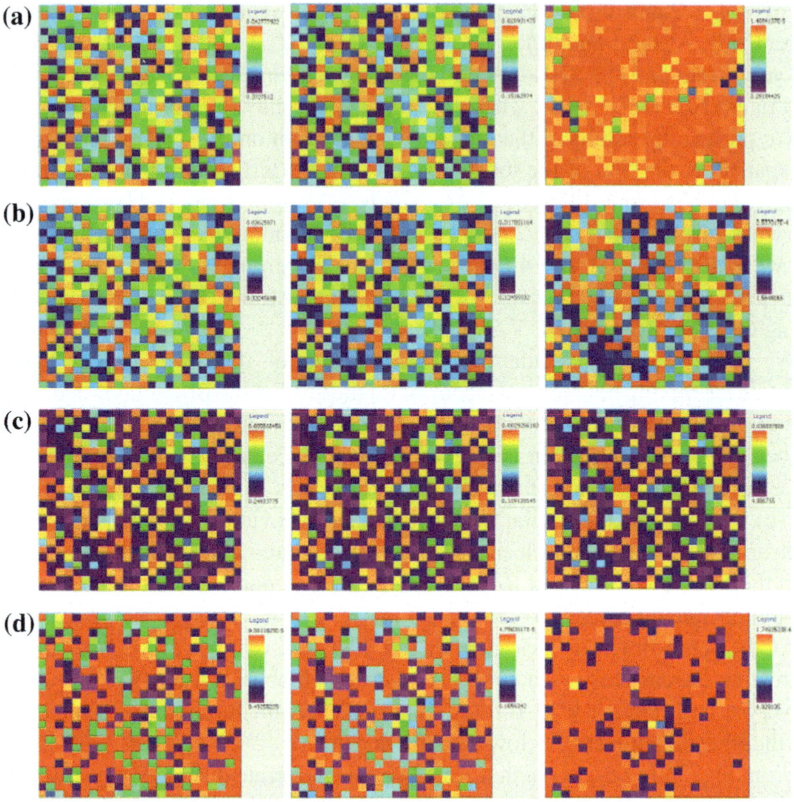

Fig. 13.1 Spatial distribution of the local electron density (left column), electron gap Δ/t (right column), and mixing of holes and electrons (middle column) in four different cases, namely: **a** for weak Hubbard interaction and weak disorder($|U|/t = 2$, $V/t = 2$), **b** for moderate Hubbard interaction and weak disorder ($|U|/t = 4$, $V/t = 2$), **c** for strong Hubbard interaction and weak disorder($|U|/t = 10$, $V/t = 2$), and **d** for strong Hubbard interaction and strong disorder ($|U|/t = 10$, $V/t = 10$). The averaged electron density $n = 0.15$ is fixed in all the cases [6]

pairs and unpaired electrons with the formation of bosonic droplets of different sizes within the unpaired fermionic matrix.

Now, let us proceed to the next important problem related to the localization of pairs in a "dirty" bosonic metal in the framework of the 2D disordered Hubbard model. To describe the relative localization of bielectron pairs, it is instructive to analyze the behavior of the pair–pair correlation function $\langle c_i^\dagger c_i^\dagger c_j c_j \rangle \propto \langle \Delta_i^\dagger \Delta_j \rangle$, which is in fact responsible for the transfer of a pair to the distance $|\mathbf{r}_i - \mathbf{r}_j|$ on the lattice, or, in other words, for the correlation in the relative spatial distribution of the pairs.

Note that the establishment of the long-range order in the system corresponds to a nonzero constant value of this correlation function at very large distances $|\mathbf{r}_i - \mathbf{r}_j| \to \infty$ and low temperatures [8]. On the lattice, the correlation function should approach

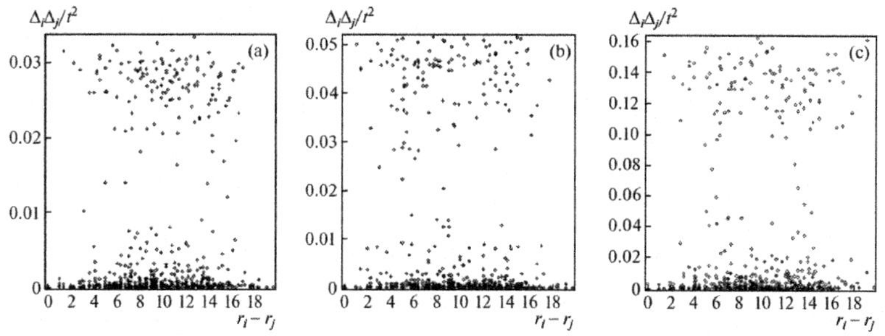

Fig. 13.2 Dependence of the mathematical expectation of the dimensionless correlation function $\Delta_i^\dagger \Delta_j/t^2$ (t is the hopping integral on the square lattice), which corresponds to the correlation in the relative spatial distribution of the pairs at the dimensionless distance $|\mathbf{r}_i - \mathbf{r}_j|/d$ (d is the intersite distance) between them. The values of the Hubbard attraction $U = -6t$ and disorder $V = 8t$ are fixed for all the three plots presented in panels **a**, **b**, and **c**, whereas the electron density is varied within a broad range from very low to moderate levels and is equal to $n = 0.07$ **a**, $n = 0.3$ **b**, and $n = 0.4$ **c** [6]

the constant value for $|\mathbf{r}_i - \mathbf{r}_j| \to L$, where L is the lattice size. Thus, for the square lattice with 24×24 sites considered in [6], the lattice size is $L = 24d$, where d is the intersite distance.

In Fig. 13.2, for three different electron densities, $n = 0.07$ (Fig. 13.2a), $n = 0.3$ (Fig. 13.2b), and $n = 0.4$ (Fig. 13.2c) at fixed values of the moderately strong Hubbard attraction $U = -6t$ and the relatively strong disorder $V = 8t = W$, we present the results for the mathematical expectation of the dimensionless ratio $\Delta_i^\dagger \Delta_j/t^2$, where t is the hopping integral on the square lattice, as function of the dimensionless distance between the pairs $|\mathbf{r}_i - \mathbf{r}_j|/d$ (d is the intersite distance).

Note that, in the calculations corresponding to all the three plots presented in Fig. 13.2, we do not perform the averaging over different disorder configurations. Thus, all the results shown in Fig. 13.2 are obtained for one particular disorder configuration.

There is an important observation regarding the broad density range in the plots of Fig. 13.2. Namely, we can see the formation of the long-range order for all plots in panels (a), (b), and (c) since the pairs are located at the maximum distance from one another (comparable to the lattice size L) prevail in our system.

We can see in Fig. 13.2 that many values of .the correlation function $\Delta_i^\dagger \Delta_j/t^2$ correspond to the same distance $|\mathbf{r}_i - \mathbf{r}_j|/d$, or, in other words, to the same spatial configuration for the relative distribution of the electron pairs. Thus, the system does not exhibit a clear dependence on the correlation function from the mutual spatial arrangement of the pairs.

As a result, the behavior of the pair correlation function illustrated in Fig. 13.2 does not allow us to conclude that within this range of parameters, our system is always in the insulating state. Quite opposite, we demonstrate in the next subsections that Fig. 13.2 corresponds to very different limiting cases from extremely low electron density $n = 0.07$ (a) to intermediate electron density $n = 0.40$ (c).

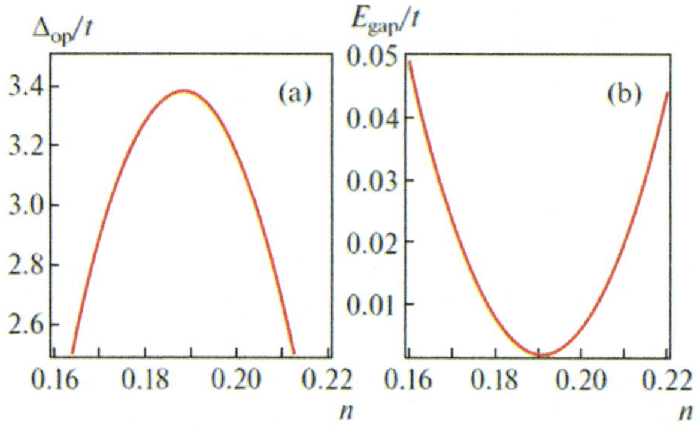

Fig. 13.3 Dependence of the average dimensionless pairing parameter Δ_{op}/t (**a**) and the dimensionless gap E_{gap}/t (**b**) measured in the units of hopping integral t on the electron density n within the density range from $n = 0.18$ to $n = 0.22$ for the strong Hubbard attraction $|U| = 10t$ and moderate disorder $V = 2t$. The calculations are performed for temperature $T = 0.01t$ on the square lattice with 24×24 sites [6]

Finally, concluding this subsection let us present the results of [6] concerning the density dependence of the average pairing parameter Δ_{op} and the gap E_{gap}. Remind that in numerical calculations, we define E_{gap} as the smallest possible eigenvalue of energy. In Fig. 13.3a, we show how the increase in the charge carrier density from $n = 0.18$ to $n = 0.22$ affects the dimensionless pairing parameter Δ_{op}/t scaled in the units of the hopping integral t. In Fig. 13.3b, we demonstrate how this increase affects the dimensionless gap E_{gap}/t. The results for the order parameter and the gap are averaged over five spatial configurations. We can see the formation of a pronounced peak in the density dependence of the averaged order parameter and a minimum in the density dependence of the gap.

13.5 Percolation Phase Transition for the Electron Density Close to $n = 0.3$

In the beginning of this section, it is instructive at first to present the calculations for the two-dimensional distributions of the local electron density and the order parameter for the moderately small averaged electron density $n = 0.17$, intermediate Hubbard attraction $|U| = 4t = W/2$, and strong disorder $V = 8t$ (see Fig. 13.4).

In Fig. 13.4, we can see that at the electron density $n = 0.17$, the droplets of the order parameter with a large number of electron pairs are formed. Moreover, the droplets play a dominant role in the spatial distribution of one-electron local density as well. The peaks of the two-dimensional distribution of electron density and order parameter touch each other but do not overlap.

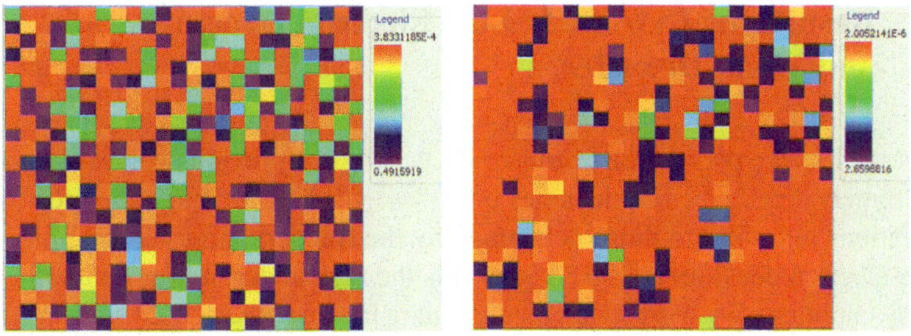

Fig. 13.4 Two-dimensional distribution of the local electron density (left panel) and order parameter (right panel) for the set of parameters: $n = 0.17$, $|U| = 4t$, and $V = 8t$ on the square lattice with 24×24 sites [6]

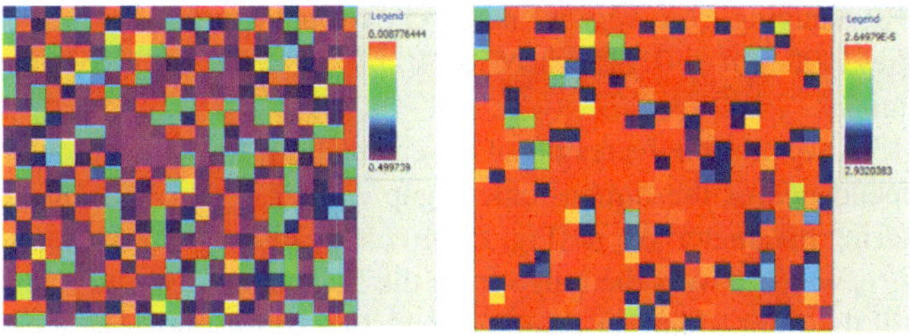

Fig. 13.5 Two-dimensional distribution of the local electron density (left panel) and order parameter (right panel) for the set of parameters: $n = 0.3$ (close to the percolation threshold), $|U| = 4t$, and $V = 8t$ on the square lattice with 24×24 sites [9]

Let us concentrate now on the electron density range close to $n = 0.3$, corresponding to Fig. 13.2b. Here, the droplets of the order parameter merge into a network of paired chains (see Fig. 13.5) and as a result, we have a percolation phase transition from the granular to SC state that takes place in the system. The percolation transition manifests itself in the formation of the large percolation bosonic cluster in the system. Preliminary estimates show that the phase transition corresponds to the critical concentration $n_c \approx 0.31$.

In such a cluster, the network of paired chains forms a tree-like pattern with a lot of branches, which in fact do not exhibit discontinuities along their length. Effectively, for $n = 0.3$, the distribution of the local order parameter corresponds to the formation of "valleys" (the regions, where the order parameter is the mostly concentrated one or where it locally has maximum values). Note that the valleys of order parameter in the right panel of Fig. 13.5 are located between the peaks of the local electron density in the left column.

13.6 Results for Extremely Low Electron Densities and Weak Disorder

In this section, we present the results of [6, 9, 10] for a previously unexplored range of extremely small electron densities $0.05 \leq n \leq 0.125$, although the idea of these calculations is similar to those performed by the Randeria group in [3–5] for much higher electron densities $n \approx 0.85$ (which is the typical density for optimally doped high-T_c cuprates). We work again on the square lattice with 24×24 sites in the limit of strong Hubbard attraction $|U| = 10t$ and low temperatures.

The results of Sect. 13.5 obtained for the moderately small electron density $n = 0.15$ and disorder amplitudes $V = 2t$ and $V = 10t$ are supplemented in this subsection by a broad spectrum of disorder values $V = 0.25t$, $V = 2t$, $V = 4t$, and $V = 10t$ (ranging from very weak to very strong disorder) for very small electron densities $0.05 \leq n \leq 0.125$ where, for $|U| \geq W = 8t$, we hope to approach the bosonic limit.

In Fig. 13.6, we present the results of the calculations for $|U| = 8t$ and $n = 0.125$ for the dependence of the dimensionless gap E_{gap}/t (Fig. 13.6a) and the dimensionless pairing parameter Δ_{op}/t (Fig. 13.6b) on the dimensionless amplitude of the disorder V/t, which covers the range from the absence of disorder (clean case) for $V/t = 0$ till the limit of strong disorder at $V/t = 10$. In Fig. 13.6, we can see the presence of maximum and minimum in the plot of the gap E_{gap}/t versus the degree of disorder. Note that the minimum at intermediate strength of disorder $V \approx 4t = W/2$ is in qualitative agreement with the results of the Randeria group [3–5].

In Fig. 13.7, we generalize the results of Fig. 13.6 to the limiting case of the extremely low electron density $n = 0.05$ (practically empty lattice) for which we still manage to get the converging iteration procedure on the lattice with 24×24 sites. We fix the strength of the Hubbard attraction $|U| = 8t$ and present the results of the calculations for the dependence of the dimensionless gap E_{gap}/t (Fig. 13.7a) and

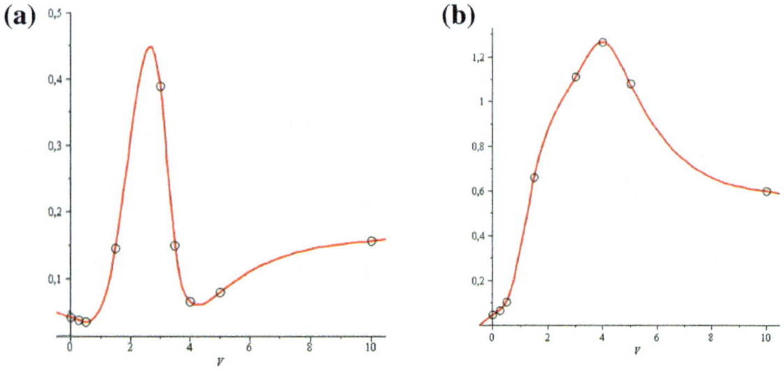

Fig. 13.6 Dependence of the dimensionless gap E_{gap}/t (**a**) and the dimensionless pairing parameter Δ_{op}/t (**b**) on the dimensionless amplitude of disorder V/t for the electron density $n = 0.125$ and Hubbard attraction $|U| = 8t = W$ on the square lattice with 24×24 sites [10]

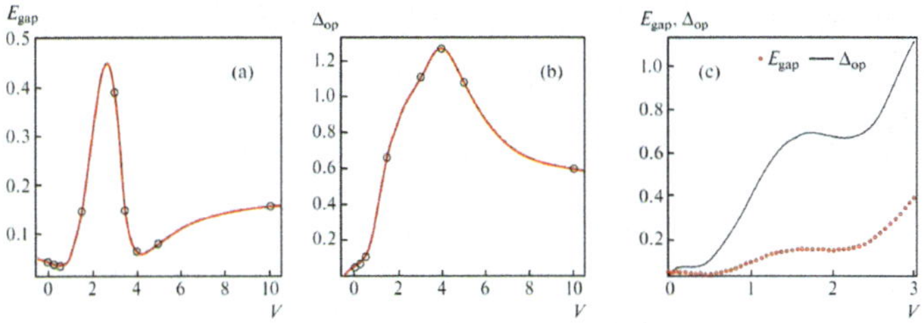

Fig. 13.7 Dependence of E_{gap}/t (**a**) and Δ_{op}/t (**b**) on the dimensionless amplitude of the disorder V/t for the strong Hubbard attraction $|U| = 8t$ and extremely low electron density $n = 0.05$ for the square lattice with 24×24 sites. We present both plots in the same panel (**c**) in the narrow range from $V/t = 0$ until $V/t = 3$ for the weak disorder [6]

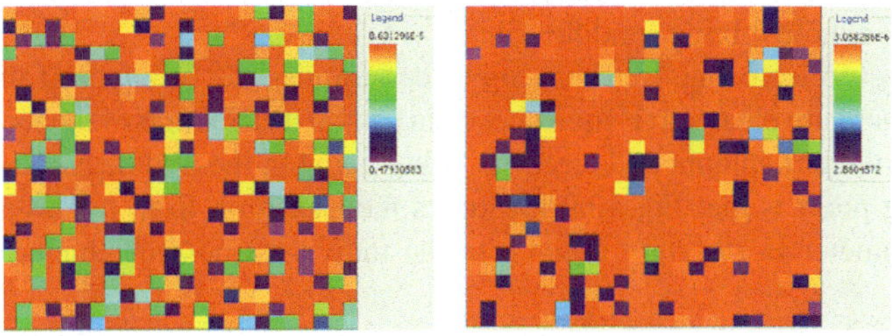

Fig. 13.8 Two-dimensional distributions of the local electron density (left panel) and the order parameter (right panel) on the square lattice with 24×24 sites for low the electron density $n = 0.11$ at the intermediately strong Hubbard attraction $|U| = 6t$ and strong disorder $V = 8t$ [6]

the dimensionless pairing parameter Δ_{op}/t (13.7b) on the dimensionless amplitude of disorder in the range from $V/t = 0$ until $V/t = 10$. In Fig. 13.7c, in the narrow range from $V/t = 0$ until $V/t = 3$ of the weak disorder, we plot both dependence of the gap and of the order parameter in the same figure. We see again the presence of the pronounced peak in the dependence of the gap on the disorder strength and a minimum for $V \approx 4t$ in Fig. 13.7a in agreement with [3–5].

Finally, concluding this section, let us present the calculations (see Fig. 13.8) generalizing the results of Fig. 13.1 for local electron density and order parameter to the case of lower electron density $n = 0.11$ at the intermediately strong Hubbard attraction $|U| = 6t$ and strong disorder $V = 8t$.

For the low electron density $n = 0.11$, individual pairs (bielectrons) or double pairs of electrons (quartets) can be observed in the adjacent cells of Fig. 13.8.

13.7 Calculations for Moderately Low and Intermediate Electron Densities in the Broad Range of Disorder Strengths

In this section, we consider moderately small and intermediate electron densities (which are close to the percolation phase transition) at $0.15 \leq n \leq 0.32$ for the broad range of the disorder strengths $0.5t \leq V \leq 10t$.

In Fig. 13.9, we generalize the results of our calculations for the moderately small electron density $n = 0.15$ presented in Fig. 13.1. Namely, for the intermediately strong Hubbard attraction $|U| = 6t$ in three cases of weak disorder $V/t = 0.5$, intermediate disorder $V/t = 6$, and strong disorder $V/t = 10$, we show the spatial distribution for the local electron density (left column), local SC gap (right column), and electron–hole mixing (middle column).

We can see from Fig. 13.9 that, up to certain disorder strength of the order of $V \approx 5t$, there are clear correlations between spatial distribution of the local electron density in the left column and electron–hole mixing in the middle column. Another interesting point is that Fig. 13.9 illustrates the transition from the droplets of the order parameter in the right column to the individual bosonic (bielectron) pairs,

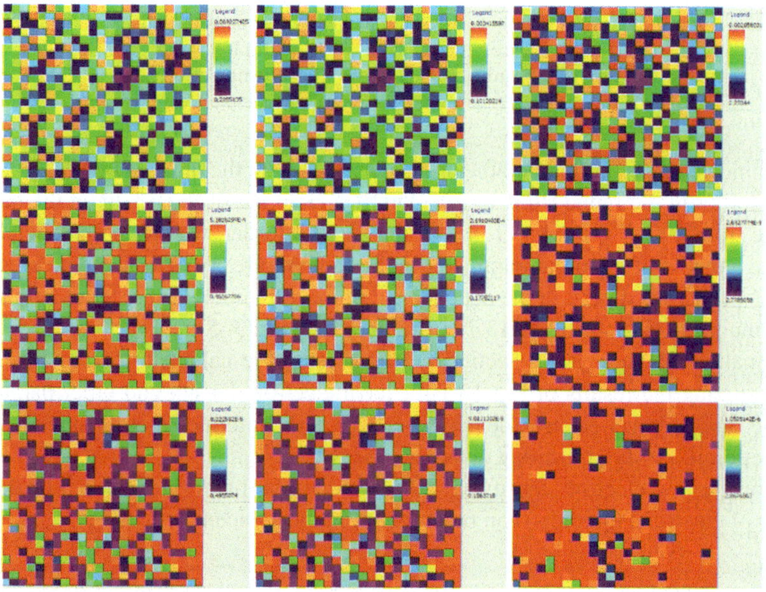

Fig. 13.9 Spatial distribution of the local electron density (left column), electron gap Δ/t (right column) and mixing of holes and electrons (middle column) in three different cases: **a** upper line, for weak disorder $V/t = 0.5$; **b** middle line, for intermediate disorder $V/t = 6$; and **c** bottom line, for strong disorder $V/t = 10$. We fix the averaged electron density $n = 0.15$ and intermediately strong Hubbard attraction $|U| = 6t$ in all three cases [6]

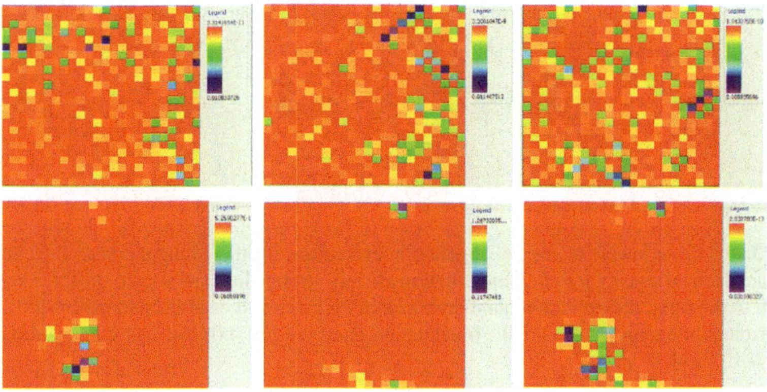

Fig. 13.10 First three Bogoliubov excitations are shown for the two sets of parameters: the Hubbard attraction $|U| = 6t$, disorder strength $V = 0.5t$, temperature $T = 0.01t$, electron density $n = 0.3$ (top line); $|U| = 6t$, $V = 4t$, $T = 0.01t$, and $n = 0.15$ (bottom line). From the left to the right, the excitation number for a given value of the Hubbard U increases yielding the excitation energies $E_1 = 0.000088t$, $E_2 = 0.004876t$, and $E_3 = 0.013679t$ for the top line and the energies $E_1 = 0.102606t$, $E_2 = 0.172969t$, and $E_3 = 0.217828t$ for the bottom line, respectively [6]

when we increase the strength of disorder (see the right corner of the bottom line in Fig. 13.9c).

At the same time, an important question concerning the critical value of the disorder strength V/t separating the domains of bosonic metal and bosonic insulator remains unresolved and deserves a more careful investigation both in case of bosonic droplets containing a large number of individual bielectron pairs. In principle, we cannot exclude the situation when we have balanced or imbalanced mixture of localized (majority) and delocalized (minority) pairs.

Finally, at the end of this section, let us deepen our insight into eigenvectors and eigenvalues of the BdG equations and calculate the local values for $|u_n(\mathbf{r}_i)|^2 + |v_n(\mathbf{r}_i)|^2$ for the positive energy values $E_n > 0$ considering the first three Bogoliubov excitations (E_1, E_2, E_3). In Fig. 13.10, we present the results for the excitation energies fixing the intermediately strong Hubbard attraction $|U| = 6t$ and considering two sets of parameters, namely the intermediate electron density $n = 0.3$ and weak disorder $V/t = 0.5$ (upper line) and moderately small electron density $n = 0.15$ and intermediate amplitude of disorder $V/t = 4$ (bottom line).

It is useful to compare the results of the top line in Fig. 13.10 (obtained for electron density $n = 0.3$) with the similar calculations for three lowest in energy Bogoliubov excitations presented in Fig. 13.11 considering a slightly lower electron density $n = 0.28$, the same strength of the Hubbard attraction $|U| = 6t$, and strong disorder $V = 8t$.

The results on the Bogoliubov excitations presented in Fig. 13.11 together with the calculations of the two-dimensional distributions of local density and order parameter presented in Fig. 13.5 and the pair–pair correlation function shown in Fig. 13.2b

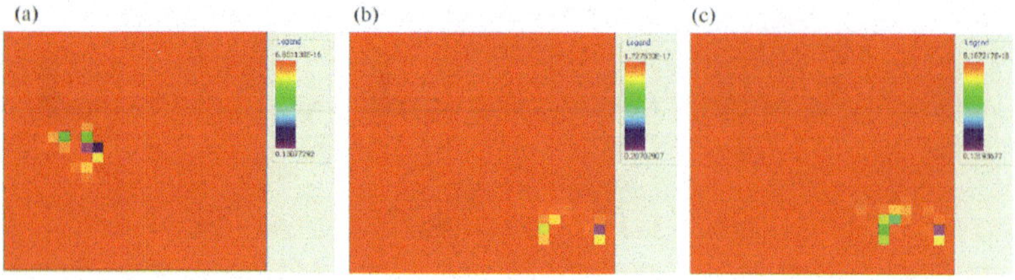

Fig. 13.11 Three lowest in energy Bogoliubov excitations for the strong disorder $V = 8t$, intermediate electron density $n = 0.28$, and intermediately strong Hubbard attraction $|U| = 6t$. From the left to the right, the excitation number increases for given values of the parameters $|U|$, V, and n: **a** excitation energy $E_1 = 0.070t$, **b** excitation energy $E_2 = 0.0840t$, and **c** excitation energy $E_3 = 0.107t$ [6]

provide a better qualitative understanding of the nature of the percolation phase transition close to the intermediate electron density $n \sim 0.3$, nevertheless, more thorough investigations of the situation occurring near the critical point $n_c \approx 0.31$ are needed.

13.8 Discussion and Conclusions

Summarizing the numerical results presented in Sects. 13.4–13.7, let us emphasize that the two-dimensional spatial distributions of the order parameter at different electron densities shown in Figs. 13.1, 13.4, 13.5, 13.8, and 13.9 yield the following qualitative picture.

At very low averaged electron densities $n = 0.11$ (see Fig. 13.8), the spatial distribution of the order parameter corresponds to the regions with individual electron pairs (bielectrons) and doubled electron pairs (quartets).

At moderately small electron densities $n = 0.15$ in Figs. 13.1 and 13.9 and $n = 0.17$ in Fig. 13.4, the spatial distribution of the order parameter is controlled by droplets containing a large number of electron pairs.

Close to the percolation threshold at electron densities $n = 0.3$ (Fig. 13.5), first, droplets of the order parameter merge in chains and then, the network of paired chains form a large percolation cluster resembling a tree with a lot of branches, which practically do not exhibit discontinuities along their lengths. Preliminary estimates for the percolation transition yield the critical concentration $n_c \approx 0.31$.

It is interesting to stress that the spectral gap in the one-particle density of states does not disappear even in the limit of large disorder despite the growing number of spatial nodes, where the local order parameter $\Delta(\mathbf{r}) \to 0$. We get a better understanding of this effect from the analysis of the spatial variation of the eigenfunctions $u_n(\mathbf{r})$ and $v_n(\mathbf{r})$ of the BdG equations, which determine $\Delta(\mathbf{r})$ in Eq. (13.12). We think that this effect should be revisited in the framework of the model of the inhomogeneous Fermi–Bose mixture with the coexistence of electron pairs and single unpaired electrons.

Let us emphasize that the new results in this section are mainly related to the strong coupling limit $|U|/t \gg 1$ of the Hubbard model for low electron density $n \ll 1$ and strong disorder $V/t \gg 1$. The limiting values of our calculations correspond to the situation, in which the strength of Hubbard attraction and the amplitude of disorder $|U|/t = V/t = 10$ for the averaged electron density $n = 0.05$ (nearly empty lattice). Note that, for the square lattice in 2D, the bandwidth is $W = 8t$. Hence, we have $|U| = V > W$ in this case.

Let us estimate the size of an electron pair in the low density and strong coupling limit. If in the absence of disorder (for $V/t = 0$) and low electron density (e.g., for $n = 0.125$) in the strong coupling limit $|U|/t = 10$, we can make simple estimates (see [2, 6, 11]), and find the values $E_F \sim 0.8t$ for the Fermi energy and $E_b \sim 0.2E_F \sim 0.16t$ for the binding energy of the pair. If we further decrease the density down to the limiting value $n = 0.05$ fixing the strength of the Hubbard attraction, then we find $E_F \sim 0.32t$ and $E_b \sim 0.5E_F \sim 0.16t$. Thus, even at such extremely low-density limit $E_b < 2E_F$, the chemical potential is positive

$$\mu = E_F - |E_b|/2 = 0.75E_F > 0\,. \tag{13.16}$$

Hence, the pairing takes place near the Fermi surface, and we are still at the BCS side of the BCS–BEC crossover.

Note that if we introduce convenient notations for the binding energy $E_b = 1/ma^2$, where a is the effective size of a pair, $m = 1/(2td^2)$ is the band mass, d is the intersite distance, and $E_F = p_F^2/2m$ for the Fermi energy, where p_F is the Fermi momentum, then the inequality $E_b < 2E_F$ means that the important parameter

$$ap_F = (2E_F/|E_b|)^{1/2} > 1\,, \tag{13.17}$$

and the size a of a pair exceeds the mean distance between electrons $1/p_F$.

At the same time, at extremely low density $n = 0.05$, we have $ap_F \sim 2$, which is much less than the corresponding parameter $p_F\xi_0 \sim 1000$ in conventional low-T_c s-wave superconductors, where the typical coherence length is $\xi_0 \sim (3-5)10^3$ Å.

Thus, we can conclude that although we are at the BCS side of the crossover, the pairs are rather compact and almost touch each other. In fact, here we are very close to the limit, when the pairs start to overlap and "crush" each other. This limit effectively corresponds just to the formation of the Fermi–Bose mixture with the coexistence of compact pairs and unpaired single electrons.

The presence of disorder makes the situation even more interesting and complicated and leads to the formation of the inhomogeneous Fermi–Bose mixture of pairs and single electrons, where a part of the one-particle and two-particle states should be localized even in the 2D system of a finite size depending upon the degree of disorder V/t.

Note that important feature of the quasi-2D systems described by the Hubbard–Anderson type of models with the local onsite Hubbard interaction and diagonal disorder is related to the fact that their global phase diagrams can exhibit a "direct" transition from the superconducting to insulating localized state in addition to the conventional phase transition from a superconductor to normal metal. This possibility is elucidated in theoretical papers [12–15] and confirmed by the experimental studies of superconductivity and localization in dirty thin films [16, 17].

Another important motivation for our study is related to recent experimental efforts to create SC flux cubits using quantum circuits with high impedances in granular superconductors of reduced dimensionality [1]. These motivations promote the calculations presented in this chapter and make them highly relevant.

References

1. L. Grünhaupt, M. Spiecker, N. Gusenkova, D. Maleeva, S.T. Skacel, I. Takmakov, F. Valenti, P. Winkel, H. Rotzinger, W. Wernsdorfer, A.V. Ustinov, I.M. Pop, Nat. Mater. **18**(8), 816 (2019)
2. M.Yu. Kagan, R. Frésard, M. Capezzali, H. Beck, Phys. Rev. B **57**, 5995 (1998)
3. A. Ghosal, M. Randeria, N. Trivedi, Phys. Rev. Lett. **81**, 3940 (1998)
4. A. Ghosal, M. Randeria, N. Trivedi, Phys. Rev. B **65**, 014501 (2001)
5. Y. Loh Lee, N. Trivedi, in *Conductor-Insulator Quantum Phase Transitions* (Oxford University Press, 2012)
6. M.Yu. Kagan, E.A. Mazur, J. Exp. Theor. Phys. **132**, 596 (2021)
7. M.Yu. Kagan, R. Frésard, M. Capezzali, H. Beck, Physica B **284–288**, 447 (2000)
8. E.M. Lifshitz, L.P. Pitaevskii, *Statistical Physics, Part II* (Pergamon, Oxford, 1980)
9. E.A. Mazur, R.S. Ikhsanov, M.Yu. Kagan, J. Phys. Conf. Ser. **2036**(1), 012019 (2021)
10. M.Yu. Kagan, E.A. Mazur, J. Phys. Conf. Ser. **1740**, 012045 (2021)
11. M.Yu. Kagan, JETP Lett. **103**(11), 728 (2016)
12. M. Feigel'man, L. Ioffe, V. Kravtsov, E. Cuevas, Ann. Phys. **325**(7), 1390 (2010)
13. M.A. Skvortsov, M.V. Feigel'man, J. Exp. Theor. Phys. **117**(3), 487 (2013)
14. I.S. Burmistrov, I.V. Gornyi, A.D. Mirlin, Phys. Rev. Lett. **108**, 017002 (2012)
15. E.Z. Kuchinskii, I.A. Nekrasov, M.V. Sadovskii, Phys. Usp. **55**(4), 325 (2012)
16. A.M. Goldman, N. Marković, Phys. Today **51**(11), 39 (1998)
17. D.B. Haviland, Y. Liu, A.M. Goldman, Phys. Rev. Lett. **62**, 2180 (1989)

Chapter 14
Droplets Formation, BEC and Superconductivity in Quantum Gases, Metallic Hydrogen and Excitonic Systems

Abstract We analyze different droplets, which arise in superconducting and superfluid (Bose-condensed) systems ranging from imbalanced and dipole Fermi gases, Fermi–Bose mixtures with dominant number of bosons, metallic hydrogen droplets near the phase boundary between atomic and molecular liquid and crystalline phases, as well as in excitonic systems in the self-similar collapse regime. We study also long-range charge oscillations and formation of the inhomogeneous phases related to acoustic plasmons in MgB_2.

14.1 Introduction. Phase Separation in Imbalanced Fermi Gas

The problem of inhomogeneous states and phase separation in liquid and solid solutions of ^3He in ^4He (see e.g. [1]) attracts the interest of the low-temperature community already quite a long time since the classical paper of Bardeen, Baym, and Pines [2].

With the discovery of Bose–Einstein condensation in ultracold Bose gases and Feshbach experiments [3–6] with ultracold Fermi gases and Fermi–Bose (as well as Fermi–Fermi and Bose–Bose) mixtures, the subject of phase separation was generalized from quantum liquids and crystals to quantum gases. The special attention in particular since the experiments of Ketterle group [7] was focused on the imbalanced (spin-polarized) Fermi gases in magnetic traps of different dimensionality.

The authors of [7] were the first to experimentally observe the phase separation in imbalanced ultracold Fermi gases. To be more specific, they detected the formation of the balanced Bose–Einstein condensate of compact local pairs $f_\uparrow f_\downarrow$ (consisting of two fermions with opposite spins) in the center of the 3D (spherically symmetric) trap. This unpolarized state corresponds to equal densities $n_\uparrow = n_\downarrow$ of "up" and "down" spins. At the same time, they observed imbalanced (strongly polarized) mixture of spins (with the prevailing "up" spins, $n_\uparrow > n_\downarrow$) at the periphery of the trap.

M. Yu. Kagan et al., *Electronic Phase Separation in Magnetic and Superconducting Materials*, Springer Series in Solid-State Sciences 201,
https://doi.org/10.1007/978-3-031-55467-4_14

It is very interesting to note that the situation is strictly reversed in quasi-one-dimensional imbalanced Fermi gases [6]. Here, the balanced (unpolarized) state of paired fermions is located at the periphery of the quasi-1D (cigar-shaped) magnetic trap, while the imbalanced strongly polarized mixture of spins (in which $n_\uparrow > n_\downarrow$) is concentrated at the trap trap center.

The first experimental results on the phase separation in the imbalanced quasi-two-dimensional Fermi gases belong to the Thomas group [7]. The experimental data of [7] demonstrate that the situation in quasi-2D geometry of the disc-shaped magnetic trap is qualitatively similar to the 3D case. Namely, the unpolarized balanced state of paired fermions is formed at the trap center.

Note that the degree of imbalance (or more precisely, the polarization degree $\alpha = (n_\uparrow - n_\downarrow)/(n_\uparrow + n_\downarrow)$) in the spin-polarized phase of the phase-separated (inhomogeneous) Fermi gas is larger that the polarization degree of the initial homogeneous gas and depending upon the dimensions of the system, density and the effective parameters of the trap can even reach 100% polarization. Note also that the spin-polarized phase of the inhomogeneous system is very interesting in itself and can be the subject of, e.g., the FFLO state [8, 9] at small polarization degrees or enhanced triplet p-wave superconductivity of the Kohn–Luttinger type [10–12] at larger polarization degrees. Moreover, the critical temperature of p-wave pairing can be strongly increased and will have a pronounced peak in the 3D case at polarization degree α =48% and in the 2D case at 60% of polarization [10–12]. Correspondingly, $n_\uparrow/n_\downarrow = 4$ in this optimum situation in 2D leading to very extended and broad peak and rather high T_c in this case, which can be only slightly less $T_{c1} \propto E_{F\uparrow} \exp(-2) \sim E_{F\uparrow}/7$ than the Fermi energy of "up" spins [11].

There is also an interesting experimental effort of the Hulet group [13] to match the quasi-1D systems with the systems of higher dimensionalities (D = 2 and 3) by creating the arrays of cigar-shaped quasi-one-dimensional tubes containing spin-polarized ultracold Fermi gas.

Theoretically [14], we can describe the 1D–3D crossover in this system in terms of the model of the ladders or coupled chains with increasing number of legs (one-leg, two-leg, three-leg ladders, etc.) [8] and investigating their phase diagrams in the limit of small number of legs N ($N = 1$), and then in the limit $N \to \infty$ (or of coupled chains in other words).

The theoretical analysis of this problem was performed in [14], where the phase diagrams of spin-polarized ultracold Fermi gases on the ladders with strong onsite Hubbard attraction between fermions were considered. In this model, for small number of legs $w = 1, 2$, and 3, we can construct the phase diagrams which contain the regions of equal density (ED) phase $n_\uparrow = n_\downarrow$, partially polarized (PP) phase with $n_\uparrow > n_\downarrow$, and fully polarized (FP) phase where $n = n_\uparrow$ (and $n_\downarrow = 0$).

In Fig. 14.1, we show the grand canonical phase diagrams of the attractive-U Hubbard model for $U = -7t$ and varying number of legs ($w = 1, 2$, and 3) of the ladder systems. The vertical axes in Fig. 14.1 correspond to the chemical potential of the system $\mu = \partial E/\partial n_{tot}$ where $n_{tot} = n_\uparrow + n_\downarrow$ is the total density of both spin components and E is the total energy. The horizontal axes correspond to the effective

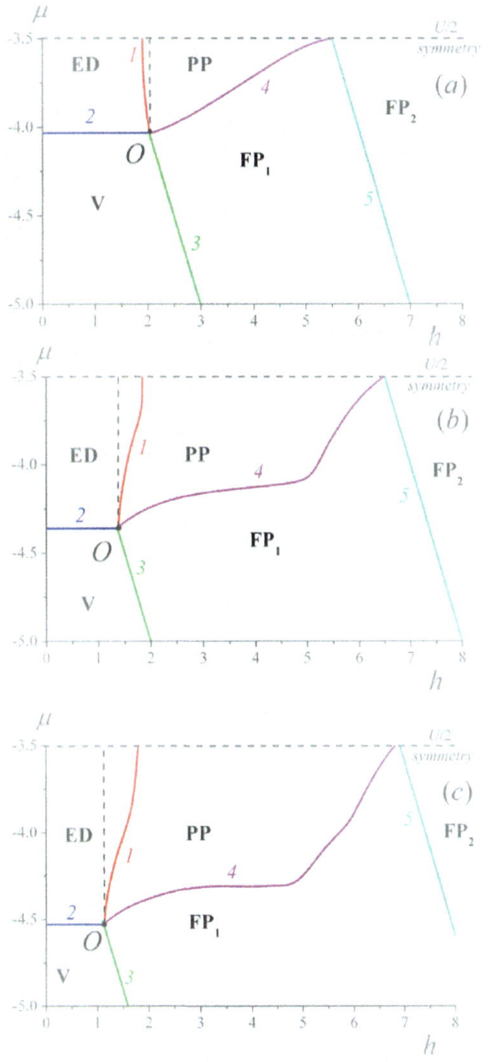

Fig. 14.1 Grand-canonical phase diagrams of the attractive-U Hubbard model for $U = -7t$ and varying number of legs w. We only show the $\mu < U/2, h > 0$ part of the phase diagram since the $h < 0$ part is symmetric with respect to the $h \to -h$ symmetry, and $\mu > U/2$ is symmetric with respect to the particle–hole symmetry of the Hubbard model. Solid lines are simulation results, the vertical dashed line is guide to an eye only. See the text for the discussion. **a** $w = 1, L = 40$; **b** $w = 2, L = 40$; **c** $w = 3, L = 40$ [14]

magnetic field $h = n_\uparrow - n_\downarrow$. In the presence of confinement potential of the trap $V(x)$, in the local density approximation, as usual we can replace $h \to h - V(x)$.

One very interesting feature for all phase diagrams on Fig. 14.1 in [14] is the presence of a multicritical point in them with the coexistence of four phases ED, PP, FP and vacuum (V) phase, where $n_\uparrow = n_\downarrow = 0$. Moreover, we have rather nontrivial phase boundaries between ED and PP phases (line 1), as well as between PP and FP phases (line 4) in Fig. 14.1. These curves are nonmonotonic (have different slopes) for two-leg and three-leg ladders. In [15] it was shown, however, that it is purely lattice effect due to filling of the higher bands at finite values of the filling fraction

$p = n_{tot}d = Nd/Lw$, where n_{tot} is the total density, N is total number of particles on the ladder, L is the length of the ladder, and d is the intersite distance. Thus, $N_{sites} = Lw/d$ is the total number of sites of the lattice and w is the number of legs.

Let us stress that experimentally the most relevant parts of the phase diagrams shown in Fig. 14.1 correspond to the vicinity of a multicritical point since it describes a continuum limit where the filling fraction is small. The qualitative feature in this domain is the slope of line 1 which separates ED and PP phases. It is positive for two- and three-leg ladders $\partial\mu/\partial h > 0$ and negative for one-leg ladder (1D spin chain) $\partial\mu/\partial h < 0$, where both experiments [6] and exact Bethe ansatz results [16] illustrate that the PP phase occupies the center of the trap, while ED phase occupies the wings of the atomic cloud.

Thus, the change of the sign of the derivative $\partial\mu/\partial h$ from negative to positive one for two- and three-leg ladders just reflects the fact that in 2D and 3D traps the situation is opposite to the 1D, and ED phase occupies the center of the trap while PP phase is on the wings of the cloud. Hence, we can say that qualitatively the increase of the number of legs correctly describes 1D to 2D and 3D crossover.

14.2 Phase Separation in Fermi–Bose Mixtures

We know quite well that the maximum solubility of ^{3}He in superfluid ^{4}He in ^{3}He–^{4}He Fermi–Bose mixture corresponds to 6.4% of ^{3}He concentration at zero pressure [17]. This maximum solubility concentration can be further increased (up to 9.5% of ^{3}He concentration or even higher) when we increase the pressure or spin-polarize the mixture [1]. At the same time, in solid He-4 already at very low concentrations of ^{3}He an inhomogeneous state is formed, namely the state with ^{3}He clusters in the matrix of solid ^{4}He.

Returning back to the Fermi–Bose mixtures of alkali isotopes ^{6}Li–^{7}Li or ^{40}K–^{87}Rb of ultracold quantum Fermi and Bose gases in magnetic traps (often in the regime of Feshbach resonance [18]) note that the optimum conditions for phase separation and formation of large droplets appear again in the strongly imbalanced case when, e.g., the bosonic density is larger than the fermionic density $n_B > n_F$.

In the case of equal densities of fermions and bosons $n_B = n_F$ in a Fermi–Bose mixture with attractive interaction between fermions and bosons, a more favorable is the creation of the composite fermion $f_\sigma b$ [19] with the subsequent creation of quartets (or local singlet pairs $f_\uparrow b f_\downarrow b$), formed by two composite holes with opposite spins [20]).

However, in the case of strong imbalance $n_B > n_F$, large bosonic droplets (containing five and more particles) are formed and experimentally measured in the system [21]. Their size again (like a FM polaron radius in Chaps. 1 and 2) is governed by the simple variational calculations, which take into account the delocalization kinetic energy and the energy of their interactions.

Namely, the limitation on the number of particles ($N > 4$) in the droplet is related in the resonance situation to the logarithm of the ratio between the large resonance

s-wave scattering length a and the small radius r_0 of the hard-core repulsive interaction. In the 2D Bose gas or strongly imbalanced 2D Fermi–Bose mixture (with $n_B \gg n_F$), this condition reads (see [22, 23])

$$N \leq N_{max} \approx 0.9 \ln \frac{a}{r_0}\,, \tag{14.1}$$

where the scattering length a determines the two-boson binding energy $|E_b|$

$$|E_b| = \frac{1}{ma^2}\,. \tag{14.2}$$

Here, m is the boson mass. Note that while the typical radius of the hard-core repulsion is $r_0 \sim 3 - 4$ Å, the scattering length in the resonance situation can reach the values as high as $a \sim 3000$ Å or even more.

Correspondingly the radius of a N-particle droplet is given by the relation

$$R_N \sim ae^{-N} > r_0 \tag{14.3}$$

and should be larger, of course, then the hard-core radius r_0. Thus, the binding energy of a N-particle droplet reads

$$|E_N| = \frac{1}{mR_N^2} \sim |E_b|e^{2N} < \frac{1}{mr_0^2}\,. \tag{14.4}$$

Large bosonic droplets, as we already mentioned, were experimentally observed by Modugno et al. [21] in the imbalanced Fermi–Bose mixture of ^{40}K (fermions) and ^{87}Rb (bosons) with the bosonic dominance. Note that large bosonic droplets were also observed [24] in the Bose gas of ^{164}Dy with a strong dipole–dipole attraction. The authors of [24] investigate the stability criterium or the inverse compressibility $\kappa^{-1} \propto \frac{d\mu}{dn} = \frac{d^2E}{dn^2} > 0$ and estimate the droplet radius R_N and bosonic density n_0 in the center of the droplet considering the typical energy functional

$$\Delta E = -Un^2 + \alpha n^{5/2} + \gamma n^3\,, \tag{14.5}$$

where n is the bosonic density. In Eq. (14.5), the first term represents the mean-field dipole–dipole attraction, while the second term (beyond the mean-field approximation) corresponds to the Lee et al. [25] repulsive corrections due to quantum fluctuations. Finally, the third term describes the three-body interaction.

These type of functionals are widely used for droplet formation in liquid ^{4}He and in the nuclear matter [26]. The presence of the second and third repulsive terms in the functional stabilizes the collapse in the attractive Bose gas. The most important is the contribution proportional to $\alpha\sqrt{n}$ from quantum fluctuations [25] to the inverse compressibility, while the contributions γn from three-body interactions are subdominant. The estimates in [24] yield $n_0 \sim 10^{14}$ cm^{-3} for the central bosonic density.

14.3 Droplets of the Metallic Phase on the Phase Boundary Between Crystalline and Liquid Metallic Hydrogen and Its Metallic Alloys

14.3.1 The State of the Art. Room Temperature SC

Rapid growth of superconducting critical temperature in hydrogen metallic alloys H_2S, H_3S, and LaH_{10} from 192 K till 203 K and, then, till 250–260 K at megabar pressures [27, 28] and finally almost room temperature 288 K (or 15 °C) at pressures $P = 2.6$ Mbar obtained in a carbonaceous sulfur hydride $[H_2\,S]_{1-x}[CH_4]_x H_2$ (which is chemically synthesized) from the H_2S+H_2+CH_4 precursor mixture) elucidates a way from high temperature to the room-temperature superconductivity [29]. On the other hand, the first convincing experimental evidence of the possible transition to metallic phase for pure metallic hydrogen at pressures $P = 5$ Mbar was demonstrated in the synchrotron infrared spectroscopic measurements [30].

14.3.2 A Bunch of Theoretical and Experimental Ideas on Metallic Hydrogen

These achievements resulted in a huge interest in both of physical and chemical communities as well as of the research community working in energetics and of the material research community in proper understanding of the gross phase diagram of the normal and superconducting phases of the metallic hydrogen and its alloys.

The most important in this context is to find a way both in theory and in experiment to reduce the pressure without reducing the SC critical temperature or, in other words, to synthesize experimentally or to predict theoretically the long-living metastable phase of the metallic hydrogen, which can serve as a local minimum of the thermodynamic potential even at very low pressures. This strategy was theoretically tried in [31]. However only the decrease of the pressure till 0.1 Mbar was obtained for the most promising metastable phase of metallic hydrogen.

One of the possible roots here according to Silvera et al. [54] and Ashcroft [32] is to investigate more carefully the first-order phase boundary (and its vicinity) between solid (crystalline) and liquid metallic hydrogen or its metallic alloys. In this region of the phase diagram, we can expect a very interesting behavior of our system which probably will have some similarities with first-order phase boundary between quantum solid and quantum liquid.

Note that this type of phase boundary was a subject of intense investigation in a low-temperature community working on the physics of superfluid and solid ^{4}He and ^{3}He (in a liquid phase which is described by Landau Fermi–Bose theories for quantum fluids) and nowadays searching a supersolid behavior [33, 34] in quantum crystals of helium.

Note that theoretically predicted in 1970-ties in [35, 36] low-dimensional phases of metallic hydrogen such as a planar phase (resembling graphite or graphene) and quasi-one-dimensional phase (resembling A.A. Abrikosov vortex lines [37]) with proton chains embedded in the Fermi liquid of the conduction electrons exhibit a lot of similarities with quantum crystals of solid ^{4}He.

These similarities with respect to, e.g., De Boer parameter of quantum degeneracy [38] or quantum melting parameter of Lindemann [39] as well as the possibility to form different Bose–Einstein condensates, namely SC Bose–Einstein condensate of Cooper pairs in electron subsystem as well as superfluid Bose–Einstein condensate of biproton pairs in ionic subsystem were indicated in [40, 41].

Note that the formation of the droplets of one phase in the matrix of another phase (liquid droplets in solid or solid droplets in metal) is a main feature of all first-order phase transitions not only in quantum systems but in classical systems as well. The quantum nature of solid and liquid phases yield additional possibilities here when we consider, for instance, the quantum diffusion (or the quantum motion) and mobility of a neutral impurity (such as an ^{3}He atom) or a charged impurity (electron and ion) in superfluid or solid ^{4}He. Here, we know quite well that either the bubble or a piece of ice is formed around the charge depending upon the sign of the charge (for negative and positive charges).

14.3.3 Eliashberg Theory for SC in Metallic Hydrogen

Let us emphasize that the critical temperature of the SC pairing in the electron subsystem is governed by the famous Eliashberg strong-coupling theory for electron–phonon interaction in metallic hydrogen. The careful numerical solution of the full set of Eliashberg equations both on real [44–46] and imaginary [42, 43] axes yield the SC critical temperature $T_c = 221.7$ K for the most promising (with respect to experimental realization) I41/AMD phase of metallic hydrogen at pressures $P = 5$ Mbar [42, 43].

In Figs. 14.2, 14.3, and 14.4, we present the results of the calculations [43] performed for the Eliasberg function $\alpha^2 F(\omega)$, electronic density of states $N_0(E)$, and order parameter $\Delta_0(T)$ as a function of temperature for the following set of parameters: the Coulomb pseudopotential $\mu^* = 0.1$, chemical potential $\mu = 6.8$ eV, and the characteristic phonon (Debye) frequency $\omega_D = 0.34$ eV.

The results are the following: $T_c = 221.7$ K (with the correction to the chemical potential of electrons), $T_c = 220.7$ K (without the correction). Although for this compound the difference between the results obtained with and without correction for the chemical potential of electrons is small, for other substances it can be quite

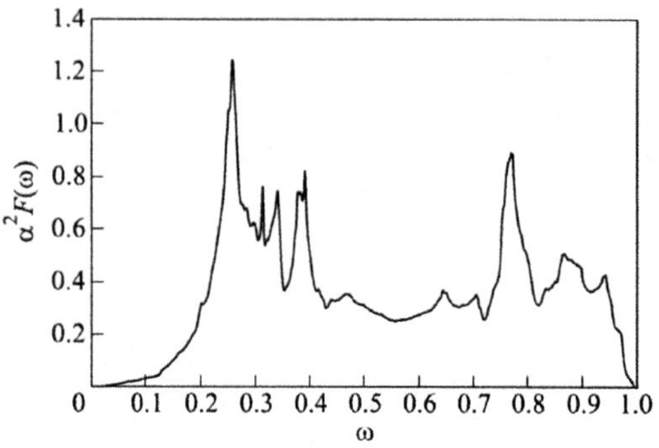

Fig. 14.2 The Eliashberg function $\alpha^2 F(\omega)$ for I41/AMD phase of metallic hydrogen under a pressure of 500 GPa [43]

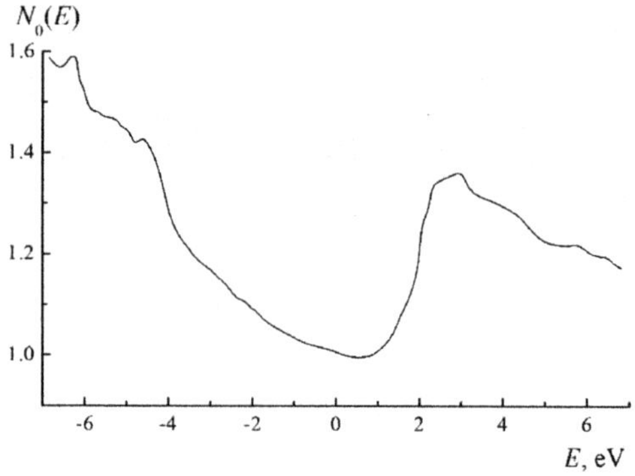

Fig. 14.3 Dimensionless electronic density of states $N_0(E)$ for I41/AMD phase of metallic hydrogen under a pressure of 500 GPa for the Debye phonon frequency $\omega_D = 0.34$ eV [43]

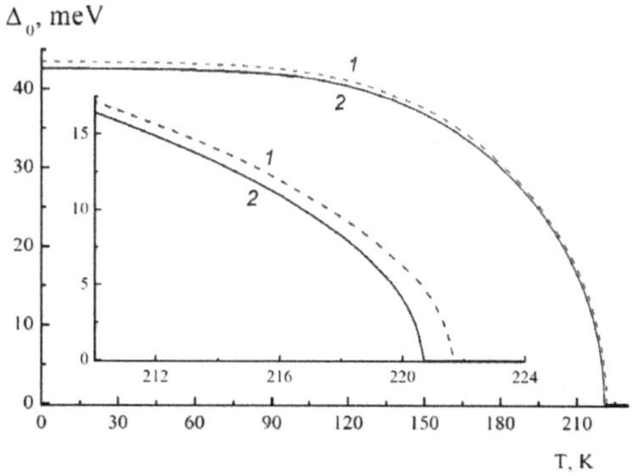

Fig. 14.4 Order parameter $\Delta_0(T)$ as a function of temperature for the I41/AMD phase of metallic hydrogen under a pressure of 500 GPa (1—calculations with the correction to the chemical potential of electrons; 2—without the correction) [43]

significant. We also get the following values for the important parameters describing the SC transition in this phase: the electron–phonon mass enhancement parameter $\zeta = 1.68$, the Sommerfeld coefficient $\gamma = 10.2$ mV/(mole K^2), the SC gap at $T = 0$: $\Delta(0) = 42.5$ meV, the thermodynamic critical magnetic field at $T = 0$ $H_c(0) = 61.8$ T, the discontinuity jump of the specific heat at the transition temperature $\Delta C(T_c) = 6.36$ J/(mole K), and the power of the isotope effect $\beta = 0.487$. From this set of parameters, we can see that the SC transition in metallic hydrogen is indeed of the phonon origin and can be described by the strong-coupling Eliashberg theory for electron–phonon interaction.

Note that the attempts to calculate the temperature of the superconducting transition of metallic hydrogen I41/AMD phase under pressure of 500 GPa have also been made earlier in [44–46], the SC temperature was calculated by solving the system of Eliashberg equations for the real axis, and proved to be 217 K. These results are in a good agreement with more recent results obtained by solving Eliashberg equations on the imaginary axis [43]. Of course, as a natural generalization of the results presented in this section, the model based on the Eliashberg equations could be extended to take into account the effect of impurities and the multiband nature of some compounds (see, e.g., [47]).

Note also that quite recently in [48], the authors presented the design principles and calculations of the SC transition temperatures for seven triple compounds with hydrogen-based alloy backbone at various pressures (see Table 14.1).

Let us emphasize that in these substances, the metallic hydride is stabilized by light elements such as B and Be. This design principle is, in contrast to previously studied hydrides such as LaH_{10}, CeH_9, and YH_6, stabilized by rare-earth elements La, Ce, and Y.

As we mentioned previously, the hopes to reduce pressure in SC phases and in the same time to increase T_c are related to the more complicated chemical composition of triple metallic hydride: H-C-S in [29] (with $T_c = 288$ K—experimental record to date) or the triple alloys theoretically considered in [48].

For all substances, for which the Eliashberg functions are given in [48], Ikhsanov et al. [43] recalculated T_c by two methods (see Table 14.1): using the standard Allen–Dynes algorithm [49] (column A-D) [43], and the solution of the full system of Eliashberg equations on the imaginary axis with an account of the corrections to the chemical potential (column El-1) [43]. T_c obtained in [48] by using the Allen-Dynes modified McMillan [50] equation and by solving the Eliashberg equations are also given in the table (columns A-D-2 [48] and El-2 in [48], respectively).

As can be seen from Table 14.1, first, the results of the calculations of [43, 48] with the full set of Eliashberg equations mainly coincide with an accuracy of 1 K (the discrepancy is obviously due to the error in digitizing the graphs of the Eliashberg functions). Second, the calculations by the Allen–Dynes algorithm, in this case, give a very good approximation to the superconducting transition temperature (no worse than 2 K for most materials). On the contrary, using the Allen–Dynes modified McMillan equation in A-D-2 gives T_c underestimated by several tens of degrees.

Table 14.1 Results for the superconducting transition temperatures of various substances with hydrogen-based alloy backbone at corresponding pressures (GPa) obtained in [43, 48] from the solution of the full set of Eliashberg equations (El-1 are results of [43] and El-2 are results of [48]), from standard Allen–Dynes approximation A-D-1 in [43] and Allen–Dynes modified McMillan equation A-D-2 in [48]

Substance	P (GPa)	T_c (K)			
		A-D-1	A-D-2	El-1	El-2
$LaBeH_8$	50	194	167	192	191
$LaBH_8$	70	160	144	159	160
$LaAlH_8$	100	145	130	145	144
$CaBeH_8$	210	307	254	303	302
$CaBH_8$	100	241	212	239	238
$YBeH_8$	100	250	215	248	249
$SrBH_8$	150	206	163	204	200

We can see also high critical temperatures (of the order of the room one, $T_c \sim 302-303$ K) for $CaBeH_8$ at pressures $P \approx 2.1$ Mbar. Note that even a higher critical temperature was predicted in [51] for Li_2MgH_{16} at pressures of about 2.5 Mbar. However, this last result should be checked thoroughly.

14.3.4 Electroneutrality Restrictions on Droplet Formation in Two-Component Charged Neutral Plasma

Note that in liquid metallic phase of pure hydrogen, the proton lattice is melted and we effectively have degenerate cold Coulomb plasma with two sorts of charged mobile particles—protons and electrons. In our language, we can say that we have a Fermi–Fermi mixture of light particles—electrons and very heavy particles—protons interacting with each other via long-range Coulomb potentials. It seems very appealing here to use the ideas of Chap. 10 on electron polaron effect for an additional increase of an effective heavy mass (due to dressing of heavy particle in the virtual cloud of soft electron–hole pairs of light particles) and an instability with respect to phase separation at large mismatch between densities and masses of the components. Unfortunately, the charge electroneutrality condition is very restrictive here leading to the tough condition of equal local densities of electrons and protons $n_{el} = n_p$, which excludes the phase separation in the Fermi–Fermi mixture with two species (two sorts) of oppositely charged particles such as protons and electrons. The situation can be milder for three and more sorts of fermions in hydrogen metallic alloys with one or two sorts of both electrons and ions.

14.3.5 Phase Diagram and Droplets Formation in Metallic Hydrogen

It is well known from the astrophysical observations (see e.g., [52, 53]) and shock-wave experiments [54, 55] that in the deep interior of some planets of the solar system such as Jupiter and Neptune (which contain hydrogen as a main constituent) at high pressure $P \geq 1.4$ Mbar the hydrogen molecules are so close to each other that they start to overlap strongly and crush each other thus losing their identities. As a result, we get the phase transition from molecular hydrogen to ionized metallic hydrogen.

Moreover, the phase diagram of hydrogen in the $P - T$ plane contains at least four different phases (see Fig. 14.5 and [56]) in the high-pressure domain. Namely, at high pressure $P \gtrsim 1.4$ Mbar [57] and low temperatures $T \leq 1000$ K, we get the phase transition between solid molecular and solid metallic hydrogen. At the same, time at high pressures and high temperatures $T \gtrsim 1000 - 2000$ K, we have the transition from liquid molecular hydrogen to liquid metallic hydrogen. Thus, the high temperature physics of liquid hydrogen phases at high pressures is closer to the ideas of Fermi–Fermi mixture of ionized protons and electrons mentioned briefly in the previous subsection (Sect. 14.3.4) of this chapter.

Nevertheless, all phase boundaries between solid molecular and solid metallic hydrogen, liquid molecular and liquid metallic hydrogen as well as the melting line between solid and liquid metallic hydrogen [56, 58] are very interesting and deserve a very careful investigation both from experimental and theoretical points of view. The similar analysis can be also generalized on the phase diagram and phase boundaries between molecular and metallic deuterium [55, 59–63].

Let us stress that the melting lines between solid and liquid phases of both molecular and metallic hydrogen correspond to the first-order phase transition and thus the droplets formation is natural for these phase boundaries. The same story holds for liquid droplets and solid (ice) droplets in helium.

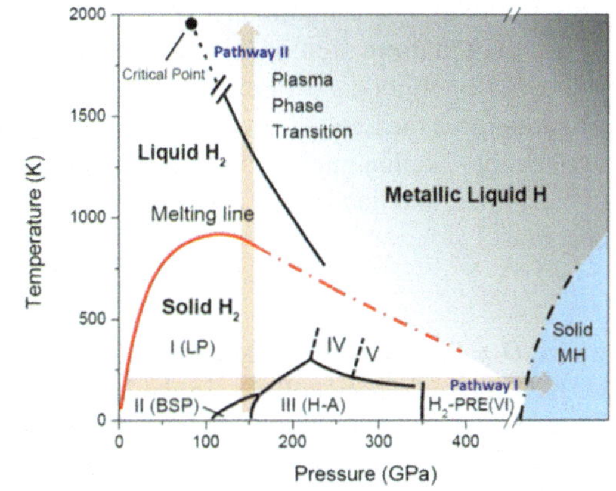

Fig. 14.5 $T - P$ phase diagram of molecular and metallic hydrogen at high pressures. The phase diagram contains (among the others) four phases of molecular and metallic hydrogen, namely two phases of solid molecular and solid metallic hydrogen at high pressures and low temperatures and two phases of liquid molecular hydrogen and liquid metallic hydrogen at high pressures and high temperatures [56]

At the same time, there are two schools of thought in relation to the liquid–liquid transition (instantaneous or the first order again) between molecular and metallic phases. From our viewpoint, however, the first-order phase transition is also more natural here. Note that while the liquid molecular phase of hydrogen is inherently insulating, the liquid phase of ionized hydrogen is metallic.

Thus, close to the phase boundary between the liquid insulating and metallic phases, we should expect the formation of the metallic droplets (and then finally a large percolation cluster) of ionized metallic hydrogen in the insulating matrix of molecular hydrogen. The same picture should be valid also for molecular and metallic deuterium. Indeed, the shock-wave experiments give us some evidence in favor of this picture with metallic droplets in the insulating matrix at least in deuterium systems [60].

14.4 Excitonic Droplets and a Fermi–Bose Mixture of Excitons and Electrons

14.4.1 *Basic Ideas of Keldysh, Halperin, and Rice on Exciton Physics*

Since the pioneering papers of Keldysh et al., Halperin-Rice, and Kohn [64–70], the phase diagrams of many semiconducting materials (having, for example, congruent electron and hole bands), especially in the situation of reduced dimensionality, can be understood on the basis of excitonic physics (or the physics of bound electron–hole pairs).

An exciton is a bosonic quasiparticle with zero charge and spin, which at certain conditions, can exhibit the Bose–Einstein condensation (BEC) phenomenon [71, 72]. In a balanced system with equal densities of electrons and holes $n_h = n_e = n$, the possibility of BEC in the Bose gas of excitons is governed by the sign of the (residual) exciton–exciton interaction.

For the repulsive sign of this interaction and at low density, we have a slightly nonideal Bogoliubov gas of excitons with the acoustic spectrum at small momenta, which satisfies the Landau criterion for superfluidity. At the same time, for the attractive exciton–exciton interaction and larger densities, we can have at first the creation of biexcitons, and finally, when the binding energy of exciton is of the order of Fermi energy,

$$|E_b| \sim E_F \propto n^{2/D}/m\,, \tag{14.6}$$

where $D = 2, 3$ is the dimensionality of the system, excitons strongly overlap, and we have the real-space collapse in the Bose gas (more precisely in the Bose liquid) of excitons with the formation of large excitonic droplets.

The typical size of these droplets can be nicely described on the level of exact soliton-type solutions in the Gross–Pitaevskii [73, 74] or nonlinear Schrödinger equation (NLSE). The basic results for the collapse here belong to Zakharov and Kuznetsov [75, 76]. Note that the same type of nonlinear Schrödinger equations also describes the self-focusing of light in nonlinear electrodynamics [77–79].

The situation is even more interesting in imbalanced semiconductors, where $n_h \neq n_e$. Here, we can have again 2-particle excitons (eh) with the binding energy $|E_b|$ and 4-particle biexcitons (eh, eh) with the binding energy E_4.

However, on top of that, the creation of three-particle complexes, or trions (see e.g. [80]), consisting of two electrons and one hole (eeh)or two holes and one electron (hhe), with the binding energy E_3, as well as large droplets [81–84] with $N > 4$ particles and binding energy E_N is also posible. Note that the similar formation of the large droplets we have discussed already in Sect. 14.1 considering imbalanced Fermi–Bose mixtures of ultracold quantum gases ^{40}K and ^{87}Rb, or more generally a Fermi–Bose mixture with the attraction between fermions and bosons [20].

Note that the phase diagram of the attractive Fermi–Bose mixture is crucially dependent upon the hierarchy of the 2-particle, 3-particle, and 4-particle binding energies and the chain of the corresponding Saha temperatures, which determine the dynamic equilibrium between the binding and ionization in the system.

Analogous approach based on the concept of Fermi–Bose mixture of excess electrons (dressed by the polaron coat) and excitons can be utilized in the physics of semiconductors at much higher temperatures [80, 85]. Moreover, in layered systems, such as transition-metal dichalcogenides (MoS_2, $MoSe_2$, WS_2, and WSe_2), we can even have a space-separated Fermi–Bose mixture of excitons (bosons) and polarons (fermions) at separate layers [86, 87]—the situation resembling the physics of $KBiO_3$ bismuthates considered in Chap. 12.

Besides that, the model of the hybrid 2D gas of excitons in the vicinity of the 2DEG considered in [87] can serve as a good platform to study the super-solidity [34] in exciton matter (as in quantum crystal of helium [33]) due to exciton-electron interactions and softening of the roton minimum in this system [17].

14.4.2 Nonlinear Schrödinger Equation and Self-similar Approach to the Analysis of Droplet Clouds

Note that in his review article [88], Keldysh formulated the concept of the coherent state of excitons. For this state, we can introduce the macroscopic wave function similar to the phenomenological wave function of the Ginzburg–Pitaevskii type [89] of superfluid Bose liquid. Formally, we can describe the system on the level of a nonlinear Schrödinger equation (NLSE) for the macroscopic Ψ function of the coherent excitonic system in a close similarity with the description of the kinetics of the formation of the Bose–Einstein condensates (see e.g. [90–92]).

Note that this formalism can be applied not only to the Bogolibov repulsive Bose gas of excitons but also in the attractive case when we deal with collapse in the excitonic system [75, 76].

A self-similar approach of Zakharov–Kuznetsov [75, 76] developed originally for the quasiclassical solutions in the collapse problem was later on generalized by Kuznetsov, Kagan, and Turlapov [93–95] for the opposite case of Bose and Fermi gas expansion into vacuum, when we switch off the trapping potential.

The idea of the calculations is the following. Let us consider the NLSE in the form

$$i\hbar\frac{\partial\Psi}{\partial t} = -\frac{\hbar^2}{4m}\Delta\Psi + \mu(n)\Psi\,, \tag{14.7}$$

where $\mu(n)$ is the chemical potential. In the case of the Bose gas of compact molecules (excitons with the mass $2m$ for equal effective masses of electrons and holes). The chemical potential in Eq. (14.7) reads

$$\mu(n) = g|\Psi|^2 = gn\,, \tag{14.8}$$

where the coupling constant can be written as

$$g = 4\pi\hbar^2 a_s/m\,, \tag{14.9}$$

and a_s is the s-wave scattering length. For the positive sign of the scattering length ($a_s > 0$), we have a repulsive slightly nonideal Bogoliubov gas of excitons, while for the negative sign ($a_s < 0$) we have the attraction, and the excitonic condensate is unstable toward the collapse in the system. We introduce in the standard way the amplitude and phase of the macroscopic Ψ function according to

$$\Psi = \sqrt{n(\mathbf{r},t)}\exp i\varphi(\mathbf{r},t)\,. \tag{14.10}$$

Then, separating real and imaginary parts in the NLSE (14.7), we obtain two hydrodynamic equations, namely, the continuity equation for the density n and the modified Euler equation for the velocity $\mathbf{v} = \nabla\varphi$ (we assume the absence of vortices $\nabla\times\mathbf{v} = 0$). These equations have the form

$$\frac{\partial n}{\partial t} + \mathrm{div}(n\nabla\varphi) = 0 \tag{14.11}$$

and

$$\frac{\partial\varphi}{\partial t} + (\nabla\varphi)^2/2 + \mu(n) + T_{QP} = 0\,. \tag{14.12}$$

Quantum pressure T_{QP}, which enters Eq. (14.12), is defined as

$$T_{QP} = \Delta\sqrt{n}/2\sqrt{n}\,. \tag{14.13}$$

Note that in the quasiclassical approximation to the Gross–Pitaevskii equation, we neglect the quantum pressure in Eq. (14.12) assuming that

$$|\mu(n)| \gg T_{QP} \,. \tag{14.14}$$

As a result, the equations of motion transform into the hydrodynamic equations [93–95] for the potential flow of ideal gas

$$\frac{\partial n}{\partial t} + \mathrm{div}(n\nabla\varphi) = 0 \tag{14.15}$$

and

$$\frac{\partial \phi}{\partial t} + (\nabla\phi)^2/2 + \mu(n) = 0 \,. \tag{14.16}$$

In the quasiclassical approximation, the system of equations (14.15) and (14.16) is usually solved in the self-similar ansatz. In this ansatz, the density reads [93–95]

$$n = \frac{1}{a_x a_y a_z} f\left(\frac{x}{a_x}, \frac{y}{a_y}, \frac{z}{a_z}\right) \,. \tag{14.17}$$

In Eq. (14.17), we assume that three scaling parameters $a_x(t)$, $a_y(t)$ and $a_z(t)$ are functions of time, and f is a scaling function which depends upon self-similar variables $\xi_x = x/a_x(t)$, $\xi_y = y/a_y(t)$, and $\xi_z = z/a_z(t)$. Note that this ansatz conserves the total number of particles N.

It should be emphasized that the system of equations (14.15) and (14.16) preserve all the symmetries and integrals of motion of the problem, including the virial theorem first formulated for the 2D NLSE with cubic non-linearity by Vlasov, Petrishev, and Talanov [79]. The virial theorem has the form

$$2m\frac{d^2}{dt^2}\left(\int r^2|\Psi|^2 d\mathbf{r}\right) = 4H = 4E \,, \tag{14.18}$$

where $H = E = const$ is the total energy of the system.

For the nonideal Bose gas of excitons with the mass $2m$ in the quasiclassical approximation (where we neglect the quantum pressure), the Hamiltonian yields

$$H = \int \left(\frac{\hbar^2}{4m}|\nabla\Psi|^2 + \frac{g}{2}|\Psi|^4\right) d\mathbf{r} \approx \frac{\hbar^2}{2m}\int \left(n\frac{(\nabla\varphi)^2}{2} + 2\pi a_s n^2\right) d\mathbf{r} \,. \tag{14.19}$$

Let us stress that the cloud size r^2 in Eq. 14.18 is determined by the Pythagorean theorem $r^2 = a_x^2 + a_y^2 + a_z^2$, where in the isotropic (spherically symmetric) case, the scaling parameters are equal: $a_x = a_y = a_z = a$, and

$$r^2 = 3a^2 \,. \tag{14.20}$$

It is interesting that the scaling parameter $a(t)$ in the isotropic case is governed by the equation of motion (the Newtonian law) [93–95]

$$\ddot{a} = \lambda/a^3\,, \tag{14.21}$$

where λ is the scaling constant, which determines the behavior of scaling function f. In the isotropic case, the scaling function depends only on the self-similar radial variable $\xi^2 = \xi_x^2 + \xi_y^2 + \xi_z^2$, and for the Bogoliubov gas with the positive chemical potential $\mu(n) = gn > 0$ has the form

$$f(\xi) = 1 - \lambda\xi^2/4\,. \tag{14.22}$$

Correspondingly, the total energy E for the self-similar solution is described by an elegant expression

$$E = \frac{3}{2}\left(\dot{a}^2 + \lambda/a^2\right)\,, \tag{14.23}$$

which contains the potential energy

$$U(a) = \lambda/a^2\,, \tag{14.24}$$

corresponding to the fall of a quantum particle to the center.

Since the total number of particles $N = const$ is also conserved in our problem

$$N = \int |\Psi|^2 d\mathbf{r}\,, \tag{14.25}$$

the virial theorem can be written as

$$2mN\frac{d^2(R^2)}{dt^2} = 4E\,, \tag{14.26}$$

where the total energy E is given by Eq. (14.23). In the spherically symmetrical case, we have

$$R^2 = \langle r^2\rangle = \langle 3a^2\rangle\,, \tag{14.27}$$

for the averaged cloud dimension squared. Moreover,

$$NR^2 = \int r^2|\Psi|^2 d\mathbf{r}\,. \tag{14.28}$$

Integrating twice over time t the left and right hand sides of expression (14.18) describing the virial theorem, we obtain the nice quadratic in time expression for the averaged cloud dimension squared

$$2mNR^2 = 2Et^2 + C_1 t + C_2\,, \tag{14.29}$$

where C_1 and C_2 are two additional integrals of motion introduced long time ago by Ermakov [96].

In the case of repulsion, for the coupling constant $g > 0$ (positive s-wave scattering length $a_s > 0$), the energy $E > 0$ is positive in Eqs. (14.18), 14.19), and (14.27) (defocusing nonlinearity in optical terminology), and thus at large times $t \to \infty$, the cloud dimension (the droplet size) is

$$R^2 = \langle 3a^2 \rangle \approx \frac{E}{mN} t^2 . \tag{14.30}$$

Hence the droplet size $R \propto t$. Thus, the expansion rate for the droplet tends to the constant value (a ballistic regime)

$$v_\infty = \sqrt{E/mN} . \tag{14.31}$$

At the same time, in the Bose gas with attraction (for the negative scattering length $a_s < 0$), the total energy in Eqs. (14.18) and (14.19) can become negative $E < 0$ in some range of parameters. In this case, the condensate is unstable with respect to a collapse. Note that in nonlinear optics, this situation corresponds to the self-focusing of light [78].

This means that naively at some intermediate times, when the Hamiltonian and the total energy $H = E \approx 0$ in Eq. (14.19), for positive values of the Ermakov integrals $C_1 > 0$ and $C_2 > 0$, the size of the droplet squared can be proportional to

$$R^2 \propto C_1 t + C_2 \quad \text{for} \quad t \to \infty . \tag{14.32}$$

However, both theoretical investigations of weak self-similar collapse reported in [75, 76], as well as experimental results suggest another possibility, namely when $C_1 < 0$ and at the same time, $C_2 > 0$. In this case, the droplet size squared at intermediate times can be given by

$$R^2 \propto (t_0 - t) , \tag{14.33}$$

and becomes very small $R \propto (t_0 - t)^{1/2} \to 0$ (collapses) at some intermediate time t_0.

Note that according to [97], the Ψ function of the NLSE in the self-similar regime can be represented in the self-similar form

$$\Psi = \frac{A}{(t_0 - t)^{1/2 + i\alpha}} \chi \left(\frac{r}{(t_0 - t)^{1/2}} \right) . \tag{14.34}$$

Close to the collapse region at $t \to t_0$, this character of the Ψ function manifests itself in singularities of the type

$$|\Psi|^2 = \frac{C}{r^2} \tag{14.35}$$

in the density profile of the system.

Note also that in the regime of strong quasiclassical collapse for the time dependence of the collapsing droplet size at $t \to t_0$, we have according to [75, 76] a slightly different exponent

$$R \propto (t_0 - t)^{2/5} \,. \tag{14.36}$$

14.4.3 Dipole-Active Excitons

Note that throughout this section, we mostly consider the condensation of excitons, which practically do not interact with light. If we consider the dipole-active excitons, their condensation leads to the generation of the coherent electromagnetic fields. In this case, our system is described by the coupled the NLSE and Maxwell equations.

From the full Hamiltonian of particles and fields, we find according to Keldysh [88] the following set of equations for amplitudes of the coherent states

$$i\hbar\frac{\partial \Phi}{\partial t} + \frac{\hbar^2}{2(2m)}\Delta\Phi + (\hbar\omega - E_0)\Phi - \mu|\Phi|^2\Phi = \mathbf{E}\mathbf{d} \,, \tag{14.37}$$

$$\nabla \times (\nabla \times \mathbf{E}) + \frac{\tilde{\varepsilon}}{c^2}\left(\frac{\partial}{\partial t} - i\omega\right)^2 \mathbf{E} = -\frac{4\pi}{c}\mathbf{d}\left(\frac{\partial}{\partial t} - i\omega\right)^2 \Phi \,, \tag{14.38}$$

where ω is the mean frequency of the field, $\mathbf{d}$ is the matrix element of the dipole moment for the excitonic transition, $\tilde{\varepsilon}$ is the dielectric constant of the crystal containing all the contributions besides the excitonic one under discussion, and $\mathbf{E}(x, t)$ is the complex amplitude of the field, which is related to the real electric field as

$$\frac{1}{2}[\mathbf{E}(x, t)\exp(-i\omega t) + \mathbf{E}^*(x, t)\exp(i\omega t)] \,. \tag{14.39}$$

The system of equations (14.37) and (14.38) contains the effects of the frequency and real space dispersion, as well as the effects of the nonlinear polarizability [88].

14.5 Acoustic Plasmons, Superconductivity, BEC, and the Formation of Droplets

14.5.1 Introduction: Acoustic Plasmons in Condensed Matter

While the standard gapped plasmon modes (often called the Vlasov or Lindhard plasmons [98]) lie in the typical energy range of 20 eV [98–100], the gapless acoustic plasmon modes (predicted in seminal papers of Pines and Nozieres [99, 100] and studied in detail by Silkin et al. [101–103]) have the typical energies of the order of 0.5

eV or lower. Thus, acoustic plasmons can serve as a natural bridge between condensed matter physics (including superconductivity and superfluidity) and nanoplasmonics.

Moreover, the ideas of L.D. Landau specified by the Fermi–Bose liquid theory can be fully applied to the acoustic plasmon modes. We can even estimate the small (power-law in temperature) contribution of the acoustic plasmon modes to the thermodynamics and to the kinetics of the system.

Note that in the dielectric formalism used in the Dolgov, Kirzhnits, and Maksimov approach [104], which is based on the important identities and sum rules for the dielectric permittivity, the plasmon contribution to thermodynamics and effective screening is governed by the real part of inverse dielectric permittivity $\mathrm{Re}1/\epsilon(q,\omega)$, while the contribution to the scattering time τ in the Drude formula for electrical conductivity is proportional to the imaginary part, $\mathrm{Im}1/\epsilon(q,\omega)$ (dielectric loss function).

To be more specific, we can expect that acoustic plasmon modes should play an important role in nonphonon mechanisms of superconductivity and at some specific conditions (on their density and dimensionality, first of all), they can even exhibit the Bose–Einstein condensation in the case of repulsive interaction between plasmonic modes or form large plasmonic droplets in real space (especially on the surfaces and in monolayers) in the case of the attractive interaction between them.

Note that in the mentioned above Pines and Nozières papers, acoustic plasmons (with linear energy spectrum) should arise in the two-band electron liquid with substantially different Fermi velocities and effective masses of "light" and "heavy electrons". In this, case the "sound" velocity for acoustic plasmons lies between the two Fermi velocities v_{Fh} and v_{FL} of heavy and light electrons, respectively

$$v_{Fh} < v < v_{FL}\,. \tag{14.40}$$

Later on, Silkin et al. generalized the Pines–Nozieres results and demonstrated the existence of the acoustic plasmon mode and mixed plasmon–phonon droplets at the metallic surface (at the 2D Tamm levels) or in the reduced dimensionality (including graphene and bismuth monolayers). First paper on superconductivity based on acoustic plasmons appeared in the middle of the sixties and belong to Fröhlich [105]. Formally, acoustic plasmons play the role of acoustic phonons in the Fröhlich mechanism; however, in the preexponential factor in the BCS-like formula for T_c enters now the typical electronic energy corresponding to a heavy electron band instead of the Debye frequency. Of course, the plasmon mechanism of superconductivity, similar to the phonon mechanism intrinsically, is a frequency dependent one and relies on the retardation effects. In the RPA approximation, this means that the full frequency dependence of dielectric function $\mathrm{Re}\epsilon(q,\omega)$ enters to the effective (screened Coulomb) interaction for the Cooper channel for heavy and light electrons, respectively

$$V_{eff}(q,\omega) = \mathrm{Re}\frac{4\pi e^2}{q^2\epsilon(q,\omega)}\,. \tag{14.41}$$

Nearly at the same time, as we already discussed in Chap. 10, Kohn and Luttinger [106] proposed another important mechanism of nonphonon superconductivity, in which only static Lindhard function [107] for the polarization operator $\Pi(q, \omega)$ and static dielectric function $\mathrm{Re}\epsilon(q, \omega = 0)$ containing the Kohn anomaly and Friedel oscillations [108, 109], is important for the Cooper problem. Later on, the results of Kohn and Luttinger were generalized by Fay and Lazer [110] and then by Kagan and Chubukov [10–12], who demonstrated the possibility to obtain a reasonable T_c in the two-band and spin-polarized case, as well as in the 2D or layered systems.

These two basic mechanisms of nonphonon superconductivity (of Kohn–Luttinger and Fröhlich) compete with each other in different low-density systems including bismuth and graphene, as well as in other semimetals, superlattices, and heterostructures [111, 112].

In high-density electron Coulomb plasma in the 3D case [113, 114] and in the absence of the lattice, the authors of [114] claim that frequency-dependent plasmon mechanism is more efficient. On the lattice in the Hubbard-type models the Kohn–Luttinger mechanism often prevails leading to the p-wave or d-wave pairing [41, 115] in contrast to the standard s-wave pairing in the plasmon mechanism.

Note that in the last several years, a more standard phonon mechanism of superconductivity, specified by the Eliashberg strong-coupling theory [116], proved to be dominant in MgB_2 and in metallic hydrogen alloys [27, 28, 32] under high pressures. However even in these systems electron–electron Kohn–Luttinger and Fröhlich mechanisms can get important corrections to the results of purely electron–phonon coupling [117, 118].

Note that the existence of the acoustic plasmon mode was explicitly demonstrated in the intensively studied (and important for applications) two-band superconductor MgB_2 [103]. The spectrum of the acoustic plasmon mode reads (see *ab initio* calculations of Silkin et al. [101] and Fig. 14.6)

$$\omega_{ac}^2 = B \sin^2(q_z d) , \tag{14.42}$$

and is practically linear close to the two points in momentum space for $q_z d \to 0$ and π.

At the same time, the Lindhard gapped plasmons have the spectrum

$$\omega_{Lind}^2 = \omega_{pl}^2 + A(1 - \cos(q_z d)) , \tag{14.43}$$

where z is the direction perpendicular to the layers (c direction). It is important to note that acoustic plasmon mode in Eq. (14.42) has a linear sound-like dispersion law not only for $q_z = 0$ but also for $q_z = Q = \pi/d$, where d is the interlayer spacing in MgB_2. Thus, both of these regions are important for acoustic plasmon contributions to the thermodynamics of the system. At the same time, the Lindhard plasmon specified by Eq. (14.43) has a large gap, which is determined by the Langmuir plasmon frequency

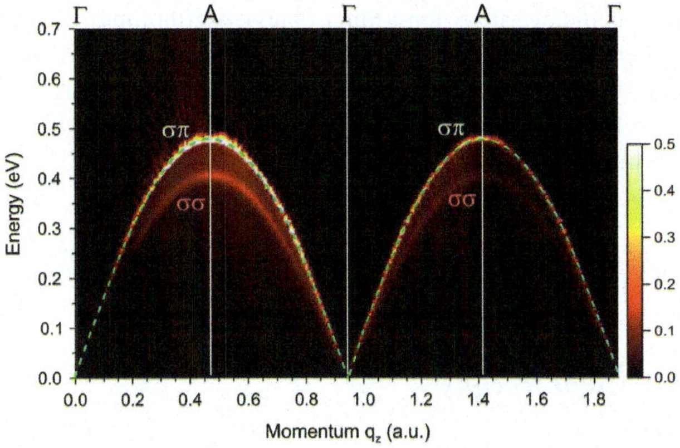

Fig. 14.6 Acoustic plasmon spectrum $\omega^2_{\sigma\pi} = B\sin^2(q_z d)$ in the layered MgB_2 metal in the z direction perpendicular to the layers. The spectrum is nearly linear close to points $q_z d \to n\pi$, where $n = 0, 1, ...$ and d is the interlayer distance [101]

$$\omega^2_{pl} = \frac{4\pi n e^2}{m_{eff}}. \tag{14.44}$$

Typically, this gap corresponds to the energy range of 20 eV and far exceeds the melting temperatures of metals. Thus, the contribution of the gapped Lindhard plasmon to the thermodynamics is usually negligibly small.

It will be interesting to estimate the contribution of acoustic plasmon mode to the reduction of the Coulomb pseudopotential μ^* [119] and the increase in T_c in MgB_2 using the results of full density functional theory.

14.5.2 Long-Range Charge Oscillations in MgB_2

Another interesting problem here is related to the behavior of the dielectric function not at zero, but also at finite (large) wave vectors q close to the nesting vector $Q = \pi/d$ and, at the same time, at small frequencies of soft acoustic plasmon modes.

This investigation is inspired by paper [120], where exciton condensation [64, 65, 67–70] was assumed to exist in transition metal dichalcogenides close to the charge density wave (CDW) state. Note that the CDW instability also manifests itself as a divergence in the effective interaction $V_{eff}(q, \omega)$ given by Eq. (14.41), but now in the particle–hole channel at small frequencies $\omega \to 0$ and finite momenta $q = Q$.

The similar situation occurs in MgB_2. Here, the structure of the dielectric function $\epsilon(q, \omega)$ due to the sound-like plasmon mode [101–103] leads to the singularity in the screened Coulomb interaction in momentum space at wave vectors q close to the "nesting" vector $Q = \pi/d$ and at frequencies close to the acoustic plasmon frequency given by Eq. (14.42).

To be more specific, we study long-range charge oscillations in the singular part of the screened Coulomb interaction in the real space $V_{eff}(z)$. We demonstrate that the effective interaction, in addition to a regular part has a singular part, which oscillates similarly to the Friedel (or RKKY) interaction in the 2D case behaving in the c direction perpendicular to the layers as

$$V_{eff}(z) = const \frac{\cos Qz}{(Qz)^2} . \tag{14.45}$$

14.5.3 Analytical Results

Here, we investigate the behavior of the dielectric function $\epsilon(q, \omega)$ in the RPA approximation for the two-band material such as MgB_2

$$\mathrm{Re}\epsilon(q, \omega) = 1 + \frac{4\pi e^2}{q^2} \left[\Pi_{LL}(q, \omega) + \Pi_{hh}(q, \omega) \right] , \tag{14.46}$$

where two polarization operators $\Pi_{LL}(q, \omega)$ and $\Pi_{hh}(q, \omega)$ correspond to two bands (π and σ bands) in MgB_2, which we for simplicity specify as heavy (h) and light (L) ones, respectively.

The zeroes of the dielectric function in Eq. (14.46) and correspondingly the poles of the effective (screened Coulomb) interaction $V_{eff}(q, \omega)$ in Eq. (14.41) in the hydrodynamic approximation provide us with two branches of plasmon spectra for the gapped (Lindhard) and acoustic plasmon modes with the spectra given by Eqs. (14.43) and (14.42), respectively.

To get a deeper insight into this result, let us consider first, for simplicity, the case of two parabolic bands of heavy and light electrons in a 3D metal. Then, in the hydrodynamic approximation, the polarization operators $\Pi_{LL}(q, \omega)$ and $\Pi_{hh}(q, \omega)$ take the form

$$\begin{aligned} \Pi_{LL}(q, \omega) &= -N_L(0) \frac{v_{FL}^2 q^2}{3(\omega^2 - v_{FL}^2 q^2/3)}, \\ \Pi_{hh}(q, \omega) &= -N_h(0) \frac{v_{Fh}^2 q^2}{3(\omega^2 - v_{Fh}^2 q^2/3)} , \end{aligned} \tag{14.47}$$

where

$$N_L(0) = \frac{m_L p_{FL}}{2\pi^2}, \qquad N_h(0) = \frac{m_h p_{Fh}}{2\pi^2} , \tag{14.48}$$

are the densities of states of light and heavy bands, respectively. In Eq. (14.48), m_L, p_{FL} and m_h, p_{Fh} are the charge carrier masses and Fermi momenta in the light and heavy bands, respectively.

As a result, after the substitution of Eqs. (14.47) and (14.48) in the relation $\mathrm{Re}\epsilon(q, \omega) = 0$, we obtain the biquadratic equation for ω^2. Two roots of this equation just correspond to the gapped Lindhard plasmon and gapless acoustic plasmon. The spectrum of Lindhard plasmons takes the form

$$\omega_{Lind}^2(q) = \omega_{pl}^2 + A v_1^2 q^2 , \tag{14.49}$$

where the gap ω_{pl}^2 in Eq. (14.49) is a combination of the Langmuir plasmon frequencies squared of heavy and light bands and v_1^2 is a combination of their Fermi velocities squared.

The gapless acoustic plasmon has a linear spectrum

$$\omega_{ac}^2(q) = v^2 q^2 , \tag{14.50}$$

where the acoustic plasmon group velocity v satisfies the requirement given by Eq. (14.40).

To get the spectra given by Eqs. (14.42) and (14.43) for the case of two-band superconductor MgB_2, it is sufficient to replace $v_{FL}^2 q^2$ and $v_{Fh}^2 q^2$ in the numerator and denominator of the hydrodynamic approximation given by Eq. (14.47) by $B_1 \sin^2 q_z d$ for the polarization operator of the light band and by $B_2 \sin^2 q_z d$ for the polarization operator of the heavy band.

After this substitution, we will get again the biquadratic equation for ω^2 with two roots corresponding to Lindhard and acoustic plasmons with the spectrum given by Eqs. (14.43) and (14.42).

Now, returning back to our calculation of the singularity in the effective interaction $V_{eff}(q, \omega)$ at finite q (close to the "nesting" vector Q), note that in this case, we should analyze the structure of the dielectric function $\mathrm{Re}\epsilon(q, \omega)$ in the resonant situation, when

$$\omega = \omega_{ac}(q) . \tag{14.51}$$

It means that the frequency in Eq. (14.51) coincides with the spectrum of acoustic plasmon given by Eq. (14.42).

The singularity in the effective interaction $V_{eff}(\mathbf{q}, \omega)$ for $q = (q_z, q_\perp)$ and q_z close to the "nesting" vector $Q = \pi/d$ has the hyperbolic form in the resonance given by

$$V_{eff}(q_z \to Q, \omega = \omega_{ac}(q)) \propto \frac{1}{|q_z - Q| - \alpha |q_\perp|} , \tag{14.52}$$

where q_z is the component of the wave vector in the direction perpendicular to the layers (in c direction), $q_\perp = (q_x, q_y)$ are two components of the wave vector inside the layer (in the ab plane), and $\alpha = const$. Note that, in fact, both wave vectors slightly larger than Q ($q_z = Q + \delta$) and slightly smaller than Q ($q_z = Q - \delta$) for small $\delta \ll Q$ are important for the character of the singularity in the effective interaction.

The Fourier transform of the effective interaction near the singularity yields the screened Coulomb interaction in real space $V_{eff}(z)$, which occurs to be similar to the 2D Friedel oscillations with the "nesting" vector Q playing the role of the Fermi sphere of diameter $2p_F$. Note that the decrease in the amplitude of the charge oscillations proportional to $1/z^2$ is related to the measure of integration in MgB_2, which, is in fact a 3D material. The measure of integration is given by $dq_z dq_\perp^2$ and defines the character of the singularity at small $\delta = |q_z - Q|$ and small $|q_\perp|$.

It is most important that both small factors (in plane, $|q_\perp|$, and along the c axis, $\delta = |q_z - Q|$) enter in the same linear fashion in the expression for $V_{eff}(z)$. This result is pretty stable with respect to the modifications of the singularity in Eq. (14.52), which conserve the same weights for both factors—along the c axis and in the ab plane. If, for instance, the singularity acquires the form

$$V_{eff}(q_z \to Q, \omega = \omega_{ac}(q)) \propto \frac{1}{\sqrt{(q_z - Q)^2 + (\alpha q_\perp)^2}}, \tag{14.53}$$

we still obtain the same result for the character of the long-range charge oscillations as in Eq. (14.45). Note that an account taken of a damping also does not change the result.

Factor α at $q_\perp$ in Eq. (14.53) enters $V_{eff}(z)$ as $1/\alpha^2$ additionally decreasing effective interaction at large values of α.

Thus, the singular (oscillating) part of the effective interaction is given by the expression

$$V_{eff}(z) = C \frac{\cos Qz}{(\alpha Q z)^2}, \tag{14.54}$$

creating some type of inhomogeneity in the system in the direction perpendicular to the layers. At the same time, the anomaly is not strong enough to produce a real CDW instability in MgB_2.

References

1. M.Yu. Kagan, Phys. Usp. **37**(1), 69 (1994)
2. J. Bardeen, G. Baym, D. Pines, Phys. Rev. **156**, 207 (1967)
3. C.A. Regal, C. Ticknor, D.S. Bohn, J.L. Jin, Nature **424**, 47 (2003)
4. M.W. Zwierlein, C.A. Stan, C.H. Schunck, S.M.F. Raupach, S. Gupta, Z. Hadzibabic, W. Ketterle, Phys. Rev. Lett. **91**, 250401 (2003)
5. Y. Shin, M.W. Zwierlein, C.H. Schunck, A. Schirotzek, W. Ketterle, Phys. Rev. Lett. **97**, 030401 (2006)
6. Y.A. Liao, A.S.C. Rittner, T. Paprotta, W. Li, G.B. Partridge, R.G. Hulet, S.K. Baur, E.J. Mueller, Nature **467**, 567 (2010)
7. W. Ong, C. Cheng, I. Arakelyan, J.E. Thomas, Phys. Rev. Lett. **114**, 110403 (2015)
8. P. Fulde, R.A. Ferrell, Phys. Rev. **135**(3A), A550 (1964)
9. A.I. Larkin, Y.N. Ovchinnikov, Sov. Phys. JETP **20**, 762 (1965)
10. M.Yu. Kagan, A.V. Chubukov, JETP Lett. **47**(10), 614 (1988)

11. M.Yu. Kagan, Phys. Lett. A **152**(5), 303 (1991)
12. M.A. Baranov, Y. Kagan, M.Yu. Kagan, JETP Lett. **64**(4), 301 (1996)
13. M.C. Revelle, J.A. Fry, B.A. Olsen, R.G. Hulet, Phys. Rev. Lett. **117**, 235301 (2016)
14. E. Burovski, R.S. Ikhsanov, A. Kuznetsov, M.Yu. Kagan, J. Phys. Conf. Ser. **1163**, 012046 (2019)
15. A.E. Feiguin, F. Heidrich-Meisner, Phys. Rev. Lett. **102**, 076403 (2009)
16. G. Orso, Phys. Rev. Lett. **98**, 070402 (2007)
17. I. Khalatnikov, *An Introduction to the Theory of Superfluidity*, 1st edn. (CRC Press, Boca Raton, 1965)
18. H. Feshbach, Ann. Phys. **19**(2), 287 (1962)
19. M.Yu. Kagan, A.V. Turlapov, Phys. Usp. **62**(3), 215 (2019)
20. M.Yu. Kagan, I.V. Brodsky, D.V. Efremov, A.V. Klaptsov, Phys. Rev. A **70**, 023607 (2004)
21. G. Modugno, G. Roati, F. Riboli, F. Ferlaino, R.J. Brecha, M. Inguscio, Science **297**(5590), 2240 (2002)
22. H.W. Hammer, D.T. Son, Phys. Rev. Lett. **93**, 250408 (2004)
23. M.Yu. Kagan, A.V. Klaptsov, I.V. Brodsky, R. Combescot, X. Leyronas, Phys. Uspekhi **49**(10), 1079 (2006)
24. I. Ferrier-Barbut, H. Kadau, M. Schmitt, M. Wenzel, T. Pfau, Phys. Rev. Lett. **116**, 215301 (2016)
25. T.D. Lee, K. Huang, C.N. Yang, Phys. Rev. **106**, 1135 (1957)
26. G.E. Volovik, *The Universe in a Helium Droplet* (Oxford University Press, 2009)
27. D. Duan, Y. Liu, F. Tian, D. Li, X. Huang, Z. Zhao, H. Yu, B. Liu, W. Tian, T. Cui, Sci. Rep. **4**, 6968 (2014)
28. A.P. Drozdov, M.I. Eremets, I.A. Troyan, V. Ksenofontov, S.I. Shylin, Nature **525**(7567), 73 (2015)
29. E. Snider, N. Dasenbrock-Gammon, R. McBride, M. Debessai, H. Vindana, K. Vencatasamy, K.V. Lawler, A. Salamat, R.P.A.U. Dias, Nature **586**, 373 (2020)
30. P. Loubeyre, F. Occelli, P. Dumas, Nature **577**, 631 (2020)
31. S.N. Burmistrov, L.B. Dubovskii, Low Temp. Phys. **43**(10), 1152 (2017)
32. N.W. Ashcroft, J. Phys. A: Math. General **36**(22), 6137 (2003)
33. E. Kim, M.H.W. Chan, Science **305**(5692), 1941 (2004)
34. A.F. Andreev, I.M. Lifshitz, Sov. Phys. JETP **29**(6), 1107 (1969)
35. E.G. Brovman, Yu. Kagan, A. Kholas, Sov. Phys. JETP **34**, 1300 (1972)
36. E.G. Brovman, Yu. Kagan, A. Kholas, V.V. Pushkarev, JETP Lett. **18**(4), 269 (1973)
37. A.A. Abrikosov, Sov. Phys. JETP **5**, 1174 (1957)
38. J. De Boer, Physica **14**(2), 139 (1948)
39. C. Kittel, *Quantum Theory of Solids* (Wiley, 1987)
40. M.Yu. Kagan, A. Bianconi, Condens. Matter **4**(2) (2019)
41. M.Yu. Kagan, JETP Lett. **103**(11), 728 (2016)
42. E.A. Mazur, R.S. Ikhsanov, M.Yu. Kagan, J. Phys. Conf. Ser. **2036**(1), 012019 (2021)
43. R.S. Ikhsanov, E.A. Mazur, M.Yu. Kagan, Software complex for the numerical solution of the isotropic imaginary-axis Eliashberg equations (2022), arXiv:2202.01452
44. N.A. Kudryashov, A.A. Kutukov, E.A. Mazur, JETP Lett. **104**(7), 430 (2016)
45. N.A. Kudryashov, A.A. Kutukov, E.A. Mazur, JETP Lett. **105**(7), 430 (2017)
46. K.S. Grishakov, N.N. Degtyarenko, E.A. Mazur, J. Exp. Theor. Phys. **128**(1), 105 (2019)
47. O.V. Dolgov, R.K. Kremer, J. Kortus, A.A. Golubov, S.V. Shulga, Phys. Rev. B **72**, 024504 (2005)
48. Z. Zhang, T. Cui, M.J. Hutcheon, A.M. Shipley, H. Song, M. Du, V.Z. Kresin, D. Duan, C.J. Pickard, Y. Yao, Phys. Rev. Lett. **128**, 047001 (2022)
49. P.B. Allen, R.C. Dynes, Phys. Rev. B **12**, 905 (1975)
50. W.L. McMillan, Phys. Rev. **167**, 331 (1968)
51. Y. Sun, J. Lv, Y. Xie, H. Liu, Y. Ma, Phys. Rev. Lett. **123**, 097001 (2019)
52. M. Zaghoo, I.F. Silvera, Proc. Natl. Acad. Sci. (USA) **114**(45), 11873 (2017)

53. M. Marley, J. Fortney, *Interiors of the Giant Planets*, 3rd edn, Chap. 33 (Elsevier, Amsterdam, 2014), pp. 743–758
54. R.P. Dias, I.F. Silvera, Science **355**(6326), 715 (2017)
55. P.M. Celliers, G.W. Collins, L.B. Da Silva, D.M. Gold, R. Cauble, R.J. Wallace, M.E. Foord, B.A. Hammel, Phys. Rev. Lett. **84**, 5564 (2000)
56. M. Houtput, J. Tempere, I.F. Silvera, Phys. Rev. B **100**, 134106 (2019)
57. S.T. Weir, A.C. Mitchell, W.J. Nellis, Phys. Rev. Lett. **76**, 1860 (1996)
58. S. Deemyad, I.F. Silvera, Phys. Rev. Lett. **100**, 155701 (2008)
59. M. Zaghoo, R.J. Husband, I.F. Silvera, Phys. Rev. B **98**, 104102 (2018)
60. M.D. Knudson, M.P. Desjarlais, A. Becker, R.W. Lemke, K.R. Cochrane, M.E. Savage, D.E. Bliss, T.R. Mattsson, R. Redmer, Science **348**(6242), 1455 (2015)
61. G.W. Collins, L.B.D. Silva, P. Celliers, D.M. Gold, M.E. Foord, R.J. Wallace, A. Ng, S.V. Weber, K.S. Budil, R. Cauble, Science **281**(5380), 1178 (1998)
62. G.W. Collins, P.M. Celliers, L.B. Da Silva, R. Cauble, D.M. Gold, M.E. Foord, N.C. Holmes, B.A. Hammel, R.J. Wallace, A. Ng, Phys. Rev. Lett. **87**, 165504 (2001)
63. V.E. Fortov, R.I. Ilkaev, V.A. Arinin, V.V. Burtzev, V.A. Golubev, I.L. Iosilevskiy, V.V. Khrustalev, A.L. Mikhailov, M.A. Mochalov, V.Y. Ternovoi, M.V. Zhernokletov, Phys. Rev. Lett. **99**, 185001 (2007)
64. L.V. Keldysh, Y.V. Kopaev, Sov. Phys. Solid State **6**, 2219 (1965)
65. B.I. Halperin, T.M. Rice, Rev. Mod. Phys. **40**, 755 (1968)
66. W. Kohn, Phys. Rev. Lett. **19**, 439 (1967)
67. A.N. Kozlov, L.A. Maksimov, Sov. Phys. JETP **21**, 790 (1965)
68. A.N. Kozlov, L.A. Maksimov, Sov. Phys. JETP **22**, 889 (1966)
69. L.V. Keldysh, A.N. Kozlov, Sov. Phys. JETP **27**, 521 (1968)
70. L.V. Keldysh, Phys Status Solidi (a) **164**(1), 3 (1997)
71. D. Snoke, Science **298**(5597), 1368 (2002)
72. J.P. Eisenstein, A.H. MacDonald, Nature **432**(7018), 691 (2004)
73. E.P. Gross, Il Nuovo Cimento **20**(3), 454 (1961)
74. L.P. Pitaevskii, Sov. Phys. JETP **13**(2), 451 (1961)
75. V.E. Zakharov, E.A. Kuznetsov, Sov. Phys. JETP **64**, 773 (1986)
76. V.E. Zakharov, E.A. Kuznetsov, Phys. Usp. **55**(6), 535 (2012)
77. E. Kuznetsov, S. Turitsyn, Phys. Lett. A **112**(6), 273 (1985)
78. V.I. Talanov, JETP Lett. **11**(6), 199 (1970)
79. S.N. Vlasov, V.A. Petrishchev, V.I. Talanov, Radiophys. Quantum Electron. **15**(8), 886 (1972)
80. M.V. Durnev, M.M. Glazov, Phys. Usp. **61**(9), 825 (2018)
81. A.E. Almand-Hunter, H. Li, S.T. Cundiff, M. Mootz, M. Kira, S.W. Koch, Nature **506**(7489), 471 (2014)
82. S. Berciaud, Nat. Photonics **13**(4), 225 (2019)
83. Y. Yu, A.W. Bataller, R. Younts, Y. Yu, G. Li, A.A. Puretzky, D.B. Geohegan, K. Gundogdu, L. Cao, ACS Nano **13**(9), 10351 (2019)
84. Z. Wang, D.A. Rhodes, K. Watanabe, T. Taniguchi, J.C. Hone, J. Shan, K.F. Mak, Nature **574**, 76 (2019)
85. M.M. Parish, F.M. Marchetti, P.B. Littlewood, Europhys. Lett. **95**(2), 27007 (2011)
86. D.K. Efimkin, E.K. Laird, J. Levinsen, M.M. Parish, A.H. MacDonald, Phys. Rev. B **103**, 075417 (2021)
87. M. Matuszewski, T. Taylor, A.V. Kavokin, Phys. Rev. Lett. **108**, 060401 (2012)
88. L.V. Keldysh, Phys. Usp. **60**(11), 1180 (2017)
89. V.L. Ginsburg, L.P. Pitaevskii, Sov. Phys. JETP **7**(5), 858 (1958)
90. Yu.M. Kagan, B.V. Svistunov, G.V. Shlyapnikov, Sov. Phys. JETP **74**(2), 279 (1992)
91. Yu. Kagan, B.V. Svistunov, Phys. Rev. Lett. **79**, 3331 (1997)
92. Yu. Kagan, B.V. Svistunov, JETP Lett. **67**(7), 521 (1998)
93. E.A. Kuznetsov, M.Yu. Kagan, A.V. Turlapov, Phys. Rev. A **101**, 043612 (2020)
94. E.A. Kuznetsov, M.Yu. Kagan, Theor. Math. Phys. **202**(3), 399 (2020)
95. E.A. Kuznetsov, M.Yu. Kagan, J. Exp. Theor. Phys. **132**(4), 704 (2021)

96. V.P. Ermakov, Appl. Anal. Discr. Math. **2**, 123 (2008)
97. V.E. Zakharov, Sov. Phys. JETP **35**(5), 908 (1972)
98. E.M. Lifshitz, L.P. Pitaevskii, *Physical Kinetics, Course of Theoretical Physics*, vol. 10 (Butterworth-Heinemann, Oxford, 1981)
99. D. Pines, P. Nozières, *Theory of Quantum Liquids*, 1st edn. (Advanced Book Classics (CRC Press, Boca Raton, 1999)
100. D. Pines, Can. J. Phys. **34**(12A), 1379 (1956)
101. V.M. Silkin, A. Balassis, P.M. Echenique, E.V. Chulkov, Phys. Rev. B **80**, 054521 (2009)
102. A. Balassis, E.V. Chulkov, P.M. Echenique, V.M. Silkin, Phys. Rev. B **78**, 224502 (2008)
103. V.P. Zhukov, V.M. Silkin, E.V. Chulkov, P.M. Echenique, Phys. Rev. B **64**, 180507 (2001)
104. O.V. Dolgov, D.A. Kirzhnits, E.G. Maksimov, Rev. Mod. Phys. **53**, 81 (1981)
105. H. Fröhlich, J. Phys. C: Solid State Phys. **1**(2), 544 (1968)
106. W. Kohn, J.M. Luttinger, Phys. Rev. Lett. **15**, 524 (1965)
107. J. Lindhard, *On the Properties of a Gas of Charged Particles* (E. Munksgaard, Copenhagen, 1954)
108. J. Friedel, Adv. Phys. **3**(12), 446 (1954)
109. W. Kohn, Phys. Rev. Lett. **2**, 393 (1959)
110. D. Fay, A. Layzer, Phys. Rev. Lett. **20**, 187 (1968)
111. J. Ruhman, P.A. Lee, Phys. Rev. B **94**, 224515 (2016)
112. J. Ruhman, P.A. Lee, Phys. Rev. B **96**, 235107 (2017)
113. A. Chubukov, N.V. Prokof'ev, B.V. Svistunov, Phys. Rev. B **100**, 064513 (2019)
114. X. Cai, T. Wang, N.V. Prokof'ev, B.V. Svistunov, K. Chen (2022), arXiv:2202.01320
115. M.Yu. Kagan, V.A. Mitskan, M.M. Korovushkin, Phys. Usp. **58**(8), 733 (2015)
116. G.M. Eliashberg, Sov. Phys. JETP **11**(3), 696 (1960)
117. D. Phan, A.V. Chubukov, Phys. Rev. B **101**, 024503 (2020)
118. D. Phan, A.V. Chubukov, Phys. Rev. B **105**, 064518 (2022)
119. H. Gao, C. Ding, W. Geng, X. Ma, Y. Li, M. Zhao, New J. Phys. **23**(7), 073036 (2021)
120. A. Kogar, M.S. Rak, S. Vig, A.A. Husain, F. Flicker, Y.I. Joe, L. Venema, G.J. MacDougall, T.C. Chiang, E. Fradkin, J. van Wezel, P. Abbamonte, Science **358**(6368), 1314 (2017)

Chapter 15
Inhomogeneous Fermi–Bose Mixture in the Aharonov–Bohm Ring with Topologically Nontrivial Superconducting Bridge

Abstract We study inhomogeneous Fermi–Bose mixtures in the interference device specified by the Aharonov–Bohm ring with SC wire in topologically nontrivial states. We analyze Majorana bound modes on the ends of the SC wire and show that the collapse of the Fano resonance in this system is connected with the increase of the length of the wire when the binding energy for the Majorana end modes tends to zero.

15.1 Introduction. Formation of the Inhomogeneous Fermi–Bose Mixture

In Chap. 11, we analyzed the manifestation of EPE and Coulomb correlations in the inhomogeneous two-band fermionic model (or in other words in the inhomogeneous Fermi–Fermi mixture) in the microcontact with four quantum dots forming the quadrupole QQD structure shown in Fig. 11.1 [1–3].

In this structure, three quantum dots QD1, QD2, and QD4 belong to the upper shoulder of the QQD structure, while QD1, QD3, and QD4 belong to the lower shoulder. Moreover, QD2 and QD3 in the central part of the QQD are coupled both by the one-particle hybridization t_0 and by the two-particle Coulomb interband interaction V.

If instead of this coupling we consider the superconducting bridge between the lower and upper shoulders, then we obtain the inhomogeneous (or space-separated) Fermi–Bose mixture with fermionic component of individual electrons at lower and upper shoulders and bosonic component of Cooper pairs in the bridge.

Moreover, if the lower and upper shoulders form the Aharonov–Bohm ring [4] and in addition, the superconducting bridge is in topologically nontrivial phase (see Fig. 15.1 and [5–7]), we have an interference device suitable to study the effects of the Aharonov–Bohm phase, different resonances [8–10], and topologically nontrivial excitations [11].

To be more specific, we show in Fig. 15.1 a SC bridge in an s-wave state in the presence of Rashba spin–orbit coupling [12, 13] and the in-plane magnetic field in the

M. Yu. Kagan et al., *Electronic Phase Separation in Magnetic and Superconducting Materials*, Springer Series in Solid-State Sciences 201,
https://doi.org/10.1007/978-3-031-55467-4_15

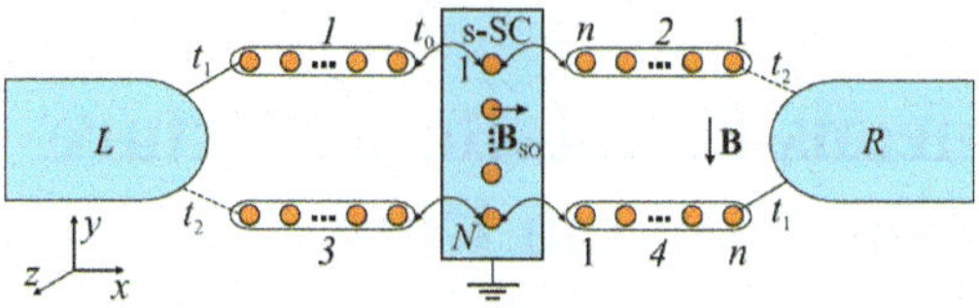

Fig. 15.1 Superconducting nanowire (SW) in the topologically nontrivial phase (s-wave SC with the Rashba spin–orbit coupling) connecting the shoulders (the arms) of the Aharonov–Bohm ring or an interference device formed by four leads: 1 (left top, LT) and 2 (right top, RT) at the upper shoulder, 3 (left bottom, LB) and 4 (right bottom, RB) at the lower shoulder. The shoulders are in the normal state and their ends are connected in parallel with metallic contacts L and R. We show the hopping integrals t_0, t_1, t_2, applied magnetic field **B**, and the effective Rashba field $\mathbf{B}_{so}$ [6]

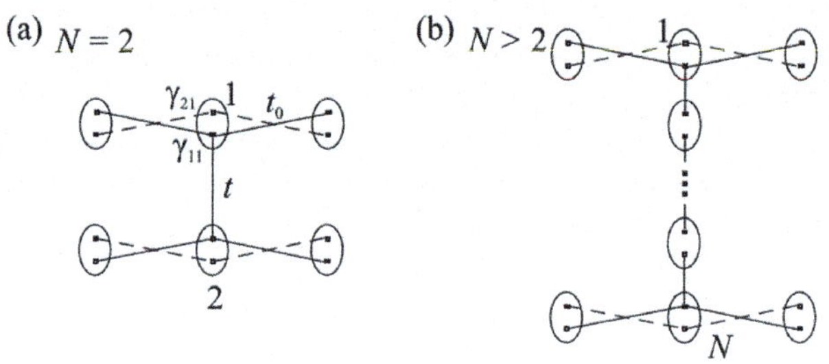

Fig. 15.2 Aharonov–Bohm ring with the topologically nontrivial SC bridge described by the Kitaev chain with $N = 2$ (**a**) or larger even numbers of sites $N = 4, 6...$ (**b**) [6]

Aharonov–Bohm ring. In this scheme, the Aharonov–Bohm ring is an interference device formed by four leads: 1 (left top, LT) and 2 (right top, RT) at the upper shoulder, 3 (left bottom, LB) and 4 (right bottom, RB) at the lower shoulder. The shoulders are in the normal state and their ends are connected in parallel with metallic contacts.

This interference device exhibits a very rich physics-related first of all to the Majorana modes [11] in SW described by the Kitaev chain model [14, 15] and additional Breit–Wigner [8] and Fano–Feshbach resonances [9, 10] in the coherent transport. We show that qualitatively these effects can be described by the toy model with four sites in normal state and even number of sites ($N = 2$ and $N = 4, 6, ...$) in the Kitaev chain (see [6] and Fig. 15.2a and b).

The formation and collapse of the different resonances in this scheme can be explained by the presence of two interacting transport channels in the system. As a result, the Fano–Feshbach resonances turn out to be related to the bound states in the continuum (BSCs). We demonstrate below that the width and position of these resonances are highly sensitive to the lowest in energy excitations in the SW, which in topologically nontrivial phase correspond to the Majorana bound states (MBS) [11] and the Andreev bound states (ABS) [16]. We investigate also the effect of the Coulomb correlations and disorder on these results.

15.2 Topological Superconductivity in Quasi-1D Nanostructures

In this section, we provide a brief introduction to the subject of topological superconductivity in quasi-1D nanostructures. Note that the topological superconductivity is a very topical field of research nowadays because of its importance for the protection of the phase in the superconducting qubits under external perturbations. The manifestation of the nontrivial topology in quasi-1D structures such as superconducting nanowire (SW) first of all is related to the presence of the Majorana bound states (MBS). As we mentioned already, the basic scenario for the formation of MBS in 1D systems deals with the combination of three factors: superconducting pairing, spin–orbit coupling, and applied magnetic field.

Often this situation occurs for a nanowire (NW) at the surface of a bulk s-wave SC. In this case, the topological superconductivity in the NW is induced by the proximity effect. In the geometry represented in Fig. 15.1 for the certain relation between the parameters, an odd number of the spin-polarized subbands arise in the normal state of the bridge. As a result, the SC pairing between electrons of the same subband takes place in it. Thus, effectively in this case, the pairing of electrons in SW is equivalent to the triplet p-wave pairing. This type of pairing with $l = S = 1$ occurs, e.g., in the superfluid ^{3}He.

Generally speaking, at zero field, there are three components of the magnetic quantum number $m = l_z$ and three components of the spin projection S_z in the isotropic B phase of ^{3}He. However, at high magnetic fields, the anisotropic triplet A phase with the pairing of two up spins at larger Fermi surface and two down spins at the smaller one is favorable. This situation is similar to the pairing of electrons belonging to the same subband in SW.

15.2.1 Model of the Kitaev Chain

Correspondingly, the SW becomes equivalent to the 1D model of the Kitaev chain describing the SC spinless fermions in the idealized 1D system. It is just in the Kitaev model [14, 15], the emergence of the MBS was first demonstrated in the particular case where the hopping integral t for electron at neighboring sites of the chain coincides with the absolute value of the SC gap

$$|\Delta| = t > 0\,, \tag{15.1}$$

and, at the same time, the chemical potential of the chain is $\mu = 0$. In this case, the MBS arise at the left and right ends of the chain.

To be more specific, we consider the Hamiltonian of 1D Kitaev chain

$$\hat{H}' = \frac{i}{2}\sum_{j=1}^{N}\left[-\mu(a_j^\dagger a_j - \frac{1}{2})\right] + \sum_{j=1}^{N-1}[-t(a_j^\dagger a_{j+1} + a_{j+1}^\dagger a_j) + \Delta^* a_j a_{j+1} + \Delta a_{j+1}^\dagger a_j^\dagger], \tag{15.2}$$

where $a_j^\dagger$ and a_{j+1} are the creation and annihilation operators of spinless fermions an sites j and $j+1$, respectively, t is the hopping integral, μ is the chemical potential, $j = 1, \ldots N$ is the number of sites in the chain, Δ is the induced SC order parameter

$$\Delta = |\Delta| \exp(i\Theta), \tag{15.3}$$

and Θ is the phase of the order parameter, which is assumed to be fixed.

After that, we can perform the transformation from fermionic operators $a_j^\dagger$ and a_{j+1} to the Majorana self-conjugated operators c_{2j-1} and c_{2j} according to the prescription

$$\begin{aligned} c_{2j-1} &= \exp(i\Theta/2)a_j + \exp(-i\Theta/2)a_j^\dagger, \\ c_{2j} &= -i\exp(i\Theta/2)a_j + i\exp(-i\Theta/2)a_j^\dagger. \end{aligned} \tag{15.4}$$

From expressions (15.4), it is easy to understand that indeed

$$c_{2j} = c_{2j}^\dagger \quad \text{and} \quad c_{2j-1} = c_{2j-1}^\dagger. \tag{15.5}$$

Correspondingly, for the anticommutators of Majorana operators at the same site, we obtain the relations

$$\{c_{2j-1}, c_{2j-1}\} = \{c_{2j}, c_{2j}\} = 2, \tag{15.6}$$

while for the operators at the neighboring sites, we have

$$\{c_{2j-1}, c_{2j}\} = 0. \tag{15.7}$$

Thus, from Eq. (15.6), we can see a striking difference between the anticommutation relations of fermions and Majorana quasiparticles at the same site.

In addition, the Majorana modes have exactly zero energies. Indeed, as it was noticed by Bogoliubov, we should satisfy two equations of motion simultaneously: $i\dot{c} = \varepsilon_0 c$ and $i\dot{c}^\dagger = -\varepsilon_0 c^\dagger$ for the Majorana operators c and $c^\dagger$. However, since $c = c^\dagger$, we have to assume that $\varepsilon_0 = -\varepsilon_0$, and hence, $\varepsilon_0 = 0$.

Moreover, we can rewrite the Hamiltonian (15.2) in terms of the Majorana operators as follows:

$$\hat{H}' = \frac{i}{2}\sum_{j=1}^{N}[-\mu c_{2j-1}c_{2j}] + \frac{i}{2}\sum_{j=1}^{N-1}[(t+|\Delta|)c_{2j}c_{2j+1} + (-t+|\Delta|)c_{2j-1}c_{2j+2}]. \tag{15.8}$$

If we consider a special point in the parameter space, where $|\Delta| = t$ and $\mu = 0$, then we obtain a simple form of Hamiltonian expressed only in terms of the Majorana operators at neighboring sites

$$\hat{H}' = it \sum_{j=1}^{N-1} c_{2j} c_{2j+1} \,. \tag{15.9}$$

Now it is convenient to perform the inverse transformation from the Majorana operators c_{2j+1} and c_{2j} in Hamiltonian (15.9) to the new fermionic operators $\tilde{a}_j$ and $\tilde{a}_j^\dagger$ assuming that they satisfy the relations

$$c_{2j} = \tilde{a}_j + \tilde{a}_j^\dagger, \quad c_{2j+1} = \frac{1}{i}(\tilde{a}_j - \tilde{a}_j^\dagger), \quad j = 1, 2, ..., N-1\,. \tag{15.10}$$

We can resolve Eq. (15.10) and find the representation for the new operators $\tilde{a}_j$ and $\tilde{a}_j^\dagger$ via the Majorana operators c_{2j+1} and c_{2j}. This yields

$$\tilde{a}_j = (c_{2j} + ic_{2j+1})/2, \quad \tilde{a}_j^\dagger = (c_{2j} - c_{2j+1})/2, \quad j = 1, 2, ..., N-1\,. \tag{15.11}$$

Then, we can easily express the product of Majorana operators $c_{2j}c_{2j+1}$ via the fermionic one

$$c_{2j}c_{2j+1} = \frac{2}{i}(\tilde{a}_j^\dagger \tilde{a}_j - \frac{1}{2})\,, \tag{15.12}$$

and, finally, diagonalize the Hamiltonian of the Kitaev chain in the real space

$$\hat{H}' = 2t \sum_{j=1}^{N-1} (\tilde{a}_j^\dagger \tilde{a}_j - \frac{1}{2})\,. \tag{15.13}$$

Note that, from Eq. (15.11), we can see that the operators $\tilde{a}_j^\dagger$ and $\tilde{a}_j$ satisfy indeed the usual fermionic anticommutation relations

$$\tilde{a}_j \tilde{a}_j = \tilde{a}_j^\dagger \tilde{a}_j^\dagger = 0, \quad \{\tilde{a}_j, \tilde{a}_j^\dagger\} = 1\,. \tag{15.14}$$

However, Hamiltonian (15.13) have one important feature, namely subscript j runs over $N-1$ sites rather than over N ones. In other words, in (15.13), some degrees of freedom are missing. They correspond to the Majorana massless fermions.

At the same time, in Eq. (15.9), where the Hamiltonian is expressed in terms of Majorana operators, the two operators at the ends of the chain c_1 and c_{2N} are also absent. Hence, the ground state is doubly degenerate. It means [17] that for a product of these missing operators, we have

$$ic_1 c_{2N} = \pm 1\,, \tag{15.15}$$

while the other products, which enter Hamiltonian (15.9) are single-valued ones

$$ic_{2j}c_{2j+1} = 1, \quad \text{for} 1 \leq j \leq N-1\,. \tag{15.16}$$

Operators c_1 and c_{2N} describe zero energy Majorana modes. Note that the two states in Eq. (15.15) with $ic_1c_{2N} = \pm 1$ are practically indistinguishable. They can be distinguished only by the parity operator, which is not local. Indeed, an operator which distinguish these states should act simultaneously on both end sites 1 and N, and thus it cannot be local. Hence, these end states form a qubit. Moreover, Eqs. (15.9)–(15.16) demonstrate that qubit states are topologically protected because they are separated from the excited states by a gap $|\Delta| = t$. We can say that the Majorana bound state (MBS) is created. It is extended in space over the chain length L.

If we go away from the special point, where $|\Delta| = t$, we obtain a similar physical picture. There is a gap in the bulk above the two nearly degenerate ground states. The ground states have the energy splitting due to the overlap of their Ψ functions at the chain of the finite length. We can say that the splitting yields the binding energy for the Majorana end states, which is proportional to

$$E_b \propto \exp\left(-N\frac{d}{\xi_0}\right) \rightarrow 0, \quad \text{for} \quad N \rightarrow \infty\,. \tag{15.17}$$

It exponentially decreases with an increase in the chain length $L = Nd$, where d is the intersite distance. In Eq. (15.17), we have the coherence length

$$\xi_0 \sim \frac{\hbar v_F}{|\Delta|}\,, \tag{15.18}$$

which is inversely proportional to the absolute value of the gap. Note that the expression (15.17) for the energy of the MBS is still valid even in a more physically relevant region, where $|\Delta| < t$.

Note also that, for arbitrary filling factors (when $\mu \neq 0$), a nontrivial topology in the Kitaev chain for $|\Delta| = t$ is determined by the so-called Majorana number

$$M = \text{sgn}(\mu^2 - 4t^2) = \pm 1\,. \tag{15.19}$$

A topologically nontrivial phase in Eq. (15.19) corresponds to the Majorana number $M = -1$. This situation takes place for the chemical potential μ located within the band. Thus, the topologically nontrivial phase corresponds to the values of the chemical potential in the $-2t < \mu < 2t$ range. For the energy gap, we have $E_{gap} \rightarrow 0$ at the border of the topologically nontrivial interval, $\mu = \pm 2t$.

Let us stress that, often in the problems of mesoscopic superconductivity, the chains with odd and even numbers of sites yield different results with respect to the total energy. In the absence of Majorana end modes, their difference in energy is just the energy cost for a single unpaired electron to appear at the chain with an odd number of sites. Thus, we have $E_{odd}(2N + 1) > E_{even}(2N)$. The formation of zero-energy Majorana modes at two ends of the Kitaev chain eliminates this difference between odd and even number of sites [17], and hence, $E_{odd}(2N+1) = E_{even}(2N)$.

We can notice also that, for an important point $\mu = 0$, we have a half-filled 1D band with an average density $n = 0.5$ in the case of spinless fermions. The numerical simulations show that for long chains in this situation, the distribution of the average density over the different sites $\langle n_j \rangle = 0.5$ is nearly constant for all $j = 1, 2, ..., N$. However, when $\mu \neq 0$ or $|\Delta| < t$, the average onsite density $\langle n_j \rangle$ drastically decreases close to the ends of the chain.

Note also that the distribution of the Bogoliubov $u - v$ coefficients over different sites $j = 1, 2, ..., N$ in the normalization condition $\sum_{j=1}^{N} [|u_j|^2 + |v_j|^2] = 1$ also changes drastically, if we work within the range of topologically nontrivial phase. Namely, for $\mu = 0$ and $|\Delta| = t$, the combination $|u_j|^2 + |v_j|^2$ in the normalization condition, for the long chains have two pronounced peaks at the two ends of the chain, for $j = 1$ and $j = N$. For both peaks, we have $u_1 = v_1 = u_N = v_n = 1/2$, and thus, $|u_j|^2 + |v_j|^2 \to 0$ for $j \neq 1, N$. At the same time, for $\mu = -2t$ on the border of the topological interval the combination $|u_j|^2 + |v_j|^2 \to 1/N$ is distributed almost uniformly over all sites of the chain including both its ends.

15.3 Importance of the Spin–Orbit Coupling. Anisotropic Aharonov–Bohm Ring

In this section, we study anisotropic AB ring with different hopping integrals $t_1 \neq t_2$ between the contacts and the shoulders of the device in Fig. 15.1. The left contact (L) has the hopping integral t_1 with upper shoulder and t_2—with the lower one. For the right contact (R), the situation is the opposite one: t_1 with the lower shoulder and t_2 with the upper one. We will show that in the anisotropic situation, new Fano resonances appear in the conductance of the ring [6]. Moreover, the width of these resonances is related with the degree of the nonlocality of the lowest energy state of the SW.

In other words, the width is governed by the energy splitting between the levels of the Majorana end states, which arise due to the overlap of their Ψ functions. If this overlap disappears and the binding energy $E_b \to 0$, the two Majorana zero modes (two Majorana fermions) become noninteracting, and the width of the Fano resonance tends to zero. According to [6], in this case, we have a collapse of the Fano resonance caused by the nonlocality of the Majorana end states.

15.3.1 Hamiltonian of the SW

The Hamiltonian of SW, which includes SC pairing, spin–orbit coupling, and the effect of the Zeeman splitting by the applied magnetic field has the form

$$\hat{H} = \sum_{j=1}^{N}\sum_{\sigma}[\xi a_{j\uparrow}^{\dagger}a_{j\downarrow} + (\Delta a_{j\uparrow}^{\dagger}a_{j\downarrow}^{\dagger} + iha_{j\uparrow}^{\dagger}a_{j\downarrow} + H.c.)] \tag{15.20}$$
$$+\frac{1}{2}\sum_{j=1}^{N-1}\sum_{\sigma}[-ta_{j\sigma}^{\dagger}a_{j+1,\sigma} + i\alpha\sigma a_{j-\sigma}^{\dagger}a_{j+1,\sigma} + H.c.],$$

where $\xi = \varepsilon_D - \mu$ is the onsite energy, ε_D is the gate voltage, μ is the chemical potential, t is the nearest neighbor hopping integral in the SW, α is the Rashba constant for the spin–orbit coupling, Δ is the s-wave SC order parameter, and h is the Zeeman energy related to the magnetic field **B** lying in the plane of the interference device. Then, the topologically nontrivial phase emerges, if the following inequalities are satisfied [18, 19]:

$$(\xi - t)^2 + \Delta^2 < h^2 < (\xi + t)^2 + \Delta^2 . \tag{15.21}$$

Let us emphasize that the Rashba constant α formally does not appear in inequalities (15.21). However, a nonzero spin–orbit coupling is essential for the formation of the MBS. Moreover, the effective Rashba field $\mathbf{B}_{so}$ should be perpendicular to the applied magnetic field **B**.

In subsection 15.3.2 and in the forthcoming sections of this chapter, we will scale all the energy parameters in the units of t assuming that $t = 1$, $\mu = 0$, $\Delta = 0.25$, and $\alpha = 0.2$.

15.3.2 Full Hamiltonian of the AB Ring with SW

Note that the upper and lower shoulders of the Aharonov–Bohm ring and their four leads: left top (LT), left bottom (LB), right top (RT), and right bottom (RB) are in the normal state. In other words, they correspond to the normal nanowires (NW) and we consider them to be identical. Their Hamiltonians $\hat{H}'_{NW(1-4)}$ can be obtained based on Eq. (15.20) by putting $\Delta = \alpha = 0$.

Finally, the coupling between SW and NW is governed by the tunneling Hamiltonian

$$\hat{H}_T = -t_0\sum_{\sigma}[(b_{Ln\sigma}^{\dagger} + b_{Rn\sigma}^{\dagger})a_{1\sigma}^{\dagger} + (d_{L1\sigma}^{\dagger} + d_{R1\sigma}^{\dagger})a_{n\sigma}^{\dagger}] + H.c., \tag{15.22}$$

where t_0 is the hopping integral between the edge sites of the SW and NW, $b_{L(R)n\sigma}^{\dagger}$ is the creation operator for the electron with spin projection σ in the left (right) upper NW (for the LT and RT leads). Correspondingly, $d_{L(R)1\sigma}^{\dagger}$ is the creation operator for the electron with spin projection σ in the left (right) lower NW (for the LB and RB leads).

At the same time, the coupling between the device (which includes both SW and NW) and the contacts is also described by the tunneling Hamiltonian. This

Hamiltonian plays here the role of the interaction operator in the Keldysh diagram technique. It is given by

$$\hat{V} = -\sum_{k\sigma} [c^{\dagger}_{Lk\sigma}(t_1 b_{L1\sigma} + t_2 d_{Ln\sigma}) + c^{\dagger}_{Rk\sigma}(t_2 b_{R1\sigma} + t_1 d_{Rn\sigma})] + H.c.\,, \tag{15.23}$$

where $c^{\dagger}_{L(R)k\sigma}$ is the creation operator for an electron with wave vector k and spin projection σ at the left (right) contact, whereas t_1 and t_2 are the hopping integrals between the contacts and the device.

Note that the Hamiltonians of the left (right) contact have the simple form:

$$\hat{H}'_{L(R)} = \sum_{k} (\varepsilon_k - \mu_{L(R)}) c^{\dagger}_{L(R)k\sigma} c_{L(R)k\sigma}\,, \tag{15.24}$$

where

$$\mu_{L(R)} = \mu \pm \frac{eV}{2} \tag{15.25}$$

is the electrochemical potential of the left (right) contact, which includes the bias voltage eV.

15.4 Evaluation of the Steady-State Current

In this section, we present a modified Nambu–Keldysh formalism for the calculation of the steady-state current (see [5–7]). Namely, it is convenient to diagonalize the full Hamiltonian of the device

$$\hat{H}'_D = \hat{H}'_{SW} + \sum_{i=1}^{4} \hat{H}'_{NW,i} + \hat{H}'_T\,, \tag{15.26}$$

utilizing the Nambu operators [20]in the site representation

$$\hat{f}_j = (f_{j\uparrow}, f^{\dagger}_{j\downarrow}, f_{j\downarrow}, f^{\dagger}_{j\uparrow})^T\,, \tag{15.27}$$

where $f_{j\sigma}$ is the annihilation operator of an electron with spin projection σ at the jth site of the SW or NW. Then, we can define the matrix Keldysh Green's function of the AB ring in the following form:

$$\hat{G}^{ab}(\tau, \tau') = -i\langle T_C \hat{\Psi}(\tau_a) \otimes \hat{\Psi}^{\dagger}(\tau'_b)\rangle\,, \tag{15.28}$$

where $a, b = +, -$, and T_C is again the ordering operator at the Keldysh time contour (see Eq. (11.14) and Fig. 11.2 in Chap. 11), which consists of the lower (superscript $+$) and upper (superscript $-$) parts. In Eq. (15.28), the Nambu operator $\hat{\Psi}$ has the dimension $4(N + 4n) \times 1$. This means that it includes the Nambu operators for both SW and four NWs. In the explicit form, it reads

$$\hat{\Psi} = (\hat{b}_{L1}...\hat{b}_{Ln}\hat{d}_{L1}...\hat{d}_{Ln}\hat{a}_1...\hat{a}_N\hat{b}_{R1}...\hat{b}_{Rn}\hat{d}_{R1}...\hat{d}_{Rn})^T \,. \tag{15.29}$$

The steady current is written again as $I = e\langle \dot{N}_L \rangle$ (see Eq. (11.10) in Chap. 11) and $N_L = \sum_{k\sigma} c^{\dagger}_{Lk\sigma} c_{Lk\sigma}$ is the operator of the number of particles (see Eq. (11.11) for the left contact. The solution of the Heisenberg equation yields for the current (in the similarity to Eq. (11.12))

$$I(t) = 2e \sum_k \mathrm{Tr}\,\{\hat{\sigma}\mathrm{Re}[\hat{t}_1^{\dagger}\hat{G}^{+-}_{r,L1}(t,t) + ... + \hat{t}_n^{\dagger}\hat{G}^{+-}_{r,Ln}(t,t)]\}\,, \tag{15.30}$$

where we put $\hbar = 1$. Note that in Eq. (15.30), we introduce the diagonal matrix

$$\hat{\sigma} = \mathrm{diag}(1, -1, 1, -1)\,. \tag{15.31}$$

At the same time, we have

$$\hat{t}_{1,n} = \frac{t_{1,2}}{2}\mathrm{diag}(\exp{(-ieVt/2)}, \exp{(ieVt/2)}, \exp{(-ieVt/2)}, \exp{(ieVt/2)})\hat{\sigma} \tag{15.32}$$

are the diagonal matrices, which depend on the time as a result of the unitary transformation of Rogovin and Scalapino [21] (see Eqs. (11.6)–(11.8) in Chap. 11). This transformation converts the voltage dependence into the interaction operator $\hat{V}$ in Eq. (15.23). Finally, in Eq. (15.30),

$$\hat{G}^{+-}_{r,L1} = i\langle \hat{b}^{\dagger}_{L1} \otimes \hat{c}_{Lk} \rangle, \quad \hat{G}^{+-}_{r,Ln} = i\langle \hat{d}^{\dagger}_{Ln} \otimes \hat{c}_{Lk} \rangle \tag{15.33}$$

are the mixed Green's functions.

It is important to stress that in the space of the Nambu operators, the Hamiltonian of the device $\hat{H}'_D$ has the form corresponding to free particles. That is why, in the averages in the definitions of the mixed Green's functions in Eq. (15.33), we should use the same prescriptions as for the averages in the T_C-ordered product of the second quantization operators [22, 23].

As a result, at $t \to 0$, Eq. (15.30) for the current can be written similarly to Eq. (11.22) as (the integration is taken over the Keldysh contour C)

$$\begin{aligned} I(t) &= 2e \int_C d\tau_1 \mathrm{Tr}\{\hat{\sigma}\mathrm{Re}[\hat{\Sigma}^{+a}_{L1,L1}(-\tau_1)\hat{G}^{a-}_{L1,L1}(\tau_1) + ... + \hat{\Sigma}^{+a}_{Ln,Ln}(-\tau_1)\hat{G}^{a-}_{Ln,Ln}(\tau_1)]\} \\ &+ 2e \int_C d\tau_1 \mathrm{Tr}\{\hat{\sigma}\mathrm{Re}[\hat{\Sigma}^{+a}_{L1,Ln}(-\tau_1)\hat{G}^{a-}_{Ln,L1}(\tau_1) + ... + \hat{\Sigma}^{+a}_{Ln,L1}(-\tau_1)\hat{G}^{a-}_{L1,Ln}(\tau_1)]\}, \end{aligned} \tag{15.34}$$

where

$$\hat{\Sigma}^{+a}_{Li,Lj}(-\tau_1) = \hat{t}_i^{+}(0)\hat{g}^{+a}_{Lk}(-\tau_1)\hat{t}_j(\tau_1) \tag{15.35}$$

are the self-energies of the left contact, $(i, j = 1, n)$, and $\hat{g}^{+a}_{Lk}(-\tau_1)$ is the bare Green's function of the left contact.

Integrating over the time τ and performing the Fourier transform, we finally obtain

$$I(t) = 2e \sum_{i,j=1,n} \int_{-\infty}^{\infty} \frac{d\omega}{\pi} \mathrm{Tr}\{\hat{\sigma}\mathrm{Re}[\hat{\Sigma}^r_{Li,Lj}(\omega)\hat{G}^{+-}_{Lj,Li}(\omega) + ... + \hat{\Sigma}^{+-}_{Li,Lj}(\omega)\hat{G}^a_{Lj,Li}(\omega)]\}, \quad (15.36)$$

where $\hat{\Sigma}^r_{Li,Lj}$ and $\hat{G}^a_{Lj,Li}$ are the block matrices of the retarded self-energies and advanced Green's functions, respectively.

In principle, it is possible to perform the further transformation of Eq. (15.36) and obtain the explicit expression for the current, which contains the terms corresponding to the local Andreev reflection and to the nonlocal transfer of charge carriers. However, for the sake of simplicity we do not present these lengthy equations in this section.

Note that the actual many-particle interactions are absent in the system considered here. That is why, the self-energies and full Green's functions in Eqs. (15.35) and (15.36) are determined taking into account all the tunneling processes between the device and the contacts. Correspondingly, the block matrices $\hat{G}^a_{Lj,Li}$ of the advanced Green's functions $\hat{G}^a$ of the whole device are determined by the following Dyson equation:

$$\hat{G}^a = [(\omega - \hat{h}_D - \hat{\Sigma}^r(\omega))^{-1}]^a, \quad (15.37)$$

where $\hat{h}_D$ is the operator of the one-particle energy in the diagonalized Hamiltonian $\hat{H}'_D$ of the device and $\hat{\Sigma}^r(\omega)$ is the matrix of the retarded self-energy. It is frequency-dependent in the general case and describes the effect of both contacts in the AB ring.

However, in our numerical calculations, we use again as in Chap. 11 a simpler approximation of wide-band contacts. In this approximation, we can neglect the real parts of the self-energies and assume the imaginary parts to be constant [24]. Then, we obtain the following explicit expressions for nonzero blocks of the retarded self-energy $\hat{\Sigma}^r$

$$\hat{\Sigma}^r_{L1,L1} = \hat{\Sigma}^r_{Rn,Rn} = -\frac{i}{2}\hat{\Gamma}_{11}, \quad \hat{\Sigma}^r_{R1,R1} = \hat{\Sigma}^r_{Ln,Ln} = -\frac{i}{2}\hat{\Gamma}_{22},$$
$$\hat{\Sigma}^r_{L1,Ln} = \hat{\Sigma}^r_{R1,Rn} = \hat{\Sigma}^r_{Ln,L1} = \hat{\Sigma}^r_{Rn,R1} = -\frac{i}{2}\hat{\Gamma}_{12}. \quad (15.38)$$

In Eq. (15.38), we introduced the matrices

$$\hat{\Gamma}_{ii} = \Gamma_{ii}\hat{I}_4, \quad (15.39)$$

where $\hat{I}_4$ is the 4×4 identity matrix, and

$$\Gamma_{ii} = 2\pi t_i^2 \quad (15.40)$$

are the functions describing the broadening of the energy levels of the device related to its interaction with the contact. Finally, ρ in Eq. (15.40) is the density of states in the contact. In expressions (15.38) for the self-energies, we introduced the notation $\Gamma_{12} = \sqrt{\Gamma_{11}\Gamma_{22}}$.

For the symmetric AB ring, we assume in the numerical calculations that $\Gamma_{11} = \Gamma_{22} = 0.01$. For the asymmetric ring, we put $\Gamma_{11} = \Gamma_{22}/2 = 0.01$.

The $\hat{G}^{+-}_{Lj,Li}$ blocks on Eq. (15.36) for the current are governed by the solution of the Keldysh equation

$$\hat{G}^{+-} = \hat{G}^{r}\hat{\Sigma}^{+-}\hat{G}^{a}\,. \tag{15.41}$$

Let us emphasize that we consider the regime where all the transient processes have been finished, and thus the bare Green's functions $\hat{g}$ of the device do not appear in Eq. (15.41) [22].

Non-zero blocks for the self-energies $\hat{\Sigma}^{+-}$ in Eq. (15.41) have the form

$$\hat{\Sigma}^{+-}_{\alpha i,\alpha j} = -2\hat{\Sigma}^{r}_{\alpha i,\alpha j}\hat{F}_{\alpha}\,, \tag{15.42}$$

where $\alpha = L, R, i, j = 1, n$, and we introduced the 4×4 diagonal matrix

$$\hat{F}_{L(R)} = \text{diag}\left(n\left(\omega \pm \frac{eV}{2}\right), n\left(\omega \mp \frac{eV}{2}\right), n\left(\omega \pm \frac{eV}{2}\right), n\left(\omega \mp \frac{eV}{2}\right)\right). \tag{15.43}$$

Note that $n\left(\omega \pm \frac{eV}{2}\right)$ in Eq. (15.43) are the Fermi–Dirac distribution functions.

15.5 Results of the Numerical Calculations

In this section, we present the results of the numerical calculations for the quantum transport. We will work again as in Chap. 11 in the regime of the linear response (small bias voltages) and consider the limit of low temperatures ($eV \to 0, k_B T \to 0$) for our system, which consists of the device and contacts shown in Fig. 15.1.

We will start by considering the limiting case of the AB ring with the minimal number of sites. To be more specific, we consider the case of six sites when $n = 1$ and $N = 2$. In Fig. 15.3a, we show the dependence of the first four excitation energies E_{1-4} on the gate field ϵ_D. At zero values of the Zeeman energy and Rashba constant ($h = \alpha = 0$), these four states are split in energy forming two pairs of the doubly degenerate states (see the thin dotted curves in Fig. 15.3a).

Moreover, since in the general case, the superconducting pairing in the ring is inhomogeneous, the gap Δ arises in the system at $\epsilon_D \neq 0$. However, at zero gate field, the energies $E_{1-4} = 0$, and that is why, the differential conductance $G = dI/dV$ exhibits the resonance behavior only at $\epsilon_D = 0$ (see the dotted line in Fig. 15.3b).

At large Zeeman energies $h > \Delta$, the gap is suppressed, and the number of zeroes in the spectrum is doubled due to the Zeeman splitting (see the dashed lines in

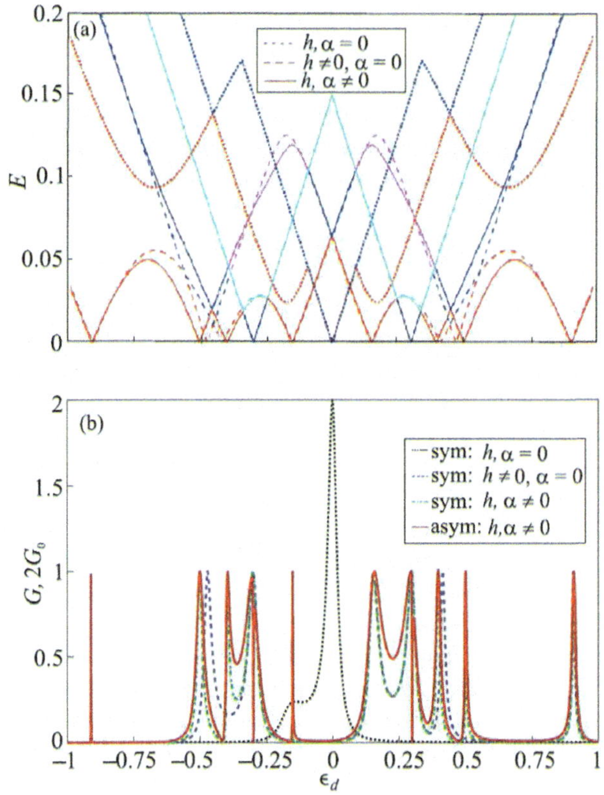

Fig. 15.3 Dependence of the excitation energies E_{1-4} (**a**) and the conductance (**b**) of the AB ring on the gate field ϵ_D. In the calculations, we use the following parameters: $n = 1$, $N = 2$ for the ring consisting of six sites, the hopping integral between NW and SW $t_0 = 0.5$, the Zeeman energy $h = 0.3$. In panel (**b**), G_0 is the conductance quantum. We scale all the parameters in the units of the hopping integral $t = 1$ between the nearest neighbor sites in the SW [6]

Fig. 15.3a). As a result, the number of peaks in the conductance increases. This fact is illustrated by the dashed curve in Fig. 15.3b.

However, not all the zeroes in the excitation energies manifest themselves as the resonances in the conductance. The cause for that is the formation of the bound states in the continuum (BSCs), which we discussed already in Chap. 11. There are several possibilities to obtain the finite lifetime of these states.

One of the possibilities is to break the spatial symmetry of the eigenstates of the ring. It can be achieved e.g. by introducing the spin–orbit coupling in the system [25]. In this case, the zeroes of the excitation spectrum related to the SW are slightly shifted and the Fano resonances emerge in the conductance (see the solid and dash-dotted curves in Fig. 15.3a and b respectively). We can conclude that to get the Dicke effect [26] in the symmetric ring with the SC central region, we need to consider the combination of the spin–orbit coupling and the applied magnetic field (the Zeeman energy).

Let us emphasize that the zero-energy states at the values of the gate fields $\epsilon_D = \pm h$ remain doubly degenerate even at non-zero values of the Rashba constant $\alpha \neq 0$. This degeneracy is also connected with the symmetry of the AB ring and leads to the emergence of the additional BSCs [27, 28]. They manifest themselves in the conductance if we consider the asymmetric case for the tunneling parameters between the device and contacts. In this case, the additional Fano resonance appears in the conductance at $\epsilon_D = \pm h$ (see the solid curve in Fig. 15.3b). The same type of effect arises, as we mentioned already in Chap. 11, if we take into account the Aharonov–Bohm phase (see [29, 30]).

Let us consider now the AB ring with a larger number of sites, namely, we use the parameters $n = 20$ and $N = 30$ for the NW and SW, respectively. According to [5], in the nontrivial topological phase of the SW when the set of inequalities (15.21) is fulfilled, and the Rashba constant $\alpha \neq 0$, we get the Dicke effect.

In Fig. 15.4, we show the pair of resonances (one is the Fano resonance, another is the Breit–Wigner one), which appear as a function of the Zeeman energy in a conductance of the symmetric AB ring with the parameters $n = 20$ and $N = 30$ for the NW and SW. As we already discussed in Sect. 15.3, the width of the Fano resonance in this system depends crucially on the degree of the localization of the Majorana states (MS) in the SW. This fact makes it possible to use our device for the detection of Majorana excitations.

Additional interesting features in the conductance related to the nonlocality of the MS appear in the asymmetric AB ring. In this case, we observe an additional narrow Fano resonance near the wide antiresonance on the dashed curve in Fig. 15.4a. We should stress, that when we increase the length of the SC bridge (namely, the length of the SW), the wide antiresonance moves closer to the narrow Fano peak.

At the same time, the narrow Fano peak collapses (its width tends to zero) and the BSC appears. We can clearly see this behavior in Fig. 15.4b. In other words, we can describe this situation as an unusual topological blockade of the Fano effect related to the asymmetry of the transport parameters of the ring. Indeed, the narrow Fano resonance disappears due to the nonlocality of the low-energy excitation in the SW.

To get a deeper insight into the mechanism controlling the collapse of Fano resonance, it is important to note that this resonance is related to the BSC emerging due to the degeneracy of the eigenstates of the closed system at zero energy. Therefore, we can assume that the vanishing of the Fano resonance can be caused by an increase in the multiplicity of the degeneracy of these states when the overlap of the Majorana wave functions becomes negligible.

15.6 Toy Model Illustrating the Collapse of Fano Resonance

To test our suggestion, let us consider a spinless toy model for the AB ring with $n = 1$ for the NW and $N = 2$ for the SW (see Fig. 15.2a). In other words, we can effectively consider the model of the Kitaev chain [14, 15] (discussed briefly in Sect. 15.2) with

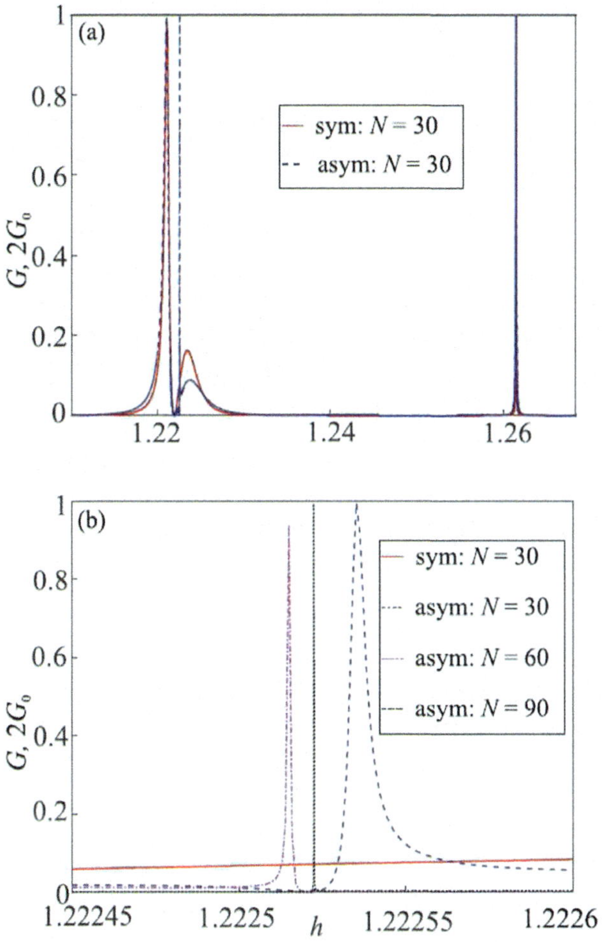

Fig. 15.4 **a** Narrow Fano resonance related to the asymmetry of the tunneling parameters of the AB ring. **b** Collapse of the narrow Fano resonance with an increase in the degree of nonlocality of the MS in the SW. The parameters for the NW and SW are: $n = 20$, $N = 30$, $t_0 = 0.1$, and $\epsilon_D = 1$ [6]

the even number of sites in the SC bridge. In this situation, the Hamiltonian of the AB ring at zero values $\epsilon_D = \mu = 0$ for the gate field and the chemical potential has the form

$$\hat{H}_D = \sum_{j=1}^{N-1} (-t a_j^\dagger a_{j+1} + \Delta a_j^\dagger a_{j+1}^\dagger) \\ - t_0 a_1^\dagger (b_{Ln} + b_{Rn}) - t_0 a_N^\dagger (d_{L1} + d_{R1}) + H.c.. \tag{15.44}$$

Diagonalizing the Hamiltonian (15.44), we get the following equation for the excitation spectrum:

$$E^4(EP_1 - 2t_0^2\delta_1^{N/2-1})(EP_2 + 2t_0^2\delta_1^{N/2-1})(EP_3 - 2t_0^2\delta_2^{N/2-1})(EP_4 + 2t_0^2\delta_2^{N/2-1}) = 0\,, \tag{15.45}$$

where

$$\delta_{1,2} = t \mp \Delta \tag{15.46}$$

and P_i, $i = 1 - 4$ are the polynomials of power $N/2$ with the property $P_{2,4} = P_{1,3}$ $(E \to -E)$, which is determined by the electron–hole symmetry.

From Eq. (15.45), it follows that for a special point of the Kitaev model [14, 15], namely for $\Delta = \pm t$, which was already mentioned in Sect. 15.2, the wave functions of the Majorana fermions do not overlap (see Eq. (15.46). Moreover, at this point, the multiplicity of degeneracy of the zero-energy states increases for $N > 2$, leading to the suppression of the narrow Fano resonance shown in Fig. 15.4b.

To develop these arguments further, we again, as in Sect. 15.2, consider our system in the Majorana representation introducing self-conjugated Majorana operators $\gamma_{ij} = \gamma_{ij}^\dagger$, $i = 1, 2$, which are related to the fermionic operators as follows:

$$\alpha_j = (\gamma_{1j} + i\gamma_{2j})/2\,. \tag{15.47}$$

In Fig. 15.2a and b, we present in the framework of our description a sketch of the device corresponding to the special point of the Kitaev model $\Delta = t$ for the two cases of $N = 2$ (Fig. 15.2a) and $N > 2$ (Fig. 15.2b), respectively. The straight lines on these figures denote the interaction between the Majorana fermions of different kinds. We can see that, for $N = 2$ in Fig. 15.2a, the upper and lower shoulders of the device remain connected due to the absence of the SC pairing in the horizontal direction.

However, for $N > 2$, the device in Fig. 15.2b is effectively divided into identical upper and lower subsystems. Each of the subsystems includes two chains of the interacting quasiparticles. Simple estimates yield for the eigenenergies of the chains with only two bonds in the horizontal direction

$$E_1 = 0, \quad E_{2,3} = \pm\frac{t_0}{\sqrt{2}}\,. \tag{15.48}$$

If we add a vertical bond, just in the same way as in the Fano–Anderson model [10, 31], then we have the following eigenenergies:

$$E_{1,2} = 0, \quad E_{3,4} = \pm\sqrt{t^2 + t_0^2/2}\,. \tag{15.49}$$

Therefore, we can say that it is just the formation of T-shape structures of the Majorana fermions, which leads to the collapse of the Fano resonance in the asymmetric AB ring.

Let us emphasize that the nonlocality of the MS does not depend on the ratio of the tunneling parameters in the NW, SW, and the contacts of the device. Thus, the effect, we study, has a universal character and manifests itself in the most general experimental situation when the tunneling parameters are different from each other. Moreover, according to Fig. 15.2, we observe that the Fano resonance is not suppressed even in the case of two noninteracting shoulders, i.e., for $t = 0$. We should stress, that the Fano resonance, which we describe here, in principle, does not emerge in the symmetric case.

15.7 Summary of the Results

Let us summarize the results presented in this section. We consider the asymmetric Aharonov–Bohm ring with different tunneling parameters. The shoulders of the ring are connected to the SC bridge in the topologically nontrivial phase. In this case, the additional narrow Fano resonance emerges in the system due to the symmetry breaking. However, when we increase the length of the SC bridge, or in other words, strongly reduce the overlap between the Majorana wave functions, this Fano resonance collapses.

To illustrate this effect, we consider a toy model of a spinless AB ring, where the Kitaev chain serves as the SC bridge. Simple analytical estimates for the excitation spectrum of this model demonstrate the increase in the multiplicity of the degeneracy of the zero-energy state for the special point $\Delta = t$ of the Kitaev model at $N > 2$. The increase in the multiplicity is due to the formation of T-shape chains of the Majorana fermions, which leads to the formation of the bound state in continuum (BSC). We can conclude that the collapse of the resonance is a direct consequence of the nonlocality of the MS.

References

1. M.Yu. Kagan, V.V. Val'kov, S.V. Aksenov, Phys. Rev. B **95**, 035411 (2017)
2. M.Yu. Kagan, V.V. Val'kov, S.V. Aksenov, J. Magn. Magn. Mater. **440**(Supplement C), 15 (2017)
3. M.Yu. Kagan, S.V. Aksenov, JETP Lett. **107**(8), 493 (2018)
4. Y. Aharonov, D. Bohm, Phys. Rev. **115**, 485 (1959)
5. V.V. Val'kov, M.Yu. Kagan, S.V. Aksenov, J. Phys.: Condens. Matter **31**(22), 225301 (2019)
6. S.V. Aksenov, M.Yu. Kagan, JETP Lett. **111**(5), 286 (2020)
7. V.V. Val'kov, M.S. Shustin, S.V. Aksenov, A.O. Zlotnikov, A.D. Fedoseev, V.A. Mitskan, M.Yu. Kagan, Phys. Usp. **65**(1), 2 (2022)
8. J.v. Neumann, E.P. Wigner, in *The Collected Works of Eugene Paul Wigner* (Springer, 1993), pp. 291–293
9. H. Feshbach, Ann. Phys. **19**(2), 287 (1962)
10. U. Fano, Phys. Rev. **124**, 1866 (1961)
11. E. Majorana, Nuovo Cim. (1924–1942) **14**, 171 (2008)

12. E. Rashba, V. Sheka, Sov. Phys. - Solid State **3**(6), 1257 (1961)
13. Y.A. Bychkov, E.I. Rashba, J. Phys. C: Solid State Phys. **17**(33), 6039 (1984)
14. A.Y. Kitaev, Phys. Usp. **44**(10S), 131 (2001)
15. A. Kitaev, Ann. Phys. **303**(1), 2 (2003)
16. A.F. Andreev, Sov. Phys. JETP **19**(5), 1228 (1964)
17. L. Fidkowski, R.M. Lutchyn, C. Nayak, M.P.A. Fisher, Phys. Rev. B **84**, 195436 (2011)
18. R.M. Lutchyn, J.D. Sau, S. Das Sarma, Phys. Rev. Lett. **105**, 077001 (2010)
19. Y. Oreg, G. Refael, F. von Oppen, Phys. Rev. Lett. **105**, 177002 (2010)
20. Y. Nambu, G. Jona-Lasinio, Phys. Rev. **122**, 345 (1961)
21. D. Rogovin, D.J. Scalapino, Ann. Phys. **86**(1), 1 (1974)
22. P.I. Arseev, Phys. Uspekhi **58**(12), 1159 (2015)
23. S. Vonsovsky, Y. Izyumov, E. Kurmaev, *Superconductivity of Transition Metals, Their Alloys and Compounds, Springer Series in Solid-State Sciences*, vol. 27 (Springer, Berlin, 1982)
24. P.I. Arseev, N.S. Maslova, V.N. Mantsevich, J. Exp. Theor. Phys. **115**(1), 141 (2012)
25. M.P. Nowak, B. Szafran, F.M. Peeters, Phys. Rev. B **84**, 235319 (2011)
26. R.H. Dicke, Phys. Rev. **89**, 472 (1953)
27. A. Volya, V. Zelevinsky, Phys. Rev. C **67**, 054322 (2003)
28. A.F. Sadreev, E.N. Bulgakov, I. Rotter, Phys. Rev. B **73**, 235342 (2006)
29. P.A. Orellana, M.L. Ladrón de Guevara, F. Claro, Phys. Rev. B **70**, 233315 (2004)
30. H. Lu, R. Lü, B.F. Zhu, Phys. Rev. B **71**, 235320 (2005)
31. P.W. Anderson, J. Phys. C: Solid State Phys. **3**(12), 2436 (1970)

Chapter 16
Disorder Effects and Phase Separation in Lattice Models, 2DEG, and Weyl Semimetals

Abstract We study electronic phase separation in cuprates in the framework of the repulsive-U Hubbard model and the $t - J$ model. We present the basic theoretic ideas important for the physics of stripes as well as recent experimental papers on phase separation in cuprates and iron-based SC. We discuss phase separation in 2DEG and disorder effects in Weyl systems related to the diffusive metal–semimetal phase transition. We comment briefly on the interesting subject of the chiral anomaly, which determines longitudinal magnetoresistance of topological materials and its possible suppression by large concentration of impurities in the ultraquantum limit.

16.1 Introduction

The subject of the different types of phase separation both in electronic and ionic subsystems of superconducting materials became very popular soon after the discovery of high-T_c superconductivity in copper oxides by J. Bednorz and A. Müller in 1986 [1–3].

We have already briefly discussed some aspects of electronic phase separation in cuprates and iron-based superconductors in Sect. 5.6 of Chap. 5 and Sect. 8.5 of Chap. 8. In this chapter, we will concentrate mostly on the one-band lattice models for cuprates and other high-T_c materials such as famous Hubbard and $t - J$ models.

The different types of phase separation in electronic subsystems at small scale were firstly emphasized in the classical papers by Schrieffer, Wen, Zhang [4], Kampf, Schrieffer [5], (spin polarons or bags) and in Bulaevskii, Nagaev, Khomskii [6], Brinkman, Rice [7] (magnetic strings). At large scale, the classical results belong here to Emery, Kivelson, and Lin [8, 9] (large clusters), Zaanen, Gunnarsson [10] (stripes), and Castellani, Di Castro, Grilli [11, 12].

M. Yu. Kagan et al., *Electronic Phase Separation in Magnetic and Superconducting Materials*, Springer Series in Solid-State Sciences 201,
https://doi.org/10.1007/978-3-031-55467-4_16

16.2 Electronic Phase Separation in Hubbard and $t - J$ Models

Note that the magnetic string oscillators (or just magnetic strings) of Bulaevskii–Nagaev–Khomskii–Brinkman–Rice [6, 7] were first predicted in magnetic insulators and semiconductors [13] in the years 1968–1970, that is substantially earlier than the discovery of high-T_c superconductivity, and only after the papers of Trugman [14] (see also the book [15]) were generalized on high-T_c cuprates in the framework of the famous $t - J$ [16] or Hubbard type of models [17].

Let us emphasize that the typical values of the important ratio of $t/J = 2 - 3$ and electron spin $S = 1/2$ in high-T_c cuprates favor string oscillators in SC materials, while in magnetic materials, such as CMR manganites $LaSrMnO_3$, where parameter $\alpha = t/(JS^2) > 75$ for spin $S = 3/2$, standard magnetic polarons of Nagaev–Mott–Kasuya type [18–20], (namely, ferrons considered in Chaps. 1–9 and [21, 22]) are more favorable.

Note also that in magnetic materials as it was emphasized in Chaps. 1–4, the FM polarons always correspond to the inhomogeneous phase-separated state with small nanoscale FM metallic droplets (containing extra holes or conduction electrons) embedded in an insulating AFM matrix as raisins in a pudding. Moreover, they appear as a consequence of the instability of the homogeneous canted state with respect to EPS governed by the negative compressibility $\kappa^{-1} = d\mu/dn < 0$.

The situation in SC materials is not so easy. Spin bags of Schrieffer, Kampf et al. [4, 5] correspond to the inhomogeneous phase-separated state. At moderate values of U/t in the Hubbard models and at optimally doped case $n \sim 0.85$ they can correspond to the homogeneous case similarly to electron polarons in the two-band Hubbard model considered in Chap. 10 on EPE. The same is true for string polarons introduced by Bulaevskii, Nagaev, Khomskii, Brinkman, Rice in [6, 7] (see also nice illustration in monograph [15] and more detailed picture for the formation of the string on Fig. 13.30 in the monograph [22]).

Only in the so-called difficult corner of the phase diagram of the Hubbard or $t - J$ models corresponding to the electron densities $n \approx 1$ close to half filling and large values of $U/t \gg 1$ (or small values of $J/t \ll 1$), we have the Nagaoka theorem [23] for one extra hole on top of AFM background, and as a consequence of it the phase separation on metallic hole rich and insulating hole-poor phases for the finite hole concentration $\delta = 1 - n$.

As it was shown by Visscher [24], in this case for $U/t \gg 1$, it is more beneficial again (as in manganites) to form small FM droplets embedded in AFM matrix with the droplet size controlled by U/t ratio and dimensionality of the system: $R_{drop} \propto a(U/t)^{1/5}$ in the 3D case and $R_{drop} \propto a(U/t)^{1/4}$ in 2D, where a is the intersite distance and R_{drop} is the droplet radius.

At the same time, as it was shown by Emery, Kivelson, Lin in [8] at small electron densities $n \to 0$ ($n \ll 1$)corresponding to strongly overdoped limit in cuprates the EPS in the 2D $t - J$ model takes place only for large physically unrealistic values of $J/t > 1$ ($J_{ps} = 3.8t$). In this case, according to the results of [8], we have phase separation on two large clusters, namely metallic PM with small electron density $n \to 0$ and AFM insulating cluster with electron density close to half filling $n \to 1$ (see also Chap. 13 and Fig. 13.28 for the phase diagram of the SC state in the 2D $t - J$ model in the monograph [22] and in [25]).

For realistic values of $J/t = 1/2 - 1/3$ and small and moderate electron densities $n < 0.85$, we get in agreement with the results of [25] the phase diagram of the SC state with the regions of p-wave and d-wave pairings. Moreover, at optimum densities $n = 0.85$ the d-wave pairing with the symmetry $d_{x^2-y^2}$ is dominantly yielding reasonable T_c of the order of 100 K for $J/t = 1/2$.

Note that in the two-band Hubbard model with one narrow band considered in Chap. 10 the tendency toward the instability of the homogeneous state (toward its negative compressibility) leading to EPS, manifests itself only at strong imbalance between the densities and masses of the two bands [26].

Note also that in a one-band Shubin–Vonsovskii [27] or extended Hubbard ($t - U - V$ model) with strong onsite Hubbard interaction $U \gg t$ and strong intersite Coulomb repulsion $V \gg t$ as it was shown in [28, 29] besides EPS close to half filling $n = 1$, there are additional regions of EPS close to the quarter filling $n = 1/2$ with nanoscale metallic droplets embedded in the charge ordered matrix (see also Chap. 4 of this monograph and our review [21]).

16.3 Theoretical Ideas Lying in the Basis of Physics of Stripes

There are many nicely written review articles (see, e.g., [47,48]) and even monographs (see, e.g., book [3] and references therein) on the phase separation and physics of stripes in high-T_c superconductors. Here we will only briefly remind the basic ideas related to the physics of stripes in copper-based [30–32] and iron-based [33, 34] superconductors.

First important theoretical results here were presented in classical papers [9, 10]. Note that in [9], the mean-field solution was obtained for a doped Mott insulator on a square lattice in the intermediate coupling regime, which corresponds to a state with Mott-insulating AFM domains separated by a regular "stripe" array of AFM domain walls (trapping the doped holes).

A complementary idea was introduced in [10]. According to [10], the system of doped holes and Cu spins has a tendency toward phase separation. Moreover, the long-range Coulomb interaction can lead to a nontrivial compromise in the system resulting in a formation of conducting stripe-like textures.

Later on the ideas of [9, 10] were generalized in [35, 36]. In this context we should mention first of all the important results of the Rice group in [9] on a very close competition between the uniform d-wave pairing and stripe phase in the underdoped and optimally doped 2D $t - J$ model. Stripe order-enhanced superconductivity in the Hubbard model for intermediate values of $U \sim W$ was thoroughly investigated by Jiang and Kivelson in [36].

It became clear soon that the stripes in high-T_c materials can stay metallic at very low temperatures and may be superconducting according to [30, 31, 35, 37]. The nature of the stripe phase however is still a subject of intense debate. In particular it is not totally clear yet whether the stripes are static or dynamic in different families of copper-based SC (see, e.g., the discussion in [31]).

16.4 First Experimental Results on EPS in Cuprates

The first experimental results on the physics of stripes in copper-based SC were obtained by Tranquada et al. [30, 37]. Later on, his group in collaboration with theorists got very interesting results [32, 38, 39] on the charge ordering and stripe formation near the structural phase transition close to the doping level $x = 1/8$ in $La_{2-x}Ba_xCuO_4$). This phase transition in ionic subsystem at x = 1/8 leads to a sharp anomaly related to the strong suppression of the SC critical temperature (see [40, 41]). Note that the incommensurate CO was recently established also in other families (YBCO, BSCCO, and HBCO) of copper-based high-T_c superconductors.

Let us stress here one important point. Usually, the common wisdom tells us that SC and CDW instabilities compete with each other. In physics of SC stripes, however, the different concepts are used such as intertwined SC and charge orders [42], which appear in the system, where charge stripes are made up of paired holes in some similarity with two-leg ladder systems [43] described by the Luther–Emery liquid [44, 45] or pair density wave state [31, 42]. In the case of intertwined orders, superconductivity can be enhanced by the formation of stripes [36].

At the end of this section, we can mention also the basic experimental papers by Jorgensen et al. [46], which reveal the important interconnection between the optimum inhomogeneity with respect to oxygen concentration and the maximum T_c in YBCOs. To be more specific, the authors of [46] investigated the role of time-dependent structural phenomena (namely, site disorder and the degradation phenomena) related to the local oxygen ordering upon quenching in $YBa_2Cu_3O_{6+y}$.

16.5 Recent Papers on EPS in Cuprates and Iron-Based SC

Among more recent papers on phase separation in copper-based high T-c SC we can mention a theoretical paper of Krasnoyarsk team by V.A.Gavrichkov et al. [47] on magnetic interaction in doped 2D perovskite cuprates and experimental paper of Tromp et al. [48] on the puddle formation, persistent gaps and non-mean-field breakdown of superconductivity in overdoped bismuth family of cuprates. Review article of Sherman [49] considers the problem of phonon-assisted phase separation in strongly correlated electron systems and contains rather disputable statement of the possibility of electronic phase separation in one-band Hubbard model close to half filling in the atomic representation. The paper by Zhang et al. [50] is more close to the subject of phase separation in ionic subsystem and discusses the possibility of the appearance and the evolution of the novel micron-scaled ribbon phase in optimally doped bismuth 2212 family $Bi_2Sr_2CaCu_2O_{8+\delta}$ at high pressure, which can possibly serve as a link between lattice inhomogeneity and high-T_c SC. Here, we can mention also a paper by Lorenz et al. [51] on magnetism, superconductivity and phase separation in $RuSr_2EuCu_2O_8$ rutheno-cuprates.

Among papers on different families of high-T_c superconductors, we can cite the article by Masi et al. [52] and the article by Imai et al. [53] on superconductivity and phase separation in $FeSe_{1-x}Te_x$, as well as articles [33, 54] on the intrinsic phase separation in pnictides $Fe_2K_{2-y}As_x$. Note also the article [55] on stripe-like nanoscale structural phase separation in plumbates bismuthates $BaPb_{1-x}Bi_xO_3$, which we already mentioned in Chap. 11. An important contribution to the physics of stripes in different families of high-T_c materials is related to the results of Bianconi's group in Rome [33, 34] (see also Kagan, Bianconi [56]). Finally, it is interesting to mention the shock-wave experiments in Livermore where the conducting droplets of metallic hydrogen (in the insulating matrix) were observed close to the first-order phase boundary between metallic liquid and metallic crystalline phases (for the brief review see, Chap. 14 and [57]).

16.6 Phase Separation in Two-Dimensional Electron Gas

The phase separation is a common feature not only in cuprates and other high-T_c superconductors but also in two-dimensional electron gas (2DEG) that forms at the interface of two insulators or semiconductors [58, 59] and in highly crystalline ultrathin films, like ZrNCl [60]. In general, the inhomogeneous states in the 2DEG arise due to electrostatic interaction between electrons in the gas with substrate. However, the details of this interaction are different for different heterostructures with 2DEG. We briefly consider here two typical cases. First, the 2DEG that forms at the interface of two insulating oxides, like $LaAlO_3/SrTiO_3$ (or $LaTiO_3/SrTiO_3$) hereafter referred to as LXO/STO (and, possibly, in the ultrathin films of ZrNCl, $TiSe_2$, MoS_2, etc.). Second, the 2DEG at the interface of two semiconductors GaAs/AlGaAs. Both systems exhibit a number of nontrivial phenomena, e.g., a gate-tunable metal–to–superconductor transition [61, 62], inhomogeneous magnetic response [63–65], and very intriguing inhomogeneous superconductivity [60, 66, 67].

The most intense research on 2DEG was implemented using the silicon-based heterostructures, for example, the silicon field effect transistor (Si-MOSFET). The 2DEG appears at an energy level arising at the nearly triangular potential well at the interface between the semiconductor and oxide layers. The electron density at this level is controlled by the applied gate voltage. This type of two-dimensional electron structures was employed for several decades as a research platform paving to the discovery, for example, of the integer and fractional Hall effects and other fascinating phenomena. The problem of phase separation and of emerging inhomogeneous states in these structures was a subject of numerous studies A detailed review of the theoretical and mostly experimental aspects of such studies can be found in the review article [68].

The physics of 2DEG in LXO/STO structures has recently attracted considerable interest (see, e.g., a review [69]), and the electronic phase separation is a key phenomena in that system [59]. This interest is highly driven by experimental discovery of an unusual inhomogeneous superconducting states in these systems [66, 67]. Now, two mechanisms of the phase separation are debated for LXO/STO, both associated with a high dielectric constant of $SrTiO_3$ [70].

According to a detailed investigation of the stability of 2DEG in LXO/STO [59], the fluctuation of the electron density in 2DEG gives rise to a high polarization of the $SrTiO_3$ substrate. As a result, electrons are captured in the self-consistent potential well, whose depth in LXO/STO is sufficient to form quantum bound states. Naturally, the compressibility of such a system is negative and the phase separation occurs in 2DEG. The second mechanism is more tricky but also rather natural. The inhomogeneous distribution of electron density is accompanied by an inhomogeneous distribution of the electric field confining the electrons at the interface. In turn, this inhomogeneous transverse electric field (with large gradients) induces an inhomogeneous Rashba spin-orbit coupling giving rise to capturing of the electrons in 2DEG by a self-consistent potential and, again, to the arising quantum bound states, negative compressibility, and the phase separation [70].

The existence of the inhomogeneity and/or multi-component phases in LXO/STO systems is reliably confirmed by a number of experiments. In particular, the magnetotransport measurements reveal the presence of high- and low-mobility charge carriers in LTO/STO, and the superconductivity develops when high-mobility charge carriers arise in 2DEG [71, 72]. Inhomogeneities are observed in magnetic experiments [63, 73, 74] and in the tunneling spectra [75]. Moreover, a percolative insulator-to-metal transition with the increase of the gate-tuned electron density, which is commonly observed in 2DEG, has a natural explanation in the framework of the picture with the electronic phase separation.

In the 2DEG in GaAs/AlGaAs heterostructures, the electronic phase separation can be also attributed to an electrostatic electron–substrate interaction but the nature of this interaction is different as compared to LXO/STO. It is commonly accepted that the inhomogeneous state in GaAs/AlGaAs arises due to disordered electrostatic potential produced by donor (or acceptor) atoms in the substrate [76–80]. The gate voltage rips out the electrons (or holes) from the substrate and tune the electron (hole) density in the 2DEG. This electron redistribution produces a disorder of the charge density in the substrate. Interplay of the kinetic energy of the 2D Fermi gas, disordered potential, and Coulomb repulsion between charge carriers in the 2DEG gives rise to the formation of the self-consistent electron (hole) droplets. This mechanism is studied in detail in the cited papers. However, in the case of GaAs/AlGaAs, we deal with an initially inhomogeneous system, which is out of the scope of the present book.

At the end of this section let us mention briefly several other interesting possibilities of phase separation in 2DEG. In [81] by Zhang et al., the size-dependent metal–insulator transition was studied in $KTaO_3$-based 2DEG. The width of the conductive channel decreasing into micrometer scale due to the coexistence and separation of metallic and insulating phase. Moreover, the authors of [81] observed hysteretic resistance-temperature relations, which are tunable by the magnetic field due to the competition between the phases.

Another possibility is related to the phase separation in the 2D electron liquid in MOSFETs [58]. In this paper, Spivak predicts an appearance of the intermediate phase between the Fermi-liquid and Wigner crystal which resembles to some extent a super-solid phase predicted some time ago by Andreev, Lifshitz [82].

Finally note a conjecture of Chang and Canningham [83] about the possible phase separation for 2DEG in the regime of fractional quantum Hall effect (FQHE). Near the boundary in the case of slowly increasing boundary potential there are transport evidence that 2DEG can phase separate into spatial regions of two fractional quantum Hall fluids with different filling factors ν.

16.7 Disorder Effects, Chiral Anomaly, and Phase Separation in Weyl Semimetals

Very recently it was quite extensive research (see, e.g., [84, 85]) on disorder effects on the evolution of the density of states, semimetal–diffusive metal transition and possible phase separation in Weyl systems.

Note that in disordered 2D graphene, a finite density of states appears already at weak disorder. Similar calculations suggest the formation of nonzero density of states in Weyl systems at a finite disorder strength. The question whether the semimetal-diffusive metal transition corresponds to a quantum critical point (QCP) at some critical strength of disorder or this transition is a continuous one (as it was shown in the renormalization group (RG) approach) is still a subject of intense debates.

Let us stress that according to [85, 86], the rare nonperturbative events (which are not accounted in the perturbative RG approach) can have a singular effect and produce a small density of states in Weyl systems even at weak disorder.

Another important topic here is the investigation of the disorder effects on the transport and thermodynamics of the Weyl systems. One of the authors of the present manuscript (Kugel et al. [87]) contributed in this field by studying the disorder effects on the specific heat and transverse magnetoresistance in Weyl semimetals under the electron scattering by the random potential of the impurities (in the limits of both zero-range and finite-range impurity potential).

The influence of the disorder effects can be very important also for the problem of chiral anomaly (specified by the appearance of the EB term) in the negative longitudinal magnetoresistance and its possible suppression by the large concentration of impurities in Weyl semimetals and topological insulators such as $Bi_{1-x}Sb_x$ close to the quantum critical point at concentration of Sb $x = 3\%$.

The most interesting here for the layered systems is the quantum limit of strong quantizing magnetic field perpendicular to the layer, low temperatures, and small impurity concentrations (when the effective chemical potential is larger than temperature). In this limit, only lower Landau level is occupied and weakly washed out by the impurities. The quantum limit is in contrast to the classical limit of large impurity concentration, high temperatures, and weak quantizing magnetic field when all Landau levels are completely washed out by the impurities and temporal effects.

However, according to the ideology of Spivak et al. [88–90], in both these limits, we can observe the interesting effects related to the infrared divergences in the expressions for the density of states, magnetoconductivity, particle number in different valleys (in the multi-valleys case and so on). This conjecture of Spivak is close to the ideology of one of the authors of this book (M.Yu.K) who together with A.F. Andreev et al. [91–93] proposed some time ago a possible mechanism relates to the washing out of the Landau levels for the suppression of chiral anomaly in superfluid He-3A in aerogel (in the presence of strong disorder).

Note that in accordance with the elegant idea of Volovik et al. [94, 95] the appearance of chiral anomaly in clean superfluid He-3A is due to the beautiful analogy between Bogoliubov–de Gennes equations for fermionic quasiparticles in He-3A in the texture of the vector of the orbital momentum (of the anisotropic p-wave Cooper pairs) and the Dirac equation for massless electrons in a strong quantizing magnetic field.

References

1. J.G. Bednorz, K.A. Müller, Zs. Phys. B **64**(2), 189 (1996)
2. J.G. Bednorz, K.A. Müller, Rev. Mod. Phys. **60**(3), 585 (1988)
3. K.A. Müller, G. Benedek, *Phase Separation in Cuprate Superconductors* (World Scientific, Singapore, 1993)
4. J.R. Schrieffer, X.G. Wen, S.C. Zhang, Phys. Rev. B **39**, 11663 (1989)
5. A. Kampf, J.R. Schrieffer, Phys. Rev. B **41**, 6399 (1990)
6. L.N. Bulaevskii, E.L. Nagaev, D.I. Khomskii, Sov. Phys. JETP **27**, 836 (1968)
7. W.F. Brinkman, T.M. Rice, Phys. Rev. B **2**, 1324 (1970)
8. V.J. Emery, S.A. Kivelson, H.Q. Lin, Phys. Rev. Lett. **64**, 475 (1990)
9. V. Emery, S. Kivelson, Physica C Superconductivity **209**(4), 597 (1993)
10. J. Zaanen, O. Gunnarsson, Phys. Rev. B **40**, 7391 (1989)
11. C. Castellani, C. Di Castro, M. Grilli, Phys. Rev. Lett. **75**(7), 4650 (1995)
12. C. Castellani, C. Di Castro, M. Grilli, J. Phys. Chem. Solids **59**(10–12), 1694 (1998)
13. E.L. Nagaev, *Physics of Magnetic Semiconductors* (Mir Publishers, 1983)
14. S.A. Trugman, Phys. Rev. B **37**, 1597 (1988)
15. P. Fulde, *Electron Correlations in Molecules and Solids*. Springer Series in Solid-State Sciences, vol. 100, 3rd edn. (Springer-Verlag, Berlin, Heidelberg, 2002)
16. P.W. Anderson, Science **235**(4793), 1196 (1987)
17. J. Hubbard, Proc. R. Soc. A **205**, 238 (1963)
18. E.L. Nagaev, JETP Lett. **6**, 18 (1967)

19. N.F. Mott, E.A. Davis, *Electronic Processes in Non-crystalline Materials*. Monographs on Physics (Oxford University Press, 1971)
20. T. Kasuya, A. Yanase, T. Takeda, Solid State Commun. **8**(19), 1543 (1970)
21. M.Yu. Kagan, K. Kugel, A. Rakhmanov, Phys. Rep. **916**, 1 (2021)
22. M.Yu. Kagan, *Modern Trends in Superconductivity and Superfluidity*. Lecture Notes in Physics, vol. 874 (Springer, Netherlands, 2014)
23. Y. Nagaoka, Phys. Rev. **147**, 392 (1966)
24. P.B. Visscher, Phys. Rev. B **10**, 943 (1974)
25. M.Yu. Kagan, T.M. Rice, J. Phys. Condens. Matter **6**(20), 3771 (1994)
26. M.Yu. Kagan, V.V. Val'kov, J. Exp. Theor. Phys. **113**(1), 156 (2011)
27. S. Schubin, S. Wonsowsky, Proc. R. Soc. A **145**, 159 (1934)
28. M.Yu. Kagan, D.V. Efremov, M.S. Marienko, V.V. Val'kov, JETP Lett. **93**(12), 725 (2011)
29. M.Yu. Kagan, V.A. Mitskan, M.M. Korovushkin, Phys. Usp. **58**(8), 733 (2015)
30. V.J. Emery, S.A. Kivelson, J.M. Tranquada, Proc. Natl. Acad. Sci. USA **96**(16), 8814 (1999)
31. B. Keimer, S.A. Kivelson, M.R. Norman, S. Uchida, J. Zaanen, Nature **518**, 179 (2015)
32. J.M. Tranquada, M.P.M. Dean, Q. Li, J. Phys. Soc. Jpn. **90**(11), 111002 (2021)
33. A. Ricci, N. Poccia, B. Joseph, G. Arrighetti, L. Barba, J. Plaisier, G. Campi, Y. Mizuguchi, H. Takeya, Y. Takano et al., Supercond. Sci. Technol. **24**(8), 082002 (2011)
34. G. Campi, A. Bianconi, Condens. Matter **6**(4) (2021)
35. P. Corboz, T.M. Rice, M. Troyer, Phys. Rev. Lett. **113**, 046402 (2014)
36. H.C. Jiang, S.A. Kivelson, Proc. Natl. Acad. Sci. USA **119**(1), e2109406119 (2022)
37. J.M. Tranquada, B.J. Sternlieb, J.D. Axe, Y. Nakamura, S. Uchida, Nature **375**(6532), 561 (1995)
38. M. Hücker, M.V. Zimmermann, G.D. Gu, Z.J. Xu, J.S. Wen, G. Xu, H.J. Kang, A. Zheludev, J.M. Tranquada, Phys. Rev. B **83**, 104506 (2011)
39. E. Berg, E. Fradkin, S.A. Kivelson, J.M. Tranquada, New J. Phys. **11**(11), 115004 (2009)
40. J.D. Axe, A.H. Moudden, D. Hohlwein, D.E. Cox, K.M. Mohanty, A.R. Moodenbaugh, Y. Xu, Phys. Rev. Lett. **62**, 2751 (1989)
41. S.J.L. Billinge, G.H. Kwei, A.C. Lawson, J.D. Thompson, H. Takagi, Phys. Rev. Lett. **71**, 1903 (1993)
42. E. Fradkin, S.A. Kivelson, J.M. Tranquada, Rev. Mod. Phys. **87**, 457 (2015)
43. E. Dagotto, T.M. Rice, Science **271**(5249), 618 (1996)
44. V.J. Emery, *Theory of the One-Dimensional Electron Gas* (Springer, US, Boston, MA, 1979), pp.247–303
45. A. Luther, Phys. Rev. B **14**, 2153 (1976)
46. J. Jorgensen, S. Pei, P. Lightfoor, H. Shi, A. Paulikas, B. Veal, Physica C **167**(5), 571 (1990)
47. V.A. Gavrichkov, S.I. Polukeev, Condens. Matter **7**(4), 57 (2022)
48. W.O. Tromp, T. Benschop, J.F. Ge, I. Battist, K.M. Bastiaans, D. Chatzopoulos, A. Vervloet, S. Smit, E. van Heumen, M.S. Golden, Y. Huang, T. Kondo, Y. Yin, J.E. Hoffman, M.A. Sulangi, J. Zaanen, M.P. Allan, Nature Materials **22**, 703 (2023)
49. A. Sherman, Int. J. Mod. Phys. B **36**(16), 2250081 (2022)
50. J. Zhang, Y. Ding, C.C. Chen, Z. Cai, J. Chang, B. Chen, X. Hong, A. Fluerasu, Y. Zhang, C.S. Ku, D. Brewe, S. Heald, H. Ishii, N. Hiraoka, K.D. Tsuei, W. Liu, Z. Zhang, Y.Q. Cai, G. Gu, T. Irifune, H.K. Mao, J. Phys. Chem. Lett. **9**(15), 4182 (2018)
51. B. Lorenz, Y. Xue, C. Chu, arXiv:0211492 (2002)
52. A. Masi, C. Alvani, A. Augieri, M. Bellusci, G. Celentano, G. De Marzi, F. Fabbri, C.F. Zignani, A. La Barbera, F. Padella, M. Pentimalli, E. Silva, A. Vannozzi, F. Varsano, IEEE Trans. Appl. Supercond. **28**(4), 1 (2018)
53. Y. Imai, Y. Sawada, F. Nabeshima, A. Maeda, Proc. Natl. Acad. Sci. USA **112**(7), 1937 (2015)
54. Y. Chen, J. Jiang, H. Yang, P. Dudin, A. Barinov, Z. Liu, H. Wen, L. Yang, Y. Chen, Nano Research **14**(3), 823 (2021)
55. P. Giraldo-Gallo, Y. Zhang, C. Parra, H. Manoharan, M. Beasley, T. Geballe, M. Kramer, I. Fisher, Nat. Commun. **6**(1), 8231 (2015)
56. M.Yu. Kagan, A. Bianconi, Condens. Matter **4**(2) (2019)

57. M.Yu. Kagan, S.V. Aksenov, A.V. Turlapov, R.S. Ikhsanov, K.I. Kugel, E.A. Mazur, E.A. Kuznetsov, V.M. Silkin, E.A. Burovski, JETP Lett. **117**(10), 755 (2023)
58. B. Spivak, Phys. Rev. B **67**, 125205 (2003)
59. N. Scopigno, D. Bucheli, S. Caprara, J. Biscaras, N. Bergeal, J. Lesueur, M. Grilli, Phys. Rev. Lett. **116**, 026804 (2016)
60. G. Dezi, N. Scopigno, S. Caprara, M. Grilli, Phys. Rev. B **98**, 214507 (2018)
61. N. Reyren, S. Thiel, A.D. Caviglia, L.F. Kourkoutis, G. Hammerl, C. Richter, C.W. Schneider, T. Kopp, A.S. Rüetschi, D. Jaccard, M. Gabay, D.A. Muller, J.M. Triscone, J. Mannhart, Science **317**(5842), 1196 (2007)
62. A. Camjayi, K. Haule, V. Dobrosavljević, G. Kotliar, Nat. Phys. **4**(12), 932 (2008)
63. X. Ariando, G. Wang, J. Baskaran, Z.Q. Liu, J.B. Huijben, A. Yi, A.R. Annadi, A. Barman, S. Rusydi, Y.P. Dhar, J. Feng, H. Ding, T. Hilgenkamp, Venkatesan. Nat. Commun. **2**(1), 188 (2011)
64. N. Teneh, A.Y. Kuntsevich, V.M. Pudalov, M. Reznikov, Phys. Rev. Lett. **109**, 226403 (2012)
65. V.M. Pudalov, M.E. Gershenson, JETP Lett. **111**(4), 225 (2020)
66. A.E.M. Smink, M.P. Stehno, J.C. de Boer, A. Brinkman, W.G. van der Wiel, H. Hilgenkamp, Phys. Rev. B **97**, 245113 (2018)
67. G. Singh, A. Jouan, G. Herranz, M. Scigaj, F. Sánchez, L. Benfatto, S. Caprara, M. Grilli, G. Saiz, F. Couëdo, C. Feuillet-Palma, J. Lesueur, N. Bergeal, Nat. Mater. **18**(9), 948 (2019)
68. V.M. Pudalov, JETP Lett. **116**(8), 286 (2022)
69. Y.Y. Pai, A. Tylan-Tyler, P. Irvin, J. Levy, Rep. Prog. Phys. **81**(3), 036503 (2018)
70. N. Bovenzi, S. Caprara, M. Grilli, R. Raimondi, N. Scopigno, G. Seibold, J. Phys. Chem. Solids **128**, 118 (2019)
71. J. Biscaras, N. Bergeal, S. Hurand, C. Grossetête, A. Rastogi, R.C. Budhani, D. LeBoeuf, C. Proust, J. Lesueur, Phys. Rev. Lett. **108**, 247004 (2012)
72. C. Bell, S. Harashima, Y. Kozuka, M. Kim, B.G. Kim, Y. Hikita, H.Y. Hwang, Phys. Rev. Lett. **103**, 226802 (2009)
73. L. Li, C. Richter, J. Mannhart, R.C. Ashoori, Nat. Phys. **7**(10), 762 (2011)
74. J.A. Bert, B. Kalisky, C. Bell, M. Kim, Y. Hikita, H.Y. Hwang, K.A. Moler, Nat. Phys. **7**(10), 767 (2011)
75. Z. Ristic, R.D. Capua, G.M.D. Luca, F. Chiarella, G. Ghiringhelli, J.C. Cezar, N.B. Brookes, C. Richter, J. Mannhart, M. Salluzzo, EPL (Europhysics Letters) **93**(1), 17004 (2011)
76. J. Shi, X.C. Xie, Phys. Rev. Lett. **88**, 086401 (2002)
77. M.M. Fogler, Phys. Rev. B **69**, 121409 (2004)
78. V. Tripathi, M.P. Kennett, Phys. Rev. B **74**, 195334 (2006)
79. V. Tripathi, K. Dhochak, B.A. Aronzon, V.V. Rylkov, A.B. Davydov, B. Raquet, M. Goiran, K.I. Kugel, Phys. Rev. B **84**, 075305 (2011)
80. V. Tripathi, K. Dhochak, B.A. Aronzon, B. Raquet, V.V. Tugushev, K.I. Kugel, Phys. Rev. B **85**, 214401 (2012)
81. J. Zhang, J. Zhang, D. Cui, L. Ye, S. Gong, Z. Wang, Z. Wu, C. Ma, J. Gao, Y. Zhao, Y. Jiang, arXiv:2211.14464 (2022)
82. A.F. Andreev, I.M. Lifshitz, Sov. Phys. JETP **29**(6), 1107 (1969)
83. A.M. Chang, J.E. Cunningham, Phys. Rev. Lett. **69**, 2114 (1992)
84. N.P. Armitage, E.J. Mele, A. Vishwanath, Rev. Mod. Phys. **90**, 015001 (2018)
85. J.H. Pixley, D.A. Huse, S.D. Sarma, Phys. Rev. X **6**, 021042 (2016)
86. R. Nandkishore, D.A. Huse, S.L. Sondhi, Phys. Rev. B **89**, 245110 (2014)
87. Ya.I. Rodionov, K.I. Kugel, B.A. Aronzon, F. Nori, Phys. Rev. B **102**, 205105 (2020)
88. D.T. Son, B.Z. Spivak, Phys. Rev. B **88**, 104412 (2013)
89. B.Z. Spivak, A.V. Andreev, Phys. Rev. B **93**, 085107 (2016)
90. A.V. Andreev, B.Z. Spivak, Phys. Rev. Lett. **120**, 026601 (2018)
91. A.F. Andreev, M.Yu. Kagan, Sov. Phys. JETP **66**(3), 504 (1987)
92. M.Yu. Kagan, D.V. Efremov, J. Low Temp. Phys. **158**, 749 (2010)
93. M.Yu. Kagan, A.V. Turlapov, Phys. Usp. **62**(3), 215 (2019)
94. A. Balatskii, G. Volovik, V. Konyshev, Sov. Phys. JETP **63**, 1194 (1986)
95. G.E. Volovik, *The Universe in a Helium Droplet* (Oxford University Press, 2009)

Chapter 17
Conclusions. Some Additional Problems

Abstract Here, we briefly underline the list of unresolved problems important to the physics of phase separation in the strongly correlated matter.

We have demonstrated that during the past half-century, the field of electronic phase separation has evolved from an exotic effect manifesting itself in some magnetic semiconductors to a very broad range of phenomena inherent to all strongly correlated electron systems and not only to them. This research field is still rapidly progressing. This book touches upon mostly the topics, to which we have contributed ourselves. Of course, some important issues related to the formation of spin and charge inhomogeneities turn out to be outside the scope of our analysis. Let us list some of them.

A very popular and well-studied topic concerns the dynamic and static stripe structures in cuprates, nickelates, and other oxide compounds. There is a vast body of literature concerning this topic, see, e.g., comprehensive reviews [1–3]. In addition to oxides, the existence of stripes has been also discussed recently for iron-based superconductors (where the stripe structures are closely related to the electron nematicity [4]) and in borides [5, 6]. We only briefly discuss this topic in Sects. 16.1–16.5 of Chap. 16.

Another large and important issue is the formation of charge inhomogeneities owing to a random potential generated by impurities. This line of research has been especially active in connection to the two-dimensional electron gas in semiconductor heterostructures [7–9]. We consider 2DEG electron gas in Sect. 16.6 of Chap. 16. Such disorder-induced inhomogeneities are conceptually related to the long-standing problem of the insulator–superconductor transition accompanied by the formation of superconducting islands within a disordered insulator [10–13]. One of the possible realizations of direct superconductor-insulator transition is discussed in Chap. 13 for the case of low-density electron system with an interparticle attraction in the presence of a strong random potential.

Note that quite recently, the formation of the superconducting droplets of the local bielectron pairs in the normal matrix of the unpaired electron states in the 2D attractive-U Hubbard model in the quasi-bosonic limit of strong onsite attraction

M. Yu. Kagan et al., *Electronic Phase Separation in Magnetic and Superconducting Materials*, Springer Series in Solid-State Sciences 201,
https://doi.org/10.1007/978-3-031-55467-4_17

$|U| \gg W$, low electron densities $n \ll 1$, and in the presence of strong diagonal disorder $V \gg W$ (W and V are the widths of the "clean" and "dirty" bands, respectively) was demonstrated in [14, 15], see also Chap. 13.

A very interesting topic concerns the phase separation under nonequilibrium conditions (see, e.g., [16]), which has attracted a vivid interest of chemists and biologists [17, 18].

Note also that the recent discovery of the unconventional superconductivity in the twisted bilayer graphene at small twist angles [19, 20] has added a new impetus to the studies of inhomogeneous states in graphene-based systems (see Chap. 9).

Among more exotic possibilities, we could mention the phase separation in the spin-imbalanced and dipole Fermi gas [21, 22] as well as inhomogeneous states in the Fermi–Bose mixtures in quantum gases in magnetic traps and in the Aharonov–Bohm ring with topologically nontrivial superconducting bridge [15, 23, 24].

Excitonic droplets and inhomogeneous plasmonic structures are important for the physics of BEC, as well as for the self-similar superfluid hydrodynamics of gas cloud expansion and collapse in novel bosonic systems [15, 25].

Recent experimental discovery [26] of the low-energy acoustic plasmons (so-called Pines' demon) in superconducting ruthenates Sr_2RuO_4 and observation of overdamped plasmons in numerical simulations [27] in bismuth family of high-T_c cuprates (with an account taken of full electron spectrum and diagonal hops in CuO layers [28]) establish a close link between nanoplasmonics and unconventional superconductivity.

Another quite extensive field of research (see, e.g., [29–36]) is the investigations of the chiral anomaly and disorder effects and their role in the evolution of the density of states, phase separation, thermodynamics and transport phenomena in Weyl semimetals and topological insulators.

The last but not the least issue is the application of the phase-separation concept to the nuclear matter including such scenarios as nuclear pasta-like structures [37, 38] and inhomogeneities in rotating neutron stars [39] (similar to the situation with the vortices in superfluid helium [40]).

We see that the problem of the phase separation is so vast and multifaceted that it cannot be covered within one, even large, treatise. Nevertheless, we believe that our book provides a comprehensive overview of an important part of this problem.

References

1. A. Bianconi, N.L. Saini (eds.), *Stripes and Related Phenomena* (Springer, US, 2013)
2. E. Berg, E. Fradkin, S.A. Kivelson, J.M. Tranquada, New J. Phys. **11**(11), 115004 (2009)
3. M. Vojta, Adv. Phys. **58**(6), 699 (2009)
4. R.M. Fernandes, A.V. Chubukov, J. Schmalian, Nat. Phys. **10**(2), 97 (2014)
5. S.V. Demishev, V.N. Krasnorussky, A.V. Bogach, V.V. Voronov, N.Yu. Shitsevalova, V.B. Filipov, V.V. Glushkov, N.E. Sluchanko, Sci. Rep. **7**(1), 17430 (2017)

6. N. Sluchanko, A. Bogach, N. Bolotina, V. Glushkov, S. Demishev, A. Dudka, V. Krasnorussky, O. Khrykina, K. Krasikov, V. Mironov, V.B. Filipov, N. Shitsevalova, Phys. Rev. B **97**, 035150 (2018)
7. V. Tripathi, K. Dhochak, B.A. Aronzon, V.V. Rylkov, A.B. Davydov, B. Raquet, M. Goiran, K.I. Kugel, Phys. Rev. B **84**, 075305 (2011)
8. N. Teneh, A.Yu. Kuntsevich, V.M. Pudalov, M. Reznikov, Phys. Rev. Lett. **109**, 226403 (2012)
9. L.A. Morgun, A.Yu. Kuntsevich, V.M. Pudalov, Phys. Rev. B **93**, 235145 (2016)
10. A.M. Goldman, N. Marković, Phys. Today **51**(11), 39 (1998)
11. Y. Dubi, Y. Meir, Y. Avishai, Nature **449**, 876 (2007)
12. Y. Dubi, Y. Meir, Y. Avishai, Physica C **468**(4), 354 (2008)
13. S. Poran, T. Nguyen-Duc, A. Auerbach, N. Dupuis, A. Frydman, O. Bourgeois, Nat. Commun. **8**(11), 14464 (2017)
14. M.Yu. Kagan, E.A. Mazur, J. Phys. Conf. Ser. **1740**, 012045 (2021)
15. M.Yu. Kagan, S.V. Aksenov, A.V. Turlapov, R.S. Ikhsanov, K.I. Kugel, E.A. Mazur, E.A. Kuznetsov, V.M. Silkin, E.A. Burovski, JETP Lett. **117**(10), 755 (2023)
16. F. Randi, I. Vergara, F. Novelli, M. Esposito, M. Dell'Angela., V.A.M. Brabers, P. Metcalf, R. Kukreja, H.A. Dürr, D. Fausti, M. Grüninger, F. Parmigiani, Phys. Rev. B **93**, 054305 (2016)
17. B. Trefz, S.K. Das, S.A. Egorov, P. Virnau, K. Binder, J. Chem. Phys. **144**(14), 144902 (2016)
18. R. Dickman, New J. Phys. **18**(4), 043034 (2016)
19. Y. Cao, V. Fatemi, S. Fang, K. Watanabe, T. Taniguchi, E. Kaxiras, P. Jarillo-Herrero, Nature **556**, 43 (2018)
20. Y. Cao, V. Fatemi, A. Demir, S. Fang, S.L. Tomarken, J.Y. Luo, J.D. Sanchez-Yamagishi, K. Watanabe, T. Taniguchi, E. Kaxiras, R.C. Ashoori, P. Jarillo-Herrero, Nature **556**, 80 (2018)
21. F. Heidrich-Meisner, G. Orso, A.E. Feiguin, Phys. Rev. A **81**, 053602 (2010)
22. D. Mitra, P.T. Brown, P. Schauß, S.S. Kondov, W.S. Bakr, Phys. Rev. Lett. **117**, 093601 (2016)
23. A.P. Menushenkov, K.V. Klementev, A.V. Kuznetsov, M.Yu. Kagan, J. Exp. Theor. Phys. **93**(3), 615 (2001)
24. A.P. Menushenkov, A.V. Kuznetsov, K.V. Klementiev, M.Yu. Kagan, J. Supercond. Novel Magn. **29**(3), 701 (2016)
25. E. Kuznetsov, M.Yu. Kagan, J. Exp. Theor. Phys. **137**, 190 (2023)
26. A.A. Husain, E.W. Huang, M. Mitrano, M.S. Rak, S.I. Rubeck, X. Guo, H. Yang, C. Sow, Y. Maeno, B. Uchoa, T.C. Chiang, P.E. Batson, P.W. Phillips, P. Abbamonte, Nature **621**, 60 (2023)
27. V.M. Silkin, S.L. Drechsler, D.V. Efremov, J. Phys. Chem. Lett. **14**(36), 8060 (2023)
28. M.R. Norman, M. Randeria, H. Ding, J.C. Campuzano, Phys. Rev. B **52**, 615 (1995)
29. N.P. Armitage, E.J. Mele, A. Vishwanath, Rev. Mod. Phys. **90**, 015001 (2018)
30. J.H. Pixley, D.A. Huse, S. Das Sarma, Phys. Rev. X **6**, 021042 (2016)
31. R. Nandkishore, D.A. Huse, S.L. Sondhi, Phys. Rev. B **89**, 245110 (2014)
32. Ya.I. Rodionov, K.I. Kugel, B.A. Aronzon, F. Nori, Phys. Rev. B **102**, 205105 (2020)
33. D.T. Son, B.Z. Spivak, Phys. Rev. B **88**, 104412 (2013)
34. B.Z. Spivak, A.V. Andreev, Phys. Rev. B **93**, 085107 (2016)
35. A.V. Andreev, B.Z. Spivak, Phys. Rev. Lett. **120**, 026601 (2018)
36. M.Yu. Kagan, A.V. Turlapov, Phys. Usp. **62**(3), 215 (2019)
37. T. Maruyama, T. Tatsumi, D.N. Voskresensky, T. Tanigawa, S. Chiba, Phys. Rev. C **72**, 015802 (2005)
38. T. Maruyama, T. Tatsumi, D.N. Voskresensky, T. Tanigawa, T. Endo, S. Chiba, Phys. Rev. C **73**, 035802 (2006)
39. S.I. Bastrukov, V.V. Papoyan, D.V. Podgainyi, Astrophysics **42**(2), 177 (1999)
40. E.L. Andronikashvili, Yu.G. Mamaladze, Rev. Mod. Phys. **38**, 567 (1966)

Index

M. Yu. Kagan et al., *Electronic Phase Separation in Magnetic and Superconducting Materials*, Springer Series in Solid-State Sciences 201,
https://doi.org/10.1007/978-3-031-55467-4

C

E

F

M

SPRINGER NATURE

GPSR Compliance

The European Union's (EU) General Product Safety Regulation (GPSR) is a set of rules that requires consumer products to be safe and our obligations to ensure this.

If you have any concerns about our products, you can contact us on ProductSafety@springernature.com

In case Publisher is established outside the EU, the EU authorized representative is:

Springer Nature Customer Service Center GmbH
Europaplatz 3
69115 Heidelberg, Germany

The manufacturer's authorised representative in the EU is Springer Nature Customer Service Centre GmbH, Europaplatz 3, 69115 Heidelberg, Germany. If you have any concerns regarding our products, please contact ProductSafety@springernature.com

Printed and bound by CPI Group (UK) Ltd, Croydon, CR0 4YY
01/10/2026
02225866-0014